RECENT ADVANCES IN GLOBAL OPTIMIZATION

PRINCETON SERIES IN COMPUTER SCIENCE
David R. Hanson and Robert E. Tarjan, Editors

RECENT ADVANCES IN GLOBAL OPTIMIZATION

Christodoulos A. Floudas
and Panos M. Pardalos, Editors

PRINCETON UNIVERSITY PRESS

Copyright © 1992 by Princeton University Press

Published by Princeton University Press, 41 William Street, Princeton,
New Jersey 08540
In the United Kingdom: Princeton University Press, Oxford

All Rights Reserved

Library of Congress Cataloging-in-Publication Data

Recent advances in global optimization / Christodoulos A. Floudas and Panos M.
Pardalos, editors.
        p. cm. – (Princeton series in computer science).
        Papers presented at a conference held at Princeton University,
        May 10-11, 1991.
        Includes bibliographical references.
        ISBN 0-691-08740-7 (cl : acid-free paper)
        ISBN 0-691-02527-4 (pbk.: acid-free paper)
        1. Mathematical optimization. 2. Nonlinear programming.
I. Floudas, Christodoulos, A. II. Pardalos, P. M. (Panos M.). III. Series.
QA402.5.R42 1992
519.3–dc20   91-35290

The publisher would like to acknowledge the authors of this volume for
providing the camera–ready copy from which this book was printed

Princeton University Press books are printed on acid–free paper, and meet the
guidelines for permanence and durability of the Committee on Production
Guidelines for Book Longevity of the Council on Library Resources

Printed in the United States of America

          10 9 8 7 6 5 4 3 2 1
(Pbk.)   10 9 8 7 6 5 4 3 2 1

Contents

Preface
Christodoulos A. Floudas & Panos M. Pardalos — ix

On Approximation Algorithms for Concave Quadratic Programming
Stephen A. Vavasis — 3

A New Complexity Result on Minimization of a Quadratic Function with a Sphere Constraint
Yinyu Ye — 19

Hamiltonian Cycles, Quadratic Programming, and Ranking of Extreme Points
Ming Chen & Jerzy A. Filar — 32

Performance of Local Search in Minimum Concave–Cost Network Flow Problems
G. M. Guisewite & P. M. Pardalos — 50

Solution of the Concave Linear Complementarity Problem
Joaquim J. Júdice & Ana M. Faustino — 76

Global Solvability of Generalized Linear Complementarity Problems and a Related Class of Polynomial Complementarity Problems
Aniekan A. Ebiefung & Michael M. Kostreva — 102

A Continuous Approach to Compute Upper Bounds in Quadratic Maximization Problems with Integer Constraints
A. Kamath & N. Karmarkar — 125

A Class of Global Optimization Problems Solvable by Sequential Unconstrained Convex Minimization
Hoang Tuy & Faiz A. Al–Khayyal — 141

A New Cutting Plane Algorithm for a Class of Reverse Convex 0–1 Integer Programs
Sihem BenSaad — 152

GLOBAL OPTIMIZATION OF PROBLEMS WITH POLYNOMIAL FUNCTIONS IN ONE VARIABLE
    V. Visweswaran & C. A. Floudas   165

ONE DIMENSIONAL GLOBAL OPTIMIZATION USING LINEAR LOWER BOUNDS
    Matthew Bromberg & Tsu–Shuan Chang   200

OPTIMIZING THE SUM OF LINEAR FRACTIONAL FUNCTIONS
    James E. Falk & Susan W. Palocsay   221

MINIMIZING AND MAXIMIZING THE PRODUCT OF LINEAR FRACTIONAL FUNCTIONS
    Hiroshi Konno & Yasutoshi Yajima   259

NUMERICAL METHODS FOR GLOBAL OPTIMIZATION
    Yu. G. Evtushenko, M. A. Potapov, & V. V. Korotkich   274

INTEGRAL GLOBAL OPTIMIZATION OF CONSTRAINED PROBLEMS IN FUNCTIONAL SPACES WITH DISCONTINUOUS PENALTY FUNCTIONS
    Quan Zheng & Deming Zhuang   298

RIGOROUS METHODS FOR GLOBAL OPTIMIZATION
    Ramon Moore, Eldon Hansen, & Anthony Leclerc   321

GLOBAL OPTIMIZATION OF COMPOSITE LAMINATES USING IMPROVING HIT AND RUN
    Zelda B. Zabinsky, Douglas L. Graesser,
    Mark E. Tuttle, & Gun–In Kim   343

STOCHASTIC MINIMIZATION OF LIPSCHITZ FUNCTIONS
    Regina Hunter Mladineo   369

TOPOGRAPHICAL GLOBAL OPTIMIZATION
    Aimo Törn & Sami Viitanen   384

LIPSCHITZIAN GLOBAL OPTIMIZATION: SOME PROSPECTIVE APPLICATIONS
    János Pintér   399

PACKET ANNEALING: A DETERMINISTIC METHOD FOR GLOBAL MINIMIZATION. APPLICATION TO MOLECULAR CONFORMATION
    David Shalloway   433

MIXED–INTEGER LINEAR PROGRAMMING REFORMULATIONS FOR SOME
NONLINEAR DISCRETE DESIGN OPTIMIZATION PROBLEMS
    I. E. Grossmann, V. T. Voudouris, & O. Ghattas   478

MIXED–INTEGER NONLINEAR PROGRAMMING ON GENERALIZED NETWORKS
    Soren S. Nielsen & Stavros A. Zenios   513

GLOBAL MINIMA IN ROOT FINDING
    Angelo Lucia & Jinxian Xu   543

HOMOTOPY–CONTINUATION ALGORITHM FOR GLOBAL OPTIMIZATION
    Amy C. Sun & Warren D. Seider   561

SPACE–COVERING APPROACH AND MODIFIED FRANK–WOLFE ALGORITHM FOR
OPTIMAL NUCLEAR REACTOR RELOAD DESIGN
    Zhian Li, P. M. Pardalos, & S. H. Levine   593

A GLOBAL OPTIMIZATION APPROACH TO SOFTWARE TESTING
    Roberto Barbagallo, Maria Cristina Recchioni,
    & Francesco Zirilli   616

Preface

Global optimization is concerned with the characterization and computation of global minima or maxima of unconstrained nonlinear functions and constrained nonlinear problems. Such problems are widespread in mathematical modeling of real world systems for a very broad range of applications. Applications include structural optimization, engineering design, VLSI chip design and database problems, image processing, computational chemistry, molecular biology, nuclear and mechanical design, chemical engineering design and control, economies of scale, fixed charges, allocation and location problems, quadratic assignment and a number of other combinatorial optimization problems such as integer programming and related graph problems (e.g. maximum clique problem).

From the complexity point of view global optimization problems belong to the class of NP-hard problems. This means that as the input size of the problem increases the computational time required to solve the problem is expected to grow exponentially.

Although standard nonlinear programming algorithms will usually obtain a local minimum or a stationary point to the global optimization problem, such a local minimum will only be global when certain conditions are satisfied such as the objective function is quasi-convex and the feasible domain is convex. In general, several local minima may exist and the corresponding function values may differ substantially. The problem of designing algorithms that obtain global solutions is very difficult, since in general there is no local criterion for deciding whether a local solution is global.

A large number of publications have appeared during the past three decades on the subject of global optimization or related problems. These papers discuss a variety of deterministic and stochastic methods for finding global solutions to nonlinear optimization problems. Deterministic methods include enumerative techniques, cutting planes, branch and bound, decomposition based approaches, bilinear programming, interval analysis, homotopy methods, interior point methods, and approximate algorithms for large-scale problems. Stochastic methods include simulating annealing, pure random search techniques, and clustering methods.

The collection of papers in this book highlights a wide spectrum of creativity and richness of ideas that belong to the area of global optimization methods. All papers in this book were regularly refereed and the majority of them were presented at the conference (organized by C. A. Floudas and P. M. Pardalos) on "Recent Advances in Global Optimization" held at Princeton University, May 10-11, 1991. The conference presented current research in global optimization and related applications in science and engineering.

We would like to take this opportunity to thank all the contributing authors and anonymous referees, Air Force Office of Scientific Research, Pennsylvania State University, and Princeton University for their help and support of the conference. Finally, we would like to thank Lilya Lorrin, the mathematical sciences editor of Princeton University Press and the editorial board of Princeton University Press for accepting this book to be published in this series.

                                        Christodoulos A. Floudas
                                        Panos M. Pardalos

                                        Summer 1991

RECENT ADVANCES IN GLOBAL OPTIMIZATION

On approximation algorithms for concave quadratic programming

Stephen A. Vavasis*
Department of Computer Science
Cornell University
Ithaca, NY 14853

June 3, 1991

Abstract

We consider ϵ-approximation schemes for concave quadratic programming. Because the existing definition of ϵ-approximation for combinatorial optimization problems is inappropriate for nonlinear optimization, we propose a new definition for ϵ-approximation. We argue that such an approximation can be found in polynomial time for fixed ϵ and k, where k denotes the number of negative eigenvalues. Our algorithm is polynomial in $1/\epsilon$ for fixed k, and superexponential in k for fixed ϵ.

1 Concave quadratic programming

Quadratic programming is a nonlinear optimization problem of the following form:
$$\begin{array}{ll} \text{minimize} & \frac{1}{2}\mathbf{x}^T H \mathbf{x} + \mathbf{h}^T \mathbf{x} \\ \text{subject to} & W\mathbf{x} \geq \mathbf{b}. \end{array} \tag{1}$$

In this formulation, $\mathbf{x}$ is the n-vector of unknowns. The remaining variables stand for data in the problem instance: H is an $n \times n$ symmetric matrix, $\mathbf{h}$ is an n-vector, W is an $m \times n$ matrix, and $\mathbf{b}$ is an m-vector. The relation '$\geq$' in the constraint $W\mathbf{x} \geq \mathbf{b}$ is the usual componentwise inequality.

Quadratic programming, a generalization of linear programming, has applications in economics, planning, and many kinds of engineering design. In

---
*Supported by the Applied Mathematical Sciences Program (KC-04-02) of the Office of Energy Research of the U.S. Department of Energy under grant DE-FG02-86ER25013.A000 and in part by the National Science Foundation, the Air Force office of Scientific Research, and the Office of Naval Research, through NSF grant DMS 8920550. Revision work was done while the author was on leave at Sandia National Laboratories, supported by U.S. Department of Energy contract DE-AC04-76DP00789.

addition, more complicated kinds of nonlinear programming problems are often simplified into quadratic programming problems.

No efficient algorithm is known to solve the general case of (1). The lack of an efficient algorithm is not surprising, since quadratic programming is known to be NP-hard, a result due to Sahni [1974]. An NP-hardness proof appears in this paper (it follows from the discussion in Section 3). More recently, Vavasis [1990] showed that the decision version of the problem lies in NP, and hence is NP-complete.

Many avenues for addressing (1) have been pursued in the literature. For example, efficient algorithms are known for the special case in which H is positive semidefinite, known as the *convex* case. See Kozlov, Tarasov and Hačijan [1979] for the first polynomial-time algorithm for the convex case. See Kapoor and Vaidya [1986] or Ye and Tse [1989] for efficient interior point algorithms for this problem. Active set methods (see Gill, Murray and Wright [1981]), a commonly-used class of methods for (1), are a combination of local search and heuristics.

A very successful way to address NP-hard combinatorial optimization problems has been approximation algorithms. In light of the importance of quadratic programming, it is surprising the ϵ-approximation algorithms have so far not been pursued. In this report, we investigate ϵ-approximation algorithms for the concave case, i.e., the case that H is negative semidefinite. Our techniques at present are not able to address the more general indefinite case (in which H has a mixture of positive and negative eigenvalues); see Section 4 for further discussion of the indefinite case. The concave case, like the general case, is NP-hard.

First it is necessary to give a definition of ϵ-approximation. To our knowledge, this concept has not been previously defined for nonlinear optimization in the literature (but see below), so it is necessary to come up with our own definition. We have the following proposal:

Definition 1 *Consider an instance of quadratic programming written in the form* (1). *Let* $f(\mathbf{x})$ *denote the objective function* $\frac{1}{2}\mathbf{x}^T H \mathbf{x} + \mathbf{h}^T \mathbf{x}$. *Let* $\mathbf{x}^*$ *be an optimum point of the problem. We say that* $\mathbf{x}^\diamond$ *is an* ϵ-*approximate solution if there exists another feasible point* $\mathbf{x}^\#$ *such that*

$$f(\mathbf{x}^\diamond) - f(\mathbf{x}^*) \leq \epsilon[f(\mathbf{x}^\#) - f(\mathbf{x}^*)].$$

Notice that we may as well take $\mathbf{x}^\#$ in Definition 1 to be the point where the objective function is maximized. Thus, another way to interpret this definition is as follows. Let Π denote the feasible region, and let interval $[a, b]$ be $f(\Pi)$. Then $f(\mathbf{x}^\diamond)$ should lie in the interval $[a, a + \epsilon(b - a)]$.

In the case that the objective function has no upper bound on the feasible region, the maximizer is of course undefined. Our definition loses its value in this situation because any feasible point is an ϵ-approximation for any $\epsilon > 0$.

The definition fails in the case that the objective function has no lower bound or in the case that no feasible points exist. In these cases, (1) is said to be *unbounded* or *infeasible* respectively. We should expect our approximation algorithm to return an indicator that the problem is unbounded or infeasible.

Finally, observe that any feasible point is a 1-approximation by this definition, and only the optimum is a 0-approximation. Thus, the definition makes sense only for ϵ in the interval $(0, 1)$.

This definition has some useful properties. First, it is insensitive to translations or dilations of the objective function. In other words, if the objective function $f(\mathbf{x})$ is replaced by a new function $g(\mathbf{x}) = af(\mathbf{x}) + b$ where $a > 0$, a vector $\mathbf{x}^\circ$ that was previously an ϵ-approximation will continue to have that property.

A second useful property is that ϵ-approximation is preserved under transformations of the feasible region. The most general kind of transformation that preserves the format of (1) is an affine linear transformation. Let $\theta : \mathbb{R}^n \to \mathbb{R}^n$ be an affine linear transformation, i.e., a function of the form $\theta(\mathbf{x}) = V(\mathbf{x} - \mathbf{x}_0)$ where V is an invertible $n \times n$ matrix. Then the problem of minimizing $f(\theta(\mathbf{y}))$ subject to $W\theta(\mathbf{y}) \geq \mathbf{b}$ is still a quadratic program, and, moreover, it is not hard to see that if $\mathbf{x}^\circ$ is an ϵ-approximate solution of the original problem, then $\theta^{-1}(\mathbf{x}^\circ)$ is an ϵ-approximate solution to the transformed problem.

We now state the main theorem of this paper.

Theorem 2 *Consider the concave case of* (1), *i.e., the case that H is negative semidefinite. Let k denote the rank of H. There is an algorithm to find ϵ-approximate solution to* (1) *in time*

$$O\left(\left\lceil\sqrt{\frac{k}{\epsilon}}\right\rceil^k \ell\right)$$

steps. In this formula, ℓ denotes the time to solve a linear programming problem of the same size as (1).

We remark that ℓ grows polynomially with the size of the input. This fact was first proved by Hačijan [1979]. The best known asymptotic bound for ℓ is due to Vaidya [1989].

The algorithm we propose is similar to algorithms that have appeared in the literature. In particular, it is very reminiscent of algorithms described in Pardalos and Rosen [1987]. Our contribution is to define a formal meaning for approximation algorithm, and then show that the algorithm achieves this bound.

The remainder of this paper is organized as follows. In Section 2 we provide the algorithm and prove the main theorem. In Section 3 we indicate why polynomial dependence on $1/\epsilon$ and exponential dependence on k is expected. In Section 4 we discuss open questions raised by this work.

The definition of approximation proposed for combinatorial optimization differs from our definition and is usually stated as follows. A feasible point $\mathbf{x}^\circ$ is an ϵ-approximation if

$$|f(\mathbf{x}^\circ) - f(\mathbf{x}^*)| \leq \epsilon \cdot f(\mathbf{x}^*).$$

See, for example, Papadimitriou and Steiglitz [1982] for an extensive discussion of approximation for combinatorial optimization. This definition does not work for nonlinear optimization because it is not preserved when a constant is added to the objective function. In particular, the definition becomes useless in the case that $f(\mathbf{x}^*) \leq 0$.

Work by Katoh and Ibaraki [1987] addresses the problem of approximation algorithms for certain kinds of quasiconcave optimization problem. Their work generalizes the kinds of allowable objective functions (i.e., they do not restrict attention to quadratic programming), but they make a number of restrictions concerning the feasible region and range of the objective function. Their results do not seem to be directly comparable to ours.

2 Proof of the main theorem

Our starting point is a problem of the form (1) in which H is negative semidefinite with rank k. We can perform change of basis to diagonalize H, either via an eigenvalue computation or an LDL^T factorization of H (see Golub and Van Loan [1989]). In either case we end up with a problem of the form

$$\begin{array}{ll} \text{minimize} & \frac{1}{2}\mathbf{y}^T D\mathbf{y} + \mathbf{c}^T\mathbf{y} + \mathbf{f}^T\mathbf{z} \\ \text{subject to} & A\mathbf{y} + B\mathbf{z} \geq \mathbf{b} \end{array} \quad (2)$$

where, in this new formulation, D is a $k \times k$ negative definite diagonal matrix, $\mathbf{y}$ is a k-vector of unknowns, and $\mathbf{z}$ is an $n - k$-vector of unknowns. Thus, the negative-definite part of the problem is confined to k variables.

Now we define the function

$$\phi(\mathbf{y}) = \min\{\mathbf{f}^T\mathbf{z} : \mathbf{z} \in \mathbb{R}^{n-k}, A\mathbf{y} + B\mathbf{z} \geq \mathbf{b}\}.$$

For a fixed choice of $\mathbf{y}$, if the system $A\mathbf{y} + B\mathbf{z} \geq \mathbf{b}$ has no feasible choice for $\mathbf{z}$, then we adopt the convention that $\phi(\mathbf{y}) = \infty$. Similarly, in the case that $\mathbf{f}^T\mathbf{z}$ has no lower bound, we say that $\phi(\mathbf{y}) = -\infty$.

Thus, the original problem can now be expressed simply as minimizing

$$q(\mathbf{y}) + \phi(\mathbf{y}) \quad (3)$$

where
$$q(\mathbf{y}) = \frac{1}{2}\mathbf{y}^T D \mathbf{y} + \mathbf{c}^T \mathbf{y}. \tag{4}$$

In this formulation there are no constraints on $\mathbf{y}$. This is equivalent to (2). In particular, for any $\mathbf{y}_1$ feasible for (2), $q(\mathbf{y}_1) + \phi(\mathbf{y}_1)$ will be the value of the minimum possible objective function value among all feasible vectors of the form $(\mathbf{y}_1, \mathbf{z})$.

Lemma 3 *The set $C = \{\mathbf{y} \in \mathbb{R}^k : \phi(\mathbf{y}) < \infty\}$ is convex, and on this domain, ϕ is a convex function.*

PROOF. First, notice that the set C in the lemma can be written:

$$C = \{\mathbf{y} \in \mathbb{R}^k : \text{There exists } \mathbf{z} \in \mathbb{R}^{n-k} \text{ such that } A\mathbf{y} + B\mathbf{z} \geq \mathbf{b}.\}.$$

This domain is convex, as the following argument shows. If $\mathbf{y}', \mathbf{y}'' \in C$, then there are $\mathbf{z}', \mathbf{z}''$ such that $A\mathbf{y}' + B\mathbf{z}' \geq \mathbf{b}$ and $A\mathbf{y}'' + B\mathbf{z}'' \geq \mathbf{b}$. This means that for all $\lambda \in [0, 1]$,

$$\begin{aligned}A((1-\lambda)\mathbf{y}' + \lambda\mathbf{y}'') + B((1-\lambda)\mathbf{z}' + \lambda\mathbf{z}'') &= (1-\lambda)(A\mathbf{y}' + B\mathbf{z}') \\ &\quad + \lambda(A\mathbf{y}'' + B\mathbf{z}'') \\ &\geq (1-\lambda)\mathbf{b} + \lambda\mathbf{b} \\ &\geq \mathbf{b}.\end{aligned}$$

This shows that $(1-\lambda)\mathbf{y}' + \lambda\mathbf{y}'' \in C$.

To show that ϕ is convex, consider $\phi(\mathbf{y}')$ and $\phi(\mathbf{y}'')$ for $\mathbf{y}', \mathbf{y}'' \in C$. We know by assumption that $\phi(\mathbf{y}'), \phi(\mathbf{y}'') < \infty$. We also assume (see the lemma below) that $\phi(\mathbf{y}'), \phi(\mathbf{y}'') > -\infty$. Then the problem of computing $\phi(\mathbf{y}')$ is equivalent to solving a linear program, and we know from linear programming theory that the minimum is achieved, say at a vector $\mathbf{z}'$. In other words, there exists a $\mathbf{z}'$ such that $\phi(\mathbf{y}') = \mathbf{f}^T\mathbf{z}'$ and $A\mathbf{y}' + B\mathbf{z}' \geq \mathbf{b}$. The same holds for $\mathbf{y}''$, i.e., there is a $\mathbf{z}''$ such that $\phi(\mathbf{y}'') = \mathbf{f}^T\mathbf{z}''$ and $A\mathbf{y}'' + B\mathbf{z}'' \geq \mathbf{b}$. Then we conclude as above that

$$A((1-\lambda)\mathbf{y}' + \lambda\mathbf{y}'') + B((1-\lambda)\mathbf{z}' + \lambda\mathbf{z}'') \geq \mathbf{b}$$

so that

$$\phi((1-\lambda)\mathbf{y}' + \lambda\mathbf{y}'') \leq \mathbf{f}^T((1-\lambda)\mathbf{z}' + \lambda\mathbf{z}'')$$

because ϕ is defined to be the minimum of expressions such as the one on the right-hand side. This last inequality is the same as

$$\phi((1-\lambda)\mathbf{y}' + \lambda\mathbf{y}'') \leq (1-\lambda)\phi(\mathbf{y}') + \lambda\phi(\mathbf{y}'').$$

∎

We now show how to determine whether the original problem is unbounded.

Lemma 4 *Problem* (2) *is unbounded if and only if*

1. *For every* $\mathbf{y} \in C$, $\phi(\mathbf{y}) = -\infty$, *or*

2. *Region C is unbounded.*

PROOF. First, we prove that either of the two conditions imply that (2) is unbounded. The first condition trivially implies that (2) is unbounded. Indeed, the existence of any $\mathbf{y}$ such that $\phi(\mathbf{y}) = -\infty$ means that (2) is unbounded, as we see from examining (3).

Suppose for the other case that C is unbounded. Then it must contain a ray (because it is convex). Let us write the ray in the form $\{\mathbf{y}_1 + t\mathbf{y}_2 : t \geq 0\}$ where $\mathbf{y}_2$ is a nonzero vector. For large enough t, the function $\phi(\mathbf{y}_1 + t\mathbf{y}_2)$ must be linear in t. This statement is proved by noting that the solution to the linear program

$$\text{minimize} \quad \mathbf{f}^T \mathbf{z}$$
$$\text{subject to} \quad B\mathbf{z} \geq \mathbf{b} - A\mathbf{y}$$

implicit in the definition of ϕ may be taken to be a basic feasible solution, and for large enough t we can assume that the choice of basis columns is independent of t.

Examining (3), we see that ϕ behaves linearly along the ray for large enough t, but the term $\frac{1}{2}(\mathbf{y}_1 + t\mathbf{y}_2)^T D(\mathbf{y}_1 + t\mathbf{y}_2)$ is quadratic in t with a negative leading term. Thus, the quadratic term dominates for t large enough and the objective function tends to $-\infty$ along the ray.

Now we prove that if the original problem is unbounded, at least one of the two conditions holds. Let $\{(\mathbf{y}_1, \mathbf{z}_1) + t(\mathbf{y}_2, \mathbf{z}_2) : t \geq 0\}$ be a ray along which the objective function of (2) tends to $-\infty$. Either $\mathbf{y}_2$ or $\mathbf{z}_2$ must be nonzero. Suppose $\mathbf{y}_2$ is nonzero; this implies that C contains the ray $\mathbf{y}_1 + t\mathbf{y}_2$ and hence is unbounded.

The other case is that $\mathbf{y}_2 = \mathbf{0}$, in which case $\mathbf{z}_2 \neq \mathbf{0}$. The ray can be written $(\mathbf{y}_1, \mathbf{z}_1 + t\mathbf{z}_2)$. We can see that $B\mathbf{z}_2 \geq \mathbf{0}$ (otherwise the ray would become infeasible for t large enough). Also, $\mathbf{f}^T \mathbf{z}_2 < 0$ because the objective function decreases along the ray by assumption. We claim that the existence of this $\mathbf{z}_2$ implies that $\phi(\mathbf{y}) = -\infty$ for any $\mathbf{y} \in C$. Let $\mathbf{z}$ be the vector such that $A\mathbf{y} + B\mathbf{z} \geq \mathbf{b}$; then $A\mathbf{y} + B(\mathbf{z} + t\mathbf{z}_2) \geq \mathbf{b}$ for any $t > 0$, and $\mathbf{f}^T(\mathbf{z} + t\mathbf{z}_2)$ is arbitrarily negative. ∎

Now we begin to describe our algorithm. The first task of our algorithm is to determine the smallest possible rectangle R such that $C \subset R$. The term *rectangle* refers to the k-fold product of compact closed intervals. Let us write

$$R = [l_1, u_1] \times \cdots \times [l_k, u_k].$$

We now explain how l_1 is determined (the other $2k - 1$ interval bounds are determined in a similar fashion). We define

$$l_1 = \min\{y_1 : \text{There exists } y_1, y_2, \cdots, y_k, \mathbf{z} \text{ such that } A\mathbf{y} + B\mathbf{z} \geq \mathbf{b}\}.$$

Thus, we see that l_1 is found by solving a linear program. From the previous lemma, we see that if this linear program is unbounded, then the original problem is unbounded.

In this manner we can determine the remaining l_i's and u_i's. We assume that $l_i < u_i$ for all i, because otherwise y_i is uniquely determined and may be dropped from the problem.

We now assume that the coordinates of the vector $\mathbf{y}$ are translated and rescaled so that the rectangle R is the k-dimensional unit cube. This simplifies the notation in the upcoming paragraphs. Note that this rescaling does not change the format of (2). Let us state this formally:

Assumption. *Without loss of generality, we assume that $R = [0, 1]^k$.*

For the next step of our algorithm we define integer m to be $\sqrt{k/\epsilon}$ rounded up; this choice is explained below. Next, we insert an $(m+1) \times \cdots \times (m+1)$ grid of points in the cube R. The coordinates of the points have the form:

$$(i_1/m, i_2/m, \cdots, i_k/m)$$

where $i_1, \ldots, i_k$ are integers satisfying $0 \leq i_1, \ldots, i_k \leq m$.

The grid partitions R into m^k smaller cubes. We now construct *linear underestimating functions* for each subcube. Focus on a particular subcube S, say the one given by

$$[r_1, s_1] \times \cdots \times [r_k, s_k].$$

By assumption, $s_i - r_i = 1/m$ for each i. For each i between 1 and k, there is a unique affine linear function of one variable, say the function $\lambda_i(y) = a_i y + b_i$ satisfying the two equations

$$\lambda_i(r_i) = d_{ii} r_i^2 / 2 + c_i r_i,$$
$$\lambda_i(s_i) = d_{ii} s_i^2 / 2 + c_i s_i.$$

Here, d_{ii} denotes the ith diagonal entry of matrix D in (2), and c_i denotes the ith entry of vector $\mathbf{c}$.

Consider the function from $\mathbb{R}^k$ to $\mathbb{R}$ given by:

$$\mu(y_1, \ldots, y_k) = \lambda_1(y_1) + \cdots + \lambda_k(y_k).$$

(Note that μ depends implicitly on the choice of S.) We claim that $\mu(\mathbf{y}) = q(\mathbf{y})$ at all vertices of of S, where $q(\mathbf{y})$ was defined by (4). This follows from the construction in the previous paragraph, since function q is separable and λ_i was constructed to agree with the components of q at boundary values.

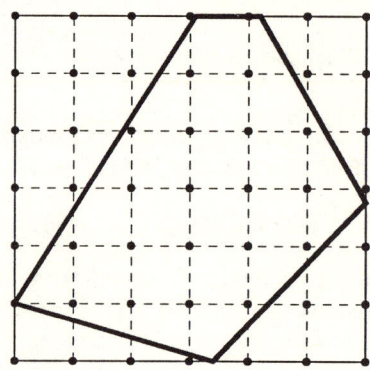

Figure 1: An example of the mesh and the subcubes.

Moreover, we claim that:

$$q(\mathbf{y}) \leq \mu(\mathbf{y}) \leq q(\mathbf{y}) + \frac{\gamma k}{8m^2} \quad (5)$$

for $\mathbf{y}$ in subcube S, where γ is the opposite of the most negative diagonal entry of D. First, we need to write down an explicit formula for λ_i, which follows from its definition above:

$$\lambda_i(y) = (d_{ii}(r_i + s_i)/2 + c_i)y - d_{ii}r_i s_i/2.$$

Then it can be derived that:

$$d_{ii}y^2/2 - c_i y - \lambda_i(y) = \frac{-d_{ii}}{2}(y - r_i)(s_i - y).$$

Thus, we see that if y is between r_i and s_i then the right hand is positive, and, as a concave quadratic function of one variable, its maximum is achieved at $(r_i + s_i)/2$. This maximum value is $(-d_{ii}/2)(1/4m^2)$. Therefore, if we let γ be the maximum value of $-d_{ii}$, $i = 1, \ldots k$, we establish (5).

See Figure 1 for an example of R, C, the grid and subcubes in the case $k = 2$.

We remark that in general it is not possible to construct a single affine linear function that agrees with a quadratic function at all vertices of a cube. It is possible in our case only because the quadratic function is separable. If the quadratic function had not been separable, then we could have proceeded by partitioning each subcube into into k-dimensional simplices. There are standard ways to triangulate cubes; Kuhn's triangulation, for example, divides each subcube into $k!$ simplices (see Todd [1976]).

The next step of the algorithm is to compute the affine linear function described in the preceeding paragraphs for each of the m^k subcubes constructed above. Define $t = m^k$. Let the subcubes be denoted $S_1, \ldots, S_t$. Let $\mu_1, \ldots, \mu_t$ be the affine functions constructed above.

For each i, we minimize $\mu_i(\mathbf{y}) + \phi(\mathbf{y})$ on subcube S_i. This is a linear programming problem. In particular, it is equivalent to the following optimization problem:

$$\begin{aligned} \text{minimize} \quad & \mu_i(\mathbf{y}) + \mathbf{f}^T \mathbf{z} \\ \text{subject to} \quad & A\mathbf{y} + B\mathbf{z} \geq \mathbf{b}, \\ & \mathbf{y} \in S_i. \end{aligned} \qquad (6)$$

This problem can be solved in time ℓ.

Finally, return as the approximate solution to the original problem (2) the vector $(\mathbf{y}^\circ, \mathbf{z}^\circ)$ that achieves the minimum objective function value among all t linear programs enumerated in (6) above.

Now we prove bounds to show that this is an ϵ-approximate solution. The simpler bound is the gap between the objective function at $(\mathbf{y}^\circ, \mathbf{z}^\circ)$ and the optimal point $(\mathbf{y}^*, \mathbf{z}^*)$.

Lemma 5 *Let $(\mathbf{y}^\circ, \mathbf{z}^\circ)$ be computed as above. Let $(\mathbf{y}^*, \mathbf{z}^*)$ be an optimum for (2). Then*

$$[q(\mathbf{y}^\circ) + \phi(\mathbf{y}^\circ)] - [q(\mathbf{y}^*) + \phi(\mathbf{y}^*)] \leq \gamma k/(8m^2).$$

PROOF. Let μ be the piecewise linear function on R defined so that μ agrees with μ_i on subcube S_i. Then the difference $q - \mu$ is between 0 and $\gamma k/(8m^2)$ on all of R, as proved earlier.

Note that $\mathbf{y}^\circ$ is a minimum for function $\mu + \phi$, and therefore,

$$\mu(\mathbf{y}^\circ) + \phi(\mathbf{y}^\circ) \leq \mu(\mathbf{y}^*) + \phi(\mathbf{y}^*).$$

Moreover,

$$\mu(\mathbf{y}^*) + \phi(\mathbf{y}^*) \leq q(\mathbf{y}^*) + \phi(\mathbf{y}^*)$$

since $\mu \leq q$ by construction of q. Finally,

$$q(\mathbf{y}^\circ) + \phi(\mathbf{y}^\circ) - \gamma k/(8m^2) \leq \mu(\mathbf{y}^\circ) + \phi(\mathbf{y}^\circ)$$

because of the properties of μ. Combining the last three inequalities gives the result. ∎

We have established an upper bound on the gap between the objective function at $(\mathbf{y}^\circ, \mathbf{z}^\circ)$ and optimum. Now we establish a lower bound on the gap between the maximum objective function value and the minimum objective value.

To find a point $(\mathbf{y}^\#, \mathbf{z}^\#)$ as in the definition of ϵ-approximation, let us recall that γ is the absolute value of the most negative entry of D. Say this entry is in the (i, i) position. Recall by construction of R that there exists a feasible point $(\mathbf{y}^0, \mathbf{z}^0)$ such that the ith entry of $\mathbf{y}^0$ is equal to 0. Similarly, there exists a point $(\mathbf{y}^1, \mathbf{z}^1)$ such that the ith entry of $\mathbf{y}^1$ is 1. We define

$$(\mathbf{y}^\#, \mathbf{z}^\#) = \frac{1}{2}(\mathbf{y}^0, \mathbf{z}^0) + \frac{1}{2}(\mathbf{y}^1, \mathbf{z}^1).$$

This point is feasible for (2) since it is the convex combination of two feasible points.

We now establish a bound on how far this point lies above optimal.

Lemma 6 *The following bound holds:*

$$\frac{1}{2}\mathbf{y}^{\#T}D\mathbf{y}^\# + \mathbf{c}^T\mathbf{y}^\# + \mathbf{f}^T\mathbf{z}^\# \geq \frac{1}{2}\left[q(\mathbf{y}^0) + \phi(\mathbf{y}^0) + q(\mathbf{y}^1) + \phi(\mathbf{y}^1)\right] + \gamma/8.$$

PROOF. First notice that

$$q(\mathbf{y}^0) + \mathbf{f}^T\mathbf{z}^0 \geq q(\mathbf{y}^0) + \phi(\mathbf{y}^0)$$

by definition of ϕ, and similarly,

$$q(\mathbf{y}^1) + \mathbf{f}^T\mathbf{z}^1 \geq q(\mathbf{y}^1) + \phi(\mathbf{y}^1).$$

Next, we have the following chain of equations.

$$\begin{aligned}
\frac{1}{2}\mathbf{y}^{\#T}D\mathbf{y}^\# + \mathbf{c}^T\mathbf{y}^\# + \mathbf{f}^T\mathbf{z}^\# &= \frac{1}{8}(\mathbf{y}^0 + \mathbf{y}^1)^T D(\mathbf{y}^0 + \mathbf{y}^1) + \frac{1}{2}\mathbf{c}^T(\mathbf{y}^0 + \mathbf{y}^1) \\
&\quad + \frac{1}{2}\mathbf{f}^T(\mathbf{z}^0 + \mathbf{z}^1) \\
&= \left(\frac{1}{4}\mathbf{y}^{0T}D\mathbf{y}^0 + \frac{1}{2}\mathbf{c}^T\mathbf{y}^0 + \frac{1}{2}\mathbf{f}^T\mathbf{z}^0\right) \\
&\quad + \left(\frac{1}{4}\mathbf{y}^{1T}D\mathbf{y}^1 + \frac{1}{2}\mathbf{c}^T\mathbf{y}^1 + \frac{1}{2}\mathbf{f}^T\mathbf{z}^1\right) \\
&\quad - \frac{1}{8}\mathbf{y}^{0T}D\mathbf{y}^0 + \frac{1}{4}\mathbf{y}^{0T}D\mathbf{y}^1 - \frac{1}{8}\mathbf{y}^{1T}D\mathbf{y}^1 \\
&= \frac{1}{2}(q(\mathbf{y}^0) + \mathbf{f}^T\mathbf{z}^0) + \frac{1}{2}(q(\mathbf{y}^1) + \mathbf{f}^T\mathbf{z}^1) \\
&\quad - \frac{1}{8}(\mathbf{y}^0 - \mathbf{y}^1)^T D(\mathbf{y}^0 - \mathbf{y}^1).
\end{aligned}$$

Comparing this chain of equations to the previous paragraph, we see that we could prove the lemma if we could get a positive lower bound on $-(\mathbf{y}^0 - \mathbf{y}^1)^T D(\mathbf{y}^0 - \mathbf{y}^1)/8$. Let us express this term in terms of coordinates:

$$-\frac{1}{8}(\mathbf{y}^0 - \mathbf{y}^1)^T D(\mathbf{y}^0 - \mathbf{y}^1) = \frac{1}{8}\sum_{j=1}^{k}(y_j^0 - y_j^1)^2 \cdot (-d_{jj})$$

where the diagonal entries of D are denoted $d_{11}, \cdots, d_{kk}$. Notice that all the terms of this summation are nonnegative. Focusing on the ith term, we observe that $(y_i^0 - y_i^1)^2 = 1$ by choice of $\mathbf{y}^0$ and $\mathbf{y}^1$. Moreover, $-d_{ii} = \gamma$, the 2-norm of matrix D. Thus, we get a lower bound of $\gamma/8$ needed for the lemma. ∎

Corollary 7 *The objective function value of the point* $(\mathbf{y}^\#, \mathbf{z}^\#)$ *is at least* $\gamma/8$ *above optimal.*

PROOF. This follows from the inequality established in the previous lemma after we notice that

$$q(\mathbf{y}^0) + \phi(\mathbf{y}^0) \geq q(\mathbf{y}^*) + \phi(\mathbf{y}^*)$$

and similarly for $\mathbf{y}^1$. ∎

Thus, we have proved that we can find a point $(\mathbf{y}^\circ, \mathbf{z}^\circ)$ that is within $\gamma k/(8m^2)$ of optimum. On the other hand, there exists at least one point that is at least $\gamma/8$ above optimal. Thus, $(\mathbf{y}^\circ, \mathbf{z}^\circ)$ is an ϵ-approximation to optimum provided that

$$\frac{\gamma k}{8m^2} \leq \epsilon \cdot \frac{\gamma}{8},$$

i.e.,

$$m \geq \sqrt{\frac{k}{\epsilon}}.$$

If we choose m to be the first integer larger than this number, we have established that $(\mathbf{y}^\circ, \mathbf{z}^\circ)$ is an ϵ-approximate solution. The running time is proportional to the time needed to solve m^k linear programming problems, i.e.,

$$\left\lceil \sqrt{\frac{k}{\epsilon}} \right\rceil^k \ell$$

steps.

This proves the main theorem. We can summarize the steps of the algorithm as follows. First, H is diagonalized to isolate the negative definite variables. Next, $l_1, \ldots, l_k$ and $u_1, \ldots, u_k$ are found. Then the problem is rescaled so that $R = [0, 1]^k$. This set R is divided into m^k subcubes, and a linear program is solved for each subcube. The minima among all of these linear programs is taken to be the ϵ-approximate solution.

In practice, a more efficient version of the algorithm would construct a hierarchy of triangulations of R. For example, suppose that m is divisible by 2. Then we could partition R into a mesh with $m/2 + 1$ grid points in each dimension and carry out the algorithm of this section. The point returned by the algorithm from the coarse grid would not necessarily be an ϵ-approximation, but the results would give good upper and lower bounds on the value of the objective function in each cube. We could then use these bounds to determine which grid cells need to be further subdivided. Applying these ideas in a hierarchical fashion would lead to a branch-and-bound style algorithm. See Pardalos and Rosen [1987] for algorithms of this style; see Papadimitriou and Steiglitz [1982] for a general description of branch and bound.

3 Dependence on ϵ and k

In this section we give some indication why the dependence on ϵ and k that we obtained in the main theorem might be expected. First, we address the dependence on ϵ. To simplify this discussion, let us restrict attention to the case $k = 1$, that is, a quadratic objective function of the form $cy^2 + \mathbf{f}^T\mathbf{z}$, where y is a scalar unknown and $c < 0$. Suppose that we were able to obtain an approximation algorithm with running time dependence on ϵ better than $1/\epsilon$. Suppose, for example, that we had an approximation algorithm whose running time were polynomial in $|\log \epsilon|$. It is known that if a point is sufficiently close to optimal for an instance of CQP—in particular, within $2^{-O(L)}$ of optimal—then an exact optimum may be found in polynomial time. Here, L denotes the number of bits needed to write the problem. If there were an approximation algorithm whose running time were polynomial in $|\log \epsilon|$, then we set $\epsilon = 2^{-O(L)}$ and in polynomial time come up with an exact solution to the concave problem with objective function $cy^2 + \mathbf{f}^T\mathbf{z}$.

This would seem to contradict recent results by Pardalos and Vavasis [1990], who proved that problems of this form (i.e., CQP problems with $k = 1$) are NP-hard. Thus, assuming $P \neq NP$, polynomial dependence on $|\log \epsilon|$ is not expected, and therefore polynomial dependence on $1/\epsilon$ seems like it might be the best possible.

Next, we investigate the dependence on k. In our bound the dependence on k is exponential. Suppose there were an approximation algorithm whose running time were polynomial in k and $1/\epsilon$. We claim that the existence of such an algorithm would imply that $P = NP$. In particular, consider the following transformation from the boolean 3-satisfiability problem to CQP. Let Q be an instance of 3SAT, i.e., a collection of m clauses each with three literals, and a total of n variables, $x_1, \ldots, x_n$. We can make a CQP instance out of Q as follows. First, let there be one variable y_i for each boolean variable x_i. The objective function is:

$$\sum_{i=1}^{n} y_i(1 - y_i).$$

This is a concave quadratic objective function. There are $2n + m$ constraints. The first $2n$ constraints are:

$$0 \leq y_1 \leq 1,$$

$$\vdots$$
$$0 \leq y_n \leq 1.$$

These constraints imply the following facts: the objective function is nonnegative on the feasible region, and, moreover, the objective function is zero iff every y_i is assigned to 0 or 1.

The remaining m constraints express the boolean clauses. For example, a clause of the form $x_5 \vee x_7 \vee \bar{x}_9$ would be expressed as a constraint:

$$y_5 + y_7 + (1 - y_9) \geq 1.$$

We claim that the boolean formula is satisfiable iff the minimum of the objective function on the feasible region is 0. This is proved using standard arguments.

Suppose now that we had a $3/(4n)$-approximate solution to this problem. First, observe that under the assumption of an approximation algorithm running in polynomial time in $1/\epsilon$ and k (in this case, $k = n$), a $3/(4n)$-approximate solution could be found in polynomial time. Observe that the maximum possible value of the objective function is $n/4$ (obtained by setting each y_i to $1/2$). In other words, the gap between minimum and maximum objective values is no more than $n/4$. Thus, a $3/(4n)$ approximate solution could be no more than $3/16$ above optimum. If the boolean formula were satisfiable, the approximation algorithm would have to return an assignment to $(y_1, \ldots, y_n)$ such that the objective function is at most $3/16$. In particular, each term of the objective function is at most $3/16$, i.e., each y_i is either in the interval $[0, 1/4]$ or $[3/4, 1]$. But we claim that from such an assignment to the y_i's, we can in fact read off a satisfying assignment to the original boolean formula. In particular, whenever y_i is between $[0, 1/4]$, set $x_i = 0$, else set $x_i = 1$.

Thus, from a $3/(4n)$-approximate solution we can determine whether the original formula is satisfiable (by checking whether the $3/(4n)$-approximate solution has an objective function value less than or equal to $3/16$).

Accordingly, we would expect that an algorithm polynomial in $1/\epsilon$ and k does not exist.

4 Open questions

Perhaps the most interesting open question is whether this work extends to indefinite quadratic programming (IQP) problems. As an example of the difficulties with the indefinite case, Murty and Kabadi [1987] show (among other things) that it is NP-hard to tell whether an indefinite quadratic program is unbounded or not. (The corresponding problem for CQP is solvable in polynomial time, and in fact our approximation algorithm detects unboundedness as indicated in Lemma 4.) It seems unreasonable that an approximation algorithm would be unable to detect unboundedness in a problem.

The difficulty mentioned in last paragraph is avoided by assuming that the feasible region of the problem is compact, in which case the problem cannot be unbounded. Even in this case, our techniques are only partially successful. In particular, our bounds concerning the point $(\mathbf{x}^\#, \mathbf{y}^\#)$ fail in the indefinite case. These bounds were necessary to establish the main theorem.

Finding more efficient approximation algorithms is another interesting question. For example, could a branch-and-bound algorithm of the sort described in Section 2 give better worst case bounds? Could randomization help?

We showed that polynomial dependence on both $1/\epsilon$ and k is not likely. We were not able to obtain the following stronger result: Polynomial dependence on k for fixed ϵ would imply that P=NP. We conjecture that this statement is true.

Another interesting question concerns the relationship between CQP and combinatorial problems in NP. Since concave quadratic programming is NP-hard, any combinatorial problem in NP can be expressed as an instance of CQP. However, CQP is general enough that many combinatorial problems can be expressed as a CQP instance without the need for complicated "gadgets." This suggests that there could be a connection between approximation algorithms for CQP and approximation for a variety of combinatorial problems.

A final question concerns the relation between our definition of ϵ-approximate and the standard combinatorial definition mentioned in the introduction. Although the combinatorial definition does not extended to nonlinear problems, our definition *does* extend to combinatorial problems.

In particular, for any combinatorial optimization problem there is a feasible set S and a real-valued objective function f defined on S. Then clearly the inequality of Definition 1 can be applied to this case. The definition does not always match up with the standard combinatorial definition. For example, it is unclear whether the approximation algorithms for problem $\triangle TSP$ given in Papadimitriou and Steiglitz are approximation algorithms in our sense. Perhaps these algorithms return the worst possible feasible solution. It would be interesting to investigate these possibilities.

Acknowledgements

Thanks to Rajeev Motwani of Stanford for originally suggesting to me the problem of approximation algorithms for nonlinear programming. Thanks to Panos Pardalos of Penn State for valuable discussions on this subject. Thanks to Mike Todd of Cornell for indicating how to improve the running time bound.

References

PHILIP E. GILL, WALTER MURRAY, AND MARGARET H. WRIGHT [1981], *Practical Optimization*, Academic Press, London.

GENE H. GOLUB AND CHARLES F. VAN LOAN [1989], *Matrix Computations, Second Edition*, Johns Hopkins University Press, Baltimore.

L. G. HAČIJAN [1979], A polynomial algorithm in linear programming, *Dokl. Akad. Nauk SSSR* 244:1093–1096. Translated in *Soviet Math. Dokl.* 20:191–194.

SANJIV KAPOOR AND PRAVIN M. VAIDYA [1986], Fast algorithms for convex quadratic programming and multicommodity flows, *Proc. 18th Annual ACM Symposium on Theory of Computing*, 147–159.

NAOKI KATOH AND TOSHIHIDE IBARAKI [1987], *Discrete Applied Mathematics* 17:39-66.

M. K. KOZLOV, S. P. TARASOV AND L. G. HAČIJAN [1979], Polynomial solvability of convex quadratic programming, *Dokl. Akad. Nauk SSSR* 248:1049–1051. Translated in *Soviet Math. Dokl.* 20:1108–1111.

KATTA G. MURTY AND SANTOSH N. KABADI [1987], Some NP-complete problems in quadratic and nonlinear programming, *Math. Progr.* 39:117–129.

CHRISTOS H. PAPADIMITRIOU AND KENNETH STEIGLITZ [1982], *Combinatorial Optimization: Algorithms and Complexity*, Prentice-Hall, New Jersey.

P. M. PARDALOS AND J. B. ROSEN [1987], *Constrained Global Optimization: Algorithms and Applications*, Springer Verlag Lecture Notes in Computer Science vol. 268, Berlin.

PANOS M. PARDALOS AND STEPHEN A. VAVASIS [1990], Quadratic programming with one negative eigenvalue is NP-hard, Technical Report 90-1160, Department of Computer Science, Cornell University, Ithaca, New York.

SARTAJ SAHNI [1974], Computationally related problems, *SIAM J. Comp.* 3:262–279.

MICHAEL J. TODD [1976], *The Computation of Fixed Points and Applications*, Springer, Berlin.

PRAVIN M. VAIDYA [1989], Speeding-up linear programming using fast matrix multiplication (extended abstract), *Proc. 30th Symp. Foundations of Computer Science,* 332–337.

STEPHEN A. VAVASIS [1990], Quadratic programming is in NP, *Inf. Proc. Lett.* 36:73-77.

YINYU YE AND EDISON TSE [1989], An extension of Karmarkar's projective algorithm for convex quadratic programming, *Math. Progr.* 44:157–179.

A New Complexity Result on Minimization of a Quadratic Function with a Sphere Constraint

Yinyu Ye [†]

Abstract

It has been shown that the problem of minimizing a general quadratic function with a sphere (or ellipsoid) constraint can be solved in $O(\log(1/\epsilon))$ iterations where ϵ is the error tolerance and each iteration solves a system of linear equations. In this paper a hybrid algorithm, combining Newton's method and a binary search, is developed to solve the problem in $O(\log(\log(1/\epsilon)))$ iterations. Thus, the problem with specified error is in the complexity class NC. This work is based on Smale's criterion for using Newton's method and is related to Renegar's result on approximating roots of polynomials.

1 Introduction

Given a matrix $Q \in R^{n \times n}$ and a vector $c \in R^n$, we consider the problem: Find an $x \in R^n$ for which there exists $\mu \geq 0$ such that

$(SQ) \qquad (Q + \mu I)x = c, \quad \|x\|^2 = 1, \quad \text{and} \quad Q + \mu I \text{ is PSD.}$

Here, PSD denotes "positive semi-definite" and $\|.\|$ designates the l_2 norm. This problem is popular since it essentially represents the optimality conditions for a sphere (or ellipsoid) constrained quadratic optimization problem

SQO minimize $q(x) = x^T Q x/2 - c^T x$
subject to $x \in S = \{x \in R^n : \|x\|^2 \leq 1\}$.

The Levenberg-Marquardt trust region method for nonlinear programming (e.g., Dennis and Schnabel [3], Gay [4], Goldfeld et al. [5], Moré

[†] Department of Management Sciences, The University of Iowa, Iowa City, IA 52242. Research supported in part by NSF Grant DDM-8922636.

[10], and Sorensen [15] among others) is based on sequentially solving SQO problems. SQO also plays an important role in interior-point algorithms for convex quadratic programming (e.g., Kapoor and Vaidya [8] and Ye and Tse [18] among others). SQO was recently used by Karmarkar et al. [9], Pardalos et al. [12] and Ye [17] for nonconvex quadratic programming, which is known to be an NP-hard problem.

If $c = 0$, then the problem becomes finding the least eigenvalue of Q, which is known an "easy" problem (e.g., Ben-Or et al. [1], Neff [11], and Renegar [13]). Thus, without loss of generality, we assume that $c \neq 0$ and $\|c\| = 1$, i.e., the system (SQ) is "normalized" for μ in some sense. We have proposed to use a binary search method for solving (SQ) [17]. The method generates an ϵ-approximate solution in $O(\log(1/\epsilon))$ iterations and each iteration solves a system of linear equations in $O(n^3)$ arithmetic operations. Thus, the arithmetic complexity of (SQ) is $O(n^3 \log(1/\epsilon))$. The bit complexity of (SQ) is analyzed by Vavasis and Zippel [16]. They further obtained a strong result that the decision problem associated with (SQ) can be solved in polynomial time. The decision problem is: Given an $n \times n$ integral matrix Q and an $n \times 1$ integral vector c, and a rational number r, does there exist an $x \in R^n$ such that $\|x\|^2 \leq 1$ and $q(x) \leq r$?

The goal of this paper is to develop a new algorithm, a hybrid of Newton's method and the binary search, to solve the problem in $O(\log(\log(1/\epsilon)))$ iterations. Thus, the arithmetic complexity of (SQ) is substantially improved over the previous result. This work is based on Smale's criterion for using Newton's method [14] and Renegar's result on approximating roots of polynomials [13]. Using Newton's method for SQO was previously discussed, for example, see Dennis and Schnabel [3] and Moré [10].

For a moment, we assume that

(A) $\qquad Q$ is positive semi-definite in (SQ).

If Q is not positive semi-definite, then let $\underline{\lambda} < 0$ be the least eigenvalue of Q, and let $\hat{Q} = Q - \underline{\lambda} I$ and $\hat{\mu} = \mu + \underline{\lambda}$. Then, (SQ) becomes

$$(\hat{Q} + \hat{\mu} I)x = c, \quad \hat{\mu} \geq 0, \quad \text{and} \quad \|x\|^2 = 1$$

where $\hat{Q}$ is positive semi-definite. Thus, if $\underline{\lambda}$ is known, the addition of Assumption (A) is without loss of generality. Even if $\underline{\lambda}$ is unknown, we will remove this assumption later using Renegar's algorithm and our algorithm to approximate $\underline{\lambda}$, and μ and x simultaneously. Note that one can form the characteristic polynomial of Q in $O(n^4)$ arithmetic operations and calculate all ϵ-approximate roots of the n- polynomial in

$O(n^2(\log n)(\log\log(1/\epsilon)) + n^3(\log n))$ arithmetic operations using Renegar's algorithm.

We now reformulate the problem (SQ). For any given $\mu \geq 0$, denote by $x(\mu)$ a solution x of the system of linear equations

$$(Q + \mu I)x = c. \tag{1}$$

Furthermore, let

$$\phi(\mu) = \|x(\mu)\|^2 - 1.$$

Then the problem is actually to find a root $\bar{\mu} \geq 0$ such that $\phi(\bar{\mu}) = 0$.

2 Some Properties of $\phi(\mu)$

Note that for $\mu \in R^{++} := (0, \infty)$, $x(\mu)$ is unique and continuous so that $\phi(\mu)$ is also continuous. Moreover, we have

Lemma 1. For $\mu \in R^{++}$
(i) $\phi(\mu) = c^T(Q + \mu I)^{-2}c - 1 > -1$ is continuous and infinitely differentiable. It is a monotonically decreasing function and the limit of $\phi(\mu)$ is -1 as $\mu \to \infty$.
(ii) For $k \geq 1$

$$\phi^{(k)}(\mu) = (-1)^k(k+1)! c^T(Q + \mu I)^{-(k+2)}c.$$

(iii) $\phi''(\mu) > 0$, so that $\phi(\mu)$ is a convex function.

Proof. To see (i) it is convenient to use the orthogonal decomposition of Q. Let

$$P^T Q P = \Lambda$$

where $P^T P = I$ and Λ is the diagonal matrix whose diagonal components are eigenvalues λ_j, $j = 1, 2, ..., n$, of Q. Let $\bar{c} = P^T c$. (Note that $\|\bar{c}\| = 1$ since $\|c\| = 1$.) Then the system of linear equations in (SQ) can be written as

$$P(\Lambda + \mu I)P^T x = c \quad \text{or} \quad (\Lambda + \mu I)P^T x = P^T c = \bar{c}.$$

For $\mu > 0$, we have

$$\phi(\mu) = \|x\|^2 - 1 = \|P^T x\|^2 - 1 = \|(\Lambda + \mu I)^{-1}\bar{c}\|^2 - 1 = \bar{c}^T(\Lambda + \mu I)^{-2}\bar{c} - 1$$

or

$$\phi(\mu) = \sum_{j=1}^n \frac{\bar{c}_j^2}{(\mu + \lambda_j)^2} - 1. \tag{2}$$

Obviously, $\phi(\mu)$ is continuous and infinitely differentiable in R^{++}.

From (2), we immediately see (ii) and (iii). $\square$

From Lemma 1, we see that $\phi(\mu)$ is an *analytic* function in R^{++}. Moreover, there exists at most one root $\bar{\mu}$ of ϕ in R^{++}. In fact, if $\bar{\mu} > 0$ is a root for ϕ, then

$$1 = \|x(\bar{\mu})\|^2 = c^T(Q + \bar{\mu}I)^{-2}c \leq \|c\|^2/\bar{\mu}^2$$

or

$$\bar{\mu} \leq \|c\| = 1. \tag{3}$$

We now study the behavior of $\phi(\mu)$ at $\mu = 0$. If Q is positive definite, then Lemma 1 extends to $\mu \in R^+ := [0, \infty)$. Suppose Q has some zero eigenvalues. For simplicity let $\lambda_1 = \lambda_2 = ... = \lambda_k = 0$ for some $k < n$ and $\lambda_j > 0$ for $j = k+1, k+2, ..., n$. Then, from (2) if

Case 1 $\qquad\qquad \bar{c}_j \neq 0 \quad$ for some $\quad j = 1, 2, ..., k$

$$\lim_{\mu \to 0^+} \phi(\mu) = \infty.$$

Otherwise,

Case 2 $\qquad\qquad \bar{c}_i = 0 \quad$ for all $\quad j = 1, 2, ..., k$

$$\lim_{\mu \to 0^+} \phi(\mu) = \sum_{j=k+1}^{n} \frac{\bar{c}_j^2}{\lambda_j^2} - 1.$$

In Case 1 the system

$$Qx = c \tag{4}$$

has no solution, and in Case 2 it has multiple solutions. In Case 2 denote the unique minimal-norm solution of (4) by Q^+c; then

$$\sum_{j=k+1}^{n} \frac{\bar{c}_j^2}{\lambda_j^2} = \|Q^+c\|^2,$$

where A^+ is the so-called pseudoinverse of A. Define $\phi(0) = \|Q^+c\|^2 - 1$ in Case 2; then $\phi(\mu)$ is continuous in R^+ too.

Note that Q^+c can be obtained using the standard Gauss elimination technique in $O(n^3)$ arithmetic operations. If Q and c are integral, then Q^+c is a rational vector whose size is bounded by a polynomial in L, the bit size of Q and c. This means that either

$$\|Q^+c\|^2 - 1 \geq 2^{-L} \quad \text{or} \quad \|Q^+c\|^2 - 1 \leq 0.$$

The following lemma is useful for our complexity analysis. A similar version of the lemma was given in [17].

Lemma 2. Let Q be a real symmetric matrix, and denote by $\lambda(Q)$ (or simply λ) an eigenvalue of Q. Moreover, let a_j's be the coefficients of the characteristic polynomial of Q: $\lambda^n + a_{n-1}\lambda^{n-1} + \ldots + a_1\lambda + a_0$, and let k be the smallest index for which $a_k \neq 0$. Then

$$|\lambda(Q)| \leq n \max_{i,j} |q_{ij}|,$$

and

either $\lambda(Q) = 0$ or $|\lambda(Q)| > \dfrac{|a_k|}{\max(1, |a_j|) + |a_k|}.$

Proof. The first inequality can be found from Golub and Van Loan [7]. For the second one, note that

$$\lambda^n + a_{n-1}\lambda^{n-1} + \ldots + a_1\lambda + a_0 = 0,$$

and a_j's are all summations of the minors of Q. Then if $\lambda \neq 0$ and $|\lambda| < 1$,

$$\begin{aligned}
|a_k| &= |\lambda^n + a_{n-1}\lambda^{n-1} + \ldots + a_{k+1}\lambda^{k+1}| \\
&\leq \max(1, |a_j|)|\lambda|^{k+1}(|\lambda|^{n-k-1} + |\lambda|^{n-k-2} + \ldots + 1) \\
&< \max(1, |a_j|)|\lambda|^{k+1}/(1 - |\lambda|) \\
&< \max(1, |a_j|)|\lambda|/(1 - |\lambda|).
\end{aligned}$$

Thus,

$$|\lambda| > \frac{|a_k|}{\max(1, |a_j|) + |a_k|}. \qquad \square$$

In the following, let the error ϵ satisfy

$$0 < \epsilon \leq \frac{|a_k|}{\max(1, |a_j|) + |a_k|}. \tag{5}$$

Furthermore, in Case 1 let ϵ satisfy

$$0 < \epsilon^2 \leq \sum_{j=1}^{k} \bar{c}_j^2 \tag{6}$$

where Q has zero eigenvalues λ_j, $j = 1, 2, \ldots, k$. Again, one can verify that if Q and c are integral, then the numbers on the right-hand sides of (5) and (6) are rationals, and their sizes are bounded by polynomials in L.

Now, we develop

Lemma 3. If the root $\bar{\mu} > 0$, i.e., the solution to the system (4) does not exist (Case 1), or a solution exists (Case 2) but $\phi(0) = \|Q^+c\|^2 - 1 \geq \epsilon$, then
$$\phi(\epsilon^3) > 0,$$
which implies that $\bar{\mu} > \epsilon^3$.

Proof. In the first case, the solution to the system (4) has no solution. Thus from (6), $\sum_{j=1}^{k} \bar{c}_j^2 \geq \epsilon^2$ where $\lambda_j = 0$ for $j = 1, 2, ..., k$. Hence, from (2)

$$\phi(\epsilon^3) > \sum_{j=1}^{k} \frac{\bar{c}_j^2}{\mu^2} - 1 \geq \epsilon^2/\epsilon^6 - 1 > 0.$$

In the second case, the system (4) has a solution but $\phi(0) = \|Q^+c\|^2 - 1 \geq \epsilon$. Note

$$\phi(0) - \phi(\epsilon^3) = \sum_{j=k+1}^{n} \frac{\bar{c}_j^2}{\lambda_j^2} - \sum_{j=k+1}^{n} \frac{\bar{c}_j^2}{(\epsilon^3 + \lambda_j)^2}$$

$$= \sum_{j=k+1}^{n} \left(\frac{\bar{c}_j^2}{\lambda_j^2} - \frac{\bar{c}_j^2}{(\epsilon^3 + \lambda_j)^2} \right)$$

$$= \sum_{j=k+1}^{n} \frac{\bar{c}_j^2}{\lambda_j^2} \left(1 - \frac{\lambda_j^2}{(\epsilon^3 + \lambda_j)^2} \right)$$

$$= \sum_{j=k+1}^{n} \frac{\bar{c}_j^2}{\lambda_j^2} \left(\frac{2\lambda_j \epsilon^3 + \epsilon^6}{(\epsilon^3 + \lambda_j)^2} \right)$$

$$\leq \sum_{j=k+1}^{n} \frac{\bar{c}_j^2}{\lambda_j^2} \left(\frac{2\lambda_j \epsilon^3 + \epsilon^6}{\lambda_j^2} \right)$$

$$= \sum_{j=k+1}^{n} \frac{\bar{c}_j^2}{\lambda_j^2} \left(\frac{2\epsilon^3}{\lambda_j} + \frac{\epsilon^6}{\lambda_j^2} \right)$$

$$\leq \sum_{j=k+1}^{n} \frac{\bar{c}_j^2}{\lambda_j^2} (2\epsilon^2 + \epsilon^4) \qquad \text{(By Lemma 2 and (5))}$$

$$= (2\epsilon^2 + \epsilon^4) \sum_{j=k+1}^{n} \frac{\bar{c}_j^2}{\lambda_j^2} = (2\epsilon^2 + \epsilon^4)(\phi(0) + 1).$$

Thus,

$$\phi(\epsilon^3) \geq (1 - 2\epsilon^2 - \epsilon^4)\phi(0) - (2\epsilon^2 + \epsilon^4) \geq (1 - 2\epsilon^2 - \epsilon^4)\epsilon - (2\epsilon^2 + \epsilon^4) > 0$$

for ϵ small enough. □

Based on these lemmas, we have used a binary search algorithm to solve the problem (SQ). It consists of two phases.

Phase 1. Check to see if the system $Qx = c$ has a solution. If not, go to Phase 2; otherwise compute Q^+c. If $\|Q^+c\| - 1 \geq \epsilon$ then go to Phase 2; otherwise the problem is solved by assigning $x = Q^+c + v$ where v is a homogenous solution to $Qv = 0$ and v satisfies

$$(1-\epsilon)^2 < \|Q^+c\|^2 + \|v\|^2 = \|Q^+c + v\|^2 < (1+\epsilon)^2.$$

(Note that $v^T Q^+ c = 0$ here.) In this case the number of the solutions for the problem (SQ) equals the number of zero-eigenvalues of Q.

Phase 2. Apply the standard binary search for the unique root $\bar{\mu}$ of ϕ over the interval $[\epsilon^3, 1]$, which is guaranteed to contain $\bar{\mu}$.

Clearly, Phase 1 will be completed in $O(n^3 + \log\log(1/\epsilon))$ arithmetic operations, and Phase 2 will be completed in $O(\log(1/\epsilon))$ bisection steps to achieve $|\mu - \bar{\mu}| \leq \epsilon$, and each step solves the system of linear equations (1) to evaluate ϕ. We see that the complexity of the problem (SQ) mainly depends on the complexity of Phase 2. This is why we are interested in developing a new approach to improve the complexity of Phase 2.

3 Smale's Criterion for Using Newton's Method

Before describing the new algorithm, we will study Smale's criterion for using Newton's method. A point μ^0 is called an approximate root of ϕ if the Newton sequence with starting point μ^0, i.e.,

$$\mu^{i+1} = \mu^i - \frac{\phi(\mu^i)}{\phi'(\mu^i)}, \quad i = 0, 1, 2, ...,$$

satisfies the property (see Proposition 4.1 of Renegar [13])

$$|\mu^{i+1} - \bar{\mu}| \leq 8(1/2)^{2^i} |\mu^0 - \bar{\mu}|.$$

Obviously, the sequence $|\mu^{i+1} - \bar{\mu}|$ converges the error ϵ in $O(\log(\log(|\mu^0 - \bar{\mu}|/\epsilon)))$ iterations. Smale further developed a checkable criterion for qualification of an approximate root ([14], also see Goldstein [6]).

Theorem 1 (Smale). Let ϕ be an *analytic* function. Then, if μ in the domain of ϕ satisfies

$$\sup_{k>1} \left| \frac{\phi^{(k)}(\mu)}{k!\phi'(\mu)} \right|^{1/(k-1)} \leq (1/8) \left| \frac{\phi'(\mu)}{\phi(\mu)} \right|.$$

μ is an approximate root of ϕ.

Using Smale's theorem, we develop the following customized theorem.

Theorem 2. Let ϕ be an analytic function in R^{++} and let ϕ be convex and monotonically decreasing. Furthermore, for $\mu \in R^{++}$ and $k > 1$ let

$$\left|\frac{\phi^{(k)}(\mu)}{k!\phi'(\mu)}\right|^{1/(k-1)} \leq \alpha\mu^{-1}$$

for some constant $\alpha > 0$. Then, if the root $\bar{\mu} \in [\hat{\mu}, (1+1/8\alpha)\hat{\mu}] \subset R^{++}$, $\hat{\mu}$ is an approximate root of ϕ.

Proof. Noting that ϕ is a monotonically decreasing function, we must have $\phi'(\mu) < 0$ in R^{++}. Assume $\phi(\hat{\mu}) > 0$, since otherwise $\hat{\mu}$ is the exact root of ϕ. Let

$$\mu^+ = \hat{\mu} - \frac{\phi(\hat{\mu})}{\phi'(\hat{\mu})}. \tag{7}$$

Then, we have $\mu^+ > \hat{\mu}$ because $\phi(\hat{\mu}) > 0$ and $\phi'(\hat{\mu}) < 0$. Also since ϕ is a convex function we must have

$$\phi(\mu^+) \geq \phi(\hat{\mu}) - \phi'(\hat{\mu})(\mu^+ - \hat{\mu}) = 0,$$

which implies

$$(1+1/8\alpha)\hat{\mu} \geq \bar{\mu} \geq \mu^+.$$

Thus, from (7)

$$\left|\frac{\phi'(\hat{\mu})}{\phi(\hat{\mu})}\right| = \frac{1}{\mu^+ - \hat{\mu}}$$
$$\geq \frac{1}{(1+1/8\alpha)\hat{\mu} - \hat{\mu}}$$
$$= 8\alpha\hat{\mu}^{-1}.$$

Therefore, for all $k > 1$

$$(1/8)\left|\frac{\phi'(\hat{\mu})}{\phi(\hat{\mu})}\right| \geq \alpha\hat{\mu}^{-1} \geq \left|\frac{\phi^{(k)}(\hat{\mu})}{k!\phi'(\hat{\mu})}\right|^{1/(k-1)},$$

which indicates that $\hat{\mu}$ is an approximate root of ϕ due to Theorem 1. □

In Theorem 2, we see that the interval $[\hat{\mu}, (1+1/8\alpha)\hat{\mu}]$, where Newton's method can prevail, is enlarged as $\hat{\mu}$ increased. Also, the conditions of ϕ being convex and decreasing may not be restrictive. For example, if ϕ is convex and increasing, we can let $y := 1/x$. Then, $\psi(y) := \phi(1/y)$ will be convex and decreasing.

We now develop the following corollary.

Corollary 1. Let ϕ be given by (2) for our problem. Then, given any $\hat{\mu} \in R^{++}$, if the root $\bar{\mu} \in [\hat{\mu}, (1+1/12)\hat{\mu}]$, $\hat{\mu}$ is an approximate root of ϕ.
Proof. From Lemma 1, ϕ is an analytic, monotonically decreasing, and convex function. Moreover, for $\mu \in R^{++}$ and $k > 1$

$$\left|\frac{\phi^{(k)}(\mu)}{k!\phi'(\mu)}\right|^{1/(k-1)} = \left|\frac{(k+1)!c^T(Q+\mu I)^{-(k+2)}c}{2k!c^T(Q+\mu I)^{-3}c}\right|^{1/(k-1)}$$

$$= (\frac{k+1}{2})^{1/(k-1)} \left(\frac{c^T(Q+\mu I)^{-(k+2)}c}{c^T(Q+\mu I)^{-3}c}\right)^{1/(k-1)}$$

$$\leq (1+\frac{k-1}{2})^{1/(k-1)}(\mu^{-k+1})^{1/(k-1)}$$

$$\leq 1.5\mu^{-1}.$$

Using $\alpha = 1.5$ in Theorem 2, we have the desired result. □

4 The Hybrid Algorithm

Based on Corollary 1 we propose a hybrid algorithm to replace Phase 2. The basic idea of the algorithm is as follows: First, we may construct a sequence of points in $[\epsilon^3, 1]$:

$$\hat{\mu}^0 = \epsilon^3, \ \hat{\mu}^1 = (1+1/12)\hat{\mu}^0, ..., \text{ and } \hat{\mu}^j = (1+1/12)\hat{\mu}^{j-1}, ...$$

until $\hat{\mu}^j = \hat{\mu}^J \geq 1$. Obviously the total number, J, of these points is bounded by $O(\log(1/\epsilon))$. Moreover, define a sequence of intervals

$$I^j = [\hat{\mu}^{j-1}, \hat{\mu}^j] = [\hat{\mu}^{j-1}, (1+1/12)\hat{\mu}^{j-1}].$$

We immediately see that if the root $\bar{\mu}$ of ϕ is in any one of these intervals, say I^j, then the front point $\hat{\mu}^{j-1}$ of I^j is an approximate root of ϕ because of Corollary 1 and, therefore, Newton's method prevails. Obviously, the length of each of these intervals is less than 1 so that Newton's method generates an μ with $|\mu - \bar{\mu}| \leq \epsilon$ in $O(\log\log(1/\epsilon))$ iterations, where each iteration solves a system of linear equations to evaluate ϕ and ϕ'.

Now the question is how to identify the interval that contains $\bar{\mu}$. This is easy since we can again apply the binary search; but this time we bisect the number of intervals. For example, the first step of the bisection is to single out $\hat{\mu}^{\hat{j}}$ where $\hat{j} = [J/2]$, and decide whether $\bar{\mu} \leq \hat{\mu}^{\hat{j}}$ or not. Thus, each bisection step reduces the total number of the intervals, each of which possibly contains the root $\bar{\mu}$, by a half. Since the total number of intervals is $O(\log(1/\epsilon))$, in at most $O(\log\log(1/\epsilon))$ bisection steps we shall locate the interval that contains $\bar{\mu}$. Again, each step solves a system of linear equations to evaluate ϕ.

New Phase 2:

Step 1. Let $\beta = 1 + 1/12$. Compute a sequence of numbers

$$b(k) = \beta^{2^k}$$

for $k = 0, 1, ..., K$, where K is the smallest integer such that

$$\beta^{2^K} \geq 1/\epsilon^3,$$

i.e.,

$$K = \lceil \log\log(1/\epsilon^3) - \log\log\beta \rceil.$$

This step uses K multiplications and K comparisons.

Let $\hat{x} = \epsilon^3$ and $\hat{k} = K$.

Step 2. Evaluate ϕ at $b(\hat{k}-1)\hat{x}$ by solving (1). Note that $b(\hat{k}-1)$ is the second number from the last in the sequence computed in Step 1. Step 2 uses one multiplication and one evaluation.

Step 3. If $\phi(b(\hat{k}-1)\hat{x}) > 0$ then $\hat{x} := b(\hat{k}-1)\hat{x}$. Let $\hat{k} := \hat{k} - 1$ and go to Step 2. This step uses one comparison.

Obviously, the loop between Step 2 and Step 3 will terminate in K steps. The total number of operations in Steps 1, 2, and 3 is $4K$. The resulted $\hat{x}$ should be an approximate root of ϕ, since $\bar{x} \in [\hat{x}, \beta\hat{x}]$.

Step 4. Use Newton's method with $\hat{x}$ as the starting point. It generates an x with $|x - \bar{x}| \leq \epsilon^3$ in $O(\log\log(1/\epsilon^3))$ iterations, and each iteration uses two function evaluations (ϕ and ϕ'), one division, and one substraction.

To summarize, we have

Theorem 3. Let Q be positive semi-definite. Then, the hybrid algorithm solves the problem (SQ) in $O(\log\log(1/\epsilon))$ iterations and each iteration uses $O(n^3)$ arithmetic operations.

In solving the system of linear equations (1) for varying values of μ, one may first symmetrically reduce Q to tridiagonal form with Householder transformations (e.g., see Golub and Van Loan [7]), Thus, in $O(n^3)$ arithmetic operations one can write $Q = ZVZ^T$ where Z is orthogonal and V is symmetric tridiagonal. Then $Q + \mu I = Z(V + \mu I)Z^T$. Thus, the system of (1) can be solved in $O(n^2)$ arithmetic operations in each iteration of the hybrid algorithm. Therefore, the arithmetic complexity of (SQ) can be further reduced to $O(n^2(\log\log(1/\epsilon)))$. This approach was pointed out by Vavasis in our private communication.

Alternatively, the system of (1) can be solved in $O(\log^2 n)$ time on $O(n^4)$ processors (e.g., Csanky [2]), assuming each arithmetic operations requires one unit time. Therefore, we have

Corollary 2. Let Q be positive semi-definite. Then, the parallel complexity of the problem (SQ) is $O((\log^2 n)(\log\log(1/\epsilon)))$ time and $O(n^4)$ processors, i.e., the problem with specified error tolerance is in the complexity class NC.

We now remove Assumption (A). Note that our previous lemmas and theorems still hold when Q and μ are replaced by $Q - \underline{\lambda} I$ and $\mu + \underline{\lambda}$, respectively. We modify Phase 1 in the algorithm of Section 2 as follows:
New Phase 1: First, formulate the characteristic polynomial of Q in $O(n^4)$ arithmetic operations. Then use Renegar's algorithm to calculate an approximate least eigenvalue of Q, $\hat{\lambda}$, such that $|\hat{\lambda} - \underline{\lambda}| \leq \epsilon^4$ in $O(n^2(\log n)(\log\log(1/\epsilon)) + n^3(\log n))$ arithmetic operations. Now, we have $-\hat{\lambda} + \epsilon^4 \geq -\underline{\lambda}$, and can tell whether $\bar{\mu} + \underline{\lambda} = 0$ or not by checking whether $\phi(-\hat{\lambda} + \epsilon^4) \leq 0$ or not. If $\phi(-\hat{\lambda} + \epsilon^4) \leq 0$, then $\phi(-\underline{\lambda} + \epsilon^3) \leq 0$, since $-\underline{\lambda} + \epsilon^3 \geq -\hat{\lambda} - \epsilon^4 + \epsilon^3 \geq -\hat{\lambda} + \epsilon^4$ and $\phi(\mu)$ is monotonically decreasing for $\mu \geq -\underline{\lambda}$. Thus, we must have $\bar{\mu} + \underline{\lambda} = 0$ from Lemma 2. In this case, we repeat the same procedure in Phase 1 to obtain an approximate $\hat{\mu} = -\hat{\lambda} + \epsilon^4$ and its corresponding $x(\hat{\mu})$. Note that

$$0 \leq \hat{\mu} + \underline{\lambda} \leq 2\epsilon^4.$$

Otherwise, use $\hat{\lambda} - \epsilon^4$ to approximate $\underline{\lambda}$ and go to the new Phase 2. Overall, we have
Theorem 4. The hybrid algorithm solves the problem (SQ) in

$$O(n^2(\log n)(\log\log(1/\epsilon)) + n^4)$$

arithmetic operations.

It is known that the (real) eigenvalue problem is in the complexity class NC (e.g., Ben-Or et al. [1] and Neff [11]). Therefore, we have
Corollary 3. The problem (SQ) with specified error tolerance is in the complexity class NC.

Acknowledgement. I would like to thank Jim Renegar and Steve Vavasis for their very helpful discussions and comments on this paper.

References

[1] M. Ben-Or, E. Feig, D. Kozen, and P. Tiwari, "A fast parallel algorithm for determining all roots of a polynomial with real roots," *SIAM J. Comput.* **17** (1988) 1081-1092.

[2] L. Csanky, "Fast parallel matrix inversion algorithms," *SIAM J. Comput.* **4** (1976) 618-623.

[3] J. E. Dennis, Jr. and R. B. Schnabel, *Numerical Methods for Unconstrained Optimization and Nonlinear Equations*, Prentice-Hall, Englewood Cliffs, New Jersey (1983).

[4] D. M. Gay, "Computing optimal locally constrained steps," *SIAM J. Sci. Statis. Comput.* **2** (1981) 186-197.

[5] S. M. Goldfeld, R. E. Quandt, and H. F. Trotter, "Maximization by quadratic hill climbing," *Econometrica* **34** (1966) 541-551.

[6] A. A. Goldstein, "A modified Kantorovich inequality for the convergence of Newton's method", manuscript (1989), to appear in *Contemporary Mathematics*.

[7] G. H. Golub and C. F. Van Loan, *Matrix Computations* (Second Edition), Johns Hopkins University Press, Baltimore (1989).

[8] S. Kapoor and P. Vaidya, "Fast algorithms for Convex quadratic programming and multicommodity flows," *Proc. 18th Annual ACM Symp. Theory Comput.* (1986) 147-159.

[9] N. Karmarkar, M. G. C. Resende and K. G. Ramakrishnan, "An interior point algorithm to solve computationally difficult set covering problems," Technical Report, AT&T Bell Laboratories (Murray Hill, NJ, 1989).

[10] J. J. Moré, "The Levenberg-Marquardt algorithm: implementation and theory," *Numerical Analysis* (G. A. Watson, ed.), Springer-Verlag, New York, New York (1977).

[11] C. A. Neff, "Specified precision polynomial root isolation is in NC," manuscript, Modeling Science Project Manufacturing Research, IBM Thomas J. Watson Center (Yorktown Heights, NY, 1990).

[12] P. M. Pardalos, C. Han and Y. Ye, "Computational aspects of an interior point algorithm for nonconvex quadratic programming," Working Paper, Department of Computer Science, The Pennsylvania State University (University Park, PA, 1990).

[13] J. Renegar, "On the worst-case arithmetic complexity of approximating zeros of polynomials," *Journal of Complexity* **3** (1987) 90-113.

[14] S. Smale, "Newton's method estimates from data at one point," in *The Merging of Disciplines: New Directions in Pure, Applied and Computational Mathematics* (R. Ewing, K. Gross, and C. Martin, Eds.), Springer-Verlag, New York (1986).

[15] D. C. Sorensen, "Newton's method with a model trust region modification," Report ANL-80-106, Argonne National Laboratory (Argonne, Il, 1980).

[16] S. A. Vavasis and R. Zippel, "Proving polynomial-time for sphere-constrained quadratic programming," Technical Report 90-1182, Department of Computer Science, Cornell University (Ithaca, NY, 1990).

[17] Y. Ye, "On the affine scaling algorithm for nonconvex quadratic programming," manuscript (1988), to appear in *Mathematical Programming*.

[18] Y. Ye and E. Tse, "An extension of Karmarkar's projective algorithm for convex quadratic programming," *Mathematical Programming* **44** (1989) 157-179.

Hamiltonian Cycles, Quadratic Programming, and Ranking of Extreme Points

Ming Chen and Jerzy A. Filar
Department of Mathematics and Statistics
University of Maryland Baltimore County
U. S. A.

ABSTRACT. In this paper we derive new characterizations of the Hamiltonian cycles of a directed graph. Our results are obtained via an embedding of this combinatorial optimization problem in a suitably ϵ-perturbed controlled Markov chain.

The frequency space of these controlled Markov chains is a known polyhedron $X(\epsilon)$ whose extreme points correspond to deterministic controls. We show that the deterministic controls whose frequency vectors satisfy just one more linear constraint are precisely the Hamiltonian cycles of the graph. From this and earlier results of Filar and Krass [2], it follows that the problem of finding a Hamiltonian cycle reduces to that of minimizing a certain indefinite quadratic over a polyhedral set $M(\epsilon)$ that is simpler than the feasible region used in [2].

In addition, we show that a simple linear objective function can be used to classify all extreme points of $X(\epsilon)$. In particular that function can take at most N distinct values, where N is the number of nodes, at the extreme points of $X(\epsilon)$. The Hamiltonian cycles are in one-to-one correspondence with those extreme points at which the linear function takes on its second lowest possible value.

The above results suggest a number of new computational approaches to the well-known Hamiltonian Cycle Problem. These approaches are currently under investigation.

1. Introduction

In this paper we consider the Hamiltonian Cycle Problem (HCP). It would be impractical to supply a complete bibliography of works on this problem, instead we refer the reader to the book of Papadimitriou and Steiglitz [5]. We begin with a brief description of only one version of the HCP.

In graph theoretic terms, the problem is to find a simple cycle of N arcs, that is a Hamiltonian cycle, in a directed graph G with N nodes and with arcs (i,j), or determine that none exist.

In this paper we follow the approach of Filar and Krass [2] to the HCP:

Consider a moving object tracing out a directed path on the graph G with its movement "controlled" by a function f mapping the nodes N into the arcs A. This function induces a "zero-one" $N \times N$ Markov matrix $P(f)$ whose positive entries correspond to the arcs "selected" by f at the respective nodes (see Section 2 for more detail). Suppose further, that this motion continues forever, and we regard $P(f)$ as a Markov Chain, and consider its "stationary distribution", contained in its *limit Cesaro-sum matrix*:

$$P^*(f) := \lim_{T \to \infty} \frac{1}{T} \sum_{t=1}^{T} P^{t-1}(f), \text{ where } P^0(f) := I_N.$$

In [2] the relationship between the ergodic class/transient state structure of such Markov Chains, and the possible cycles in the graph was studied. It was shown in [2] that, after a suitable perturbation introduced in Section 2 below, from the limiting behavior of these chains it is possible to derive a new characterization of Hamiltonian cycles.

In particular, it was shown that the Hamiltonian cycles of a graph correspond precisely to the global minima (with objective function equal to 0) of an indefinite quadratic program of the general form:

$$\begin{aligned} \text{minimize } & [\mathbf{x}^T Q \mathbf{x}] \\ \text{Subject to}: A_1 \mathbf{x} &= \mathbf{b}_1 \\ A_2 \mathbf{x} &\geq \mathbf{b}_2(\epsilon) \\ \mathbf{x} &\geq \mathbf{0}, \end{aligned}$$

where Q is a block-diagonal matrix with each block $Q_i = J_i - I_i$, and where the matrices J_i and I_i are the matrix of all ones and the identity matrix respectively. If a is the number of arcs in the graph G, then the dimensions of both A_1 and A_2 are $(N+1) \times a$. These matrices (and the vectors $\mathbf{b}_1$ and $\mathbf{b}_2(\epsilon)$) are from the formulation in Filar and Krass [2]. In sequel we shall require only A_1 and $\mathbf{b}_2(\epsilon)$ which are determined by constraints (C1)–(C2) in Section 2 below.

In Section 3 we show that the feasible region of the above quadratic program can be simplified by replacing $A_2 \mathbf{x} \geq \mathbf{b}_2(\epsilon)$ with a single equality constraint (see Theorem 3.2). In the process we classify all the extreme points of the "frequency space" $X(\epsilon)$ of the controlled ϵ-perturbed Markov chain problem in which the graph has been embedded. This classification is based on the fact that a certain linear objective function can take at most N distinct values at the extreme points of $X(\epsilon)$ (see Remark 3.3). The Hamiltonian cycles of G are in one-to-one correspondence with those extreme points at which the linear objective function takes on its second lowest value (see Remark 3.3). A byproduct of our approach is a new linear programming formulation for the problem of finding a simple cycle with the minimal number of arcs.

Of course, it is well known that the HCP is an NP-complete problem. At this stage it is not clear whether new computational approaches based on our Theorem 3.2, and the observations contained in Remarks 3.1–3.3 will lead to significant algorithmic improvements. The purpose of Remarks 3.2–3.3 is to suggest that there may be algorithms that move from one extreme point of $X(\epsilon)$ to another and are guided by the values of a convex objective function.

2. Definitions and Preliminaries

A discrete Markovian decision process Γ is observed at discrete time points $t = 1, 2, \ldots,$. The *state space* is denoted by $E = \{1, 2, \ldots, N\}$. With each state $i \in E$, we associate a finite set $A(i)$ of "actions". At any time point t the system is in one of the states and an action has to be chosen by the decision maker. If the system is in state i and action $a \in A(i)$ is chosen, then an immediate reward r_{ia} is earned and the process moves to a state $j \in E$ with transition probability p_{iaj}, where $p_{iaj} \geq 0$ and $\sum_{j=1}^{N} p_{iaj} = 1$.

Henceforth, the process Γ will be synonymous with the four-tuple $\langle E, A, r, p \rangle$, where $A = \{A(i) \mid i \in E\}$, $r = \{r_{ia} \mid (i, a) \in E \times A(i)\}$ and $p = \{p_{iaj} \mid (i, a, j) \in E \times A(i) \times E\}$. Sometimes p will be referred to as the *law of motion* of Γ.

A *decision rule* f^t at time t is a function which assigns a probability to the event that action a is taken at time t. In general, a *policy* f is a sequence of decision rules: $f = (f^1, f^2, \ldots, f^t \ldots)$, where f^t may depend on all realized states up to and including time t, and on all realized actions up to time t. A policy is called *stationary* if all its decision rules are identical and depend only on the current state. A *deterministic policy* is a stationary policy with nonrandomized decision rules. In particular, we shall denote a stationary policy f by the collection of probability vectors $\mathbf{f}(i) = (f(i,1), f(i,2), \ldots, f(i,n_i))$, where $n_i = |A(i)|$ for $i = 1, \ldots, N$. Here $f(i,k)$ is the probability that action k is chosen in state i whenever that state is visited. If f is deterministic, each $f(i,a) \in \{0,1\}$ and hence we shall write $f = (f(1), \ldots, f(N))$, where $f(i)$ now denotes the action chosen whenever state i is visited.

Let X_t be the state at time t, Y_t be the action at time t, and $P_f(X_t = j, Y_t = a \mid X_1 = i)$ be the conditional probability that at time t the state is j and the action taken is a, given that the initial state is i and the decision maker uses a policy f. Now if R_t denotes the reward at time t, then for any policy f and initial state i the expectation of R_t is given by

$$E_f(R_t, i) = \sum_{j \in E} \sum_{a \in A(j)} P_f(X_t = j, Y_t = a \mid X_1 = i) r_{ja}. \qquad (2.1)$$

The manner in which we aggregate the resulting stream of expected rewards $\{E_f(R_t, i); t = 1, 2, \ldots\}$ defines the Markov Decision Process discussed in the sequel:

Average Reward Markovian Decision Process (AMD): Here the corresponding overall reward is defined by

$$\phi_i(f) = \liminf_{T \to \infty} \frac{1}{T} \sum_{t=1}^{T} E_f(R_t, i).$$

A policy, f^*, is called *optimal* if for every $i \in E$

$$\phi_i(f^*) = \max_f \phi_i(f). \qquad (2.2)$$

We shall assume that the *initial distribution* on the states of Γ is the given vector $\gamma = (\gamma_1, \ldots, \gamma_N)^T$, with $\gamma_i = P(X_1 = i)$ and $\sum_{i=1}^{N} \gamma_i = 1$.

Given a stationary policy f, let $p_{ij}(f) := \sum_{a \in A(i)} p_{iaj} f(i, a)$. It is now clear that f defines a Markov Chain with the probability transition matrix

$$P(f) := \left(p_{ij}(f)\right)_{i,j=1}^{N}. \tag{2.3}$$

For any policy f, initial distribution γ, $j \in E$ and $a \in A(j)$, define

$$x_{ja}^T(f) := \frac{1}{T} \sum_{t=1}^{T} \sum_{i=1}^{N} \gamma_i P_f(X_t = j, Y_t = a \mid X_1 = i). \tag{2.4}$$

Further, let $X(f)$ denote the set of all limit points of the vectors $\{\mathbf{x}^T(f) \mid T = 1, 2, \ldots\}$, where $\mathbf{x}^T(f)$ is a $\sum_{i=1}^{N} |A(i)|$-dimensional vector with entries given by (2.4). If $X(f) = \{\mathbf{x}(f)\}$, a singleton, then the entries $x_{ja}(f)$ of $\mathbf{x}(f)$ can be interpreted as the *long-run expected state-action frequencies* induced by f. Similarly, the long-run expected frequencies of visits to any state $j \in E$ under f are given by

$$x_j(f) := \sum_{a \in A(j)} x_{ja}(f). \tag{2.5}$$

A Markov Decision Process is called *unichain* if for any deterministic policy f, the Markov chain induced by $P(f)$ has one ergodic set plus a (perhaps empty) set of transient states.

Consider the following linear program (LP1):

$$\max \sum_{i \in A} \sum_{a \in A(i)} r_{ia} x_{ia}$$

subject to:

(C1) $\quad \sum_{i \in E} \sum_{a \in A(i)} (\delta_{ij} - p_{iaj}) x_{ia} = 0, \quad j \in E$

(C2) $\quad \sum_{i \in E} \sum_{a \in A(i)} x_{ia} = 1$

(C3) $\quad x_{ia} \geq 0, \quad i \in E, a \in A(i),$

where δ_{ij} is the Kronecker delta. Let X denote the feasible region of the above program, and $C(S)$ denote the class of stationary strategies of the unichain MDP. Now consider the map $T : X \to C(S)$, where $T(\mathbf{x}) = f_x$ is defined by

$$f_x(i,a) = \begin{cases} \dfrac{x_{ia}}{x_i}, & \text{if } x_i = \sum_{a \in A(i)} x_{ia} > 0 \\ 1, & \text{if } x_i = 0 \text{ and } a = a_1 \\ 0, & \text{if } x_i = 0 \text{ and } a \neq a_1, \end{cases} \quad (2.6)$$

for every $i \in E$ and $a \in A(i)$, where a_1 denotes the first available action in a given state according to some fixed ordering. Also consider the map $\hat{T} : C(S) \to X$ where $\hat{T}(f) = \mathbf{x}(f)$ is defined by (consistently with (2.4) and (2.5)):

$$x_{ia}(f) = p_i^*(f) f(i,a), \quad i \in E, \quad a \in A(i). \quad (2.7)$$

In the above, $p_i^*(f)$ is the i-th entry of the unique fixed probability vector (stationary distribution vector) of $P(f)$. The transformations T and $\hat{T}$ have been studied by a number of authors (e.g., see Derman [1], Kallenberg [4]). Those of their properties that we shall require in the sequel are summarized in the following result which can be reconstructed with the help of [1] and [4].

Theorem 2.1 *Let Γ be a unichain Markov Decision Process, and (LP1), $C(S), X, T$ and $\hat{T}$ be as defined above. Then*

(i) For all $f \in C(S)$ and any initial state $k \in E$

$$\phi_k(f) = \sum_{i \in E} \sum_{a \in A(i)} r_{ia} x_{ia}(f).$$

(ii) If $\mathbf{x}^0$ is optimal in (LP1) then $f_{\mathbf{x}^0} = T(\mathbf{x}^0)$ is optimal in Γ. Conversely, if f^0 is an optimal stationary policy in Γ, then $\hat{T}(f^0) = \mathbf{x}(f^0)$ is optimal in (LP1).

(iii) For all $\mathbf{x} \in X$, $\hat{T}(T(\mathbf{x})) = \mathbf{x}$.

(iv) If $L(S) = \{\mathbf{x}(f) \mid f \in C(S)\}$, then $X = L(S)$, and the extreme points of X correspond to those $\mathbf{x}$ for which $f_{\mathbf{x}}$ is a deterministic policy.

Embedding of a Directed Graph in an MDP:

Consider a directed weighted graph G with the vertex set $V = \{1, \ldots, N\}$, the arc set $\mathcal{A}$ and with weights c_{ij} associated with the arcs (i, j). The first MDP which we shall associate with G will be the process $\Gamma =< E, A, r, p >$, $E = \{1, 2, \ldots, N\} = $ set of vertices of G, $A(i) = \{j \in E \mid (i,j) \in \mathcal{A}\}$ for each $i \in E$ and $A = \bigcup_{i=1}^{N} A(i)$, $r = \{r_{ij} = -c_{ij} \mid i \in E, j \in A(i)\}$, and $p = \{p_{iaj} \mid i \in E, a \in A(i), j \in E\}$ with $p_{iaj} = \delta_{aj}$, the Kronecker delta. Also, we assume that 1 is the initial state. We shall say that a deterministic policy f in Γ *is a Hamiltonian cycle* in G if the set of arcs $\{(1, f(1)), (2, f(2)), \ldots, (N, f(N))\}$ is a Hamiltonian cycle in G. If the above set of arcs contains cycles of length less than N, we say that f *has subcycles in* G. Note that if f is a Hamiltonian cycle, then the Markov Chain $P(f)$ is unichain and it follows form (2.5) that $x_j(f) = \frac{1}{N}$ for every $j \in E$.

The preceding embedding of the graph G in a Markov Decision Process Γ suggests that analysis be carried out in the space X of long-run state-action frequencies, the union of $\{\mathbf{x}(f)\}$ over all policies f. This approach seems attractive since characterization of the space X as a polyhedral set is now available (e.g., see Hordijk and Kallenberg [3]). However, we point out that MDP Γ defined above is not unichain. This leads to several technical difficulties: it is known ([3]) that there are points in that space that cannot be obtained from any stationary policy via the transformation $\hat{T}$, and furthermore the long-run frequency $\mathbf{x}(f)$ is not continuous in f. These, and some other, technical difficulties would vanish if Γ were a unichain Markov Decision Process. In view of the above in Filar and Krass [2] the law of motion of Γ was perturbed to $p(\epsilon) := \{p_{iaj}(\epsilon) \mid (i, a, j) \in E \times A(i) \times E\}$ where for any $\epsilon \in (0, 1)$ we define

$$p_{iaj}(\epsilon) := \begin{cases} 1 & \text{if } i = 1 \text{ and } a = j \\ 0 & \text{if } i = 1 \text{ and } a \neq j \\ 1 & \text{if } i > 1 \text{ and } a = j = 1 \\ \epsilon & \text{if } i > 1, a \neq j, \text{ and } j = 1 \\ 1 - \epsilon & \text{if } i > 1, a = j, \text{ and } j > 1 \\ 0 & \text{if } i > 1, a \neq j, \text{ and } j > 1. \end{cases}$$

Node 1 now denotes the "home" node; for each pair of nodes i, j

(not equal to 1) corresponding to a (deterministic) arc (i,j), our perturbation replaces that arc by a pair of "stochastic arcs" $(i,1)$ and (i,j):

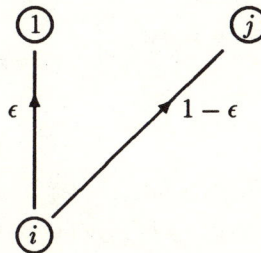

with weights ϵ and $(1-\epsilon)$ respectively ($\epsilon \in [0,1)$). This stochastic perturbation has the interpretation that a decision to move along arc (i,j) results in movement along (i,j) only with probability of $(1-\epsilon)$, and with probability ϵ it results in return to the home node 1.

Note also that the ϵ-perturbed process $\Gamma(\epsilon) =< E, A, r, p(\epsilon) >$ clearly "tends" to Γ as $\epsilon \to 0$. It has the following properties, that were established in [2].

Consider the polytope $X(\epsilon)$ defined by the constraints corresponding to (C1)-(C3) in the perturbed process $\Gamma(\epsilon)$, and a smaller polyhedron $G(\epsilon)$ defined by $2N+2$ constraints:

(C1) $\quad \sum_{i \in E} \sum_{a \in A(i)} (\delta_{ij} - p_{iaj}(\epsilon)) x_{ia} = 0, \quad j \in E$

(C2) $\quad \sum_{i \in E} \sum_{a \in A(i)} x_{ia} = 1$

(C3) $\quad x_{ia} \geq 0, \quad i \in E, a \in A(i)$

(C4) $\quad \sum_{a \in A(i)} x_{ia} \geq c(\epsilon); \quad i \in E$

(C5) $\quad \sum_{i \in E} x_{i1} = c(\epsilon),$

where $c(\epsilon) = \frac{(1-\epsilon)^{N-2}}{d(\epsilon)}$. The results listed below are from [2]:

Theorem 2.2 *(i) Let $\epsilon \in (0,1)$ and f be a deterministic policy of $\Gamma(\epsilon)$. Then f is a Hamiltonian cycle in G if and only if $\mathbf{x}(f) \in G(\epsilon)$.*

(ii) Let $\mathbf{x}$ be an extreme point of $X(\epsilon)$ which is also in $G(\epsilon)$. Then $f_\mathbf{x} = T(\mathbf{x})$ is a Hamiltonian cycle in G.

Lemma 2.1 *Let $G(\epsilon)$ be the polytope defined by constraints C(1)-C(5), and let $\mathbf{x} \in G(\epsilon)$. Then $\mathbf{x}$ satisfies the ratio constraint*

$$\frac{x_{ia}}{x_i} \in \{0,1\} \quad \forall i \in E, \, a \in A(i) \tag{2.8}$$

if and only if

$$\mathbf{x}^T Q \mathbf{x} = 0, \tag{2.9}$$

where Q is a block-diagonal matrix with its i-th block equal to Q_i for $i = 1, \ldots, N$, and Q_i is an $n_i \times n_i$ matrix with all diagonal entries equal to 0 and off-diagonal entries equal to 1, and $n_i = |A(i)|$ for $i \in E$. Set $Q_i := (0)$ if $n_i = 1$.

Remark 2.1 For $i \in E$ with $n_i > 1$, it can be easily verified that the only eigenvalues of Q_i are -1 and $n_i - 1$. Thus, Q is indeed an indefinite quadratic matrix (unless $n_i = 1$ for every i, in which case the underlying graph is trivial).

□

Theorem 2.3

1. *Suppose $f \in C(D)$ corresponds to a Hamiltonian cycle in G. Then $\mathbf{x}(f)$ is a global solution of the indefinite quadratic program*

$$\min\{\mathbf{x}^T Q \mathbf{x} | \mathbf{x} \in G(\epsilon)\}, \tag{2.10}$$

   and $\mathbf{x}(f)^T Q \mathbf{x}(f) = 0$.

2. *Conversely, let the global minimum of the quadratic program be 0, and suppose it is achieved at $\mathbf{x}^*$. Then $f_{\mathbf{x}^*} = T(\mathbf{x}^*)$ is a deterministic policy which corresponds to a Hamiltonian cycle in G.*

3. Main Results

In this Section we shall exploit the structure of the underlying graph G, and of the ϵ-perturbed embedding in the MDP $\Gamma(\epsilon)$, to differentiate between the deterministic policies which are Hamiltonian cycles and those which are not. The basis of this differentiation is the linear function on the frequency space of $X(\epsilon)$ of $\Gamma(\epsilon)$ which represents the long-run frequency of visits to the "home" state 1. That is,

$$x_1(f) := \sum_{a \in A(1)} x_{1a}(f),$$

recalling that $\mathbf{x}(f) \in X(\epsilon)$ for each $f \in C(S)$. Of course, it is natural for us to focus on state 1 since it is distinguished from others by our perturbation.

Note that with each $f \in C(\mathcal{D})$ we can associate a subgraph G_f of G defined by

$$\text{arc}(i,j) \in G_f \iff f(i) = j.$$

We shall also denote a simple cycle of length m beginning at 1 by

$$C_m^1 = \{(i_1 = 1, i_2), (i_2, i_3), \ldots, (i_m, i_{m+1} = 1)\}, \quad m = 2, 3, \ldots, N.$$

Of course, C_N^1 is a Hamiltonian cycle. If G_f contains a cycle C_m^1 we write $G_f \supset C_m^1$. We can now partition the set of deterministic policies into $C(\mathcal{D}) = G(\mathcal{D}) \cup B(\mathcal{D})$, where $G(\mathcal{D}) := \{f \in C(\mathcal{D}) \mid G_f \supset C_m^1 \text{ for some } m = 2, \ldots, N\}$ and $B(\mathcal{D}) := C(\mathcal{D}) \backslash G(\mathcal{D})$. Note that since each $f \in C(\mathcal{D})$ is a function on E, then G_f can contain at most one cycle C_m^1. With cycle C_m^1 we can associate a scalar $d_m(\epsilon) := 1 + \sum_{i=2}^{m}(1-\epsilon)^{i-2}$, where $m = 2, 3, \ldots, N$.

Proposition 3.1 *Let $\epsilon \in (0,1)$, $f \in C(\mathcal{D})$ and $\mathbf{x}(f)$ be its long-run frequency vector as defined in [2]. The frequency of visits to state 1 is given by*

$$x_1(f) = \sum_{a \in A(1)} x_{1a}(f) = \begin{cases} \frac{1}{d_m(\epsilon)}, & \text{if } G_f \supset C_m^1 \\ \frac{\epsilon}{1+\epsilon}, & \text{if } f \in B(\mathcal{D}) \end{cases}.$$

Proof:

1. Suppose that for some $m = 2, 3, \ldots, N$, $G_f \supset C_m^1 = \{(i_1 = 1, f(i_1)), (i_2, f(i_2)), \ldots, (i_m, f(i_m)) = 1)\}$ with $f(i_{k-1}) = i_k$ for $k = 2, \ldots, m$. Since we are only interested in the frequencies of visits to state 1 note that there is no loss of generality in assuming that

$$C_m^1 = \{(1, f(1)), (2, f(2)), \ldots, (m, f(m))\}$$

with $f(k-1) = k$ for $k = 2, 3, \ldots, m$. The probability transition matrix of the Markov chain induced in $\Gamma(\epsilon)$ by f is now of the form

$$P(f) = \begin{bmatrix} 0 & 1 & 0 & \cdots & 0 & 0 & | & 0 & \cdots & 0 \\ \epsilon & 0 & 1-\epsilon & \cdots & 0 & 0 & | & 0 & \cdots & 0 \\ \vdots & \vdots & \vdots & & \vdots & \vdots & | & \vdots & & \vdots \\ \epsilon & 0 & 0 & \cdots & 0 & 1-\epsilon & | & 0 & \cdots & 0 \\ 1 & 0 & 0 & \cdots & 0 & 0 & | & 0 & \cdots & 0 \\ - & - & - & - & - & - & - & - & - & - \\ & & & & & & | & & & \\ & & P_{21}(f) & & & & | & & P_{22}(f) & \\ & & & & & & | & & & \end{bmatrix}$$

where the dimension of the top left-hand block is $m \times m$. It follows that if $m < N$, then the states $m+1, m+2, \ldots, N$ are all transient. Now, if we let $\mathbf{p}^*(f) = (p_1^*(f), \ldots, p_N^*(f))$ denote the unique stationary distribution vector of $P(f)$, then $p_i^*(f) = 0$ for $i \geq m+1$. In addition $\mathbf{p}^*(f)$ must satisfy the equations $\mathbf{p}^*(f)P(f) = \mathbf{p}^*(f)$ and $\sum_{i=1}^{N} p_i^*(f) = 1$. It is now easy to verify that

$$p_i^*(f) = \begin{cases} p_1^*(f), & \text{if } i = 1, 2 \\ (1-\epsilon)^{i-2}p_1^*(f), & \text{if } i = 3, \ldots, m \\ 0, & \text{if } i \geq m+1. \end{cases}$$

Summing the above to 1. we obtain

$$\sum_{i=1}^{N} p_i^*(f) = \left[1 + \sum_{i=2}^{m}(1-\epsilon)^{i-2}\right] p_1^*(f) = 1,$$

and by recalling the definition of $d_m(\epsilon)$ we have together with (2.7) that

$$x_1(f) = p_1^*(f) = \frac{1}{d_m(\epsilon)} \quad \text{for } m = 2, 3 \ldots, N.$$

2. Suppose that $f \in B(\mathcal{D})$, then G_f must contain a directed path $i_1 = 1, i_2, \ldots, i_m, i_l$, where i_l is the first state to repeat itself, that

HAMILTONIAN CYCLES

is,

$$f(i_k) = \begin{cases} i_{k+1}, & \text{if } k = 1, 2, \ldots, m-1 \\ i_l \text{ for some } l \in \{2, 3, \ldots, m-1\} & \text{if } k = m. \end{cases}$$

with $i_k \neq i_{k+1}$ for $k = 1, \ldots, m-2$. Thus the above path begins at 1, proceeds (without repetitions) to i_m, and returns to some $i_l \neq 1$. Without loss of generality we shall assume that $i_k = k$ for each $k = 1, 2, \ldots, m$, and $l = m - 1$. Hence in $\Gamma(\epsilon)$ the policy f induces a Markov chain with the probability transition matrix of the form

$$P(f) = \begin{bmatrix} 0 & 1 & 0 & \cdots & 0 & 0 & | & 0 & \cdots & 0 \\ \epsilon & 0 & 1-\epsilon & \cdots & 0 & 0 & | & 0 & \cdots & 0 \\ \epsilon & 0 & 0 & \cdots & 0 & 0 & | & 0 & \cdots & 0 \\ \vdots & \vdots & \vdots & & \vdots & \vdots & | & \vdots & & \vdots \\ \epsilon & 0 & 0 & \cdots & 1-\epsilon & 0 & | & 0 & \cdots & 0 \\ - & - & - & - & - & - & - & - & - & - \\ & & & & & & | & & & \\ & & P_{21}(f) & & & & | & & P_{22}(f) & \\ & & & & & & | & & & \end{bmatrix}$$

where the top left hand block is of dimension $m \times m$. Following a similar argument to that in part (1), we note that states i are transient whenever $i \geq m+1$. Hence $p_i^*(f) = 0$ if $i \geq m+1$, and from the first co-ordinate of the equation $\mathbf{p}^*(f)P(f) = \mathbf{p}^*(f)$ we obtain

$$\begin{aligned} p_1^* &= \epsilon(\sum_{i=2}^{m} p_i^*(f)) \\ &= \epsilon(\sum_{i=2}^{N} p_i^*(f)) \\ &= \epsilon(1 - p_1^*(f)). \end{aligned}$$

Thus, $x_1(f) = p_1^*(f) = \frac{\epsilon}{1+\epsilon}$.

□

The above proposition lead us to the following new characterization of the Hamiltonian cycles of a directed graph.

Theorem 3.1   *1. Let $f \in C(\mathcal{D})$ be a Hamiltonian cycle. Recall that this means that G_f is a Hamiltonian cycle, then, we have (a) $\mathbf{x}(f)$ is an extreme point of $X(\epsilon)$, and (b) $x_1(f) = \frac{1}{d_N(\epsilon)}$.*

2. Conversely, suppose that $\mathbf{x}$ is an extreme point of $X(\epsilon)$, $x_1 := \sum_{a \in A(1)} x_{1a}$, and $x_1 = \frac{1}{d_N(\epsilon)}$, then $f_\mathbf{x} = T(\mathbf{x})$ is a Hamiltonian cycle.

Proof:

1. Let $f \in C(\mathcal{D})$ be a Hamiltonian cycle. From Theorem 2.1 we have that $\mathbf{x}(f)$ is an extreme point of $X(\epsilon)$. By Proposition 3.1 $x_1(f) = \frac{1}{d_N(\epsilon)}$.

2. Again from Theorem 2.1 $f_\mathbf{x} \in C(\mathcal{D})$ and $x_1(f_\mathbf{x}) = x_1 = \frac{1}{d_N(\epsilon)}$, since $\mathbf{x} = \hat{T}(T(\mathbf{x}))$. Thus $f_\mathbf{x}$ is a Hamiltonian cycle by Proposition 3.1.

□

In view of the above characterization it is possible to simplify the quadratic programming formulation of the Hamiltonian cycle problem that was given in Theorem 4.1 of Filar and Krass [2]. These authors used a feasible region $G(\epsilon)$ consisting of $X(\epsilon)$ cut by $N+1$ additional linear constraints. However, we can now define the simpler feasible polyhedral region:

$$M(\epsilon) := X(\epsilon) \cap \left\{ \mathbf{x} \mid \sum_{a \in A(1)} x_{1a} = \frac{1}{d_N(\epsilon)} \right\}$$

and establish an analogous result. That is, let Q be a block-diagonal matrix with its ith block equal to Q_i for $i = 1, 2, \ldots, N$, and Q_i is an

$n_i \times n_i$ matrix with all diagonal elements equal to 0 and off-diagonal elements equal to 1, and $n_i := |A(i)|$ for $i = 1, \ldots, N$. Also set $Q_i := (0)$ if $n_i = 1$. The next Theorem now follows immediately from Theorem 3.1.

Theorem 3.2

1. *Suppose $f \in C(\mathcal{D})$ corresponds to a Hamiltonian cycle in G. Then $\mathbf{x}(f)$ is a global solution of the indefinite quadratic program*

$$\min\{\mathbf{x}^T Q \mathbf{x} \mid \mathbf{x} \in M(\epsilon)\}, \qquad (\dagger)$$

   and $\mathbf{x}^T(f) Q \mathbf{x}(f) = 0$.

2. *Conversely, let the global minimum of the quadratic program be 0, and suppose it is achieved at $\mathbf{x}^*$. Then $f_{\mathbf{x}^*} = T(\mathbf{x}^*)$ is a deterministic policy which corresponds to a Hamiltonian cycle in G.*

Remark 3.1 It follows from the preceding Theorem that the Hamiltonian cycles of G are characterized as the global minimizers of the indefinite quadratic program. Note that if the global minimum of ($\dagger$) is positive, then no Hamiltonian cycle exists in G.

Remark 3.2 Of course, the quadratic program ($\dagger$) is algorithmically, a difficult problem since Q ia an indefinite matrix. It is , however, possible to reformulate ($\dagger$) as the problem:

$$\min \left(\sum_{a \in A(1)} x_{1a} - \frac{1}{d_N(\epsilon)} \right)^2 \qquad (\ddagger)$$

Subject to :

$$\mathbf{x} \in \text{ext}(X(\epsilon)),$$

where $\text{ext}(X(\epsilon))$ denotes the set of extreme points of $X(\epsilon)$. While this reformulation has a convex objective function, its feasible region now lacks an elegant and mathematically explicit characterization.

Remark 3.3 It is important to note, however, that Proposition 3.1 neatly partitions all the elements of $\text{ext}(X(\epsilon))$. This is because it is easy to check that

$$\frac{1}{d_m(\epsilon)} > \frac{1}{d_{m+1}(\epsilon)} > \frac{\epsilon}{1+\epsilon}$$

for all $\epsilon \in (0,1)$ and $m = 2, 3, \ldots, N-1$. Thus the polytope $X(\epsilon)$ has all its extremum points distributed among N discrete value levels of the linear function $l(\mathbf{x}) := x_1 = \sum_{a \in A(1)} x_{1a}$. In particular all the Hamiltonian cycles (if any) lie on the second lowest level of

$$l(\mathbf{x}) = \frac{1}{d_N(\epsilon)}.$$

Remark 3.4 Note that if we now solve the linear program

$$\max \left(\sum_{a \in A(1)} x_{1a} \right) \qquad (\ddagger)$$

Subject to:

$$\mathbf{x} \in X(\epsilon),$$

and if we obtain an optimal solution $\mathbf{x}^0$ with $\sum_{a \in A(1)} x_{1a}^0 = \frac{1}{d_m(\epsilon)}$ for some $m = 2, 3, \ldots N$, then m is the number of arcs of a shortest (if all arcs have equal length) simple cycle in G beginning at 1. In particular if $f(\mathbf{x}^0) = T(\mathbf{x}^0)$, then $G_{f(\mathbf{x}^0)}$ identifies such a shortest simple cycle.

4. Example

We now show how in the simple example given in [2] we obtain the simplified quadratic program of our Theorem 3.2. Consider the following complete graph G on four nodes (with no loops):

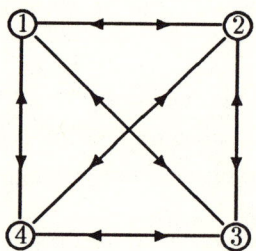

and think of the nodes as the states of our original MDP, Γ, and of the arcs emanating from a given node as of actions available at that state. The Hamiltonian cycle $\tau_1 : 1 \to 2 \to 3 \to 4 \to 1$ corresponds to the deterministic policy $f_1 : \{1,2,3,4\} \to \{2,3,4,1\}$, where $f_1(2) = 3$ corresponds to the controller choosing arc (2,3) in state 2 with probability 1. The Markov chain induced by f_1 is given by the transition matrix

$$P(f_1) = \begin{pmatrix} 0 & 1 & 0 & 0 \\ 0 & 0 & 1 & 0 \\ 0 & 0 & 0 & 1 \\ 1 & 0 & 0 & 0 \end{pmatrix}$$

which is *irreducible*, that is, all the states belong to one ergodic class. On the other hand, the union of two sub-cycles: $1 \to 2 \to 1$ and $3 \to 4 \to 3$ corresponds to the policy $f_2 : \{1,2,3,4\} \to \{2,1,4,3\}$ which identifies the Markov chain transition matrix

$$P(f_2) = \begin{pmatrix} 0 & 1 & 0 & 0 \\ 1 & 0 & 0 & 0 \\ 0 & 0 & 0 & 1 \\ 0 & 0 & 1 & 0 \end{pmatrix}$$

containing two distinct ergodic classes.

Recall that our perturbation changes the Markov control problem Γ, to an ϵ-perturbed problem $\Gamma(\epsilon)$ that is unichain. For instance, the policy f_2 now has the Markov chain matrix

$$P_\epsilon(\pi_2) = \begin{pmatrix} 0 & 1 & 0 & 0 \\ 1 & 0 & 0 & 0 \\ \epsilon & 0 & 0 & 1-\epsilon \\ \epsilon & 0 & 1-\epsilon & 0 \end{pmatrix}$$

The reader is invited to verify that the simplified quadratic program of Theorem 3.2 is now of the form

$$minimize\ [\mathbf{x}^T Q \mathbf{x}]$$
$$Subject\ to: A\mathbf{x} = \mathbf{b}$$
$$\mathbf{x} \geq \mathbf{0},$$

where the vectors and matrices making up the above system are given below:

$$Q = \begin{pmatrix} Q_1 & 0 & 0 & 0 \\ 0 & Q_2 & 0 & 0 \\ 0 & 0 & Q_3 & 0 \\ 0 & 0 & 0 & Q_4 \end{pmatrix},$$

where for $i = 1, 2, 3, 4$ $Q_i = \begin{pmatrix} 0 & 1 & 1 \\ 1 & 0 & 1 \\ 1 & 1 & 0 \end{pmatrix}$

The vector $\mathbf{b}^T = (0, 0, 0, 0, 1, .2695)$, and with ϵ fixed at 0.1. The coefficient matrix A, with the same epsilon value is

$$\begin{pmatrix} 1 & 1 & 1 & -1 & -.1 & -.1 & -1 & -.1 & -.1 & -1 & -.1 & -.1 \\ -1 & 0 & 0 & 1 & 1 & 1 & 0 & -.9 & 0 & 0 & -.9 & 0 \\ 0 & -1 & 0 & 0 & -.9 & 0 & 1 & 1 & 1 & 0 & 0 & -.9 \\ 0 & 0 & -1 & 0 & 0 & -.9 & 0 & 0 & -.9 & 1 & 1 & 1 \\ 1 & 1 & 1 & 1 & 1 & 1 & 1 & 1 & 1 & 1 & 1 & 1 \\ 1 & 1 & 1 & 0 & 0 & 0 & 0 & 0 & 0 & 0 & 0 & 0 \end{pmatrix}$$

It is now easy to check that one global optimum of the above quadratic program is attained at

$$\hat{\mathbf{x}}^T = (.2695, 0, 0, 0, .2695, 0, 0, 0, .2425, .2183, 0, 0)$$

which induces the Hamiltonian cycle τ_1 via the transformation T.

References

[1] Derman, C. (1970). *Finite Markovian Decision Processes.* Academic Press, New York.

[2] Filar, J. A. and Krass, D. (1990), Hamiltonian Cycles, Markov Chains, and Entropy Like Functions. *Submitted to Math. of Op. Res.*

[3] Hordijk, A. and Kallenberg, L. C. M. (1984). Constrained Undiscounted Stochastic Dynamic Programming. *Math. Oper. Res.*, 9 pp. 276-289.

[4] Kallenberg, L. C. M. (1983). *Linear Programming and Finite Markovian Control Problems.* Mathematical Centre Tracts #148, Amsterdam.

[5] Papadimitriou, C. and Steiglitz, K. (1982). *Combinatorial Optimization: Algorithms and Complexity.* Prentice Hall, New Jersey.

Performance of Local Search In Minimum Concave-Cost Network Flow Problems

G.M. Guisewite[1] P.M. Pardalos[2]

Abstract

The performance of several local search algorithms for the single source, uncapacitated (SSU) version of the minimum concave cost network flow problem (MCNFP) is investigated. The local search algorithms considered include Gallo and Sodini's search of adjacent extreme flows, and a new technique based on a larger search neighborhood. Computational results and implementation details are provided for sparse and dense networks, serial and parallel architectures, and for various local search techniques. Sparse problems with up to 1,000 nodes and 10,000 arcs, and dense problems with up to 150 nodes and 22,000 arcs are solved. Worst case performance examples for the algorithms are presented.

Keywords

Concave-cost network flow, uncapacitated, single source, global optimization, local optimization, parallel processing.

1. Introduction

The single-source uncapacitated (SSU) version of the minimum concave-cost network flow problem (MCNFP) requires establishing a minimum

[1]HRB Systems, State College, PA
[2]Pennsylvania State University, University Park, PA

cost flow from a single generating source to a set of sinks, through a directed network. All arcs are uncapacitated, indicating that the entire source flow can pass through any arc. The SSU MCNFP can be stated formally as follows:

Given a directed network $G = (N_G, A_G)$ consisting of a set N_G of n nodes and a set A_G of m ordered pairs of distinct nodes called arcs, coupled with a n-vector (demand vector) $d = (d_i)$ with $d_1 < 0$ and $d_i \geq 0, i = 2, \ldots, n$, and a concave cost function for each arc, $c_{ij}(x_{ij})$, then solve

$$global\ min \sum_{(i,j) \in A_G} c_{ij}(x_{ij})$$

subject to

$$\sum_{(k,i) \in A_G} x_{ki} - \sum_{(i,k) \in A_G} x_{ik} = d_i, \forall i \in N_G \quad (1)$$

and

$$0 \leq x_{ij}, \forall (i,j) \in A_G. \quad (2)$$

All demands are assumed to be integral. The requirement that only $d_1 < 0$ corresponds to the single source case. The lack of an upper bound for the x_{ij} gives rise to the uncapacitated case.

The SSU MCNFP is a concave optimization problem over a convex polyhedron. The concave case differs from the convex case in that if a finite optimal solution exists, then an extreme point of the feasible domain exists that is optimal [2]. In addition, for the concave case a local optimum need not be a global optimum. For the SSU case, an extreme flow (corresponding to an extreme point) is a tree [15]. The leaves of the solution tree correspond to a subset of the sink nodes. The integral demands and the network constraints give rise to extreme flows of integral value.

A SSU MCNFP has a finite optimal solution if it contains no negative cost-oriented cycles, and all sinks are reachable from the source (i.e. there exists a directed path from the source to each sink). The latter requirement is necessary for the existence of a feasible flow. The presence of a negative cost-oriented cycle would imply an unbounded negative cost solution; the absence of such a cycle guarantees a finite solution [8].

In this paper, cases of the SSU MCNFP with arc flow costs that are non-negative, non-decreasing, and concave, are considered. This property of objective functions accurately reflects cost functions for models of real world problems in areas such as production planning and transportation analysis. For example, in a production setting, decreasing concave arc cost functions would exclude the influence of demand on production. Numerous application areas for SSU MCNFP are discussed in [6].

Efficient algorithms for the SSU MCNFP have been found only for a small set of structured problems [12], [13], [15], [16]. This is not surprising, as the general global search problem for the SSU MCNFP is known to be NP-hard [5], [7], [8]. In order to avoid the excessive computations required by an exact global search, the performance of local search for the SSU MCNFP is investigated. A recent survey of both local and global search algorithms for the general MCNFP can be found in [6].

This paper develops and evaluates a new local search algorithm for SSU MCNFP. The new algorithm searches for a local minimum over a larger neighborhood than Gallo and Sodini's [4] adjacent extreme flow algorithm, but requires less computation in the test cases considered. The performance of this new algorithm is compared to existing techniques for both uniprocessor and multiprocessor test cases. Sections 2 through 3 of this paper motivate and introduce algorithms for testing if an extreme feasible flow of a SSU MCNFP is a local optimum for various definitions of local optimum. Section 4 develops algorithms to compute a local optimum, and Section 5 extends these algorithms to incorporate multiple processors. Sections 6 through 8 present the details of an implementation of these algorithms, the computer architecture on which they were executed, and processing results for numerous test cases.

2. Local Optima

A solution X to a SSU MCNFP, where X is a vector of arc flows, is locally optimal if no better solution exists in a specified neighborhood of X. Specifically, X is a local optimum for neighborhood N and (arc) cost function f if $f(X) < f(Y)$ for all Y in $N(X)$. Varying the definition of neighborhood results in different conditions for a local optimum. The standard marginal definition of local optimality defines a neighborhood of X to be

$$N_\epsilon(X) = \{X' | X' \text{ satisfies (1) and (2) and } ||X - X'|| < \epsilon\}$$

for a specified vector norm and $\epsilon > 0$. Local search based on N_ϵ for concave optimization is explored by Minoux [9] and Yaged [14]. For the single commodity case with fixed-charge arc costs, all extreme points are local optima. This led to the development of the following generalized definition of neighborhood by Gallo and Sodini [4]:

$$N_{AEF}(X) = \{X' | X' \text{ satisfies (1) and (2) and }$$
$$X' \text{ is an adjacent extreme flow to } X\}.$$

Here, X' is an adjacent extreme flow to an extreme solution X if X' is an extreme flow, and $X \cup X'$ contains a single undirected cycle.

An even more relaxed definition of neighborhood is the following:

$N_{AF}(X) = \{X'|X'$ satisfies (1) and (2) and X' is an adjacent flow to $X\}$.

Here, X' is an adjacent flow to an extreme solution X if X' results from rerouting a single subpath of flow within X. This concept of neighborhood was developed by Guisewite and Pardalos [7] for single source problems, and independently by Plasil and Chlebnican [11] for the multicommodity case.

It is obvious from the definitions that $N_{AEF} \subseteq N_{AF}$. It is also true that if X is a local optimum over N_{AEF}, then X is a local optimum over N_ϵ. The quality (cost) of a local optimum improves with the size of the neighborhood. That is, if $N_A \subseteq N_B$, with X_A a local optimum over N_A and X_B a local optimum over N_B, then the cost of X_B is less than or equal to the cost of X_A.

3. Algorithms to Test for a Local Optimum

Checking if an extreme flow X is a local optimum over neighborhoods N_{AEF} and N_{AF} can be accomplished by enumerative search over the neighborhoods. Both neighborhoods, however, can contain an exponential number of elements. This is true even for very simple networks. In this section, characterizing the elements in both N_{AEF} and N_{AF} leads to efficient algorithms for checking local optimality. Implementations that efficiently process and utilize memory for sparse and dense networks are described. In Section 4 these algorithms are employed in a Simplex-like search algorithm to locate a local optimum.

Gallo and Sodini [4] characterize adjacent extreme flows for SSU MCNFP as follows:

Let X be an extreme flow with induced tree $T_X = (N_{T_X}, A_{T_X})$, where $N_{T_X} = \{i | X_{ij} > 0 \text{ or } X_{ij} > 0\}$ and $A_{T_X} = \{(i,j) | X_{ij} > 0\}$. An extreme flow X', with induced tree $T_{X'} = (N_{T_{X'}}, A_{T_{X'}})$, is adjacent to X if and only if the arcs in $A_{T_{X'}}$ and not in A_{T_X} constitute a path that connects two nodes in N_{T_X} and does not contain any other node in N_{T_X}.

This characterization implies that only augmenting paths from n_o to n_d, denoted by p'_{n_o,n_d}, that satisfy $p'_{n_o,n_d} \cap T_X = \{n_o, n_d\}$ need be considered. Without loss of generality, assume that flow is being rerouted to node n_d with flow magnitude $IN(n_d)$. If otherwise, the rerouted flow would be less than $IN(n_d)$ and an undirected cycle would result (at node n_d). The resulting augmenting path can be described as $p'_{S,n_d} = p_{S,n_o} + p'_{n_o,n_d}$, where $p_{S,n_o} \subseteq T_X$ and the replaced path is $p_{S,n_d} \subseteq T_X$.

Implementing the above search directly would require solving up to $(|N_{T_X}| - 1)^2$ single source shortest weighted path problems. One problem could result for each $n_d \in N_{T_X} - \{S\}$ and $n_o \in N_{T_X} - \{n_d\}$, and corresponds to finding the minimum cost augmenting flow of size

$IN(n_d)$ from n_o to n_d. Processing can be reduced by noting that all augmenting paths to n_d originate at the source, travel some $p_{S,n_o} \subseteq T_X$, then take a path p'_{n_o,n_d} to the destination node n_d. If all incoming arcs to nodes in $N_{T_X} - \{n_d\}$ except for arcs in A_{T_X} are removed, then $p'_{n_o,n_d} \cap T_X = \{n_o, n_d\}$ is enforced for any node pair n_o and n_d. This implies one single source shortest path problem is sufficient for each destination node n_d, for a total of $|N_{T_X}| - 1$ shortest path problems.

The resulting algorithm, extended to return the best adjacent extreme flow X_{BEST} to an extreme flow X, is summarized in Figure 1. The algorithm is a variation of the algorithm presented in [4]. The stated algorithm avoids adding new nodes and arcs to the original network G, simplifying the implementation. The following notation is used in the algorithm:

S is the source node
$G = (N_G, A_G)$ is the original network
X is the original extreme solution
$T_X = (N_{T_X}, A_{T_X})$ is the tree induced by X
p_{S,n_i} is the unique path from S to n_i in T_X
$IN(n_i)$ is the flow into node n_i for extreme solution X
BIG is a sufficiently large cost.

The corresponding algorithm for search neighborhood N_{AF} is the same as Algorithm 1 except step (3) is omitted, the main loop is over a subset of linebreak $N_{T_X} - \{S\}$, and the resulting flow may be nonextreme. For this case, the modified flow X' induces a network $G_{X'}$, which is possibly not a tree. Recalling that $N_{AF}(X)$ consists of adjacent flows to X, where an adjacent flow X' with induced network $G_{X'} = (N_{X'}, A_{X'})$ satisfies $G_{X'} = T_X - p_{S,n_d} + p'_{S,n_d}$, with $p_{S,n_d} \subseteq T_X$ and p'_{S,n_d} any path from S to n_d in G. Noting that p'_{S,n_d} is unrestricted indicates that only a single problem need be solved for each unique flow magnitude occurring on the paths to each sink. That is, if $p'_{S,n_d} = p'_{S,n_i} + p'_{n_i,n_d}$ where each arc in p'_{n_i,n_d} has flow $IN(n_d)$, then the set of all subpaths from S to n_d in G includes all paths of the form p'_{S,n_i} in G. Unique flow magnitudes occur at branch points and sinks, where branch points are nodes in T_X with out-degree greater than one. For extreme flows, at most (2 * (number of sinks)) + 1 unique flow magnitudes exist. Denoting the set of breakpoints in T_X by BP_{T_X} and the set of sinks in G by $SINK_G$, results in the algorithm in Figure 2.

The motivation for considering the neighborhood N_{AF} was to reduce the computation required to check if a flow was a local optimum, while increasing the size of the neighborhood searched. The fact that flows in

Algorithm 1
  (0) Initialize loop variables
    local.opt = TRUE
    best.cost = cost of solution X
    $T_{BEST} = T_X$
    $X_{BEST} = X$
  FOR {each $n_d \in N_{T_X} - \{S\}$}
    (1) Remove flow into node n_d ($T_{X'} = T_X - p_{S,n_d}$)
      $\forall (i,j) \in p_{S,n_d}, \quad x'_{ij} = x_{ij} - IN(n_d)$
      $\forall (i,j) \in A_G - p_{S,n_d}, \quad x'_{ij} = x_{ij}$
    (2) Compute weights for augmenting flow
      $\forall (i,j) \in A_G, \quad w_{ij} = C(x'_{ij} + IN(n_d)) - C(x'_{ij})$
      $\forall (i,j) \notin A_G, \quad w_{ij} = BIG$
    (3) Prevent non-extreme solutions by implicitly removing arcs
      $\forall (i,j) \ni j \in N_{T_X} - \{n_d\}$ AND $(i,j) \notin A_{T_X}, \quad w_{ij} = BIG$
    (4) Solve shortest path problem from source S to n_d
      aug.cost, p'_{S,n_d} = SHORTEST(S, n_d, W)
    (5) Compute overall cost of rerouted flow
      reduced.cost = cost of X'
      new.cost = reduced.cost + aug.cost
    (6) Check if new flow is of lower cost than current best
      IF(new.cost < best.cost)
        {local.opt = FALSE
        best.cost = new.cost
        $T_{BEST} = T_X - p_{S,n_d} + p'_{S,n_d}$
        X_{BEST} = flow induced by T_{BEST}}

Figure 1: N_{AEF} based local optimum test

Algorithm 2
(0) Initialize loop variables
    local.opt = TRUE
    best.cost = cost of solution X
    $G_{BEST} = T_X$
    $X_{BEST} = X$
FOR {each $n_d \in BP_{T_X} \cup SINK_G$}
    (1) Remove flow into node n_d $(G_{X'} = T_X - p_{S,n_d})$
        $\forall (i,j) \in p_{S,n_d}, \quad x'_{ij} = x_{ij} - IN(n_d)$
        $\forall (i,j) \in A_G - p_{S,n_d}, \quad x'_{ij} = x_{ij}$
    (2) Compute weights for augmenting flow
        $\forall (i,j) \in A_G, \quad w_{ij} = C(x'_{ij} + IN(n_d)) - C(x'_{ij})$
        $\forall (i,j) \notin A_G, \quad w_{ij} = BIG$
    (3) Solve shortest path problem from source S to n_d
        aug.cost, p'_{S,n_d} = SHORTEST(S, n_d, W)
    (4) Compute overall cost of rerouted flow
        reduced.cost = cost of X'
        new.cost = reduced.cost + aug.cost
    (5) Check if new flow is of lower cost than current best
        IF(new.cost < best.cost)
            {local.opt = FALSE
            best.cost = new.cost
            $G_{BEST} = T_X - p_{S,n_d} + p'_{S,n_d}$
            X_{BEST} = flow induced by G_{BEST}}

Figure 2: N_{AF} based local optimum test

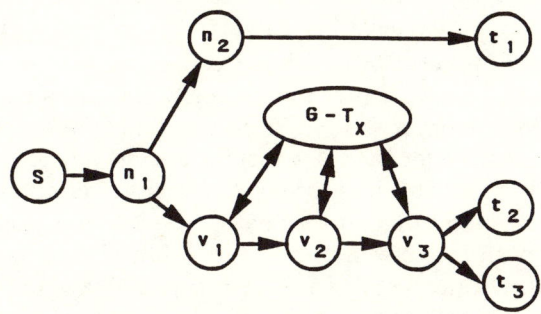

Figure 3: Sample flow problem

N_{AF} need not be extreme causes some additional complexity in using the algorithm to search for a local optimum.

It would be advantageous to improve the N_{AEF} algorithm to require the same number of shortest path problems as the N_{AF} algorithm. The following indicates that this is unlikely. Suppose that $p_{S,n_d} \subseteq T_X$ is such that $p_{S,n_d} = p_{S,n_i} + p_{n_i,n_d}$, where $p_{n_i,n_d} = (v_1, v_2) + (v_2, v_3)$ and arcs (v_1, v_2), (v_2, v_3) carry flow $IN(v_3)$. This case is summarized in Figure 3. The node v_3 is a branch point, while v_1 and v_2 are not. Suppose one attempts to identify the minimum cost adjacent extreme flow with destination v_1, v_2, or v_3 with one shortest path problem. The set of adjacent extreme flows from S to v_j includes paths $p_{S,v_i} \subseteq T_X$, $i = 1, \ldots, j-1$, augmented by a suitable path from v_i to v_j in $G - T_X$. This implies incoming and outgoing arcs for v_1 and v_2 in A_G, and incoming arcs for v_3 in A_G must remain in the modified problem. However, this allows paths of the form $p'_{S,v_1} + p'_{v_1,v_2} + p'_{v_2,v_3}$ with $p'_{v_1,v_2} \neq (v_1, v_2)$ and $p'_{v_2,v_3} \neq (v_2, v_3)$, which are not in N_{AEF}. This contradicts a result found in [10], where an N_{AEF} based algorithm requiring the same number of iterations as the above N_{AF} algorithm is presented.

4. Local Search Algorithms

The above algorithms to test if a flow is a local optimum can be used to locate a local optimum. This is achieved by using a Simplex-like search in Algorithms 3 − 6 (Figure 4). Algorithms 3 and 4 employ neighborhood N_{AEF}, and Algorithms 5 and 6 make use of neighborhood N_{AF}. In Algorithms 3 and 5, the best neighboring vertex is computed using Algorithm

1, i.e. X_{BEST}. Algorithms 4 and 6 make use of Algorithm 1 modified to terminate when any better neighboring vertex is encountered.

Gallo and Sodini [4] used Algorithm 3 combined with a variation of Algorithm 1 to locate a local optimum. Algorithm 4 provides additional options in terms of the order that adjacent solutions are evaluated. The local optimum computed for both Algorithms 3 and 4 depends on the initial solution used to initiate the process. Gallo and Sodini [4] computed the initial solution based on shortest paths to each sink using weights α_{ij} for problems with arc costs $\alpha_{ij} * x_{ij}^{\beta}$. In [7], initial solutions based on various greedy algorithms applied to the actual concave arc costs were evaluated. It was determined that the greedy initial solutions resulted in faster convergence to a local optimum for both the N_{AEF} and N_{AF} local search algorithms.

As noted earlier, if neighborhood N_{AF} is employed to define a local optimum, the computed adjacent flow X' may not be an extreme flow. This permits two choices in using Algorithm 2 to locate a local optimum:

(a) Generalize Algorithm 2 to allow for non-extreme initial flows
(b) Alter the computed X' to be extreme at each iteration.

For case (a), the number of shortest path problems required could increase exponentially, because the number of candidate leaving paths p_{S,n_d} would include choices for each cycle in X'. Case (b) requires detecting nodes in $G_{X'} = G_{BEST}$ with in-degree greater than one, and choosing a single path to each such node. At most $|N_{X'}|$ nodes can have this property. Case (b) offers the possibility of generating an adjusted flow X'' with cost less than the cost of X'. This is accomplished by selecting the paths to each node based on minimizing cost. Implementing case (b) results in Algorithms 5 and 6.

Figure 5 presents a method to alter a nonextreme flow X' to an extreme flow X''. The algorithm attempts to minimize the cost of flow X'' by selecting the minimum cost path over some spanning tree $ST_{X'}$ of $G_{X'}$. Each surplus arc (n_o, n_d) connects two nodes in $ST_{X'}$. Algorithm 7 identifies, for an existing surplus arc (n_o, n_d), the paths $p_{S,n_o} \subseteq ST_{X''}$ and $p_{S,n_d} \subseteq ST_{X''}$, where X'' is the current modified flow (originally X'). The minimum flow on each path is identified, min_o and min_d, and the appropriate units of flow are rerouted to path $p_{S,n_o} + (n_o, n_d)$ or p_{S,n_d}. This rerouting of flow alters X'', $ST_{X''}$, and the set $SURPLUS$ of surplus arcs.

To see that Algorithm 7 terminates in a finite (and small) number of iterations, note that decreasing path $p_{S,n_o} + (n_o, n_d)$ by min_o flow or p_{S,n_d} by min_d flow results in a nonempty set of arcs in X'' decreasing to 0 flow and leaving the spanning tree $SP_{X''}$. There exists one leaving

Algorithm 3 (Best First N_{AEF} Local Search):
  Find an initial extreme feasible solution X
  WHILE (X is not a local optimum over N_{AEF})
    move to the best neighboring vertex X' in N_{AEF}
    $X \leftarrow X'$

Algorithm 4 (First Better N_{AEF} Local Search):
  Find an initial extreme feasible solution X
  WHILE (X is not a local optimum over N_{AEF})
    move to the first better neighboring vertex X' in N_{AEF}
    $X \leftarrow X'$

Algorithm 5 (Best First N_{AF} Local Search):
  Find an initial extreme feasible solution X
  WHILE (X is not a local optimum over N_{AF})
    move to the best neighboring flow X' in N_{AF}
    alter X' to be extreme $\Rightarrow X''$
    $X \leftarrow X''$

Algorithm 6 (First Better N_{AF} Local Search):
  Find an initial extreme feasible solution X
  WHILE (X is not a local optimum over N_{AF})
    move to the first better neighboring flow X' in N_{AF}
    alter X' to be extreme $\Rightarrow X''$
    $X \leftarrow X''$

Figure 4: Four local search algorithms

Algorithm 7:
(1) Generate a spanning tree $ST_{X'} = (N_{ST_{X'}}, A_{ST_{X'}})$
    for $G_{X'} = (N_{X'}, A_{X'})$
(2) Identify the set of surplus arcs $SURPLUS = A_{X'} - A_{ST_{X'}}$
(3) $X'' = X'$, $ST_{X''} = ST_{X'}$
WHILE $\{\ \exists (n_o, n_d) \in SURPLUS\ \}$
    $\{$(4) Construct paths $p_{S,n_o}, p_{S,n_d} \subseteq ST_{X''}$
    (5) Compute $min_o = min\{min_{(i,j) \in p_{S,n_o}}\{X''_{ij}\}, X''_{n_o,n_d}\}$
          $min_d = min_{(i,j) \in p_{S,n_d}}\{X''_{ij}\}$
    (6) Reroute min_d units of flow onto $p_{S,n_o} + (n_o, n_d)$
        or min_o units of flow onto p_{S,n_d}
    (7) Update X'', $ST_{X''}$, $SURPLUS\}$

Figure 5: Forcing an extreme flow

arc (n_o^*, n_d^*), such that n_d^* has another incoming arc of nonzero flow, i.e. the end of the subpath removed. If (n_o^*, n_d^*) was not previously in $ST_{X''}$, then $(n_o^*, n_d^*) = (n_o, n_d)$ and the current surplus arc is removed. If (n_o^*, n_d^*) was previously in $ST_{X''}$, then another surplus arc (n', n_d^*) enters $ST_{X''}$ as a result of rerouting flow. In either case, the number of surplus arcs decreases by one at each iteration. The total number of iterations is then $|A_{X'} - A_{ST_{X'}}|$. A variation of Algorithm 7 was implemented. The implemented approach updated $ST_{X''}$ after enumerating the entire current $SURPLUS$ set of arcs, and processed the arcs in $SURPLUS$ in order of minimum distance to the root (source node). This approach gained efficiency for the cases encountered, in which the set of surplus arcs was extremely small. $ST_{X''}$ was generated using breadth first search on $G_{X'}$. The rerouting of flow in step (6) was determined by selecting the path that minimized the resulting flow.

Algorithms 3 through 6 for local search can be efficiently implemented. However, it is an open problem as to whether the number of iterations required for the WHILE loop is polynomial in the size of G. Empirical results indicate that for each case, the number of iterations is less than the number of nodes, even for randomly generated initial extreme solutions.

MAIN originally holds the network and initial solution
P_i correspond to attached coprocessors

MAIN:
- (1) Compute and broadcast subproblem tasking
  $(SP_i, i = 1, \ldots, P)$
- (2) Broadcast G and X
- (3) Gather $best.cost(i), i = 1, \ldots, P$
- (4) Identify j such that $best.cost(j)$ is minimum
- (5) Broadcast selected processor j
- (6) Receive T_{BEST} (or G_{BEST})

P_i:
- (1) Receive/distribute subproblem tasking $(SP_i, i = 1, \ldots, P)$
- (2) Receive/distribute G and X
- (3) Apply Algorithm 1 or 2 (FOR each $n_d \in SP_i$)
- (4) Return $best.cost_i$/distribute $best.cost_j, j > i$
- (5) Receive/distribute selected processor j
- (6) IF (processor j) return T_{BEST} (or G_{BEST})
  ELSE IF (processor $i < j$)
     receive/distribute T_{BEST} (or G_{BEST})

Figure 6: Parallel test for a local optimum

5. Multiprocessor Algorithms

A coarse grain multiprocessing approach to testing for a local optimum is to evenly distribute the FOR loop in Algorithms 1 and 2 over distinct processors. For Algorithm 1, $|N_{T_X} - \{S\}|$ subproblems, consisting of generating and solving a single source shortest path problem, can be solved independently. For Algorithm (2), $|BP_{T_X} \cup SINK_G|$ independent problems exist. Static allocation of these problems is appropriate, as each subproblem requires approximately the same amount of processing time. The resulting approach is summarized in Figure 6.

The performance of this approach is limited by the overhead to distribute and gather data across processors, and by the relationship of the number of processors P to the number of subproblems solved. The multiprocessor algorithm fits easily into the local search algorithms of Section 4. Because the number of subproblems at each iteration of local search is variable, it is difficult to estimate the number of processors to employ. The N_{AF} based approach incurs additional overhead as the algorithm to force a solution to be extreme is currently executed on a single processor. Performance of the algorithm for best first local search over N_{AEF} and N_{AF} is summarized in Section 7.

6. Implementation Details

The algorithms described in previous sections were implemented on a multiprocessor system. In some cases, performance (processing time) was sacrificed to reduce storage requirements. The main objective of the implementation was to demonstrate that large problems could be solved efficiently, and to identify the effects of various network properties on the convergence of the local search algorithms. Algorithms for both the dense network and sparse network cases were implemented.

The dense implementation uses an adjacency matrix to describe the underlying network. Structures used included arrays $ALPHA$, $BETA$, C, X, X', and X_{BEST}. Arc costs are of the form $c_{ij}(x_{ij}) = alpha_{ij} * X_{ij}^{beta_{ij}}$, where arc (i,j) is in A_G if $alpha_{ij} \neq BIG$, X stores the arc flows for the current extreme solution, X' stores the arc flows for subproblems, and X_{BEST} maintains the current best augmenting flow. This sparse representation simplifies tracking arcs in a forward direction, but is inefficient for searching arcs in a backward direction. Problems with complete networks of 150 nodes were solved.

The sparse implementation uses an adjacency list to describe the underlying network. The same structures were employed, but with size equal to the number of arcs in the network. Additional structures included $FROM$, TO, and $JINDEX$. For each node n_i, $FROM(n_i)$ and $TO(n_i)$ served as pointers into the $JINDEX$ array. Arc (i,j) is stored such that $JINDEX(l) = j$ for some l satisfying $FROM(i) \leq l \leq TO(i)$. Problems with 1000 nodes and 10,000 arcs were solved.

The single source shortest path algorithm employed was a variation of Dijkstra's algorithm [1] with the distance vector stored as a linked list. It is anticipated that large problems could gain efficiency with an alternate data structure. This approach to solving the shortest path problems is sufficient due to the assumption of non-decreasing arc flow costs. This assumption guarantees that all weights in the shortest path problems are non-negative.

Our multiprocessor system consists of one to twenty Transputer T800s. The Transputer is a microprocessor developed under the European Espirit project, and is designed to facilitate parallel processing. Each 20 MHz T800 consists of a 10 MIP fixed point processor, a 1.5 MFLOP floating point coprocessor, a 4 KByte cache, and 4 DMA (direct memory access) I/O processors, all on a single chip. Our system includes 1 MByte of memory per processor. Experience indicates that 3-5 MIPS are achievable by each processor for large general processing applications. Four processors are configured in a pipeline on a single board, with the remaining DMA links connected to the board edge through a cross-bar switch. The cross-bar is software programmable, allowing dynamic reconfiguration of the processor interconnect.

The existence of only four DMA links per processor limits processor configurations achievable by the system. Configuring the processors in a 3-tree minimizes the distance of processors to the processor originally containing the network (MAIN). However, any configuration beyond a simple linear array requires the use of the crossbar switch, which reduces the achievable data input/output (I/O) rates to 75% of the maximum. The following analysis indicates that a simple linear array configuration ($MAIN \leftrightarrow P_1 \leftrightarrow ... \leftrightarrow P_P$) minimizes I/O overhead for our coarse grain implementation as a result of the crossbar degradation:

M = number of bytes to be moved
P = number of processors
B = block size of data transfer
α = I/O startup time $\approx$ 5 microseconds
β = block data rate $\approx$ 1.7 MBytes/second

$T_{LA}(X,Y,Z)$ is the time to broadcast X bytes of data through Y processors using a data block size of Z bytes when the processors are organized in a linear array

$T_{3-tree}(X,Y,Z)$ is the time to broadcast X bytes of data through Y processors using a data block size of Z bytes when the processors are organized in a 3-tree

$$T_{LA}(M,P,B) = T_{LA}(M,1,B) + ((P-1) * T_{LA}(B,1,B))$$
$$T_{LA}(M,1,B) \approx (M/B) * T_{LA}(B,1,B)$$
$$T_{LA}(B,1,B) = (2*\alpha) + (B/\beta)$$
$$\Rightarrow T_{LA}(M,P,B) \approx ((M/B) + (P-1)) * ((2*\alpha) + (B/\beta))$$
$$T_{3-tree}(M,P,B) = T_{3-tree}(M,1,B)$$
$$\quad\quad + (1 + \lceil log_3(P-1) \rceil) * T_{3-tree}(B,3,B))$$
$$T_{3-tree}(M,1,B) \approx (M/B) * T_{3-tree}(B,3,B)$$
$$T_{3-tree}(B,3,B) = (4*\alpha) + (B/(.75*\beta))$$
$$\Rightarrow T_{3-tree}(M,P,B) \approx$$
$$\quad ((M/B) + (1 + \lceil log_3(P-1) \rceil)) * ((4*\alpha) + (B/(.75*\beta)))$$

The analysis corresponds to pipelined I/O, where the time to move (M/B) blocks of data into P processors equals the time to move all the data through processor 1, plus the time for the last block of data to propagate to the most distant processor. The .75 * β block data rate for the 3-tree case results from the decrease due to the crossbar switch. Minimizing $T_{LA}(M, P, B)$ and $T_{3-tree}(M, P, B)$ over possible block sizes $1 \leq B \leq M$, yields the optimal block sizes $B^*_{LA} \approx \sqrt{M}$ and $B^*_{3-tree} \approx 3 * \sqrt{M}$ for the twenty processor case. This indicates that the I/O times for the linear array are less than the 3-tree case for networks with more than 21 arcs ($M = 4 * |A_G|$, as 4 bytes are used to represent each arc).

Test networks are generated in a random fashion. Arcs are generated by computing two random integers uniformly distributed in $[1, 2, \ldots, n]$, where n is the number of nodes in the network. Duplicate arcs and arcs of the form (i, i) are discarded. After the specified number of arcs is successfully generated, the resulting network is tested for connectivity by solving a single source shortest path problem. If the connectivity is suitably high, then cost functions are generated for each arc. Each cost function is of the form $\alpha_{ij} x^{\beta_{ij}}$, where the α_{ij} are uniformly distributed in $[1, 2, \ldots, 100]$ and the β_{ij} are uniformly distributed in $[.1, .2, \ldots, 1]$. The test generator is implemented so that if two problems contain the same number of nodes, and the same random number seed, but have a different number of arcs, then the smaller network generated is a subset of the larger network.

7. Processing Results

Numerous tests were executed to identify the performance of the local search algorithms for a range of networks. The test problem parameters that were varied included:

(1) Network size - number of nodes, number of sinks
(2) Network density - average number of arcs/node
(3) Processing architecture - number of processors.

Each test case was executed for the algorithms and implementations described in the previous sections, and include the averaged results for ten randomly generated networks. The initial solution was obtained using a greedy algorithm based on dependent shortest weighted paths from the source node to each sink. The arc weights were computed using the required sink flows and the actual concave arc costs. This corresponds to initial solution technique (d2) in [7]. Processing times presented exclude the initial solution times, and indicate the time needed to perform local search, given the initial solution.

Figure 7 compares the performance of the sparse and dense implementations of Algorithm 3 (N_{AEF} best first search) and Algorithm 5

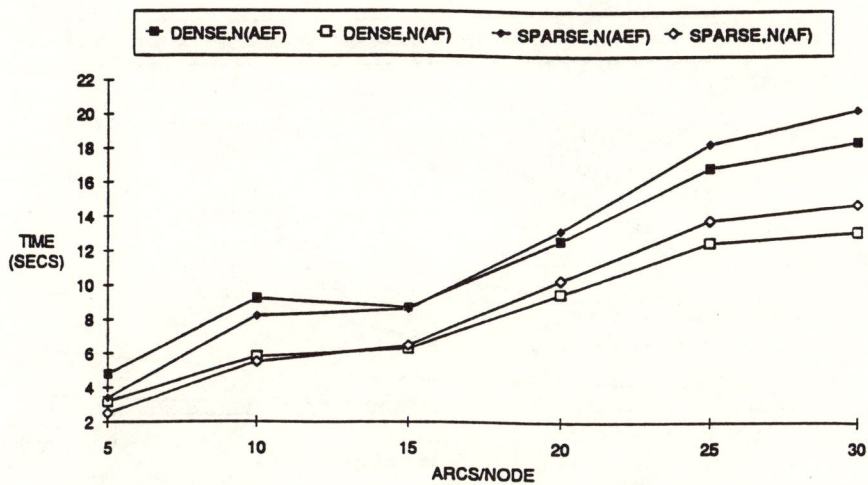

Figure 7: Sparse vs dense uniprocessor local search

(N_{AF} best first search) for locating a local optimum. Algorithm 5 outperforms Algorithm 3 for both the sparse and dense case. The sparse and dense implementations perform similarly for the case of fifteen arcs per node. The test cases contain fifty nodes, indicating the density at this point is roughly 30%. The sparse implementations perform slightly better for cases with lower density, and worse for higher densities.

Figures 8 and 9 demonstrate the processing time required for sparse and dense implementations of Algorithm 5 as a function of network size and density. For the sparse case, processing times increase roughly as a linear function of the network density and as the square of the number of nodes. The increase in time due to the number of nodes corresponds to the implementation of the shortest path algorithm, i.e. $(O(|N_G|^2)$. The dense results exhibit an anomaly for the 150 node case at the highest density. This is due to an increase in the number of iterations required for local search to converge. The irregular performance indicates that network structure is equally a key factor in determining processing time for local search.

Figures 10 and 11 demonstrate the processing required for the sparse implementation of Algorithm 5 as a function of the number of sinks. Figure 10 demonstrates processing times increase proportionate to the

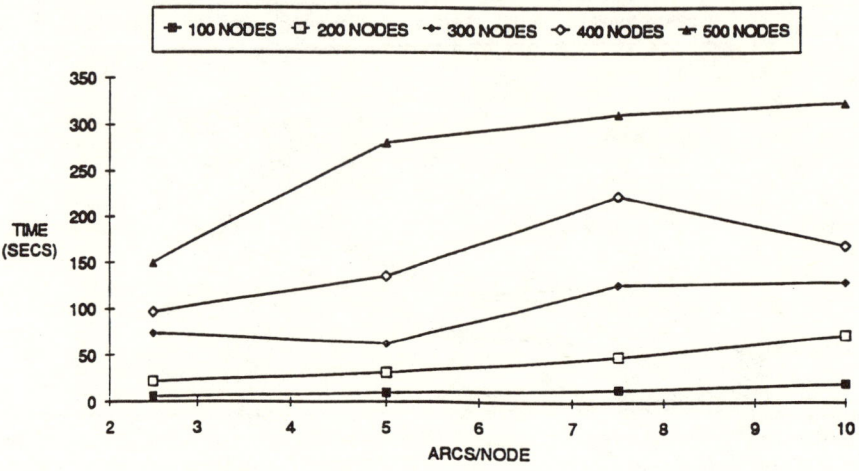

Figure 8: Sparse local search vs network size and density

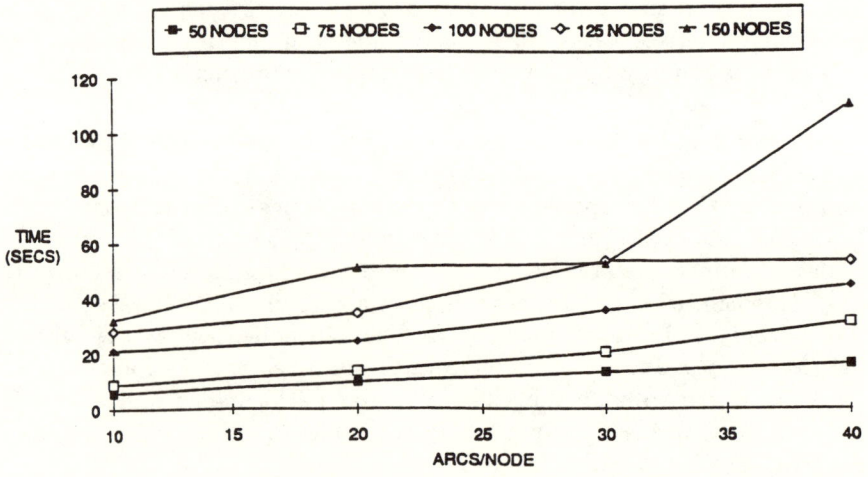

Figure 9: Dense local search vs network size and density

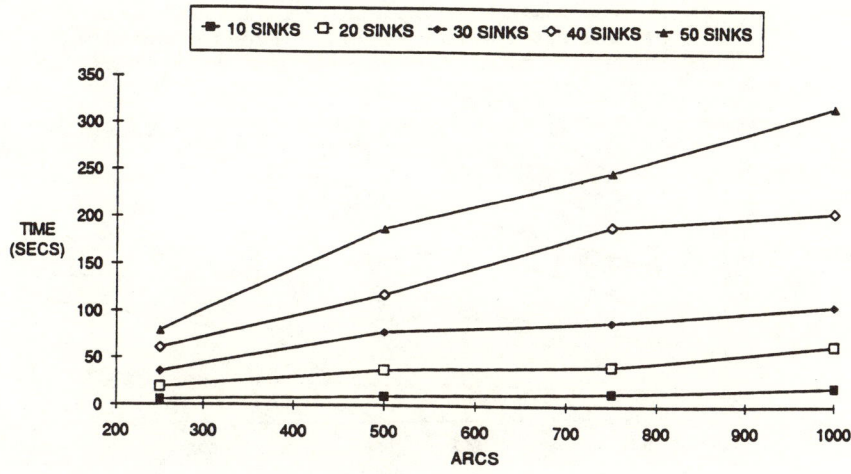

Figure 10: Sparse local search vs the number of sinks (time)

square of the number of sinks. This increase corresponds to a linear increase in the number of iterations required (Figure 11) in addition to a linear increase in the number of shortest path problems solved at each iteration. This result indicates that the number of sinks has a significant effect on processing time for local search in random networks. This is similar to the result found in [3] for the send-and-split global search technique. For send-and-split, processing time increases as an exponential function of the number of sinks.

Figures 12 and 13 indicate the effects of employing multiple processors on varying network densities for the sparse implementations of Algorithms 3 and 5. Results are presented for both processing time (Figure 12) and speedup (Figure 13). The definition of speedup employed is the processing time on a single processor divided by the processing time on multiple processors, when using the same algorithm for one and multiple processors. Both algorithms achieve significant speedups for the twenty processor case. Algorithm 3 achieves a greater speedup due to the increased number of subproblems solved at each step. The effective utilization of twenty processors was 80% for Algorithm 3, and 73% for Algorithm 5. Network density had little effect on the speedups achieved. This seems to contradict the idea that increased workloads improve mul-

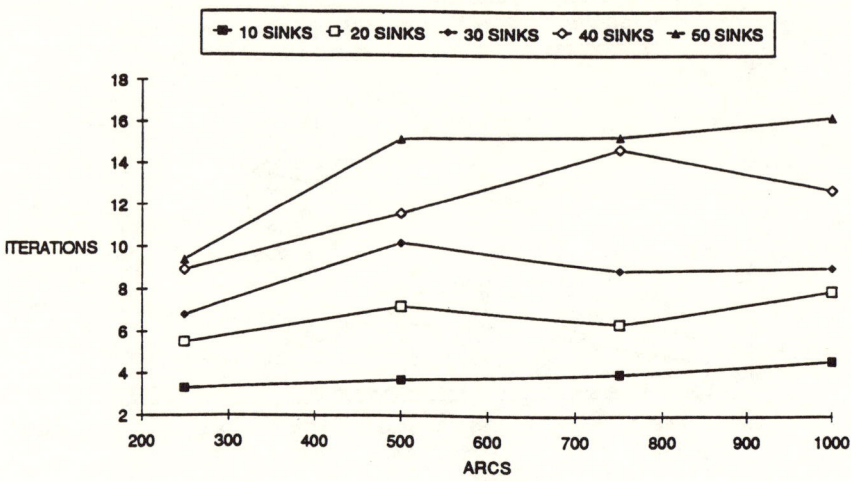

Figure 11: Sparse local search vs the number of sinks (iterations)

tiprocessor utilization. This is true for cases where I/O overhead has a significant impact. For the current implementation, processor utilization is predominantly determined by the number of shortest path problems relative to the number of processors.

Figures 14 and 15 indicate the effects of employing multiple processors as the number of sinks varies, for the dense implementations of Algorithms 3 and 5. Results are, again, presented for both processing time (Figure 14) and speedup (Figure 15). The processor utilization varies irregularly with the number of sinks for both algorithms. This is due to the number of subproblems for each iteration increasing proportionately with the number of sinks. The utilization increases until the average number of subproblems at each iteration exceeds the number of processors employed. Then the utilization drops due to the load imbalance across the processors.

To analyze the multiprocessor overhead, the effects of load balancing and other factors, such as I/O time, can be modeled as follows:

$$\text{Achieved speedup} = \frac{\text{Time}(1)}{(\text{Time}(1) * (1/R)) + \text{other}} \qquad (3)$$

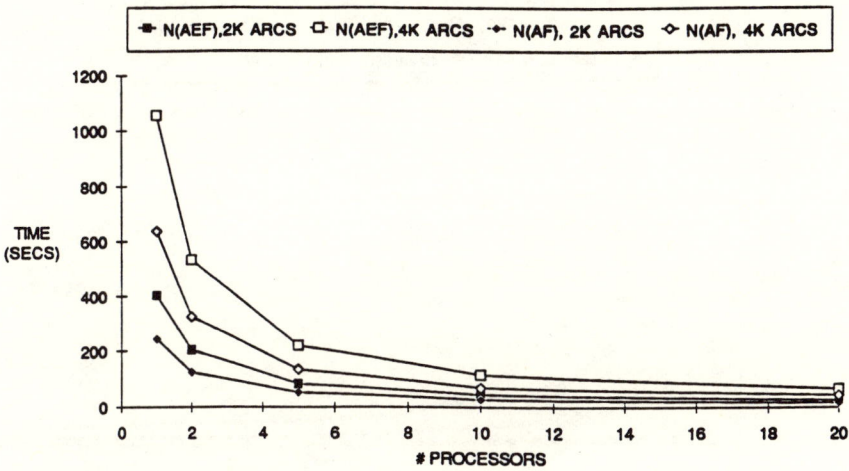

Figure 12: Sparse multiprocessor local search vs network density (time)

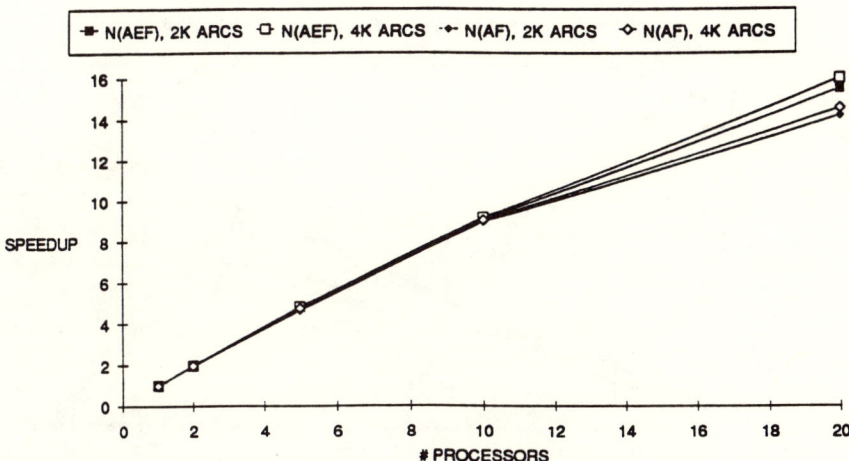

Figure 13: Sparse multiprocessor local search vs network density (speedup)

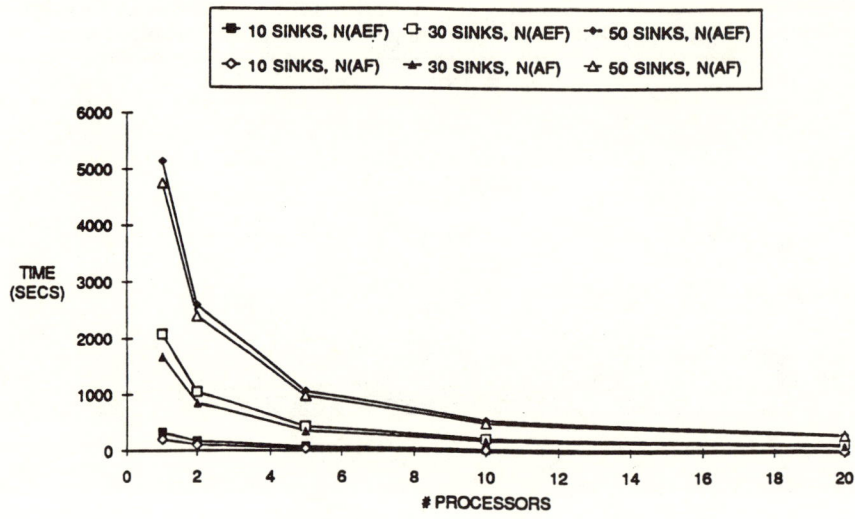

Figure 14: Dense multiprocessor local search vs #sinks (time)

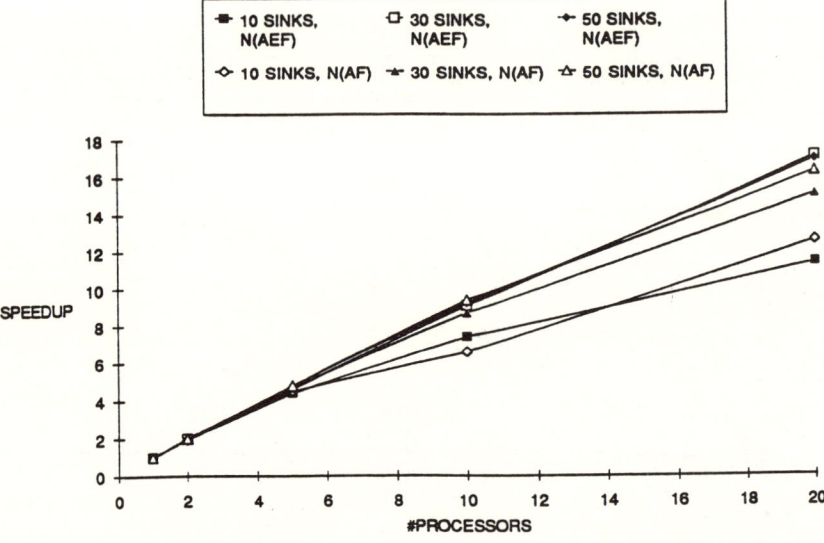

Figure 15: Dense multiprocessor local search vs #sinks (speedup)

Time(1) denotes the time to solve the problem on one processor, R denotes the speedup limitation due to the distribution of subproblems, and "other" denotes all other overhead. R can be computed directly from the total subproblems solved, N_{TOT}, and the average of the maximum number of subproblems solved at each iteration of the local search process, N_{AVG}. Then $R = N_{TOT}/N_{AVG}$. For the test cases considered, "other" overhead accounts for roughly 5% of the resulting twenty processor times.

The processing results presented indicate that local search can be applied to large cases of the SSU MCNFP. Dense problems with 150 nodes, 22,000 arcs, and 50 sinks were solved within 7 minutes on the twenty processor system. Sparse problems with 1000 nodes, 10,000 arcs, and 50 sinks were solved within 38 minutes using twenty processors. Results indicate that more processors can be efficiently utilized when the number of shortest path problems required in each iteration increases. All test results presented correspond to randomly generated networks. These random networks tend to have short overlapping paths to the sinks. Layered networks, evaluated in [7], required a larger number of shortest path problems at each iteration, with the number of problems increasing with the number of layers.

The processing time for the initial solution technique was not included in the above results. This was to avoid the additional degree of freedom imposed by the initial solution technique. In [7], it was demonstrated that an initial solution technique based on solving independent shortest path problems for each sink resulted in roughly 50% more iterations than the initial solution technique used here. The independent shortest paths could easily be distributed across multiple processors for this case. However, the resulting multiprocessor speedup, when compared to the best known uniprocessor approach, would be on the order of 66% of the maximum in the best case. For the case of dependent shortest path problems, the initial solution technique requires, at worst, time proportional to a single iteration of local search executed on one processor. For a problem requiring few iterations, this can dominate the multiprocessor time. The dependent shortest path approach can be improved on multiple processors by constructing as much of each shortest path problem as possible on independent processors, then solving the problems in the required order. Another alternative is to develop an initial solution technique that offers a median between the fully dependent and fully independent approaches. Also, in [7] a comparison of the N_{AEF} and N_{AF} based algorithms, in terms of the quality of the computed local optima and the number of shortest path subproblems solved, is provided.

8. Evaluating the Computed Local Optimum

A second criteria for comparing local search algorithms is to identify the magnitude of the computed cost. This is important if a local optimum with cost near to the global optimum is desired. For small problems, the computed local optima were evaluated by comparing them to the global optimum obtained via complete enumeration. For larger problems, local optima were compared to the best known solution generated by a global search heuristic. The global search heuristic randomly generated 200 initial solutions, then applied local search to each initial solution. The probability of obtaining a global optimum increases with the number of initial solutions for this approach.

For small sparse test problems, the local search algorithms found the global optimum a high percentage of the time. For random networks with 10 nodes, 50 arcs and 5 sinks, local search with a greedy initial solution located the global optimum in 84% of the 100 test cases. The global search heuristic located the global optimum in all cases. For problems with 25 nodes, 125 arcs, and 10 sinks, the local search algorithm found the best known solution in 71% of the 100 test cases. For problems with 50 nodes, 250 arcs, and 20 sinks, the local search algorithm found the best known solution in 54% of the 100 test cases. In all test cases, the global heuristic solution had cost less than or equal to the local search solution. Over the 300 test cases, the largest deviation between the computed local optimum and the best known solution was 30%. The average deviation of the local optima that were not global optima was 7%.

In theory, the computed local optimum can be arbitrarily large in comparison to the global optimum. This is demonstrated by the example network in Figure 16. If arc costs for this network satisfy $C_{S,t_i}(1) = a > 0$, $\forall i$, $C_{S,v}(j) = a + \epsilon$, $\epsilon > 0$, $\forall j$, and $C_{v,t_i}(1) = (a/n) > 0$, $\forall i$; then the computed local optimum has cost $n * a$, and the global optimum has cost $(2 * a) + \epsilon$. For this case, the result holds for all the greedy initial solution algorithms presented in [4] and [7].

9. Summary

Local search algorithms for the SSU MCNFP were developed and implemented on a multi-microprocessor system. Two neighborhoods were used to define the local search process. Processing results indicate that the relaxed neighborhood, although allowing non-extreme solutions at each iteration, offered an improved approach to the local search problem. This was due to a fewer number of shortest path problems requiring solution at each iteration of the local search process. This reduction of problems,

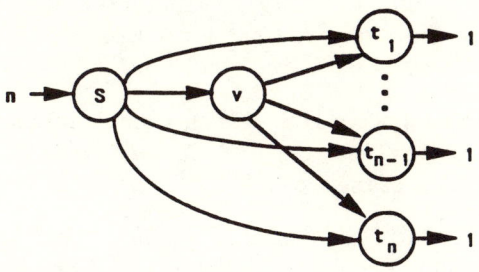

Figure 16: Worst Case Local Search Performance

however, limited the number of processors that could be used effectively in the multiprocessor approach presented. The overall processing times, for all cases, were less for the relaxed approach. The difference in processing time for the two neighborhoods became less as the number of processors increased. The locally optimal flows computed by both techniques were close to the global optimum (or best known solution) a large percentage of the time. This indicates that the local search algorithms would be useful as global search heuristics, or as starting solutions for a branch-and-bound global search algorithm. Computed local optima for small networks correspond to a global optimum a larger percentage of the time than in larger networks. This indicates conclusions for global search heuristics based on small test problems may not apply for larger networks.

References

[1] Aho, A.V., J.E. Hopcroft, and J.D. Ullman (1974), *The Design and Analysis of Computer Algorithms*, Addison-Wesley, Reading, Mass.

[2] Eggleston, H.G. (1963), *Convexity, Cambridge Tracts in Mathematics and Mathematical Physics No. 47*, Cambridge University Press, Cambridge, Mass.

[3] Erickson, R.E., Monma, C.L., and Veinott, Jr., A.F. (1987), "Send-and-Split Method for Minimum-Concave-Cost Network Flows," *Mathematics of Operations Research, Vol. 12, No. 4*, 634-664.

[4] Gallo, G. and Sodini, C. (1979), "Adjacent Extreme Flows and Application to Min Concave-Cost Flow Problems," *Networks 9*, 95-121.

[5] Garey, M.R. and Johnson, D.S. (1979), *Computers and Intractability: A Guide to the Theory of NP-Completeness*. W.H. Freeman and Company, San Francisco, CA.

[6] Guisewite, G.M. and Pardalos, P.M. (1990), "Minimum Concave-Cost Network Flow Problems: Applications, Complexity, and Algorithms," *Annals of Operations Research 25*, 75-100.

[7] Guisewite, G.M. and Pardalos, P.M. (1990), "Algorithms for the Single-Source Uncapacitated Minimum Concave-Cost Network Flow Problem," Working Paper, Department of Computer Science, Pennsylvania State University.

[8] Lozovanu, D.D.(1983), "Properties of Optimal Solutions of a Grid Transport Problem with Concave Function of the Flows on the Arcs," *Engineering Cybernetics, Vol 20*, 34-38.

[9] Minoux, M. (1976), "Multiflots De Coût Minimal Avec Fonctions De Coût Concaves," *Annals of Telecommunication, Vol 31, No 3-4*, 77-92.

[10] Pelzwerger, B.V. and Shafir, A.Y. (1990), "The Interactive Fixed Charge Inhomogeneous Flow Optimization Problem," *Annals of Operations Research 25*, 223-242.

[11] Plasil, J. and Chlebnican, P. (1990), "A New Algorithm for the Min Concave Cost Flow Problem," Working paper, Technical University of Transport and Communications, Czechoslovakia.

[12] Wagner, H.M. (1960), A Postscript to "Dynamic Problems in the Theory of the Firm," *Naval Research Logistics Quarterly, Vol. 7*, 7-12.

[13] Wagner, H.M. and Whitin, T.M. (1958), "Dynamic Version of the Economic Lot Size Model," *Management Science, Vol 5, No. 1*, 89-96.

[14] Yaged, Jr. B.(1971), "Minimum Cost Routing for Static Network Models," *Networks, Vol. 1*, 139-172.

[15] Zangwill, W.I. (1968), "Minimum Concave-Cost Flows in Certain Networks," *Management Science, Vol. 14, No. 7*, 429-450.

[16] Zangwill, W.I. (1969), "A Backlogging Model and a Multi-echelon Model of a Economic Lot Size Production System- A Network Approach," *Management Science, Vol. 15, No. 9*, 506-527.

SOLUTION OF THE CONCAVE LINEAR COMPLEMENTARITY PROBLEM

Joaquim J. Júdice *'   Ana M. Faustino *"

Abstract

The Linear Complementarity Problem (LCP) consists of finding vectors $z \in R^n$ and $w \in R^n$ such that $w = q + Mz$, $z \geq 0$, $w \geq 0$, $z^T w = 0$. The LCP is Concave if its matrix M is Negative Semi-definite (NSD). In this paper it is shown that a hybrid enumerative method simplifies and becomes much more efficient to solve medium and large scale Concave LCPs. A polynomial algorithm is also introduced for the solution of the Concave LCP when all off-diagonal elements of M are nonnegative ($M \in NSD_+$). Computational experience indicates that the algorithm is quite efficient to solve large-scale Concave LCPs with NSD_+ matrices. It is also established that $Q \cap NSD = \emptyset$, $Q_0 \cap NSD \neq \emptyset$ and $NSD_+ \subset Q_0$, where Q and Q_0 are the well-known classes of matrices related with the existence of a solution to the LCP.

The incorporation of the hybrid enumerative method in a Sequential LCP (SLCP) algorithm for finding global minima of Concave Quadratic Programs is also discussed. It is shown that the resulting SLCP method is in general capable of finding a global minimum efficiently, but it has difficulties in establishing that such a solution has been found.

Keywords: Linear Complementarity Problem, Global Concave Quadratic Optimization, Polynomial Algorithms, Enumerative Methods, Classes of Matrices.

---

* Partially supported by Instituto Nacional de Investigação Científica (INIC) under project 89/EXA/5.
' Departamento de Matemática da Universidade de Coimbra, 3000 Coimbra, Portugal.
" Departamento de Engenharia Civil da Universidade do Porto, 4000 Porto, Portugal.

1 - Introduction

The Linear Complementarity Problem (LCP) consists of finding vectors $z \in R^n$ and $w \in R^n$ such that

$$w = q + Mz, \; z \geq 0, \; w \geq 0, \; z^T w = 0 \qquad (1)$$

where $q \in R^n$ and M is an n by n real square matrix. This nonlinear optimization problem has received much interest during the past twenty years. Many direct and iterative algorithms have been developed for the solution of the LCP. These algorithms can only process the LCP (find a solution or show that no solution exists) when the matrix M satisfies certain properties. We suggest Murty's book [35] for an excellent survey of the most important algorithms and respective classes of matrices. If the matrix M does not belong to any of these classes, then the LCP can be solved either by reducing it to a nonconvex optimization problem [2, 4, 17, 37, 38] or by using an enumerative method [1, 15, 20, 26, 33, 34].

The LCP is equivalent to the following quadratic programming (QP) problem [11]

$$\text{Minimize } f(z) = q^T z + \frac{1}{2} z^T (M + M^T) z \qquad (2)$$

$$\text{subject to } z \in K$$

where

$$K = \{ z \in R^n : q + Mz \geq 0, \; z \geq 0 \} \qquad (3)$$

is the feasible set of the LCP. Since the quadratic function $f(z)$ of the QP is bounded from below on its constraint set, there are three possible cases:

(i) $K = \emptyset \Rightarrow$ LCP is infeasible

(ii) $f(\bar{z}) = \underset{z \in K}{\text{Min}} f(z) = 0 \Rightarrow \bar{z}$ is solution of the LCP

(iii) $\underset{z \in K}{\text{Min}} f(z) > 0 \Rightarrow$ LCP is feasible but has no solution

Because of the equivalence stated above, three important classes of LCPs may be distinguished, namely Convex, Concave and Nonconvex, depending on the status of the equivalent QP. Since the matrix M is Positive Semi-definite (PSD), Negative Semi-definite (NSD) and Indefinite (IND) for Convex, Concave and Nonconvex QPs respectively, then we can write

(i) LCP is Convex $\Leftrightarrow M \in$ PSD ($z^T M z \geq 0$ for all $z \in R^n$)

(ii) LCP is Concave $\Leftrightarrow M \in$ NSD ($z^T M z \leq 0$ for all $z \in R^n$)

(iii) LCP is Nonconvex $\Leftrightarrow M \notin$ PSD and $M \notin$ NSD $\qquad (4)$

The problem of the existence of a solution for the LCP is usually presented in terms of the classes of matrices Q and Q_0, where

$M \in Q \Leftrightarrow$ LCP has a solution for all $q \in R^n$

$M \in Q_0 \Leftrightarrow$ LCP has a solution for each $q \in R^n$ such that $K \neq \emptyset$. (5)
It has been shown that PSD matrices are Q_0-matrices and there are some classes of indefinite matrices that are included in the class Q_0 [35].

Another important research aspect of the LCP is related with the existence of polynomial algorithms for its solution. Recently it has been established that Convex LCPs can be solved in polynomial time [29]. Furthermore there exist some nonconvex LCPs that can also be solved in polynomial time [3, 32, 35, 40].

As stated in [35] there is a lack of important studies about the Concave LCP. This problem is interesting on its own, but it is also an important tool for finding global minimum of concave quadratic programs [37]. In this paper we investigate this problem. It is well-known that the Concave LCP is NP-Complete [10]. Hence it is expected that only some sort of enumerative technique can process the LCP in this case. In this paper we show that the hybrid enumerative method developed by Al-Khayyal [1] and improved by us [20] can be simplified for the Concave LCP and becomes powerful for the solution of such type of problem. We also investigate the solution of the Concave LCP by solving its equivalent QP (2) by a special purpose algorithm [18]. Some computational experience presented in the last section of this paper shows that the hybrid enumerative method is preferable for Concave LCPs.

We also show that the class NSD has a nonempty intersection with the class Q_0 but it is not included strictly in it. Furthermore there are no Q-matrices that are also NSD. We also introduce a suclass NSD_+ of NSD matrices in which all off-diagonal elements are nonnegative. We show that $NSD_+ \subset Q_0$ and we develop a polynomial algorithm for the solution of Concave LCPs with such matrices. Computational experience presented in this paper shows that this polynomial method performs quite well for large-scale LCPs and is more efficient than the hybrid enumerative method.

Recently we have developed a Sequential LCP (SLCP) algorithm for the solution of Bilevel Linear Programs [19, 22]. This SLCP algorithm can also be used to find global minima of Concave Quadratic Programs (CQP) by using Keller's method [27] and the hybrid enumerative algorithm described in this paper, or by exploiting the equivalence between CQPs and Bilinear Programs (BLP) [18]. In this paper we study the efficiency of the SLCP algorithm under these two alternatives for the solution of the CQP test problems described in [13]. The computer results show that in both cases the SLCP algorithm can in general find a global minimum efficiently (the BLP alternative normally performs better) but it is quite difficult to establish that such a solution has been found (the situation is worse for the BLP approach).

The organization of this paper is as follows. In Section 2 some properties of the Concave LCP are established. The hybrid enumerative method is discussed in Section 3. Section 4 is devoted to the solution of the Concave LCP by exploiting its equivalence to the QP (2). The polynomial method for NSD_+ matrices is described in Section 5. Finally computational experience with medium and large scale Concave LCPs is presented in the last section of this paper.

2 - Properties of the Concave LCP

In this section we investigate some theoretical properties of the LCP when M is a NSD or a NSD_+ matrix, where

$$M \in NSD_+ \Leftrightarrow M \in NSD \text{ and } m_{ij} \geq 0 \text{ for all } i \neq j \quad (6)$$

We start by recalling the concepts of Principal Transform and Schur Complement of a matrix. If M is a square matrix of order n, $F \subset \{1,\ldots, n\}$ and M_{FF} is a principal submatrix whose rows and columns belong to F, then there exists a permutation matrix P such that

$$P^T M P = \begin{bmatrix} M_{FF} & M_{FT} \\ M_{TF} & M_{TT} \end{bmatrix}$$

where $T = \{1,\ldots, n\} - F$. A Principal Transform of M with pivot M_{FF} is the matrix

$$\bar{M} = \begin{bmatrix} M_{FF}^{-1} & -M_{FF}^{-1}M_{FT} \\ M_{TF}M_{FF}^{-1} & (M|M_{FF}) \end{bmatrix} \quad (7)$$

where

$$(M | M_{FF}) = M_{TT} - M_{TF} M_{FF}^{-1} M_{FT} \quad (8)$$

is called the Schur Complement of M_{FF} in M.

By using proofs similar to those presented in [35] it is possible to establish a number of important properties that are gathered in the following lemma.

Lemma 1 - Let M be a square matrix of order n.

(i) $M \in NSD\ (NSD_+) \Rightarrow m_{ii} \leq 0$ for all $i = 1,\ldots, n$

(ii) $M \in NSD$ and $m_{ii} = 0 \Rightarrow m_{ij} = -m_{ji}$ for all $j \neq i$

$M \in NSD_+$ and $m_{ii} = 0 \Rightarrow m_{ij} = m_{ji} = 0$ for all $j \neq i$

(iii) $M \in NSD$ $(NSD_+) \Rightarrow M_{FF} \in NSD$ (NSD_+) for all $F \subset \{1,...,n\}$

(iv) If $\bar{M}$ is the matrix given by (7), then
$$M \in NSD \Rightarrow \bar{M} \in NSD$$

(v) If M_{FF} is nonsingular and $(M \mid M_{FF})$ is the matrix given by (8), then
$$M \in NSD\ (NSD_+) \Rightarrow (M \mid M_{FF}) \in NSD\ (NSD_+)$$

If $M \in NSD_+$, then $-M$ is an M-matrix [5], whence [5]

Lemma 2 - If $M \in NSD_+$ and is nonsingular then $M^{-1} \le 0$.

It follows from this lemma that the NSD_+ class has a non-empty intersection with the class of square inverse antitone matrices [28, 39]. It is important to add that none of these two classes contains the other one.

It has been shown [39] that square inverse antitone matrices are Q_0-matrices and cannot be Q. Furthermore it is stated in [39] without proof that Lemke's algorithm [30] cannot be applied to solve the LCP when its matrix M belongs to this class. Next, we investigate these properties for the class of NSD matrices.

Theorem 1 - (i) If $M \in NSD_+$ and $q < 0$, then LCP is infeasible.

(ii) $Q \cap NSD_+ = \emptyset$

Proof: We only prove the first result, since (ii) is a consequence of (i) and the definition of a Q-matrix. Two cases are possible and are discussed below.

a) If M is nonsingular the LCP (1) is equivalent to the LCP
$$z = -M^{-1} q + M^{-1} w \tag{9}$$
$$z \ge 0,\ w \ge 0,\ z^T w = 0$$

Since $0 \ne M^{-1} \le 0$ and $q < 0$, then $M^{-1} q > 0$ and LCP (9) is infeasible.

b) If M is singular there exists a set $F \subset \{1,...,n\}$ such that
$$M = \begin{bmatrix} M_{FF} & b & M_{FT} \\ a^T & m_{rr} & c^T \\ M_{TF} & h & M_{TT} \end{bmatrix}$$

M_{FF} is nonsingular, $T = \{1,...,n\} - (F \cup \{r\})$ and $m_{rr} - a^T M_{FF}^{-1} b = 0$. Since M_{FF} is nonsingular, we can interchange the z_i variables with the

w_i variables for $i \in F$ and obtain an equivalent LCP. This LCP takes the following form

$$\begin{bmatrix} z_F \\ w_r \\ w_T \end{bmatrix} = \begin{bmatrix} \bar{q}_F \\ \bar{q}_r \\ \bar{q}_T \end{bmatrix} + \begin{bmatrix} M_{FF}^{-1} & -M_{FF}^{-1}b & -M_{FF}^{-1}M_{TF} \\ a^T M_{FF}^{-1} & 0 & 0 \\ M_{TF}M_{FF}^{-1} & 0 & \bar{M}_{TT} \end{bmatrix}$$

But

$$\bar{q}_r = q_r - a^T M_{FF}^{-1} q_F < 0$$

since $a \geq 0$, $q < 0$ and $M_{FF}^{-1} \leq 0$. Furthermore $a^T M_{FF}^{-1} \leq 0$. Therefore there exist no positive entries in row r, whence this LCP is infeasible and the same happens with LCP (1).

Recently, Cottle [12] has established that the second result of the last theorem is valid for any NSD matrix. Because of the simplicity of the proof we present it below.

Theorem 2 - $NSD \cap Q = \emptyset$

Proof: If $M \in Q$, there exists a vector $z > 0$ such that $Mz > 0$ [31], whence $z^T Mz > 0$ for this vector and $M \notin NSD$.

It is a simple exercise to prove that $NSD \subset Q_0$ for matrices of order 2. However, it is possible to generate NSD matrices of order $n \geq 3$ such that the feasible set K is not empty and the minimum of the QP (2) is positive. So $NSD \cap Q_0 \neq \emptyset$, but NSD is not included in Q_0. In Section 5 we develop a polynomial algorithm that processes a LCP with a matrix $M \in NSD_+$, and we establish that $NSD_+ \subset Q_0$ as a consequence of the convergence of this algorithm.

As is stated in the introduction, direct and iterative methods can only process the LCP for special classes of matrices. Lemke's algorithm [30] is the one which can be applied for a larger number of classes. In this method a vector $p \geq 0$, such that $p_i > 0$ for all $q_i < 0$, is introduced to get an initial system of the form

$$w = q + z_0 p + Mz$$

where z_0 is an artificial variable. In the initial step of Lemke's algorithm a pivotal operation is performed that makes z_0 basic by changing with a variable w_r in such a way that the corresponding value $\bar{z}_0$ of z_0 satisfies $q + \bar{z}_0 p \geq 0$. If w_r is the w-variable that becomes nonbasic by changing with z_0, then z_r is the entering variable in the next iteration.

The entries of the updated column of this variable are given by

$$\bar{m}_{rr} = -\frac{m_{rr}}{p_r} \geq 0, \quad \bar{m}_{ir} = m_{ir} - \frac{m_{rr}}{p_r} p_i \geq 0, \, i \neq r$$

So the so-called termination in ray occurs and the algorithm is unable to process the LCP. Hence we have established the following result:

Theorem 3 - If $M \in NSD_+$, then Lemke's algorithm with a vector $p \geq 0$ cannot process the LCP.

This theorem is not valid for a general NSD matrix. In fact Lemke's method is able to find stationary points of Concave Quadratic Programs by solving an augmented Concave LCP and using a nonnegative vector p [35]. However, computational experience presented in the last section of this paper shows that in general Lemke's method is unable to process the Concave LCP.

3 - A Hybrid Enumerative Method for the Concave LCP

In the definition of the LCP we may distinguish the linear constraints and the complementarity condition $z^T w = 0$. Since the variables z_i and w_i are forced to be nonnegative, then this last condition is equivalent to

$$z_i w_i = 0 \text{ for all } i = 1,\ldots, n \tag{10}$$

Hence the LCP possesses a combinatorial nature similar to the 0-1 integer programming problem. By recognizing that fact, enumerative methods have been developed for the solution of the LCP [1, 15, 20, 26, 33, 34]. To date the best algorithm of this type was designed by Al-Khayyal [1] and improved by us [20]. In this algorithm a binary tree of the form

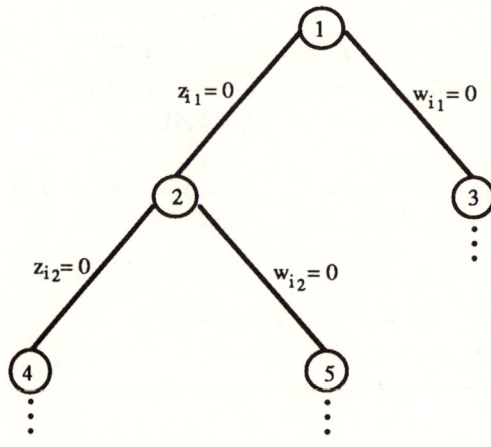

is implicitly searched. Node 1 is generated by solving the linear program

Minimize z_0
Subject to $w = q + z_0 p + Mz$
$$z \geq 0, w \geq 0, z_0 \geq 0 \qquad (11)$$

where p is a nonnegative vector such that $p_i > 0$ for all $q_i < 0$. This linear program is solved by a Modified Basis Restricted Entry Simplex (MBRES) algorithm in which there is a control of the number of basic pairs of complementary variables (z_i, w_i). The procedure only uses basic feasible solutions, that is, basic solutions of the feasible set of the LCP

$$\Gamma = \{(z, w): w = q + Mz, z \geq 0, w \geq 0\} \qquad (12)$$

Each node is generated by solving a linear program in which a variable is minimized by using the same MBRES technique. Some heuristic techniques for the choice of the node and the pair of complementary variables for branching have been developed [20] and are incorporated in the algorithm. The enumerative method also contains Al-Khayyal's Modified Reduced Gradient algorithm. This algorithm looks for a local star minimum of the function

$$g(z, w) = \sum_{i=1}^{n} z_i w_i \qquad (13)$$

that is, a basic feasible solution $(\bar{z}, \bar{w}) \in \Gamma$ such that

$$g(\bar{z}, \bar{w}) \leq g(z, w)$$

for all its adjacent basic feasible solutions $(z, w) \in \Gamma$. Next we show that this kind of algorithm can be simplified for the Concave LCP and becomes a simplex-type procedure. If $(\tilde{z}, \tilde{w})$ is an adjacent basic feasible solution of $(\bar{z}, \bar{w})$, then we can write

$$(\tilde{z}, \tilde{w}) = (\bar{z}, \bar{w}) + \mu_0 (d^z, d^w)$$

where μ_0 is the minimum ratio of the simplex method and $d = (d^z, d^w)$ is a feasible direction such that d^z and d^w are the vectors of the components corresponding to z and w respectively. But

$$g(\tilde{z}, \tilde{w}) = (\bar{z} + \mu_0 d^z)^T (\bar{w} + \mu_0 d^w)$$
$$= \bar{z}^T \bar{w} + \mu_0 [\bar{w}^T d^z + \bar{z}^T d^w] + \mu_0^2 (d^z)^T d^w$$

Since $(\tilde{z}, \tilde{w}) \in \Gamma$, then

$$\bar{w} + \mu_0 d^w = q + M(\bar{z} + \mu_0 d^z)$$

whence

and
$$\bar{w} + \mu_0 d^w = (q + M\bar{z}) + \mu_0 M d^z$$
$$d^w = M d^z$$
Therefore
$$(d^z)^T d^w = (d^z)^T M d^z \le 0$$
since $M \in$ NSD. Hence
$$g(\tilde{z}, \tilde{w}) - g(\bar{z}, \bar{w}) \le \mu_0 [\bar{w}^T d^z + \bar{z}^T d^w]$$
But $(\bar{w}^T d^z + \bar{z}^T d^w)$ is exactly the reduced-cost associated with the linear function
$$h(z, w) = \bar{z}^T w + \bar{w}^T z \tag{14}$$
of the nonbasic variable that is increased to generate $(\tilde{z}, \tilde{w})$. Hence Al-Khayyal's algorithm can be replaced by a simplex-type algorithm in which the reduced-costs of the nonbasic variables are associated with the linear function (14). The algorithm also incorporates Bland's rule [7] to avoid the occurrence of cycling in the degenerate cases. We suggest [22] for a description of the steps of the hybrid enumerative method, incorporating the procedures for generating the nodes, some important heuristic techniques and this last simplex-type procedure. The implementation of this method for large-scale LCPs is described in [20].

It is important to add that this hybrid enumerative method can also handle more general LCPs of the form
$$w = q + Mz$$
$$v = b + Rz$$
$$v, w, z \ge 0, z^T w = 0 \tag{15}$$
where $M \in$ NSD. This has important implications on the development of a Sequential LCP algorithm for the solution of Concave Quadratic Programs. This is discussed in the next section.

4 - A Sequential LCP Algorithm for Concave Quadratic Programming and Application to the Concave LCP

Consider the Concave Quadratic Programming Problem (CQP)
$$\text{Minimize } c^T x + \frac{1}{2} x^T Q x \tag{16}$$
$$\text{subject to } A x \ge b$$
$$x \ge 0$$

where $c, x \in R^\ell$, $A \in R^{m \times \ell}$, $b \in R^m$ and $Q \in R^{\ell \times \ell}$ is a symmetric NSD matrix. If this CQP has an optimal solution, then it is equivalent [23] to the following Minimum Linear Complementarity Problem (MLCP):

Minimize $c^T x + b^T u$

subject to $s = c + Qx - A^T u$ (17)
$r = -b + Ax$
$x, u, s, r \geq 0$, $x^T s = u^T r = 0$

This MLCP can be solved by the Sequential LCP (SLCP) algorithm described in [22]. In this procedure LCPs (λ_k) of the form

$$\begin{bmatrix} s \\ r \end{bmatrix} = \begin{bmatrix} c \\ -b \end{bmatrix} + \begin{bmatrix} Q & -A^T \\ A & 0 \end{bmatrix} \begin{bmatrix} x \\ u \end{bmatrix}$$

$$v_0 = \lambda_k - \begin{bmatrix} c & b \end{bmatrix}^T \begin{bmatrix} x \\ u \end{bmatrix}$$ (18)

$$v_0 \geq 0, \begin{bmatrix} x \\ u \end{bmatrix} \geq 0, \begin{bmatrix} s \\ r \end{bmatrix} \geq 0, \begin{bmatrix} x \\ u \end{bmatrix}^T \begin{bmatrix} s \\ r \end{bmatrix} = 0$$

are solved sequentially, where $\{\lambda_k\}$ is a sequence defined by

$$\lambda_{k+1} = c^T x^k + b^T u^k - \gamma_0 |c^T x^k + b^T u^k| \quad (19)$$

with γ_0 a small positive number and (x^k, u^k) a solution of LCP(λ_k). The SLCP algorithm terminates at an iteration k when the LCP(λ_k) has no solution. In this case the solution x^{k-1} of the LCP(λ_{k-1}) is a global minimum for the CQP. Since each LCP(λ_k) has the form (15) with

$$M = \begin{bmatrix} Q & -A^T \\ A & 0 \end{bmatrix} \in \text{NSD}$$

then the hybrid enumerative method can be used to solve these LCPs. The following features of the SLCP algorithm should be distinguished:

(i) The constraint $c^T x + b^T u \leq \lambda_0$ is not considered in the LCP(λ_0). Hence Keller's method [27] (or any other active-set method [6]) can be used to find a solution (x^0, u^0) for this LCP. After finding such a solution, λ_0 is set equal to $c^T x^0 + b^T u^0$.

(ii) The procedure MAXVAR described in [19] is incorporated in the SLCP algorithm to improve its efficiency.

(iii) The solution (x^{k-1}, u^{k-1}) of the LCP (λ_{k-1}) is the initial solution for the LCP (λ_k). This has a great effect on the efficiency of the SLCP method.

The SLCP algorithm can be implemented for the solution of large-scale Concave LCPs with sparse structure, by using the implementations of Keller's method, hybrid enumerative algorithm and MAXVAR procedure described in [21], [20] and [19] respectively.

Computational experience with the SLCP algorithm on the solution of some nonconvex optimization problems [18, 19, 22] has indicated that the method is usually efficient to find a global minimum, but it has difficulties in recognizing that such a solution has been achieved. In fact it is necessary to establish that the last LCP (λ_k) has no solution and this usually requires a lot of tree search in the hybrid enumerative method. These characteristics of the SLCP algorithm are highlighted in the last section of this paper, when we report computational experience with the algorithm for the solution of the CQP test problems described in [13].

As is stated in Section 1, the Concave LCP is equivalent to the CQP (2). Furthermore $\bar{z}$ is a solution of the LCP if and only if $f(\bar{z}) = 0$. Hence the last step of the SLCP algorithm has not to be performed and this contributes to a great increase on the efficiency of the algorithm to solve the Concave LCP. In the last section of this paper some Concave LCPs are solved by this approach and a comparison with the simple hybrid enumerative method is presented.

5 - A polynomial method for the Concave LCP with a NSD$_+$ matrix

Consider the set of constraints
$$w = q + Mz$$
of the LCP(1). A pivot operation is called Principal if the pivot is a diagonal element of M, that is, if a basic variable becomes nonbasic by changing with its complement.

If only principal pivot operations are performed, then the complementarity condition $z^T w = 0$ always holds. Furthermore in each iteration we have a system of the form

$$\bar{w} = \bar{q} + \bar{M}\bar{z} \tag{20}$$

where $(\bar{w}, \bar{z})$ is a permutation of (w, z) and $\bar{M}$ is a principal transform of M ($\bar{M} \in$ NSD by lemma 1). Principal Pivot Algorithms are procedures that only use principal pivot operations. Since the complementarity condition is satisfied in each iteration, then a solution of the LCP can be found when a system of the form (20) is obtained with $\bar{q} \geq 0$. These algorithms have proven to be quite successful in the solution of large-scale LCPs with special classes of matrices [24, 25]. In this section we describe a polynomial Principal Pivoting Algorithm for the solution of the Concave LCP with a NSD_+ matrix, which is closely related to Chandrasekaran's method [8]. The steps of the algorithm are as follows:

Polynomial Algorithm

Step 0 - Let $I=\{i : q_i < 0\}$, $J=\{1,...,n\} - I$, $F=\emptyset$ and $\bar{M}=M$.

Step 1 - Let
$$r = \min \{i \in I\} \tag{21}$$
and consider the set
$$A = \{j \in J - F : \bar{m}_{rj} > 0\} \tag{22}$$
If $A \neq \emptyset$ go to Step 3. Otherwise go to Step 2.

Step 2 - Set $I = I - \{r\}$. If $I \neq \emptyset$ go to Step 1. Otherwise the LCP is infeasible and stop.

Step 3 - Let $s \in A$. Set $F = F \cup \{s\}$ and perform a principal pivot operation with pivot $\bar{m}_{ss}$. Go to Step 4.

Step 4 - Let $I = \{i : \bar{q}_i < 0\}$ and $J = \{1,...,n\} - I$. If $I = \emptyset$, the vector z defined by
$$z_i = \begin{cases} \bar{q}_i & \text{if } i \in F \\ 0 & \text{if } i \notin F \end{cases}$$
is a solution of the LCP. Otherwise go to Step 1.

Next we prove that this algorithm terminates in at most n pivot operations. First of all we show that the principal pivot operation in Step 3 can always be performed. Consider the initial iteration of the algorithm. Since $M \in NSD_+$, either $m_{ss} < 0$ and the principal pivot operation can be performed or $m_{ss} = 0$. In the latter case the matrix M would have the following form

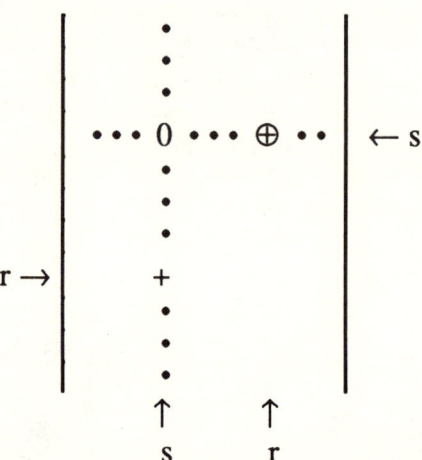

where $+$, $\oplus$ and 0 correspond to positive, nonnegative and zero entries respectively. Such a situation is impossible by lemma 1. Since only principal pivot operations are performed, then in any iteration of the algorithm we have a tableau of the form

$$\begin{array}{c|ccc} & 1 & w_F & z_T \\ \hline z_F = & \bar{q}_F & \bar{M}_{FF} & \bar{M}_{FT} \\ w_T = & \bar{q}_T & \bar{M}_{TF} & \bar{M}_{TT} \end{array} \qquad (23)$$

where $T = \{1,\ldots,n\} - F$,

$$\bar{M} = \begin{bmatrix} \bar{M}_{FF} & \bar{M}_{FT} \\ \bar{M}_{TF} & \bar{M}_{TT} \end{bmatrix}$$

is the principal transform of M given by (7) and $\bar{q}$ satisfies

$$\bar{q}_F = -M_{FF}^{-1} q_F, \quad \bar{q}_T = q_T + M_{TF}\bar{q}_F \qquad (24)$$

Because of the definition of the set A, $r \notin F$ and $s \notin F$. But $\bar{M}_{TT}$ is the Schur Complement of M_{FF} in M. Hence $\bar{M}_{TT} \in NSD_+$ and $\bar{m}_{ss} < 0$ by using a proof similar to the initial iteration.

As stated above a tableau of the form (23) can be considered in each iteration of the algorithm. Next we show that $\bar{q}_F \geq 0$. To do this, suppose that $\bar{q}_F \geq 0$ in tableau (23) and another principal pivot

operation is performed. If $\bar{m}_{ss}$ is the pivot of this operation, then $\bar{m}_{ss} < 0$, $\bar{q}_s \geq 0$ and
$$\bar{M}_{Fs} = -M_{FF}^{-1} M_{Fs} \geq 0$$
since $M_{FF}^{-1} \leq 0$ by lemma 2. As $F = F \cup \{s\}$ and the new components of the transformed of $(\bar{q}_F, \bar{q}_s)$ are given by
$$\bar{q}_F - \frac{\bar{q}_s}{\bar{m}_{ss}} \bar{M}_{Fs}, \quad -\frac{\bar{q}_s}{\bar{m}_{ss}}$$
then the result follows.

Therefore whenever a variable z_i, $i \in I$, becomes basic (i becomes an element of F), then it never changes its status until the end of the algorithm. This shows that there are at most n principal pivot operations and the algorithm is polynomial.

To establish the convergence of the algorithm it is sufficient to prove that if a pivot cannot be found then the LCP is infeasible. To do this, it is better to write tableau (23) in the following form

	1	w_F	z_I	z_T
$z_F =$	$\bar{q}_F$	$\bar{M}_{FF}$	$\bar{M}_{FI}$	$\bar{M}_{FT}$
$w_I =$	$\bar{q}_I$	$\bar{M}_{IF}$	$\bar{M}_{II}$	$\bar{M}_{IT}$
$w_T =$	$\bar{q}_T$	$\bar{M}_{TF}$	$\bar{M}_{TI}$	$\bar{M}_{TT}$

where
$$\bar{q}_F \geq 0, \bar{q}_I < 0, \bar{q}_T \geq 0, \bar{M}_{FF} = M_{FF}^{-1} \leq 0,$$
$$\begin{bmatrix} \bar{M}_{FI} \\ \bar{M}_{FT} \end{bmatrix} = -M_{FF}^{-1} \begin{bmatrix} M_{FI} \\ M_{FT} \end{bmatrix} \geq 0, \begin{bmatrix} \bar{M}_{IF} \\ \bar{M}_{TF} \end{bmatrix} = \begin{bmatrix} M_{IF} \\ M_{TF} \end{bmatrix} M_{FF}^{-1} \leq 0,$$
$$\begin{bmatrix} \bar{M}_{II} & \bar{M}_{IT} \\ \bar{M}_{TI} & \bar{M}_{TT} \end{bmatrix} = (M \mid M_{FF}) \in NSD_+$$

If the algorithm cannot reach Step 3, then we must have for all $i \in I$
$$\bar{m}_{ij} = 0 \text{ for all } j \neq i, j \notin F$$
Since $\bar{q}_i < 0$, $\bar{m}_{ii} \leq 0$ and $\bar{M}_{IF} \leq 0$, then the LCP is infeasible.

So the algorithm is polynomial. Furthermore either terminates with a solution of the LCP or the LCP is infeasible. This shows that

$NSD_+ \subset Q_0$. The inclusion is obviously strict, since there is a large number of other subclasses of Q_0-matrices [35].

In an implementation of this algorithm for large-scale Concave LCPs, it is important to note that all the information required by the algorithm can be found by solving systems with the matrix M_{FF}. If $T = \{1,\ldots, n\} - F$ then by using (24) we get the following procedure to find the updated vector $\bar{q}$:

$$\boxed{\begin{array}{l} \text{Solve} \quad M_{FF}\, \bar{q}_F = -q_F \\[4pt] \text{Compute} \quad \bar{q}_T = q_T + M_{TF}\, \bar{q}_F \end{array}} \quad (25)$$

On the other hand the elements $\bar{m}_{rj}$ in Step 1 can be found by:

$$\boxed{\begin{array}{l} \text{Solve} \quad M_{FF}\, \alpha = M_{rF} \\[4pt] \text{Compute} \quad \bar{m}_{rj} = m_{rj} - M_{Fj}^T\, \alpha \end{array}} \quad (26)$$

Since $M_{FF} \in NSD_+$ the solution of the systems by elimination without pivoting is stable [14]. If the matrix M is symmetric then the implementation described in [25] can be used to solve large-scale Concave LCPs by this polynomial algorithm. In this implementation an Analyse Phase for controlling the occurrence of too much fill-in is first performed and the updating of the LDL^T decompositions is done following the ordering achieved in the Analyse Phase. This type of implementation can be extended for the unsymmetric case if the Analyse Phase is performed for the matrix $M+M^T$ and the LDU decompositions are updated according to a scheme similar to that described in [25]. In practice it is important to choose r in criterion (21) according to the ordering achieved in Analyse Phase. It is also better to choose the index s as the first element of A in this ordering.

6 - Computational Experience

In this section we present some computational experience with the techniques discussed in this paper on the solution of medium and large scale Concave LCPs. All the tests have been performed in a CDC CYBER 180-830 of the University of Porto. Table 1 reports the experience of solving eleven test problems taken from different sources by the following algorithms:

    (i) Al-Khayyal's hybrid enumerative method [1] with the improvements presented in [20] - ALKHYBEN.

(ii) The hybrid enumerative method with the simplex-type procedure described in Section 3 instead of Al-Khayyal's Modified Reduced Gradient method - HYBEN.
(iii) Keller's method [27] using the implementation described in [21] - KELLER.
(iv) The extension of Lemke's method for quadratic programs described in [35] - LEMKE.

The eleven test problems have been generated in the following way:
(i) The matrices of the test poblems TP1, TP2 and TP3 are symmetric tridiagonal and their elements have been randomly generated.
(ii) The matrices of the test problems TP4 and TP5 are symmetric and have been randomly generated by a scheme similar to that described in [22].
(iii) The test problems TP6 and TP7 are taken from [16].
(iv) The test problems TP8, TP9, TP10 and TP11 are the Kuhn-Tucker conditions of quadratic programs of the form (16). The matrix Q and the vector c of these quadratic programs are those used in TP4 (TP8-10) and TP5 (TP11). The constraints have been generated by a technique similar to that presented in [22].

In this table and in the next three we use the following notations:
n = dimension of the LCP = order of matrix M.

ℓ = number of variables of the quadratic program.
m = number of constraints of the quadratic program.
NI = number of pivot operations.
ND = number of nodes.
T = CPU time in seconds.
UN = the method is unable to find a solution (termination in ray has occurred).
C = cycling has occurred.
NLCP = number of LCPs(k) solved by the SLCP algorithms (Tables 2 and 4).

The following conclusions can be asserted from the results presented in Table 1:
(i) The hybrid enumerative method with the modifications presented in this paper (HYBEN) is efficient to solve large-scale Concave LCPs.
(ii) The modifications recommended in this paper increase very much the efficiency of the hybrid enumerative method, since they reduce drastically the computational effort of each iteration.

(iii) In general Keller and Lemke methods are unable to solve Concave LCPs. However, there are more efficient than HYBEN when they can do the job.

(iv) Cycling may occur in Lemke's method. We should note that Bland's rule [7] is implemented in our code. This shows that, contrary to the Convex case [9], the use of Bland's rule is not sufficient to prevent the occurrence of cycling when the LCP is concave. We believe that the lexico-minimum rule described in [35] can overcome these cycling problems. However, it seems quite involved to incorporate this type of technique in an implementation for large-scale LCPs.

In Table 2 it is compared the efficiency of the HYBEN method with the SLCP algorithm (SLCPQP) described in Section 2 for the solution of medium scale Concave LCPs. Another SLCP algorithm (SLCPBLP) was also considered in our computational study. This method consists of applying the SLCP algorithm to a Bilinear Program that is equivalent to the Concave Quadratic Program [18].

We have generated the test problems of Table 2 by considering positive definite matrices taken from known sources and multiplying all their entries by -1. Pang's technique [36] is used to generate the test problems TP12 and TP13. A block tridiagonal matrix associated with the solution of elliptic differential equations by the finite-difference method has been chosen to generate TP14. Finally the tridiagonal matrix of the diagonal blocks has lead to the test problem TP15. We have randomly generated five different right-hand side vectors for each matrix M of each test problem. Table 2 contains the behavior of the algorithms by showing the best (B), average (A) and worst(W) performance in terms of pivot operations.

The total computational efforts of the SLCP methods are reported under the columns headed by TOTAL. We have followed the recommendations stated in [18, 19] and choose $\gamma_0 = 10^{-3}$ in both the SLCP algorithms. The following conclusions can be stated from the computational results:

(i) The HYBEN algorithm is more efficient than the SLCP algorithms.

(ii) It seems that SLCPBLP is more robust than SLCPQP in the sense that is smaller the difference between the best and the worst performance.

(iii) The HYBEN algorithm deals with lower dimensional LCPs. This implies that the Basis matrices for the SLCP algorithms are larger and this contributes to an increase of the time of each pivot operation.

These conclusions lead to our recommendation of using the HYBEN algorithm for large-scale Concave LCPs instead of exploiting its equivalence to a nonconvex optimization problem.

Table 3 contains the results of the experiments with the polynomial (POLYN) algorithm to solve large-scale Concave LCPs with NSD_+ matrices. The matrices of the test problems TP16 and TP17 (TP18 and TP19) come from the same sources of the test problem TP15 (TP14). Furthermore the test problems TP20 and TP21 have been randomly generated. The test problems TP16, TP17 and TP18 have also been solved by the HYBEN algorithm to compare its efficiency with the POLYN method. The computer results lead to the following conclusions:

(i) The POLYN algorithm performs quite well and is more efficient than the HYBEN method.

(ii) The POLYN algorithm tends to find solutions with more basic z-variables than the HYBEN method. To see this property we added in the results of the HYBEN method a column headed by BVZ, which represents the number of z-variables in the solution found by the HYBEN algorithm. Our conclusion follows from the fact that BVZ is always smaller than the number NI of the z-variables that are basic in the solution found by the POLYN method. We have not found an exploration for these performances so far.

(iii) For each test problem we have generated two problems which differ on the number neg of negative components of the vector q. This quantity usually influences the behavior of Principal Pivoting methods [24]. The results of Table 3 show that it is also the case for the POLYN algorithm but the increase of NI with neg is small.

Recently Floudas and Pardalos [13] have produced a list of test problems for nonconvex global optimization. In Table 4 we present the results of the solution of their Concave Quadratic Programs by the SLCP algorithm with four different forms:

(i) The LCPs(λ_k) are solved by the procedure described in Section 4 - SLCPQP.

(ii) The Concave Quadratic Programs are tranformed into their equivalent Bilinear Programs which are solved by the SLCP algorithm described in [18] - SLCPBLP.

(iii) The LOWBND algorithm [18] is first applied to find a lower-bound LVAL for the quadratic function of the quadratic program (16). Then the cut
$$c^T x + b^T u \geq LVAL$$

is introduced in the MLCP constraints and the SLCPQP is applied to this MLCP - SLCPLBQP.
(iv) The same procedure of (iii) but using the SLCPBLP - SLCPLBBLP.

In Table 4 we use the notation NS when the hybrid enumerative method cannot process a LCP(λ_k) (find a solution or establish that no solution exists) in less than 30000 pivot operations. In the columns headed by OPTIMAL and LAST we write the computational efforts required to find the best solution found by the algorithm and to show that the last LCP has no solution respectively. The performance of the algorithm LOWBND is presented in the columns headed by its name. The results presented in Table 4 lead to the following conclusions:

(i) All the four SLCP algorithms have found the global minima stated in [13] for the first six test problems. Furthermore they have been able to establish that these solutions are global minima, since they could show that the last LCP(λ_k) has no solution. The best solutions stated in [13] for the test problems 7.1 to 7.5 and 8 have also been found by the four SLCP algorithms. However, the algorithms have not been able to establish that such solutions are global minima. The only exception was the test problem 8 in which the SLCPLBQP algorithm could do it. It is also interesting to note that in the problem 7.4 the algorithms have found a better solution than the one stated in the book [13]. This solution x* is given by
$x_3^*=1.0429$, $x_{11}^*=1.7467$, $x_{13}^*=0.43147$, $x_{16}^*=4.4331$
$x_{18}^*=18.859$, $x_{20}^*=16.487$, $x_i^*=0$, $i \neq 3, 11, 16, 18, 20$
and the value of the objective function is - 4150.4102.

(ii) The efficiency of the algorithms SLCPQP and SLCPBLP to find the best solution does not seem to be much influenced by the use of the algorithm LOWBND and the respective cut. However, the performance of the hybrid enumerative method for the last LCP(λ_k) is usually improved by the use of this cut.

(iii) In many problems the SLCP algorithms are quite efficient to find the best solution (or global minimum). However, there are some test problems in which the SLCP algorithms face some difficulties in fulfilling this goal. It is important to add that a good solution for these problems can be found with a relative small effort, but the computational effort to find the best solution is quite involved.

(iv) The **SLCPBLP** algorithm is usually more efficient than the **SLCPQP** method for finding the best solution (or global minimum). The difference between the performances of the two algorithms is even big for some test problems. However, SLCPQP performs much better for the test problem 7.4.

As a final conclusion of the results presented in this table, we claim that the use of a SLCP algorithm as a global minimizer method holds some promise. However, it is still necessary to develop algorithms that can find efficiently tight lower-bounds and to speed-up the SLCP method in the last stages for some difficult problems.

This paper also shows that the Concave LCP can be solved efficiently by a hybrid enumerative method. Furthermore a polynomial algorithm for the solution of a special Concave LCP is also developed and performs even better than the hybrid enumerative method. We think that other efficient algorithms for the solution of Concave LCPs can be developed in the future with important implications on the solution of Concave Quadratic Programs.

TP	n	ℓ	m		HYBEN	ALKHYBEN	LEMKE	KELLER
1	500	500	—	NI ND T	160 53 27.8	160 53 52.5	UN	UN
2	1000	1000	—	NI ND T	325 103 92.1	325 103 263.6	UN	UN
3	2000	2000	—	NI ND T	699 243 422.4	699 243 1989.	UN	UN
4	500	500	—	NI ND T	189 65 35.9	189 65 71.	UN	UN
5	600	600	—	NI ND T	531 87 115.9	531 87 298.7	UN	UN
6	1199	799	400	NI ND T	325 23 109.	287 13 280.4	19 — 5.2	4 — 0.5
7	1499	999	500	NI ND T	124 1 42.8	124 1 152.8	C	9 — 1.2
8	520	500	20	NI ND T	136 7 33.3	136 7 87.1	7 — 1.3	30 — 1.16
9	550	500	50	NI ND T	1355 23 427.2	1354 23 1124.	307 — 38.7	668 — 72.2
10	550	500	50	NI ND T	3613 25 1091.	3709 23 2492.	C	821 — 88.7
11	650	600	50	NI ND T	1905 45 729.2	1905 45 2064.	683 — 120.9	764 — 104.

Table 1 - Solution of Large-Scale Concave LCPs

CONCAVE LINEAR COMPLEMENTARITY PROBLEM

TP	n		SLCPBLP			SLCPQP			HYBEN		
			NLCP	NI	T	NLCP	NI	T	ND	NI	T
12	100	B	1	63	1.9	1	20	.69	1	21	.67
		A	4	163	14.	3	100	9.	4	32	1.6
		W	9	255	27.6	7	332	34.3	5	54	3.
13	300	B	1	34	2.8	1	10	.88	1	9	.69
		A	2	67	10.2	2	80	19.	2	18	1.7
		W	4	132	21.7	3	189	50.6	3	36	3.8
14	100	B	7	290	26.1	4	131	11.2	3	96	4.4
		A	5	320	31.8	5	156	26.1	3	126	7.2
		W	3	363	30.3	7	198	56.1	3	138	8.7
15	100	B	1	18	.43	1	14	.36	1	3	.06
		A	2	66	5.	2	64	5.	2	8	.36
		W	4	219	20.6	5	258	23.5	7	23	1.4

Table 2 - Comparison of SLCP and HYBEN for medium-scale Concave LCPs

TP	n	neg	POLYN		HYBEN			BVZ
			NI	T	ND	NI	T	
16	500	130	484	19.6	83	589	67.4	233
		166	497	20.1	105	623	80.4	344
17	1000	305	987	81.9	195	1349	289.	528
		336	992	90.6	189	1129	261.	656
18	512	223	466	41.6	5	388	80.5	300
		251	503	57.6	5	570	268.	421
19	1008	480	987	245.4				
		498	996	284.8				
20	500	202	424	39.2				
		227	434	47.5				
21	1000	258	772	73.6				
		337	781	78.6				

Table 3 - Solution of Concave LCPs with NSD_+ matrices

Prob	ℓ	m	SLCPBLP					SLCPQP				
			OPTIMAL			LAST		OPTIMAL			LAST	
			NLCP	NI	T	NI	T	NLCP	NI	T	NI	T
1	5	6	3	22	.15	214	2.6	5	95	1.0	125	1.8
2	6	7	1	12	.03	72	.87	1	6	.02	56	.61
3	13	19	5	91	1.2	2062	69.3	3	72	.94	2295	63.4
4	6	8	1	16	.06	239	3.6	1	8	.03	185	2.1
5	10	21	1	22	.21	2078	61.0	1	11	.10	1439	27.7
6	10	15	5	271	4.3	11300	224.	6	397	5.4	4902	76.4
7.1	20	10	24	16786	467.	NS		20	31853	750.	NS	
7.2	20	10	18	18259	506.	NS		16	34142	799.	NS	
7.3	20	10	9	13594	351.	NS		12	32654	690.	NS	
7.4	20	10	5	37529	951.	NS		5	20493	492.	NS	
7.5	20	10	18	428	10.8	NS		38	11685	281.	NS	
8	24	20	7	14	3.0	NS		9	940	19.7	NS	

Prob.	ℓ	m	LOWBND		SLCPLBBLP					SLCPLBQP			
					OPTIMAL			LAST		OPTIMAL			LAST	
			NI	T	NLCP	NI	T	NI	T	NLCP	NI	T	NI	T
1	5	6	12	0.3	3	21	.15	248	3.3	4	167	2.1	165	2.3
2	6	7	22	0.5	1	12	0.3	0	0	1	6	0.1	0	0
3	13	19	33	1.4	5	91	1.2	2056	70.3	3	67	.89	2260	62.9
4	6	8	18	0.6	1	16	0.6	239	3.8	1	8	0.3	185	2.3
5	10	21	32	.28	1	22	.21	1860	54.0	1	11	.10	1372	27.2
6	10	15	34	.21	5	282	4.7	10667	223.	5	281	4.1	4723	79.1
7.1	20	10	347	3.9	24	16782	480.	NS		20	31599	763.	NS	
7.2	20	10	348	3.9	18	18258	520.	NS		16	34625	839.	NS	
7.3	20	10	346	3.9	9	13590	360.	NS		12	32877	701.	NS	
7.4	20	10	333	3.8	5	37528	974.	NS		5	20614	506.	NS	
7.5	20	10	342	3.8	18	428	11.2	NS		38	12224	300.	NS	
8	24	20	143	.95	8	125	2.2	NS		7	327	7.0	23368	517.

Table 4 – Solution of Concave Quadratic Programs from the collection [13]

References

[1] F.A. Al-Khayyal, *An implicit enumeration procedure for the general linear complementarity problem,* Mathematical Programming Studies 31 (1987) 1-20.

[2] F.A. Al-Khayyal, *Linear, quadratic and bilinear programming approaches to the linear complementarity problem,* European Journal of Operations Research 24 (1986) 216-227.

[3] F.A. Al-Khayyal, *On characterizing linear complementarity problems as linear programs,* Optimization 20 (1989) 715-724.

[4] J.F.Bard and J.E.Falk, *A separable programming approach to the linear complementarity problem,* Computers and Operations Research 9 (1982) 153-159.

[5] A.Berman and R.J.Plemmons, *Nonnegative Matrices in the Mathematical Sciences,* Academic Press, New York, 1979.

[6] M.J.Best, *Equivalence of some quadratic programming algorithms,* Mathematical Programming 30 (1984) 71-87.

[7] R.G.Bland, *New finite pivoting rules for the simplex method,* Mathematics of Operations Research 2 (1977) 103-107.

[8] R.Chandrasekaran, *A special case of the complementary pivot problem,* Opsearch 7 (1970) 263-268.

[9] Y.Y.Chang, *Least-index resolution of degeneracy in linear complementarity problems,* Technical Report 79-14, Department of Operations Research, Stanford University, 1979.

[10] S.J.Chung, *NP-completeness of the linear complementarity problem,* Journal of Optimization Theory and Applications 60 (1989) 393-399.

[11] R.W.Cottle, *Note on a fundamental theorem in quadratic programming,* Journal SIAM 12 (1964) 663-665.

[12] R.W.Cottle, private communication.

[13] C.A.Floudas and P.M.Pardalos, *A Collection of Test Problems for Constrained Global Optimization Algorithms*, Lecture Notes in Computer Science 455, Springer-Verlag, Berlin Heidelberg, 1990.

[14] R.E.Funderlic and R.J.Plemmons, *LU decompositions of M-matrices by elimination without pivoting*, Linear Algebra and its Applications 41 (1981) 99-110.

[15] G.B.Garcia and C.E.Lemke, *All solutions to linear complementarity problems by implicit search*, Proceedings of the 39th Meeting of the Operations Research Society of America, 1971.

[16] N.I.M.Gould, *An algorithm for large-scale quadratic programming*, Harwell Report CSS 219, 1989.
[17] R.Horst and H.Tuy, *Global Optimization*, Springer-Verlag, Berlin Heidelberg, 1990.
[18] J.J.Júdice and A.M.Faustino, *A computational analysis of LCP methods for bilinear and concave quadratic programming*, to appear in Computers and Operations Research.
[19] J.J.Júdice and A.M.Faustino, *A sequential LCP method for bilevel linear programming*, to appear in Annals of Operations Research.
[20] J.J.Júdice and A.M.Faustino, *An experimental investigation of enumerative methods for the linear complementarity problem*, Computers and Operations Research 15 (1988) 417-426.
[21] J.J.Júdice and A.M.Faustino, *An implementation of Keller's method for two concave optimization problems*, to appear in Investigação Operacional.
[22] J.J.Júdice and A.M.Faustino, *The solution of the linear bilevel linear programming problem by using the linear complementarity problem*, Investigação Operacional 8 (1988) 77-95.
[23] J.J.Júdice and G.Mitra, *Reformulation of mathematical programming problems as linear complementarity problems and investigation of their solution methods*, Journal of Optimization Theory and Applications 57 (1988) 123-149.
[24] J.J.Júdice and F.M.Pires, *Bard-type methods for the linear complementarity problem with symmetric positive definite matrices*, IMA Journal of Mathematics Applied in Business and Industry 2 (1988/89) 51-68.
[25] J.J.Júdice and F.M.Pires, *Direct methods for convex quadratic programs subject to box constraints*, Investigação Operacional 9 (1989) 23-56.
[26] I.Kaneko and W.P.Halman, *An enumerative algorithm for a general linear complementarity problem*, Technical Report WP 78-11, University of Wisconsin Madison, Wisconsin, 1978.
[27] E.L.Keller, *The general quadratic optimization problem*, Mathematical Programming 5 (1973) 311-337.
[28] G.J.Koehler, *A comment on Rao's class M and S matrices*, Management Science 23 (1977) 1247.
[29] M.Kojima, S.Mizuno and A.Yoshise, *A polynomial-time algorithm for a class of linear complementarity problems*, Mathematical Programming 44 (1989) 1-26.

[30] C.E.Lemke, *On complementarity pivot theory*, in "Mathematics of Decision Sciences", edited by G.B.Dantzig and A.F.Veinott Jr., American Mathematical Society, Providence, 1968, pp.95-113.

[31] C.E.Lemke, *Recent results on complementarity problems*, in "Nonlinear Programming", edited by Rosen, Mangasarian and Ritter, Academic Press, New York, 1970, pp.350-384.

[32] O.L.Mangasarian, *Simplified characterizations of linear complementarity problems solvable as linear programs*, Mathematics of Operations Research 4 (1979) 268-273.

[33] G.Mitra and G.R.Jahanshalou, *Linear complementarity problem and a tree search algorithm for its solution*, in "Survey of Mathematical Programming", edited by A.Prekopa, North-Holland, Amsterdam, 1979, pp.35-56.

[34] K.G.Murty, *An algorithm for finding all the feasible complementary bases for a linear complementarity problem*, Technical Report 72-7, Department of Industrial and Operations Engineering, University of Michigan, 1972.

[35] K.G.Murty, *Linear Complementarity, Linear and Nonlinear Programming*, Heldermann Verlag, Berlin, 1988.

[36] J.S.Pang, *A new and efficient algorithm for a class of portfolio selection problems*, Operations Research 28 (1980) 754-767.

[37] P.M.Pardalos and J.B.Rosen, *Constrained Global Optimization: Algorithms and Applications*, Lecture Notes in Computer Science 268, Springer-Verlag, Berlin Heidelberg, 1987.

[38] B.Ramarao and C.M.Shetty, *Application of disjunctive programming to the linear complementarity problem*, Naval Research Logistics Quarterly 31 (1984) 589-600.

[39] A.K.Rao, *On the linear complementarity problem*, Management Science 22 (1973) 427-429.

[40] Y.Ye and P.M.Pardalos, *A class of linear complementarity problems solvable by polynomial time*, to appear in Linear Algebra and its Applications.

Global Solvability of Generalized Linear Complementarity Problems and a Related Class of Polynomial Complementarity Problems

Aniekan A. Ebiefung [*]   Michael M. Kostreva [*]

Abstract

Characterizations of the set of Q - matrices and its complement are given for the Generalized Linear Complementarity Problem (GLCP) introduced by Cottle and Dantzig. The characterizations provide a basis for algorithms which not only solve the GLCP for a general matrix N, but also generate all complementary solutions. Also considered is a relationship between the solution set of the GLCP and that of a specially structured polynomial nonlinear complementarity problem. Existence of a solution for the polynomial complementarity problem is demonstrated in the case when N is a Q-matrix, while uniqueness is ruled out even when N is a P-matrix. Properties of the N matrix and some related systems of linear inequalities are considered. Specialization of the results to the linear complementarity problem are seemingly novel as well.

1. INTRODUCTION

The Generalized Linear Complementarity Problem of Cottle and Dantzig (Ref. 1) is: Given an m x n, m $\geq$ n, vertical block matrix N, of type $(m_1, \ldots, m_n)$ and q in R^m, find $z \in R^n$ and $w \in R^m$, such that

$$\text{GLCP(q,N):} \quad w = Nz + q, \quad w \geq 0, \quad z \geq 0. \quad z_j \prod_{i=1}^{m_j} w_i^j = 0 \quad (j=1,\ldots,n).$$

[*] Mathematical Sciences Department, Clemson University, Clemson, SC 29634-1907.

That GLCP(q, N) has a solution when N is strictly copositive or a P-matrix was shown by Cottle and Dantzig (Ref. 1). More recent work on the GLCP(q, N) can be found in (Ref. 2) where P- matrices are characterized. A useful engineering application of the Generalized Complementarity Problem is given by Oh in (Ref. 3).

After a review of the literature, we observed that there are no published results on the existence of solutions to GLCP(q, N) for a general vertical block matrix N of type $(m_1, \ldots, m_n)$.

In this paper, necessary and sufficient conditions under which problem GLCP(q, N) has a solution are provided. The characterizations given form a basis for algorithms for solving problem GLCP(q, N) for a general matrix N.

Definitions and notation used in this paper are presented in section 2. The main results, including the algorithms are stated in section 3. In section 4, we record some mathematical properties of the N matrix associated with problem GLCP(q, N). Finally, in section 5, we derive some related results for the linear complementarity problem which are discernible from the theory developed for GLCP(q, N).

2. DEFINITIONS AND NOTATION

The Linear Complementarity Problem is: Given an n x n real matrix M and an n x 1 vector q, find vectors $w \in R^n$ and $z \in R^n$ such that $w = Mz + q \geq 0$, $z \geq 0$, and $w^T z = 0$ (equivalently $w_i z_i = 0$, i=1,..., n).

Let the above problem be denoted LCP(q, M).

Definition 2.1. Let M be a real square matrix of size n. M is called a Q-matrix iff LCP(q, M) has solution for each $q \in R^n$. Similarly, GLCP(q, N) has a solution for each $q \in R^m$ iff N is a vertical block Q-matrix of type $(m_1, \ldots, m_n)$.

Definition 2.2. An n x n matrix M is said to be a P-matrix iff all its principal minors are strictly positive.

Definition 2.3. If M is a square matrix of order n, then M is said to be copositive iff $x^t M x \geq 0$ for each $x \geq 0$. The matrix M is called strictly copositive iff for all $x \geq 0$, $x \neq 0$, $x^t M x > 0$. M is copositive plus iff M is copositive and if $x^t M x = 0$ for some $x \geq 0$, $x \neq 0$, then $(M + M^t)x = 0$.

Definition 2.4. By an m x n vertical block matrix of type $(m_1, \ldots, m_n)$ is meant a matrix N, where the j-th block, N^j, is of dimension m_j x n and

$$m = \sum_{j=1}^{n} m_j .$$

The vectors q and w in R^m are also partitioned to conform to the entries in the blocks, N^j, of N.

Definition 2.5. Let N be a vertical block matrix of type $(m_1, \ldots, m_n)$. A submatrix M of N of size n is called a representative submatrix if its j-th row is drawn from the j-th block, N^j, of N. A vertical block matrix of type $(m_1, \ldots, m_n)$ has $\prod_{j=1}^{n} m_j$ representative submatrices.

A principal submatrix of N is a principal submatrix of some representative submatrix. The determinant of such a matrix is called a principal minor of N.

Definitions 2.6. Let N be a vertical block matrix of type $(m_1, \ldots, m_n)$. N is called a P-matrix if all its representative submatrices are P-matrices. The concepts of copositive, copositive plus, and strictly copositive are also defined in this sense.

Definition 2.7. A complementary solution of $w = q + Nz$ is called a proper solution. A nonnegative proper solution (w, z) solves GCLP(q, N).

A solution of $w = q + Nz$ that satisfies complementarity for all but one variable is called an almost proper solution. The corresponding basis is called an almost proper basis.

Definition 2.8. A vector q in R^n is <u>nondegenerate with respect to a matrix M</u> of order n if it is not a linear combination of any (n - 1) (or less) column vectors in (I, - M).

A vector $q \in R^m$ is said to be <u>nondegenerate with respect to an m x n vertical block matrix N of type $(m_1, \ldots, m_n)$</u> if the solution (w, z) of w=Nz+q contains at most n zero components.

Finally, the following lemma of Farkas is used often in the sequel.

<u>Lemma (Farkas's)</u>. Let A be a given m x n matrix and b a column vector of order m x 1. Then exactly one of the systems (I) and (II) below has a solution and the other system is inconsistent

(I) Ax = b, $x \geq 0$.

(II) $u^t A \leq 0$, $u^t b > 0$.

3. CHARACTERIZATIONS AND EXISTENCE OF SOLUTIONS

Cottle and Dantzig, (Ref. 1), showed that GLCP(q, N) has a solution when N is strictly copositive or a P-matrix. The theory of existence of solutions for a general vertical block matrix N is provided in this section. The first theorem is that of Cottle and Dantzig and is stated for the sake of completeness. We start with their formulation.

Let $\zeta = \{z: Nz + q \geq 0, z \geq 0\}$, the set of feasible solutions for GLCP (q, N). An extreme point of ζ is called proper or almost proper according as the corresponding basic solution is proper or almost proper.

Now let q^0 be a real number larger than the sum of the components of any extreme point of ζ. Let e_n denote a column vector of length n having all components equal to 1 and z_0 a variable with complement $w^0 = q^0 + 0 \cdot z_0 - e_n^t z$. Using the variables so defined, we transform the given GLCP(q, N) into a form so that a modification of Lemke's algorithm may be applied.

NEW PROBLEM

(NP): $\begin{pmatrix} w^0 \\ w \end{pmatrix} = \begin{pmatrix} q^0 \\ q \end{pmatrix} + \begin{pmatrix} 0 & -e_n^t \\ e_m & N \end{pmatrix} \begin{pmatrix} z_0 \\ z \end{pmatrix},$

$\begin{pmatrix} w^0 \\ w \end{pmatrix} \geq 0, \begin{pmatrix} z_0 \\ z \end{pmatrix} \geq 0.$

$z_j \left(\prod_{i=1}^{m_j} w_i^j \right) = 0, \qquad (j = 1, \ldots, n)$

Cottle and Dantzig, (Ref. 1), showed that (NP) has a solution when the modified Lemke's algorithm is applied to it, assuming nondegeneracy of the $(q^0, q)^t$ vector. A solution of (NP) in which $z_0 = 0$ gives a solution to the GLCP(q, N).

<u>Theorem 3.1</u> (Cottle and Dantzig (Ref. 1).). If the matrix N is strictly copositive or a P-matrix, then the GLCP(q, N) has a solution. If the matrix N is copositive plus and Lemke's algorithm applied to (NP) fails to produce a solution for GLCP(q, N), then there is no feasible solution.

The next two theorems provide necessary conditions under which GLCP(q, N) may or may not have a solution depending on the vector q.

<u>Theorem 3.2.</u> Let N be a vertical Block Matrix of type $(m_1, \ldots, m_n)$ and $q < 0, q \in R^m$. If GLCP(q, N) has a solution, then there exists a representative submatrix M and a non-zero nonnegative vector z such that $z_j(Mz)_j \geq 0$ for each j, $(j = 1, \ldots, n)$.

<u>Proof:</u>

Let (w, z) be a given solution to GLCP(q, N). (w, z) must satisfy

$w = Nz + q \geq 0, z \geq 0.$

$z_j \left(\prod_{i=1}^{m_j} w_i^j \right) = 0, \qquad (j = 1, \ldots, n).$

The last term implies that there exists an index i_j such that

$$z_j w^j_{i_j} = 0, \qquad (j=1,\ldots,n).$$

But $w^j = N^j z + q^j \qquad (j=1,\ldots,n)$.

Consequently,

$$z_j w^j_{i_j} = z_j \left(N^j z + q^j\right)_{i_j} = 0 \qquad (j=1,\ldots,n).$$

That is,

$$z_j \left(N^j z\right)_{i_j} = -z_j q^j_{i_j} \qquad (j=1,\ldots,n).$$

Thus there exists a representative submatrix M such that
$$z_j(Mz)_j = -z_j q_{i_j} \qquad (j=1,\ldots,n).$$

Since $z \geq 0$ and not equal to zero for all j, and $q < 0$, we must have that
$$z_j(Mz)_j \geq 0 \qquad (j=1,\ldots,n). \qquad \square$$

<u>Theorem 3.3.</u> Let N be a vertical block matrix of type $(m_1,\ldots,m_n)$. If N is not a Q-matrix, then there exists a representative submatrix M and $z \geq 0$, $z \neq 0$ such that
$$z_j(Mz)_j \leq 0 \qquad (j=1,\ldots,n).$$

<u>Proof:</u> If N is not a Q-matrix, then there exists a q, nondegenerate, such that GLCP(q, N) has no solution (such a q can be obtained through perturbation). Thus (NP) terminates with $w^0 = 0$ when Lemke's algorithm is applied to it.

Let $\left(\hat{w}, \hat{w}^0, \hat{z}, \hat{z}_0\right)$ be the almost basic feasible solution and $\left(\overline{w}, \overline{w}^0, \overline{z}, \overline{z}_0\right)$ the extreme direction associated with the solution.

Then, similar to (Ref. 4),
$$\left(\overline{w}, \overline{w}^0, \overline{z}, \overline{z}_0\right) \neq 0$$

$$\left(\frac{\overline{w}^0}{\overline{w}}\right) \geq 0, \overline{z} \geq 0, \overline{z}_0 \geq 0$$

$$\left(\frac{\overline{w}^0}{\overline{w}}\right) = \begin{pmatrix} 0 & -e_n^t \\ e_m & N \end{pmatrix} \left(\frac{\overline{z}_0}{\overline{z}}\right)$$

$$\overline{z}_0 \neq 0$$

$$\left(\frac{\overline{z}_0}{\overline{z}}\right)_j \prod_{i=1}^{m_j} \left(\frac{\overline{w}^0}{\overline{w}}\right)_i^j = 0 \qquad (j=1,\ldots,n)$$

Consequently, for all $\lambda \geq 0$ we obtain

$$\hat{w} + \lambda \overline{w} = N(\hat{z} + \lambda \overline{z}) + e_m(\hat{z}_0 + \lambda \overline{z}_0) + q \quad \text{and}$$

$$(\hat{z}_j + \lambda \overline{z}_j) \prod_{i=1}^{m_j} \left(\hat{w}_i^j + \lambda \overline{w}_i^j\right) = 0 \quad (j=1,\ldots,n).$$

The nonnegativity of the variables implies that for each j, $(j = 1, \ldots, n)$

$$\hat{z}_j \prod_{i=1}^{m_j} \overline{w}_i^j = 0.$$

$$\hat{z}_j \prod_{i=1}^{m_j} \hat{w}_i^j = 0.$$

POLYNOMIAL COMPLEMENTARITY PROBLEMS

$$\bar{z}_j \prod_{i=1}^{m_j} \bar{w}_i^j = 0.$$

$$\bar{z}_j \prod_{i=1}^{m_j} \hat{w}_i^j = 0.$$

From (NP) and the second to the last term, we have that

$$\bar{z}_j \prod_{i=1}^{m_j} \left[(N\bar{z})_i^j + \bar{z}_o \right] = 0 \quad (j=1,\ldots,n).$$

Hence there is an index i_j, such that

$$\bar{z}_j \left[(N\bar{z})_{i_j}^j + \bar{z}_o \right] = 0 \quad (j=1,\ldots,n).$$

That is,

$$\bar{z}_j (N\bar{z})_{i_j}^j = -\bar{z}_j \bar{z}_o \quad (j=1,\ldots,n).$$

But $\bar{z}_o \geq 0$, $\bar{z}_o \neq 0$. So we must obtain $\bar{z}_j (N\bar{z})_{i_j}^j \leq 0$.

Consequently, there is a representative submatrix M of N such that

$$\bar{z}_j (M\bar{z})_j \leq 0 \quad (j=1,\ldots,n).$$

□

<u>Corollary 3.1.</u> Let N be a vertical block matrix of type $(m_1,\ldots,m_n)$. If $N > 0$ and q is nondegenerate, then GLCP(q, N) has a solution.

The following results give necessary and sufficient conditions under which N is or is not a Q-matrix. The following information is needed for some of the theorems.

Let $Z^1 = \{(w, z): w = Nz + q, w \geq 0, z \geq 0\}$,

and $Z^2 = \{(w, z, s): w = Nz + q - s, w \geq 0, z \geq 0, s \geq 0\}$.

If $(\overline{w}, \overline{z}, \overline{s})$ is feasible to Z^2 with $\overline{s} = 0$, then $(\overline{w}, \overline{z})$ is feasible to Z^1. Similarly, if $(\hat{w}, \hat{z})$ is feasible to Z^1, then $(\hat{w}, \hat{z}, 0)$ is feasible to Z^2. Consequently, all points feasible to Z^1 can be obtained from Z^2. More importantly, if Z^2 is infeasible then Z^1 is infeasible. Hence it is sufficient to examine the infeasibility of GLCP(q, N) by examining the infeasibility of Z^2. If $q \geq 0$, $w = q$, $z = 0$ solves GLCP(q, N). So assume q is nonpositive. Let e be a vector of length m with all components equal to 1.

Consider the linear programming problem

(LP):
$$\min \quad e^t s$$
$$\text{(s.t.)} \quad w - Nz + s = q$$
$$w \geq 0, z \geq 0, s \geq 0.$$

Let $v_o = \min e^t s$.

If $(\overline{w}, \overline{z}, \overline{s})$ is feasible to (LP) with $v_o = 0$, then $(\overline{w}, \overline{z})$ is feasible to Z^1 and such feasible points that satisfy the complementarity conditions solve GLCP(q, N). The aftermath of this is that Z^1 is infeasible if (LP) is infeasible.

The dual of (LP) is:

(DP):
$$\max \quad q^t u$$
$$\text{(s.t.)} \quad u^t [I, -N, I] \leq e$$
$$u \text{ unrestricted.}$$

<u>Theorem 3.4.</u> Let N be a vertical block matrix of type $(m_1, \ldots m_n)$ and let $q \in R^m$.

If N is not a Q-matrix, then one or more of the following holds:

(1) There exists a vector $u \in R^m$ such that
$$q^t u > 0 \text{ and } u^t[I, -N, I] \leq 0.$$

(2) There exists no vector $u \in R^m$ such that
$$u^t[I, -N, I] \leq e.$$

(3) For all $0 \neq z \geq 0$, there is an index k such that
$$z_k \neq 0 \text{ and } (Nz)^k + q^k > 0.$$

Proof: Suppose N is not a Q-matrix. Then there exist $q \in R^m$ such that GLCP(q, N) has no solution. It is sufficient to consider the case of q with $q_j < 0$, for some j, $1 \leq j \leq m$. That GLCP(q, N) has no solution implies that any of the following possibilities may occur:

Case 1: Z^1 is empty. By the discussion before the theorem, it is enough to examine infeasibility of Z^1 by examining the infeasibility of (LP). But (LP) infeasible implies either that (DP) is infeasible or (DP) is unbounded. If (DP) is unbounded, then there exists $u \in R^m$ such that $u^t[I, -N, I] \leq 0$ and $q^t u > 0$. On the other hand, infeasibility of (DP) means the system $u^t[I, -N, I] \leq e$ has no solution for each u in R^m. Thus either (1) or (2) is true.

Case 2: Z^1 is nonempty but for each $z \neq 0$, z in Z^1, there exists an index k such that

$$z_k \prod_{i=1}^{m_k} w_i^k \neq 0, \quad 1 \leq k \leq n.$$

This implies that $z_k \neq 0$, and $w_i^k \neq 0$ \quad (i = 1, ..., m_k).
Hence, $w_k > 0$, for some k, $1 \leq k \leq n$. But $w^k = (Nz)^k + q^k$. Thus this part of the theorem follows.

Theorem 3.5. Let N be a vertical block matrix of type ($m_1, ..., m_n$) and q a vector in R^m.
Suppose one or more of the following holds:

(1) There exists a vector u in R^m such that $q^t u > 0$ and $u^t[I, -N, I] \leq 0$.

(2) There exists no vector u in R^m such that $u^t[I, -N, I] \leq e$.

(3) For all $0 \neq z \geq 0$, there is an index k such that $z_k \neq 0$ and $(Nz)^k + q^k > 0$.

Then N is not a Q-matrix.

Proof:

Suppose (1) holds: The existence of a vector u in R^m such that $q^t u > 0$ and $u^t[I, -N, I] \leq 0$ implies that (DP) is unbounded. This is equivalent to the infeasibility of (LP) and hence of Z^1. The result follows.

Assume (2) holds: In this case, (DP) is infeasible. Consequently, the (LP) is infeasible or unbounded. Noting that $v_0 = \min e^t s$ is bounded below by zero, we rule out unboundedness of the (LP). Hence we accept the only alternative - the infeasibility of the (LP). Thus Z^1 is empty.

Suppose (3) holds: Since $z \geq 0$, $z_k \neq 0$ implies that $z_k > 0$. Also there exists a block, w^k, of w, conforming to N^k such that $w^k = (Nz)^k + q^k > 0$.

This is equivalent to saying that $w_i^k \neq 0$ $(i = 1, \ldots, m_k)$.

Thus we have that for some k, $1 \leq k \leq n$, $z_k \prod_{i=1}^{m_j} w_i^k \neq 0$.

For this q, GLCP(q, N) is not solvable, so N is not a Q-matrix. □

Theorem 3.6. Let N be a vertical block matrix of type $(m_1, \ldots, m_n)$ and q a vector in R^m. N is not a Q-matrix if and only if one or more of the following hold:

(1) There exists a vector u in R^m such that $q^t u > 0$ and $u^t[I, -N, I] \leq 0$

(2) There is no vector in R^m such that $u^t[I, -N, I] \leq e$

(3) For all nonzero $z \geq 0$, there is an index k such that $z_k \neq 0$ and $(Nz)^k + q^k > 0$.

Proof: Theorems 3.4 and 3.5. □

Example 3.1: The use of Theorem 3.6 is illustrated in this example.

Let $N = \begin{bmatrix} 4 & -4 \\ 1 & 2 \\ 1 & 4 \\ -5 & 4 \end{bmatrix}$ be of type (2, 2) Choose $q = (-1, -1, -1, -1)^t$.

To show that N is not a Q-matrix, it suffices to show GLCP(q,N) has no solution for some q. Solve the system $u^t[I, -N] \leq 0$ to have $u^t = (-1, 0, 0, -1)^t$. Note that solving $u^t[I, -N] \leq 0$ is equivalent to solving $u^t[I, -N, I] \leq 0$ since the homogeneous system $w = Nz - s$ is bounded. Verify that $q^t u = 2 > 0$. Hence one of the conditions in the theorem is satisfied and so N is not a Q-matrix.

Theorem 3.7. Let N be a vertical block matrix of type($m_1, \ldots, m_n$) and $q \in R^m$. Suppose $\Gamma = \{z: Nz + q \geq 0, z \geq 0\}$ is nonempty. For each j define on the set Γ the function

$$G_j(z) = \min_{1 \leq i \leq m_j} \left\{ (Nz)_i^j + q_i^j \right\} \quad (j=1,\ldots,n).$$

Then for all $z \in \Gamma$,

$$z_j \prod_{i=1}^{m_j} \left((Nz)_i^j + q_i^j \right) = 0 \quad \text{if and only if } z_j G_j = 0 \quad (j=1,\ldots,n).$$

Proof: $z_j \prod_{i=1}^{m_j} \left[(Nz)_i^j + q_i^j \right] = 0$ if and only if $z_j \left[(Nz)_k^j + q_k^j \right] = 0$

for some k, $1 \leq k \leq m_j$, if and only if

$$\min_{1 \le i \le m_j}\left\{(Nz)_i^j + q_i^j\right\} = 0$$

if and only if $z_j \min_{1 \le i \le m_j}\left\{(Nz)_i^j + q_i^j\right\} = 0$

for any fixed j (j= 1, ..., n), if and only $z_j G_j = 0$. □

Theorem 3.8. Let N be a vertical block matrix of type $(m_1, ..., m_n)$. N is a Q-matrix if and only if for each $q \in R^m$:

(1) There is no vector $u \in R^m$ such that $u^t q > 0$ and $u^t [I, -N] \le 0$

and

(2) If p(z) = G(z) (G_j as defined in (8)) there exists $z \ge 0$ so that $p(z) \ge 0$ and the inner product (z, p) = 0.

Proof: Let $q \in R^m$ Suppose there is no vector $u \in R^m$ such that $u^t q > 0$ and $u^t [I, -N] \le 0$. By Farkas's Lemma, the system

$$[I, -N]\begin{bmatrix} w \\ z \end{bmatrix} = q, \qquad \begin{pmatrix} w \\ z \end{pmatrix} \ge 0$$

has a solution. So $Z^1 \ne \emptyset$. Moreover, suppose p(z) = G(z) where

$$G_j(z) = \min_{1 \le i \le m_j}\left\{(Nz)_i^j + q_i^j\right\}.$$

If there is a $z \ge 0$ such that $p(z) \ge 0$, then $G_j(z) \ge 0$ (j=1,...,n). Hence $z \in Z^1$. If in addition (z,p)=0, then we obtain $z_1 G_1 + ... + z_n G_n = 0$. By theorem 3.8, this implies that

$$z_j \prod_{i=1}^{m_j} w_i^j = 0 \quad (j=1, ..., n).$$

The result $N \in Q$ follows.

For the converse, suppose there is a u in R^m such that $u^t q > 0$ and $u^t[I,-N] \leq 0$. By Farkas's lemma, the system

$$[I,-N]\begin{bmatrix} w \\ z \end{bmatrix} = q, \quad \begin{pmatrix} w \\ z \end{pmatrix} \geq 0$$

has no solution. This is sufficient to show that N is not in Q.

On the other hand, if $p(z)=G(z)$ and there is no $z \geq 0$ such that $p(z) \geq 0$ and $(z,p) = 0$, then for some j, $1 \leq j \leq n$, we obtain:

<u>Case 1</u>: $z_j G_j < 0$. Consequently, GLCP(q,N) is not feasible.

<u>Case 2</u>: $z_j G_j > 0$. In this case GLCP(q,N) has no complementary feasible solution. The converse follows. □

<u>Corollary 3.2</u>. Let N be a vertical block matrix of type $(m_1,...,m_n)$ and q a vector in R^m. GLCP(q,N) has no solution if there is a vector u in R^m such that $u^t q > 0$, $u^t[I,-N] \leq 0$.

<u>Remark</u>. The piecewise linear (but nonlinear) complementarity problem is now seen to be related to the GLCP(q,N) in terms of existence theory. Szanc(1989) also investigated this connection, but his analysis was for P-matrices.

Implicit in the proofs of Theorems 3.7 and 3.8 is the the following algorithm for solving problem GLCP(q,N) for any matrix N. The algorithm terminates after a finite number of arithmetic operations.

<u>Algorithm 3.1</u>.
<u>Step 1</u>

$$\text{Let } I^j = \left\{ i: q_i^j < 0 \right\} \ (1 \leq i \leq m_j), \ (j = 1,...,n).$$

If I^j is empty for all j, then $(w,z) = (q,0)$ is a complementary feasible solution. Stop, or if more solutions are required, go to Step 3. Otherwise, go to Step 2.

<u>Step 2</u> Solve the system
$$u^t [I, -N] \leq 0.$$
If there is a u such that $u^t q > 0$, then stop, feasible set is empty. Otherwise, go to Step 3.

<u>Step 3</u> Solve the system
$$z_j G_j = 0 \ (j = 1, \ldots, n)$$

where $G_j(z) = \min_{1 \leq i \leq m_j} \left\{ (Nz)_i^j + q_i^j \right\}$

If there exists $z \geq 0$ such that $Nz + q \geq 0$, then (w,z), where $w = Nz + q$, is a complementary feasible solution. Otherwise GLCP(q,N) has no complementary feasible solution. Stop. □

We now show the relationship of the solution of GLCP(q,N) to that of a certain polynomial nonlinear complementarity problem. Polynomial functions are important tools in mathematics. They are studied in the framework of game theory by Kostreva (Ref. 5) and in a general nonlinear systems context by Li, Sauer and York (Ref. 6). In the area of complementarity, polynomials have not been explicitly considered to date.

<u>Theorem 3.9</u>.Let N be a vertical block matrix of type $(m_1,\ldots,m_n)$. Consider the polynomial function $f : R^n \longrightarrow R^n$ defined by
$$f_j(z) = \prod_{i=1}^{m_j} (N^j z + q^j)_i \quad j = 1, 2, \ldots, n.$$
If N is a Q-matrix, then for each q in R^m

(1) there is no vector u in R^m such that $u^t q > 0$ and $u^t[I,-N] \leq 0$,

and

(2) there exists a solution to the following nonlinear complementarity problem:

Find z in R^n such that for $i=1,...,n$, $z_i \geq 0$, $f_i(z) \geq 0$, and $z_i f_i(z) = 0$.

Proof : Suppose N is a Q-matrix. Then for each q in R^m, the feasible set is nonempty. By Farkas's Lemma, property (1) holds. From the definition of GLCP(q,N), the solution of GLCP(q,N) solves the nonlinear polynomial complementarity problem . □

Remark. If N is a P-matrix, if $N > 0$, or if N is strictly copositive, then N is a Q-matrix (Ref. 1). Hence, from Theorem 3.9, there exists at least one solution to the polynomial nonlinear complementarity problem. Existence results for the polynomial complementarity problem have not been previously obtained. The structure of the function is, of course, key to this finding.

Example 3.2. The following example shows that the above nonlinear complementarity problem may have more than one solution, even when N is a P-matrix.

Let $f_1(z_1,z_2) = (4z_1 - 3z_2)(3z_1 - 4z_2)$ and $f_2(z_1,z_2) = z_1 + z_2 - 7$ corresponding to the vertical block matrix N, of type (2,1).

$$N = \begin{bmatrix} 4 & -3 \\ 3 & -4 \\ 1 & 1 \end{bmatrix}$$

Then $(z_1,z_2)^t = (3,4)^t$ and $(z_1,z_2)^t = (4,3)^t$ are solutions of the nonlinear complementarity problem, since they are positive zeroes of $(f_1,f_2)^t$. The vector $q = (0,0,-7)^t$ is degenerate with respect to N and hence solving GLCP(q,N) (by Lemke's algorithm) requires extra care. Multiplicity of solutions is independent of degeneracy, however. It is easy to check that the above $(f_1,f_2)^t$ is not a P-function on the positive orthant in R^n

Theorems 3.7 and 3.9 suggest the following algorithm for solving GLCP(q,N) for any N. Again, the algorithm terminates after a finite number of arithmetic operations.

Algorithm 3.2.

Step 1 : If $q \geq 0$, then $(w,z)=(q,0)$ is a complementary feasible solution. Stop, or if more solutions are required, go to Step 3. Otherwise, continue.

Step 2 : Solve the system $u^t[\,I,-N] \leq 0$.
If there exists u such that $q^t u > 0$, then stop. Feasible set is empty. Otherwise go to Step 3.

Step 3 : Solve the system

$$z_j \prod_{i=1}^{m_j} \left[(Nz)_i^j + q_i^j \right] = 0 \quad (j=1,\ldots,n).$$

If there is a $z \geq 0$ such that $w = Nz + q \geq 0$, then (w,z) is a complementary feasible solution. Otherwise there is no complementary feasible solution. Stop. □

The system in Step 3 is a system of polynomial equations for which special purpose algorithms (Refs. 5 and 6) that can find all real solutions exist. Observe, however, that this system may also be solved by finding all the complementary basic points for all the representative submatrices. This involves solving $[(\Pi_j m_j) 2^n]$ linear nxn systems of equations where each system is independent of all the others. Hence, they may be solved in parallel. Moreover, in a small size problem, Step 2 of the algorithm may be omitted, since all complementary points are generated in Step 3. This is demonstrated in Example 3.3. However, in large problems, detecting infeasibility in Step 2 could provide a substantial savings in computation time.

Example 3.3.

Let $N = \begin{bmatrix} -2 & 1 \\ 3 & 2 \end{bmatrix}$ be a vertical block matrix of type (1,1).

Part 1: Let $q = (-1,1)^t$.

Step 1 : q is nonpositive ; go to step 2 .

Step 2 : Problem is infeasible. Stop here or go to step 3.

Step 3 : Solve the system

$$z_1(-2z_1 + z_2 - 1) = 0.$$
$$z_2(3z_1 + 2z_2 + 1) = 0.$$

The solution set is $\{(0,0)^t, (0,-1/2)^t, (-3/7,1/7)^t, (-1/2,0)^t\}$. It is obvious that GLCP(q,N) has no complementary feasible solution in this case.

Part 2 : Take $q = (1,-1)^t$.

Step 1: Skipped.

Step 2: Problem is feasible.

Step 3: Solve the system

$$z_1(-2z_1 + z_2 + 1) = 0.$$
$$z_2(3z_1 + 2z_2 - 1) = 0.$$

The solution set generated is $\{(0,0)^t, (1/2,0)^t, (3/7,-1/7)^t, (0,1/2)^t\}$. The solutions to the problem are $w = (0,1,1/2,0)^t$ and $w = (3/2,0,0,1/2)^t$ corresponding to the vectors $z = (0,1/2)^t$ and $z = (1/2,0)^t$.

Remarks: The problems solved are small but representative of data of the most general class. The matrix is not in P, not strictly copositive, not in any class of Garcia (Ref. 7) and not in Q. The new algorithm solves problems which are outside the domain of the earlier methods.

4. SOME PROPERTIES OF THE VERTICAL BLOCK MATRIX N

In this section, we give some mathematical properties of the vertical block matrix N. The first result is a generalization of a theorem by C. B. Garcia (Ref. 7).

<u>Theorem 4.1</u>. Suppose N is a vertical block matrix of type$(m_1,...,m_n)$, where $m = m_1 + ... + m_n$ and q a vector in R^m. If there is a nonzero vector x in R^m such that $q^t x < 0$ and $N^t x \leq 0$, then GLCP(q,N) is infeasible.

<u>Proof</u>: Suppose there exists an x such $q^t x < 0$ and $N^t x \leq 0$. Let $w = Nz + q$, $z \geq 0$. Then

$$w^t x = (Nz + q)^t x,$$
$$w^t x = z^t N^t x + q^t x < 0.$$

Since $x \geq 0$, there exists at least a j such that $w_j < 0$. Thus GLCP(q,N) is not feasible. □

The corollary below follows from Theorem 4.1 by choosing the vector q strictly positive.

<u>Corollary 4.1</u>. Let N be a vertical block matrix of type $(m_1,...,m_n)$.
Suppose for all $z \geq 0$, $z \neq 0$, $Nz < 0$, then N is not a Q-matrix.

If x is a non-zero nonnegative vector such that $Nx > 0$, we show, in the results that follow, that there exists a strictly positive vector depending on x that satisfies the same conditions. Such a vector is obtained by perturbing x.

<u>Theorem 4.2</u>. Let N be a vertical block matrix of type $(m_1,...,m_n)$. Suppose there is a $z \geq 0$, $z \neq 0$, in R^n such that $Nx > 0$. Then there is a vector $y > 0$ such that $Ny > 0$.

<u>Proof</u>: Let e_n be an nx1 column vector with each component equal to 1. For any $\varepsilon > 0$, $x \geq 0$, the vector $y = x + \varepsilon e_n > 0$. For a sufficiently small $\varepsilon > 0$, we obtain $Ny = N(x + \varepsilon e_n) = Nx + \varepsilon Ne_n > 0$. □

Theorem 4.3. Suppose N and x are as defined in Theorem 4.2. Then there exists a representative submatrix M of N and a vector $y > 0$ such that $y_j(My)_j > 0$, $(j=1,...,n)$.

Proof: Since $Nx > 0$, $(Nx)_i > 0$, for each i, $i=1,...,m$. Let N^k be the k-th block of N. By Theorem 4.2, $Ny > 0$. Thus for each j, there exists an index i_j with $y_j(N^j y)_{i_j} > 0$. By definition, this implies the existence of a representative submatrix M of N such that $y_j(My)_j > 0$ $(j=1,...,n)$. □

Corollary 4.2. Let N be a vertical block matrix of type $(m_1,...,m_n)$. Suppose there is a non-zero nonnegative vector x in R^n such that $Nx > 0$. Then there is a representative submatrix M such that $x_j(Mx)_j \geq 0$ $(j=1,...,n)$.

If N is a Q-matrix, we examine the types of systems of inequalities involving N and a nonnegative vector x in R^n such that a solution exists. Such systems are important since any constructive characterization of the set of Q-matrices might depend heavily on such systems. The results given, however, indicate how difficult it is to obtain a constructive characterization of Q. Note that any characterization of Q which depends on all q in R^m cannot be constructive. We start with

Theorem 4.4. Let N be a vertical block matrix of type $(m_1,...,m_n)$. If N is a Q-matrix, then the system $Nx > 0$, $x \geq 0$ has a solution.

Proof: By Theorem 3.8, the system $u^t q > 0$, $q \in R^m$, $u^t[\ I,-N\ ] \leq 0$. has no solution since N is a Q-matrix. By Farkas's lemma, the system $[I,-N]s = q$, $s \geq 0$, $s \in R^{m+n}$ has a solution for any q in R^m. In particular, if $q < 0$, then the system $Nx > 0$, $x \geq 0$ has a solution. □

Corollary 4.3. If N is a vertical block Q-matrix of type $(m_1,...,m_n)$, then the system $Ny > 0$, $y > 0$, has a solution.

Proof: Theorems 4.2 and 4.4. □

Corollary 4.4. Suppose N is a vertical block Q-matrix of type $(m_1,...,m_n)$. Then there is a representative submatrix M of N and a vector $y > 0$ such that $y_j(My)_j > 0$ $(j=1,...,n)$.

Proof: Theorems 4.2, 4.3, and 4.4. □

Corollary 4.5. Let N be a vertical block matrix of type $(m_1,...,m_n)$. If N is a Q-matrix, then the system $Nz \geq 0$, $z \geq 0$, has a solution.

Proof: Choose $q=0$ in Theorem 4.4. □

Corollary 4.6. Let N be a vertical block matrix of type $(m_1,...,m_n)$. Suppose N is a Q-matrix, then the system $Nx < 0$, $x \geq 0$, has a solution.

Proof: Choose $q > 0$ in Theorem 4.4. □

5. SOME RESULTS FOR THE LINEAR COMPLEMENTARITY PROBLEM

Presented in this section are some linear complementarity results discernible from the theory we have developed for the GLCP. To the best of our knowledge, these results do not duplicate any earlier results for the LCP. Some very recent results of Al-Khayyal (Ref.8) deal with characterizations of Q and its complement in a different way.

Theorem 5.1. Let M be a real square matrix and q a strictly negative vector in R^m. If LCP(q,M) has a solution, then there exists a vector $z \geq 0$, $z \neq 0$, such that $z_j(Mz)_j \geq 0$ $(j=1,...,n)$.

Proof: Consider M as a vertical block matrix of type $(1,...,1)$ and use Theorem 3.2. □

Theorem 5.2. Suppose M is an nxn real matrix. If M is not a Q-matrix, then there exists a vector $z \geq 0$, $z \neq 0$, such that $z_j(Mz)_j \leq 0$, $j=1,...,n$.

Proof: Use Theorem 3.3. Note that M is of type $(1,...,1)$. □

Theorem 5.3. Let M be a real square matrix of order n and q a vector in R^n. Then M is not a Q-matrix if and only if one or more of the following holds:

(1) There exists a vector u in R^n such that

$q^t u > 0$, $u^t[I, -M, I] \leq 0$.

(2) there is no vector u in R^n such that

$u^t[I, -M, I] \leq e$.

(3) for all $z \geq 0$, $z \neq 0$, there is an index k such that $z_k \neq 0$ and $(Mz)_k + q_k > 0$.

Proof: Use Theorems 3.4 and 3.5. □

Theorem 5.4. Suppose M is a real square matrix of order n.
Let G ={ z : Mz +q ≥0, z ≥0 }. M is a Q-matrix if and only if for each vector q in R^n ;

(1) There exists no vector u in R^n such that $u^t[\ I,M] \leq 0$, $u^t q > 0$ and

(2) there exists a vector z ∈ G, p=p(z,q) ≥ 0 such that the inner product (z,p) =0.

Proof: Use Theorem 3.8. □

6. CONCLUSIONS

Q - matrices corresponding to Cottle-Dantzig Generalized Linear Complementary Problem are characterized in this paper. The characterization leads to algorithms which are capable of not only processing problem GLCP(q, N) for a general matrix N but also of generating all complementary feasible solutions.

Mathematical properties of the Q-matrices are also presented as well as some results on the Linear Complementary Problem.

The research suggests algorithms for solving the generalized linear Complementarity Problem for any matrix N. While these algorithms appear computationally complex, it is important to remember that problem GLCP(q, N) itself is NP-complete in the case of a general matrix N. That this is so follows from the fact that LCP(q,M) is NP-complete (Ref. 9). The existence of an effective algorithm, for solving GLCP(q, N) for a general matrix N, which performs well on the average, is an open question.

Results of the type obtained for the polynomial complementarity problem are unusual in the sense that none of the typical monotonicity properties (P-functions or copositivity) of nonlinear functions are needed to prove existence of solutions. The utility of polynomial functions in model building makes this result of interest to the mathematical community.

7. REFERENCES

1. COTTLE, R. W., and DANTZIG, G. B., A Generalization of the Linear Complementarity Problem, Journal of Combinatorial Theory, Vol. 8, pp. 79-90, 1970.

2. SZANC, B. P., The Generalized Complementarity Problem, Rensselaer Polytechnic Institute, Troy, New York, Ph.D Thesis, 1989.

3. OH, K.P., The Formulation of the Mixed Lubrication Problem as a Generalized Nonlinear Complementarity Problem, Transactions of the American Society of Mechanical Engineers Journal of Tribology, Vol. 106, pp. 598-603, 1986.

4. BAZARAA, M.S., and SHETTY, C. M, Nonlinear Programming - Theory and Algorithms, John Wiley and Sons, New York, First Edition, pp. 445, 1979.

5. KOSTREVA, M. M., Nonconvexity in Noncooperative Game Theory, International Journal of Game Theory, Vol. 18, 247-259, 1989.

6. LI, T. Y., SAUER, T. and YORKE, J. A., Numerically Determining Solutions of Systems of Polynomial Equations, Bulletin of the American Mathematical Society, Vol. 18, pp.173-177, 1988.

7. GARCIA, C. B., Some Classes of Matrices in Linear Complementarity Theory, Mathematical Programming, Vol. 5, pp. 299-310, 1973.

8. AL-KHAYYAL, F. A., Necessary and Sufficient Conditions for the Existence of Complementary Solutions and Characterizations of the Matrix Classes Q and Q_0, to appear in Mathematical Programming.

9. CHUNG, S. J., NP-Completeness of the Linear Complementarity Problem, Journal of Optimization Theory and Applications Vol. 60, pp. 393-399, 1989.

A Continuous Approach to Compute Upper Bounds in Quadratic Maximization Problems With Integer Constraints

A. Kamath[†]   N. Karmarkar[†]

April 19, 1991

Abstract

In the graph partitioning problem, as in other NP-hard problems, the problem of proving the existence of a cut of given size is easy and can be accomplished by exhibiting a solution with the correct value. On the other hand proving the non-existence of a cut better than a given value is very difficult. We consider the problem of maximizing a quadratic function $x^T Q x$ where Q is a $n \times n$ real symmetric matrix with x an n-dimensional vector constrained to be an element of $\{-1, 1\}^n$. In the paper we propose a technique for obtaining upper bounds on solutions to the problem using a continuous approach. The method is then applied for finding lower bounds on cutsizes in graph partitioning problems and thereby proving non-existence of cuts better than the lower bound.

0. Outline

In the first section we introduce the quadratic optimization problem and give a motivation for the concepts underlying the development of the interior point approach for solving the problem. In the second section we describe the method of generating a sequence of decreasing upper bounds for the quadratic maximization problem and then introduce a minor simplification of the technique. In the third section we show how the approach has been used to find lower bounds on the cutsizes for the graph partitioning problem [4] and present a few implementation details. Conclusions and directions for future work are presented in the last section.

[†] AT&T Bell Laboratories, Murray Hill, NJ 07974

1. Introduction

Let S be the set of n-dimensional vectors defined as follows

$$S = \{\mathbf{x} = (x_1, \ldots, x_n) | x_i \in \{-1, 1\}, i = 1, \ldots, n\}.$$

Consider the quadratic optimization problem

max $f(\mathbf{x})$

s.t. $\mathbf{x} \in S$

where $f(\mathbf{x}) = \mathbf{x}^T Q \mathbf{x}$, $Q \in R^{n \times n}$ is a symmetric matrix. (1)

Let f_{max} denote the maximum value of $f(\mathbf{x})$ in problem (1).

Since Q is a real symmetric matrix all its eigenvalues are real. If all the eigenvalues are negative the problem is easy [5][14]. Otherwise problem (1) is NP-hard[3] and obtaining an upper bound on f_{max} by finding the optimum solution to (1) is difficult. We consider only the NP-hard case in which at least one eigenvalue of Q is positive [15] and we shall present in this paper a technique for finding a good upper bound on f_{max}.

Evaluating $f(\mathbf{x})$ at any $\mathbf{x} \in S$ gives us a lower bound on the solution to this problem. However to show that the optimal value of $f(\mathbf{x})$ in S, can be no more than some real number μ using a combinatorial approach, one may have to evaluate $f(\mathbf{x})$ for all $\mathbf{x}$ in S. In contrast we present here an efficient technique based on the interior point approach, for obtaining upper bounds on solutions to this maximization problem.

The main concept in an interior point approach to such problems is to embed the discrete set (S in the current case) in a continuous set T where $S \subseteq T$. Maximum of a function $f(\mathbf{x})$ over T gives an upper bound on $f(\mathbf{x})$ in S. We would like to choose set T so that not only can the maximum be found in a computationally efficient manner but also the bound obtained would be as good as possible.

A commonly used approach is to choose the continuous set to be the box X defined as follows

$$X = \{\mathbf{x} \in R^n | -1 \leq x_i \leq 1, i = 1, \ldots, n\}.$$

But optimizing a quadratic function $f(\mathbf{x})$ over X is still NP-hard [15] so instead we enclose the box in a ball B where

$$B = \{\mathbf{x} \in R^n | \mathbf{x}^T \mathbf{x} \leq n\}.$$

For maximizing $f(\mathbf{x})$ on B, we need to just find the maximum eigenvalue λ_{max} of Q. Then since

$$\frac{\mathbf{x}^T Q \mathbf{x}}{\mathbf{x}^T \mathbf{x}} \leq \lambda_{max}, \qquad \forall \mathbf{x} \in R^n - \{\mathbf{0}\},$$

$$\mathbf{x}^T Q \mathbf{x} \leq n \lambda_{max}, \qquad \forall \mathbf{x} \in B$$

and therefore we get $n\lambda_{max}$ to be an upper bound on f_{max}.

Instead of the ball B we may choose any ellipsoid E enclosing the box X and still retain the computational ease of the resulting maximization problem. In this paper we shall limit ourselves to ellipsoids $E(\mathbf{w})$ whose axes are along the co-ordinate axes. A technique for ellipsoids with some other given set of axes can be derived along similar lines.

$$\text{Let } U = \left\{ \mathbf{w} = (w_1, w_2, \ldots, w_n) \in R^n \,\bigg|\, \sum_{i=1}^{n} w_i = 1, w_i > 0, i = 1, \ldots, n \right\},$$

and $E(\mathbf{w}) = \{\mathbf{x} \in R^n \,|\, \mathbf{x}^T W \mathbf{x} \leq 1 \text{ where } W = \text{diag}(\mathbf{w}), \mathbf{w} \in U\}$.

It can easily be verified that the ellipsoid E encloses the box X.

If λ is the maximum eigenvalue of the matrix $W^{-1/2} Q W^{-1/2}$ then

$$\max_{x} \frac{\mathbf{x}^T Q \mathbf{x}}{\mathbf{x}^T W \mathbf{x}} = \max_{y} \frac{\mathbf{y}^T W^{-1/2} Q W^{-1/2} \mathbf{y}}{\mathbf{y}^T \mathbf{y}} = \lambda \qquad (2)$$

implying
$$\mathbf{x}^T Q \mathbf{x} \leq \lambda, \quad \forall \mathbf{x} \in E(\mathbf{w}).$$

This gives us a simple technique for maximizing $f(\mathbf{x})$ on $E(\mathbf{w})$.

Since $E(\mathbf{w})$ contains the set S, λ is an upper bound on f_{max}. We note that λ is a function of the weights $\mathbf{w}$. In the following section, we propose an iterative method for systematically modifying the shape of the ellipsoid by changing the weights $\mathbf{w}$ so as to minimize λ and hence obtain a better bound on f_{max}.

2. Interior Point Approach to the Problem

In our interior point approach to the problem, we start with some weights $\mathbf{w}^{(0)} = (w_1^0, w_2^0, \ldots, w_n^0) \in U$ and iteratively modify them obtaining in the process a sequence of ellipsoids $E^{(1)}, \ldots, E^{(k)}$ such that if $\mu^{(i)}$ is maximum of $f(\mathbf{x})$ over $E^{(i)}$ then

$$\mu^{(1)} > \mu^{(2)} > \ldots > \mu^{(k)}.$$

In the $(\mu, \mathbf{w})$ space, consider the region defined by

$$L = \left\{ (\mu, \mathbf{w}) \,\bigg|\, \mu \geq 0, \mathbf{w} \in U, \frac{\mathbf{x}^T Q \mathbf{x}}{\mathbf{x}^T W \mathbf{x}} \leq \mu, \quad \forall \mathbf{x} \in R^n - \{\mathbf{0}\} \right\}$$

where $W = \text{diag}(\mathbf{w})$.

It may be trivially confirmed that for any $(\mu, \mathbf{w}) \in L$ the value of μ gives a bound on $f_{\max}$. Hence to get good bounds on $f_{\max}$ we need to minimize μ over set L.

We note that since the weights in $\mathbf{w}$ are positive the condition

$$\frac{\mathbf{x}^T Q \mathbf{x}}{\mathbf{x}^T W \mathbf{x}} \leq \mu, \quad \forall \mathbf{x} \in R^n - \{\mathbf{0}\},$$

can be rephrased as

$$\mathbf{x}^T (\mu W - Q) \mathbf{x} \geq 0, \quad \forall \mathbf{x} \in R^n \quad (3)$$

implying that the matrix $\mu W - Q$ is positive semidefinite.

Lemma 1. The minimization problem given by

$$\min \mu \quad (4)$$
$$\text{s.t.} \ (\mu, \mathbf{w}) \in L$$

has convex [13] level sets.

Proof.

The level sets in the minimization problem (4) over the space $(\mu, \mathbf{w})$ can be defined as

$$L(\alpha) = \{(\mu, \mathbf{w}) | 0 \leq \mu \leq \alpha, \ (\mu, \mathbf{w}) \in L\}.$$

Consider the regions K and P in the $(\mu, \mathbf{w})$ space defined as

$K = \{(\mu, \mathbf{w}) | \mu \leq \alpha, \ \mathbf{w} \in U\}$
$P = \{(\mu, \mathbf{w}) | \mu \geq 0, w_i \geq 0, i = 1, \ldots, n, \mathbf{x}^T(\mu W - Q)\mathbf{x} \geq 0, \forall \mathbf{x} \in R^n\}$.

Obviously $L(\alpha)$ is the intersection of regions K and P. Convexity of K can be proved easily. So if we can show that P also is convex then convexity of $L(\alpha)$ would follow trivially.

Let $(\mu^{(1)}, \mathbf{w}^{(1)})$ and $(\mu^{(2)}, \mathbf{w}^{(2)})$ belong to P. To prove that P is convex, it suffices to show that for $0 \leq \beta \leq 1$, $(\beta \mu^{(1)} + (1-\beta)\mu^{(2)}, \beta \mathbf{w}^{(1)} + (1-\beta)\mathbf{w}^{(2)})$ also belongs to P. Without loss of generality we may assume $\mu^{(1)} \geq \mu^{(2)}$.

As a digression let us first show that

$$\mathbf{x}^T \left[\frac{\mu^{(1)} W^{(2)} + \mu^{(2)} W^{(1)}}{2} - Q \right] \mathbf{x} \geq 0, \quad \forall \mathbf{x} \in R^n \quad (5)$$

Since $(\mu^{(1)}, \mathbf{w}^{(1)})$ and $(\mu^{(2)}, \mathbf{w}^{(2)})$ belong to P,

$$\mathbf{x}^T \mu^{(1)} W^{(1)} \mathbf{x} \geq \mathbf{x}^T Q \mathbf{x}, \quad \forall \mathbf{x} \in R^n$$
$$\mathbf{x}^T \mu^{(2)} W^{(2)} \mathbf{x} \geq \mathbf{x}^T Q \mathbf{x}, \quad \forall \mathbf{x} \in R^n. \quad (6)$$

For each **x** one of the following cases holds

a) $x^T W^{(1)} x \geq x^T W^{(2)} x$,

or b) $x^T W^{(2)} x > x^T W^{(1)} x$.

Case (a)

$$x^T \left[\frac{\mu^{(1)} W^{(2)} + \mu^{(2)} W^{(1)}}{2} - \mu^{(2)} W^{(2)} \right] x$$

$$= \frac{\mu^{(1)} - \mu^{(2)}}{2} x^T (W^{(2)}) x + \frac{\mu^{(2)}}{2} x^T (W^{(1)} - W^{(2)}) x \geq 0.$$

Therefore from (6) we may conclude that for case (a)

$$x^T \left[\frac{\mu^{(1)} W^{(2)} + \mu^{(2)} W^{(1)}}{2} - Q \right] x \geq 0.$$

Case (b)

$$x^T \left[\frac{\mu^{(1)} W^{(2)} + \mu^2 W^{(1)}}{2} - \frac{\mu^{(1)} W^{(1)} + \mu^{(2)} W^{(2)}}{2} \right] x$$

$$= \left[\frac{\mu^{(1)} - \mu^{(2)}}{2} \right] x^T (W^{(2)} - W^{(1)}) x \geq 0.$$

Therefore from (6) we may conclude that for case (b)

$$x^T \left[\frac{\mu^{(1)} W^{(2)} + \mu^{(2)} W^{(1)}}{2} - Q \right] x \geq 0.$$

This proves the result (5) for all **x**. Hence from (5) and (6) we may conclude that for all **x**

$$x^T ((\beta \mu^{(1)} + (1-\beta) \mu^{(2)})(\beta W^{(1)} + (1-\beta) W^{(2)}) - Q) x$$

$$= \beta^2 x^T [\mu^{(1)} W^{(1)} - Q] x + (1-\beta)^2 x^T [\mu^{(2)} W^{(2)} - Q] x$$

$$+ 2\beta(1-\beta) x^T \left[\frac{\mu^{(1)} W^{(2)} + \mu^{(2)} W^{(1)}}{2} - Q \right] x$$

$$\geq 0.$$

We may now conclude that $(\beta \mu^{(1)} + (1-\beta) \mu^{(2)}, \beta w^{(1)} + (1-\beta) w^{(2)})$ belongs to P and hence P is a convex set.

Therefore the level set defined by $L(\alpha)$ is convex. Hence the problem (4) is a convex programming problem. ∎

We now propose an iterative approach for solving problem (4). We start with some vector of weights $\mathbf{w}^{(0)} \in U$. Let $\mathbf{w}^{(k)} \in U$ be the state of the weight vector at the kth iteration. The maximum value of $f(\mathbf{x})$ over $E(\mathbf{w}^{(k)})$ is denoted by $\lambda^{(k)}$ which is equal to the maximum eigenvalue of $W^{(k)-1/2} Q W^{(k)-1/2}$. Let $\mu^{(k)}$ be the value of the upper bound variable μ at the kth iteration and be such that $\mu^{(k)} > \lambda^{(k)}$. We define $M^{(k)} = \mu^{(k)} I - W^{-1/2} Q W^{-1/2}$. Also we use the notation $\lambda_i(M)$ to denote the ith eigenvalue of a real symmetric matrix M.

At each iteration we need to enforce the positive semidefiniteness condition on $\mu^{(k)} W - Q$ as given in (3). Because of the positivity of the $\mathbf{w}$ vector, condition (3) may also be phrased as

$$\mathbf{y}^T M^{(k)} \mathbf{y} \geq 0, \quad \forall \mathbf{y} \in R^n.$$

If we were to enforce the condition for each $\mathbf{y}$ in $R^n - \{\mathbf{0}\}$ it would require an infinite number of inequalities. The same effect can be achieved by imposing conditions on eigenvalues of $M^{(k)}$. Since $M^{(k)}$ is a real and symmetric matrix it has n real eigenvalues $\lambda_i(M^{(k)})$, $i = 1, \ldots, n$. For $M^{(k)}$ to be positive semi-definite its eigenvalues must be non-negative which give us the inequality constraints

$$\lambda_i(M^{(k)}) \geq 0, \quad i = 1, \ldots, n. \tag{7}$$

Since the weights also need to be non-negative, we get the inequality constraints

$$w_i \geq 0, \quad i = 1, \ldots, n. \tag{8}$$

Subject to these inequality constraints (7) and (8) we have to reduce the upper bound μ at each iteration. We note that $\lambda_i(M^{(k)})$ are the slacks of the constraints in (7). Correspondingly we can define the potential function [7] [9] [10]

$$\phi_1(\mathbf{w}) = -\ln \prod \lambda_i(M^{(k)})$$
$$= -\ln \det(M^{(k)}).$$

We note that unlike the eigenvalues, the determinant of a matrix is a smooth function which can be expressed as a polynomial in the components of the matrix [16] [17].

The collection of slack variables in (8) give us the additional term

$$\phi_2(\mathbf{w}) = -\ln \prod_{i=1}^{n} w_i.$$

So we define the potential function

$$\phi(\mathbf{w}) = \phi_1(\mathbf{w}) + \gamma \phi_2(\mathbf{w})$$
where $\gamma > 0$ is a parameter.

The potential function is defined only for points that are in the interior of $L(\mu^{(k)})$ i.e. $\lambda_i(M^{(k)}) > 0$ and $w_i > 0$, $i = 1, \ldots, n$. We shall later modify our method so that the conditions (8) on **w** do not arise, permitting us to remove the term $\phi_2(\mathbf{w})$ from the potential function.

From our experience with interior point techniques we note that by linearizing the constraints in (7) the optimization problem (4) may be transformed into a linear programming problem [8]. This suggests a method for solving the problem (4) by alternating between two steps as we do in linear programming - a potential step in which we approximately minimize the potential function $\phi(\mathbf{w})$ and an objective step in which we reduce the upper bound variable μ.

The following minimization problem has to be solved in the potential step at the kth iteration:

$$\min \phi(\mathbf{w})$$
$$\text{s.t. } \mathbf{w} \in U$$
$$\lambda_i(M^{(k)}) \geq 0, \quad i = 1, \ldots, n. \quad (10)$$

We shall now suggest a modification that will simplify the minimization problem. For that let us define

$$V = \left\{ \mathbf{d} = (d_1, d_2, \ldots, d_n) \in R^n \,\middle|\, \sum_{i=1}^{n} d_i = 0 \right\} \quad (11)$$

Lemma 2. Let $\mathbf{w} \in U$ and $\mathbf{d} \in V$, then given

$$W = \text{diag}(\mathbf{w}) \text{ and } D = \text{diag}(\mathbf{d}),$$

$$\frac{\mathbf{x}^T(Q-D)\mathbf{x}}{\mathbf{x}^T W \mathbf{x}} \leq \mu, \quad \forall \mathbf{x} \in R^n - \{\mathbf{0}\} \quad (12)$$

$$\Rightarrow \mathbf{x}^T Q \mathbf{x} \leq \mu, \quad \forall \mathbf{x} \in S.$$

Proof.

For $\mathbf{x} \in S$, we have $x_i \in \{-1, 1\}$, $i = 1, \ldots, n$ and hence

$$\mathbf{x}^T W \mathbf{x} = \Sigma w_i = 1, \quad \forall \mathbf{x} \in S.$$

Similarly from (11) we have

$$\mathbf{x}^T D \mathbf{x} = \Sigma d_i = 0, \quad \forall \mathbf{x} \in S.$$

On substituting these results in (12), it follows immediately that

$$\mathbf{x}^T Q \mathbf{x} \leq \mu, \quad \forall \mathbf{x} \in S. \quad \blacksquare$$

This suggests that we have an additional degree of freedom in minimizing the bound f_{max} on S. So we may define the optimization problem over $(\mu, \mathbf{w}, \mathbf{d})$ space. But before doing so let us consider the inequality (12) which is the only constraint that has both the $\mathbf{w}$ and $\mathbf{d}$ variables. This constraint may be rephrased as

$$\mathbf{x}^T(\mu W + D - Q)\mathbf{x} \geq 0, \quad \forall \mathbf{x} \in R^n - \{\mathbf{0}\}.$$

Let us define for $\mathbf{w} \in U$ and $\mathbf{d} \in V$

$$M(\mu, \mathbf{w}, \mathbf{d}) = \mu W + D - Q.$$

For any given $(\mu, \mathbf{w}, \mathbf{d})$ we can define $(\mu, \mathbf{w}', \mathbf{d}')$ where $\mathbf{w}' = \frac{1}{n}\mathbf{e}$ and $\mathbf{d}' = \mu \mathbf{w} + \mathbf{d} - \frac{\mu}{n}\mathbf{e}$.

It is obvious that $\mathbf{w}' \in U$ and $\mathbf{d}' \in V$ and that

$$M\left(\mu, \frac{1}{n}\mathbf{e}, \mathbf{d}'\right) = \frac{\mu}{n}I + \mu W + D - \frac{\mu}{n}I - Q$$

$$= M(\mu, \mathbf{w}, \mathbf{d}).$$

Hence it suffices to treat M as a function of only μ and $\mathbf{d}$, keeping $\mathbf{w}$ constant at $\frac{1}{n}\mathbf{e}$ and varying $\mathbf{d}$ iteratively to reduce the upper bound μ.

The weights $\mathbf{w}$ and diagonals $\mathbf{d}$ are equivalent. However, there is an advantage in working in $\mathbf{d}$ space because unlike $\mathbf{w}$, the diagonals $\mathbf{d}$ are not constrained to be positive. Hence a term corresponding to $\phi_2(\mathbf{w})$ is not required in the potential function.

For notational convenience we define $\zeta = \frac{\mu}{n}$ and $M^{(k)} = \zeta^{(k)}I + D - Q$.

The problem to be solved at the kth iteration can then be described as

$$\min \phi(\mathbf{d}) \tag{13}$$

$$\text{s.t.} \sum_{i=1}^{n} d_i = 0$$

$$\lambda_i(M^{(k)}) \geq 0, \quad i = 1,\ldots,n$$

where

$$\phi(\mathbf{d}) = -\ln \prod_{i=1}^{n} \lambda_i(M^{(k)}).$$

This is a constrained optimization problem. We shall later describe an approach that solves this problem approximately. However, before doing so, we need to prove a few lemmas.

QUADRATIC MAXIMIZATION PROBLEMS

Lemma 3. Let $\mathbf{x} = (x_1, \ldots, x_n) \in R^n$ and $M(\mathbf{x})$ be a real symmetric positive-definite matrix of size $n \times n$ whose elements depend on $\mathbf{x}$. Let $\phi(\mathbf{x}) = \ln \det M(\mathbf{x})$ and suppose for a pair of indices i and j

$$\frac{\delta M}{\delta x_i} = \mathbf{a}_i \mathbf{a}_i^T \text{ and } \frac{\delta M}{\delta x_j} = \mathbf{a}_j \mathbf{a}_j^T$$

where $\mathbf{a}_i, \mathbf{a}_j \in R^n$ are independent of $\mathbf{x}$.

Then the first and second order derivatives of $\phi(\mathbf{x})$ are given by

$$\frac{\delta \phi}{\delta x_i} = \mathbf{a}_i^T M^{-1} \mathbf{a}_i , \quad \frac{\delta \phi}{\delta x_j} = \mathbf{a}_j^T M^{-1} \mathbf{a}_j$$

and
$$\frac{\delta^2 \phi}{\delta x_i \delta x_j} = -(\mathbf{a}_i^T M^{-1} \mathbf{a}_j)^2 .$$

Proof.

In order to differentiate with respect to x_i we keep all variables except x_i constant. For analysis, let us use the notation $M_i(x_i)$ for $M(\mathbf{x})$ when all variables except x_i are unchanged.

$$\frac{\delta \phi}{\delta x_i} = \frac{1}{\det(M(\mathbf{x}))} \frac{\delta}{\delta x_i} \det(M(\mathbf{x}))$$

$$= \frac{1}{\det(M(\mathbf{x}))} \lim_{\Delta x_i \to 0} \frac{\det(M_i(x_i + \Delta x_i)) - \det(M_i(x_i))}{\Delta x_i}$$

$$= \frac{1}{\det(M(\mathbf{x}))} \lim_{\Delta x_i \to 0} \frac{\det(M_i(x_i) + \Delta x_i \mathbf{a}_i \mathbf{a}_i^T) - \det(M_i(x_i))}{\Delta x_i}$$

$$= \frac{1}{\det(M(\mathbf{x}))} \lim_{\Delta x_i \to 0} \det(M_i(x_i)) \mathbf{a}_i^T M_i^{-1}(x_i) \mathbf{a}_i$$

$$= \mathbf{a}_i^T M^{-1} \mathbf{a}_i$$

Similarly we can show that $\dfrac{\delta \phi}{\delta x_j} = \mathbf{a}_j^T M^{-1} \mathbf{a}_j$.

On differentiating $MM^{-1} = I$, we get

$$\frac{\delta M^{-1}}{\delta x_i} = -M^{-1} \frac{\delta M}{\delta x_i} M^{-1} .$$

The result for the second order derivatives now follows easily as

$$\frac{\delta^2 \phi}{\delta x_i \delta x_j} = \mathbf{a}_i^T \frac{\delta M^{-1}}{\delta x_j} \mathbf{a}_i = -\mathbf{a}_i^T M^{-1} \mathbf{a}_j \mathbf{a}_j^T M^{-1} \mathbf{a}_i$$

$$= -(\mathbf{a}_i^T M^{-1} \mathbf{a}_j)^2 . \blacksquare$$

Lemma 4. Let $\mathbf{h}$ be the gradient and H be the Hessian of the potential function $\phi(\mathbf{d})$ in (13). Let the inverse of $M(\mathbf{d}) = \zeta I + D - Q$ be denoted by $M^{-1} = [r_{ij}]$ then

$$h_i = -r_{ii} \text{ and } H_{ij} = r_{ij}^2, \quad i,j = 1, \ldots, n.$$

Proof.

Let $\mathbf{e}_i$ be an n-dimensional vector whose i^{th} co-ordinate is 1 and all other co-ordinates are 0. Then we get the results,

$$\frac{\delta M(\mathbf{d})}{\delta d_i} = \mathbf{e}_i \mathbf{e}_i^T, \quad i = 1, \ldots, n$$

hence $h_i = \dfrac{\delta \phi(\mathbf{d})}{\delta d_i} = \mathbf{e}_i^T M^{-1} \mathbf{e}_i = r_{ii}$

and $H_{ij} = \dfrac{\delta^2 \phi(\mathbf{d})}{\delta d_i \delta d_j} = -(\mathbf{e}_i^T M^{-1} \mathbf{e}_j)^2 = -(r_{ij})^2$. ∎

Let $\mathbf{d}^{(k)}$ be the state of the diagonal variables in the kth iteration and $\lambda^{(k)}$ be the maximum eigenvalue of $Q - D^{(k)}$. If we ensure that $\zeta^{(k)} > \lambda^{(k)}$, then the potential function $\phi(\mathbf{d})$ is defined for $(\zeta^{(k)}, \mathbf{d}^{(k)})$ and we can make a quadratic Taylor series approximation of $\phi^{(k)}(\mathbf{d})$ around $\mathbf{d}^{(k)}$ which is given by

$$T^{(k)}(\mathbf{d}) = \phi(\mathbf{d}^{(k)}) + \mathbf{h}^T \Delta \mathbf{d} + \frac{1}{2} \Delta \mathbf{d}^T H \Delta \mathbf{d} \qquad (14)$$

where $\Delta \mathbf{d} = \mathbf{d} - \mathbf{d}^{(k)}$

and $\mathbf{h}$ is the gradient and H is the Hessian of $\phi(\mathbf{d})$ at $\mathbf{d}^{(k)}$.

The Hessian H and gradient $\mathbf{h}$ can be computed using Lemma 4.

To obtain the new $\mathbf{d}^{(k+1)}$ that minimizes $\phi(\mathbf{d})$, we solve the problem

$$\min \; T^{(k)}(\mathbf{d}) \qquad (15)$$

$$\text{s.t. } \mathbf{e}^T \mathbf{d} = 0.$$

On applying the first-order Kuhn-Tucker optimality conditions, we get the linear system

$$H \Delta \mathbf{d} + \mathbf{h}^T = \alpha \mathbf{e}^T \qquad (16)$$

$$\mathbf{e}^T \Delta \mathbf{d} = 0.$$

We solve this linear system to obtain $\Delta \mathbf{d}$ which gives a new set of diagonals

$$\mathbf{d}^{(k+1)} = \mathbf{d}^{(k)} + \Delta \mathbf{d}.$$

… A better bound $\lambda^{(k+1)}$ is obtained with the new diagonal $\mathbf{d}^{(k+1)}$ and the process is repeated until the changes in diagonal become insignificant.

Our algorithm for minimizing the bound $f_{\max}$ on (1) can hence be described as follows:

Algorithm

Begin

1. Initialize $d^{(0)}$ to all zeros, $K = 0$
2. Compute maximum eigenvalue of Q and put the value in $\lambda^{(0)}$
3. $\zeta^{(0)} = \gamma \lambda^{(0)}$ { γ is a parameter > 1 }
4. Repeat
   begin loop
5.   Construct
$$M = \zeta^{(K)} I + D^{(K)} - Q$$
    Define
$$\phi^{(K)}(\mathbf{d}) = \ln \det \left[\zeta^{(K)} I + D - Q \right]$$
6.   Compute Hessian H and gradient $\mathbf{h}$ for $\phi^{(K)}(\mathbf{d})$
7.   Solve
$$\begin{bmatrix} H & -\mathbf{e}^T \\ -\mathbf{e}^T & 0 \end{bmatrix} \begin{bmatrix} \Delta \mathbf{d} \\ \alpha \end{bmatrix} = \begin{bmatrix} -\mathbf{h}^T \\ 0 \end{bmatrix}$$
8. $\mathbf{d}^{(K+1)} = \mathbf{d}^{(K)} + \Delta \mathbf{d}$
9.   Compute $\lambda^{(K+1)} = $ max eigenvalue of $Q - D^{(K+1)}$
10.   Update ζ
$$\zeta^{(K+1)} = \frac{\zeta^{(K)} + \lambda^{(K+1)}}{2}$$
11. $K = K + 1$
    end loop
12. Until $\|\Delta \mathbf{d}\| < \delta$
13. Return $n\lambda^{(K)}$.

End

---

3. Applying the Technique for Getting Lower Bounds on Graph Partition Cuts

We consider here the graph partitioning (GP) problem which is an *NP*-complete problem occurring in various applications [1][11][12][6]. Formally the problem could be defined as follows.

Problem Definition: [3][4][6]

Given a graph $G = (V, E)$ where $|V| = n$, partition the set of vertices V into disjoint subsets V^1 and V^2 where $|V^1| = |V^2| = n/2$ (assume n is even) so that the number of edges that have one endpoint in V^1 and the other in V^2 is minimized.

Quadratic Formulation:

This problem can be posed as a quadratic optimization problem (QP) of the form (1). The transformation of graph partitioning into a QP may be done as follows.

With each vertex $v_i \in V$ we associate a variable x_i. The variable x_i takes value 1 or -1 depending on whether vertex v_i is put in partition V^1 or V^2 respectively. Now consider the objective function

$$C(\mathbf{x}) = \sum_{(i,j) \in E} x_i x_j = \mathbf{x}^T A \mathbf{x}, \tag{17}$$

where A is the adjacency matrix for the graph.

One may check that with the above interpretation of the variables

$$C(\mathbf{x}) = 2[m - 2 \text{ (number of crossing edges)}],$$

$$\text{where} \quad \mathbf{x} \in S \text{ and } m = |E|. \tag{18}$$

Hence the objective of minimizing the size of the cut in a partition can be achieved by maximizing $C(\mathbf{x})$. We also need to ensure that there are equal numbers of nodes on either side of the partition which can be expressed by the constraint

$$\mathbf{e}^T \mathbf{x} = 0, \tag{19}$$

where $\mathbf{e}$ is the n-dimensional vector of all ones.

This constraint could be implicitly captured in the quadratic objective by modifying $C(\mathbf{x})$ to

$$GP(\mathbf{x}) = \mathbf{x}^T A \mathbf{x} - \beta \mathbf{x}^T \mathbf{e} \mathbf{e}^T \mathbf{x}. \tag{20}$$

For $\mathbf{x} \in S$ and $\beta > \dfrac{n}{2}$, it can be shown that the global maximum of $GP(\mathbf{x})$ gives us the best possible partition for the graph.

Another way of imposing (19) in the quadratic objective is by using the projection matrix P_e defined as

$$P_e = \left[I - \frac{1}{n}\mathbf{e}\mathbf{e}^T\right].$$

The new function so obtained looks like

$$GP'(\mathbf{x}) = \mathbf{x}^T P_e A P_e \mathbf{x}. \qquad (21)$$

So essentially what we have done is to transform the graph partitioning problem to one of the form (1) where $f(\mathbf{x})$ can be chosen to be $GP(\mathbf{x})$ or $GP'(\mathbf{x})$. We can hence use the scheme described previously to obtain lower bounds on the cutsize in the partitioning problem.

4. Computational Experience

We have tested our method on randomly generated instances of the graph partitioning problem with number of vertices ranging from 20 to 200. The edges in the graphs were generated with probability 0.5. We report here some implementational details and results obtained with this approach.

4.1 Implementation

To start with, the diagonals are all initialised to zero and we obtain an initial estimate on the upper bound μ. This μ is scaled by a factor slightly greater than one. At k^{th} iteration, $\mu^{(k)}$ is taken to be the arithmetic mean of $\mu^{(k-1)}$ and the new upper bound obtained in the previous iteration. At each iteration, we ensure that the current iterate is at a strictly interior point (M is positive definite).

The maximum eigenvalue and the corresponding eigenvector can be computed using some eigenvalue technique [16] like the power or the QR or the Jacobi method. At each iteration we find the maximum eigenvalue of $Q - D^{(k)}$ and using an adaptive rounding heuristic, we round the corresponding eigenvector to a ± 1 solution that gives us a partition of the graph. The cutsize corresponding to this partition is an upper bound on the optimum cutsize. This is then compared with the lower bound on the cutsize estimated using our scheme. If the two bounds differ by a value less than one then we can prove that the rounded solution corresponds to an optimum partition.

TABLE 1

Nodes	Edges	Iter	Initial Lower Bound	Improved Lower Bound	Upper Bound by Rounding
20	93	39	32.10	33.69	34
20	98	53	32.78	36.03	37
20	90	49	26.07	29.79	31
20	101	43	35.10	37.72	38
20	92	45	28.96	32.64	33
40	385	45	137.82	145.82	151
40	374	65	138.29	144.02	150
40	380	59	137.55	146.01	153
40	376	56	135.33	143.94	150
40	365	61	124.51	134.23	138
50	612	56	231.19	238.89	246
50	586	54	219.88	225.65	238
50	623	59	242.10	248.20	256
50	621	68	236.61	245.23	250
50	606	53	233.08	240.68	250
100	2487	43	1017.96	1041.76	1073
100	2436	61	991.98	1016.15	1048
100	2415	80	985.68	1008.19	1041
100	2542	56	1041.93	1072.69	1098
100	2479	62	1009.17	1039.47	1070
200	10063	92	4348.07	4424.49	4534
200	9968	81	4321.55	4378.93	4491
200	9886	73	4286.80	4331.56	4439
200	10022	90	4320.35	4409.07	4510
200	9874	93	4271.31	4339.68	4444

4.2 Results

We report in Table 1 the results obtained using our scheme, for graph partitioning problems having from 20 to 200 nodes. The Table gives the number of nodes and edges in the graph followed by the number of iterations taken by our algorithm. We then report the bound in the first iteration followed by the improved bound obtained using our technique and then compare it with the solution obtained by rounding the eigenvector corresponding to the maximum eigenvalue.

For graphs with 20 nodes the cut obtained by rounding the eigenvectors and the bounds obtained using our technique are close enough to prove optimality of the cuts. For larger problems the lower bounds were close to the cuts obtained from rounding. The gap between the two, some part of which may be attributed to the non-optimality of the rounded cut, can be seen to be small.

5. Conclusion and Directions of Future Work

In this paper, we have addressed the problem of finding an upper bound in a quadratic maximization problem with integer constraints. Combinatorial techniques usually have a hard time on such problems. We propose here a continuous approach for solving this problem. This technique is then applied for finding lower bounds on cutsizes in instances of the Graph Partitioning problem with promising results. Extensions and improvements to this method are being studied. The complexity analysis of the technique is also a subject of our current research. We believe that this method would be useful in several related applications.

REFERENCES

[1] A. Dunlop and B. Kernighan, "A Procedure for Placement of Standard-Cell VLSI Circuits," *IEEE Trans. Computer-Aided Design* **4**, 92-98, 1985.

[2] F. Gantmakher, *Theory of Matrices*, Chelsea, 1959.

[3] M. Garey and D. Johnson, *Computers and Intractability: A Guide to the Theory of NP-Completeness*, W. H. Freeman, San Francisco, 1979.

[4] M. Garey, D. Johnson and L. Stockmeyer, "Some Simplified NP-Complete Graph Problems," *Theor. Comput. Sci.* **1**, 237-267, 1976.

[5] C. Han, P. Pardalos, Y. Ye, "Computational Aspects of an Interior Point Algorithm for Quadratic Problems with Box Constraints," *Proc. of Conf. on Large-Scale Numerical Optimization*, Cornell Univ., 1989.

[6] D. Johnson, C. Aragon, L. McGeoch and C. Schevon, "Optimization by Simulated Annealing: An Experimental Evaluation, Part I (Graph Partitioning)," *Operations Research* **37**, No. 6, 865-892, 1989.

[7] A. Kamath, N. Karmarkar, K. Ramakrishnan and M. Resende, "Computational Experience with an Interior Point Algorithm on the Satisfiability Problem," *Annals of Opns. Res.*, **25**, 43-58, 1990.

[8] N. Karmarkar, "A New Polynomial Time Algorithm for Linear Programming," *Combinatorica* **4**, 373-395, 1984.

[9] N. Karmarkar, "An Interior-Point Approach to NP-Complete Problems," Math. Develop. arising from Linear Prog., Proc. of a Joint Summer Research Conf., Bowdoin College, 1988, *American Math. Soc. - Contemporary Math.* **114**, 1991.

[10] N. Karmarkar, M. Resende and K. Ramakrishnan, "An Interior Point Algorithm to Solve Computationally Difficult Set Covering Problems," *Mathematical Programming*, Special Issue on "Interior Point Methods: Theory and Practice," eds. Roos and Vial, 1991.

[11] B. Kernighan and S. Lin, "An Efficient Heuristic Procedure for Partitioning Graphs," *Bell Syst. Tech. J.* **49**, 291-307, 1970.

[12] J. Kral, "To the Problem of Segmentation of a Program," *Information Processing Machines*, 1965.

[13] D. Luenberger, *Linear and Nonlinear Programming*, Addison Wesley, 1973.

[14] R. Monteiro and I. Adler, "Interior path following primal-dual algorithms part II: convex quadratic programming," *Mathematical Programming* **44**, 43-66, 1989.

[15] P. Pardalos and S. Vavasis, "Quadratic Programming with One Negative Eigenvalue is NP-hard," to appear in *Journal of Global Optimization*, 1991.

[16] F. Rellich, *Perturbation Theory of Eigenvalue Problems*, Gordon and Breach, 1969.

[17] J. Wilkinson, *The Algebraic Eigenvalue Problem*, Clarendon Press, Oxford, 1965.

A Class of Global Optimization Problems Solvable by Sequential Unconstrained Convex Minimization

Hoang Tuy[1]   Faiz A. Al-Khayyal[2]

Abstract

We consider the problem of maximizing the sum of certain composite functions. Each term is the composition of a convex decreasing function, bounded from below, with a convex function having compact level sets. We show that this problem is equivalent to a convex maximization problem over a compact convex set and develop a specialized polyhedral annexation procedure to find a global solution for the case when the inside function is a polyhedral norm. Such a problem arises in location theory and very recently a method for finding only local solutions was proposed.

1 Introduction

In a recent paper, Idrissi et al. [5] considered the following class of global optimization problems for the case $m = 2$

$$\max_{x \in R^m} \sum_{j=1}^n q_j[h_j(x)] \qquad (P)$$

under the assumptions that: (A1) q_j is a strictly convex function, strictly decreasing on R_+, with values in R_+, for which $\lim_{t \to \infty} q_j(t) = 0$ for all j; and (A2) h_j is a convex function defined on R^m, with values in R_+, such that $\lim_{|x| \to \infty} h_j(x) = +\infty$ for all j, where $|\cdot|$ denotes an arbitrary norm.

The objective function $\varphi(x) := \sum_{j=1}^n q_j[h_j(x)]$ is generally neither convex nor concave. In [5], a method that computes only a local (instead of global) maximum is proposed. The aim of this paper is to present a method for finding a *global* maximum. The method reduces to solving a sequence of unconstrained convex minimization problems of the form

$$\min_{x \in R^m} \sum_{j=1}^n \alpha_j h_j(x)$$

---

[1] Institute of Mathematics, Hanoi, Vietnam.
[2] School of Industrial and Systems Engineering, Georgia Institute of Technology, Atlanta, Georgia, 30332-0205.

for $\alpha_j \geq 0$. The method is a specialization of the polyhedral annexation procedure earlier developed in [8] (see also [4]). The utilization of the general procedure herein demonstrates the value and versatility of polyhedral annexation approaches.

As noted in [5], for the case $m = 2$, problem (P) arises in certain single facility location problems. Specifically, consider the problem of locating a "desirable" facility (such as a community center or branch library) which is designed to serve n districts of a town or region. District j has a population located at $a_j \in R^2, j = 1, \ldots, n$, which can be interpreted as the population center of mass. Let $x \in R^2$ denote the location of the desirable facility and take $h_j(x) = |x - a_j|$ as the distance function from x to district j using an arbitrary polyhedral norm $|\cdot|$ (see, e.g., [7]). Then $q_j(|x - a_j|)$ measures the attraction of the district j population to the desirable facility when it is located at x. According to assumption (A1), the farther x is away from a_j the less attractive it looks to the district j population. Thus, problem (P) is to determine the optimal location $x \in R^2$ that maximizes the total attraction.

In general, a different norm can be associated with each district. When the distance measures between points are not symmetric (this occurs when there are one-way streets, for example) then gauges can be used for h_j instead of norms.

Problem (P) is more difficult than the familiar Fermat-Weber problem with nonlinear costs [2] where the objective is to *minimize* and the "cost" functions $q_j : R_+ \to R_+$ are assumed to be strictly increasing and convex such that $q_j(0) = 0$. Note that the Fermat-Weber problem is to minimize a convex function of a distance measure while problem (P) *maximizes* a convex function of the same measure. In fact, we show in Section 2 that problem (P) is equivalent to a convex maximization problem.

Idrissi et al. [5] develop a procedure based on solving a sequence of parameterized Fermat-Weber problems for finding only a local solution of problem (P). As in the more general case treated in [2], all local solutions of problem (P) are among the *intersection points* of the polyhedral norm. These points can be very easily characterized as follows. Let $\{v^1, v^2, \ldots, v^s\}$ denote the vertices of the "unit ball" in R^2 associated with the polyhedral norm of district j. For each $a_j, j = 1, \ldots, n$, draw the s_j half-lines emanating from a_j through v^i for all $i = 1, \ldots, s_j$. An intersection point is where two noncollinear half lines, emanating from a_k and a_ℓ ($k, \ell = 1, \ldots, n$), cross. It is easy to see that there are finitely many intersection points and that R^2 is thus partitioned into finitely many polyhedra each having vertices defined by intersection points and set directions defined by the half-lines constructed above. The polyhedra are called *elementary polyhedra* (see Figure 1). It is demonstrated

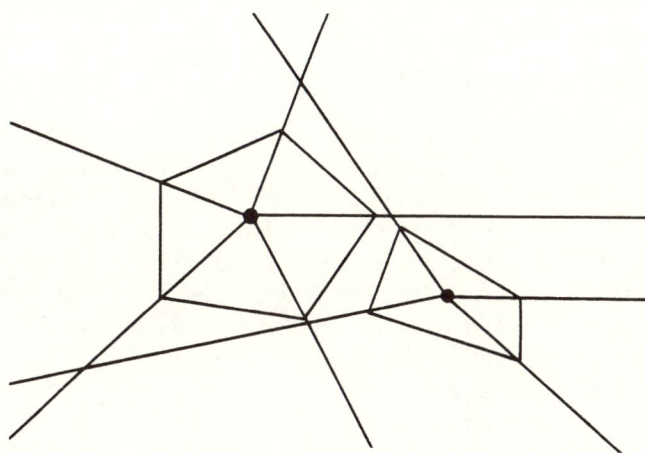

Figure 1: Intersection points of polyhedral norms and elementary polyhedra.

in [5] that the objective function of (P) is convex on every elementary polyhedron; thus, the problem can be viewed as maximizing a piecewise convex function when R^2 is tiled by elementary polyhedra. The algorithm in [5] effectively uses a parameterized partial linearization of the objective to achieve linearity over each elementary polyhedron which gives rise to a Fermat-Weber problem that is solved for a fixed parameter vector which is, in turn, successively updated. Only improving vertices of elementary polyhedra are generated until a local solution is encountered.

In contrast, our general approach is based on our observation in Section 2 that problem (P) can be reformulated as a constrained convex maximization problem which is a "classical" problem in global optimization. Consequently, our approach does not break down for $m \geq 3$ as does the algorithm in [5]. We employ an existing algorithm based on successively enlarging polyhedra and specialize this procedure in Section 3 to the problem at hand. In Section 4 we prove that every limit point of our algorithm is a global solution to problem (P). The closing Section 5 discusses some implementation considerations and techniques for computing certain quantities required by the algorithm. The treatment in the remainder of the paper assumes a basic knowledge of the fundamental techniques in deterministic global optimization, as detailed in [4], and has therefore been deliberately kept concise.

2 Basic Properties

In this section we first prove that problem (P) is equivalent to a constrained convex maximization problem that can be solved by an existing polyhedral annexation procedure. When this procedure is specialized to our problem, we show that it reduces to solving a sequence of unconstrained convex minimization subproblems.

Proposition 1 (see [5], Proposition 2.1). *Let $\varphi(x) = \sum_{j=1}^{n} q_j[h_j(x)]$. Then $\varphi(x)$ has a global maximizer on R^m. In particular the function $\varphi(x)$ is bounded on R^m.*

Proposition 2. *If $\bar{x}$ solves (P) then $(\bar{x}, \bar{t})$ with $\bar{t} = h(\bar{x})$ solves the convex maximization problem*

$$\max_{(x,t)} \{\sum_{j=1}^{n} q_j(t_j) : h_j(x) \leq t_j \ (j = 1, \ldots, n),$$

$$(x, t) \in R^{m+n}\}. \quad (Q)$$

Conversely, if $(\bar{x}, \bar{t})$ solves (Q) then $\bar{x}$ solves (P).

Proof. If (x, t) satisfies $h(x) \leq t$ and $\sum_{j=1}^{n} q_j(t_j) > \sum_{j=1}^{n} q_j[h_j(\bar{x})]$ then $\sum_{j=1}^{n} q_j[h_j(x)] \geq \sum_{j=1}^{n} q_j(t_j) > \sum_{j=1}^{n} q_j[h_j(\bar{x})]$. This proves the first assertion. To prove the second assertion, observe that if $\sum_{j=1}^{n} q_j[h_j(x)] > \sum_{j=1}^{n} q_j[h_j(\bar{x})]$ then for $t = h(x)$ we would have $\sum_{j=1}^{n} q_j(t_j) = \sum_{j=1}^{n} q_j[h_j(x)] > \sum_{j=1}^{n} q_j[h_j(\bar{x})] \geq \sum_{j=1}^{n} q_j(\bar{t}_j)$. □

Thus the problem is a convex maximization problem which can be solved by a general purpose global optimization method (see [8]). We exploit the special structure of the problem to obtain a more efficient streamlined version of this procedure.

Denote

$$\tilde{D} = \{t \in R_+^n : h(x) \leq t \text{ for some } x \in R^m\} \quad (1)$$
$$f(t) = \sum_{j=1}^{n} q_j(t_j),$$

where $h : R^m \to R_+^n$. Then problem (Q) can be rewritten as

$$\max_{t\in\tilde{D}} f(t). \qquad (\tilde{Q})$$

Let $\bar{t}$ be the best feasible solution of $(\tilde{Q})$ known at a certain stage. The core of our method is a procedure for solving the following subproblem:

$(\tilde{Q},\bar{t})$: *Determine whether $f(t) \leq f(\bar{t})$ for all $t \in \tilde{D}$ (i.e., $\bar{t}$ is globally optimal) and if not, find a feasible solution t' such that $f(t') > f(\bar{t})$.*

Let $t^o \in \tilde{D}$ be such that $f(t^o) < f(\bar{t})$. Denote $\gamma = f(\bar{t})$, $\tilde{C} = \{t \in R^n : f(t) \leq \gamma\}$, $D = \tilde{D}\setminus\{t^o\}$ and $C = \tilde{C}\setminus\{t^o\}$. Then $0 \in D \cap \text{int} C$ and $(\tilde{Q},\bar{t})$ can be reformulated as:

$(\tilde{Q},\bar{t})$: *Determine whether $D \subset C$ and if not find a point in the difference $D\setminus C$.*

For solving this problem, we observe the following properties of C and D.

Proposition 3. *C contains the orthant R^n_+.*

Proof. For any $t \in R^n_+$ we have $f(t^o + t) \leq f(t^0)$ because each $q_j(\cdot)$ is strictly decreasing on R_+. Hence $f(t^o + t) \leq \gamma$; i.e., $t^o + t \in \tilde{C}$ or $t \in C$. □

Proposition 4. *For any vector $s \in R^n_+$ we have*

$$\min_{t\in\tilde{D}} \sum_{j=1}^n s_j t_j = \min_{x\in R^m} \sum_{j=1}^n s_j h_j(x).$$

Proof. Obvious from (1) and the fact $s \in R^n_+$. □

It turns out that, due to these properties, the polyhedral annexation procedure in [8] specialized to problem $(\tilde{Q},\bar{t})$, will reduce to solving a sequence of unconstrained convex minimization subproblems. We show this in the next section.

3 Solution Method

Let C^* and D^* denote the polars of C and D, respectively. From general properties of polars, we have $D \subset C$ if and only if $C^* \subset D^*$. We now consider how to check $C^* \subset D^*$.

From Proposition 3 it follows that $C^* \subset R^n_-$, and from $0 \in \text{int} C$ it follows that C^* is compact (see, e.g., [6]).

Now consider a polytope S such that $C^* \subset S \subset R^n_-$. We are interested in knowing whether $S \subset D^*$, because if this holds then *a fortiori* $C^* \subset D^*$.

Proposition 5. *Let V be the vertex set of S. We have $S \subset D^*$ if and only if*

$$\max\{\textstyle\sum_{j=1}^n v_j t_j : t \in D\} \leq 1 \quad \text{for all } v \in V. \tag{2}$$

Proof. Since D^* is convex, we have $S \subset D^*$ if and only if $V \subset D^*$. But from the definition of polars, $v \in D^*$ if and only if (2) holds. □

Thus, to check whether or not $S \subset D^*$, we solve the subproblem $\max\{\sum_{j=1}^n v_j t_j : t \in D\}$ for each $v \in V$. Setting $s = -v$, we have $s \in R_+^n$ (because $S \subset R_-^n$). By Proposition 4, the latter subproblem is equivalent to

$$- \min_{x \in R^m} \textstyle\sum_{j=1}^n s_j [h_j(x) - t_j^o].$$

Denote by $R(s)$ the subproblem

$$\min_{x \in R^m} \textstyle\sum_{j=1}^n s_j h_j(x) \tag{R(s)}$$

and let $\mu(s)$ be the optimal value, and $x(s)$ an optimal solution of $R(s)$. It is convenient to use the notation $\langle \cdot, \cdot \rangle$ for inner product.

Corollary 1. *If $\langle t^o, s \rangle - \mu(s) \leq 1$ for all $s \in -V$ then $\bar{t}$ is a global optimal solution of $(\bar{Q})$.*

Suppose now that, for some $\tilde{s} \in -V$, we have

$$\langle t^o, \tilde{s} \rangle - \mu(\tilde{s}) > 1. \tag{3}$$

Let $\tilde{x} = x(\tilde{s})$ and $\tilde{t} = h(\tilde{x})$ (so $\tilde{t} \in \tilde{D}$ and $\langle \tilde{t}, \tilde{s} \rangle = \mu(\tilde{s})$). If $f(\tilde{t}) > f(\bar{t})$ then we have obtained a better feasible solution than the current best $\bar{t}$. Otherwise, $f(\tilde{t}) \leq f(\bar{t})$, compute

$$\theta = \sup\{\lambda : f(t^o + \lambda(\tilde{t} - t^o)) \leq \gamma\}. \tag{4}$$

Note that $\theta \geq 1$ because $f(t^o) < \gamma$ while $f(\tilde{t}) \leq \gamma$.

Proposition 6. *The cut*

$$\langle \tilde{t} - t^o, t \rangle \leq \frac{1}{\theta} \tag{5}$$

excludes $\tilde{v} = -\tilde{s}$ from S without excluding any point of C^*.

Proof. Since $\langle \tilde{t}, \tilde{s} \rangle = \mu(\tilde{s})$ it follows from (3) that $\langle \tilde{t} - t^o, -\tilde{s} \rangle = \langle t^o, \tilde{s} \rangle - \mu(\tilde{s}) > 1$; i.e., $-\tilde{s}$ violates the inequality (5). On the other hand, since from the definition of θ (see (4)), $\theta(t-t^o) \in C$, it follows that any point $t \in C^*$ satisfies (5). □

Thus, the polytope S', formed by applying the cut (5) to S, given by
$$S' = S \cap \{t : \langle \tilde{t} - t^o, t \rangle \leq \frac{1}{\theta}\},$$
does not contain $-\tilde{s}$ but still contains C^*. Consequently, the procedure can be repeated from S' in place of S.

To start, we need a polytope $S_1 \supset C^*$. This polytope can be constructed, for example, as follows.

Let $e = (1, \ldots, 1)$ be a vector of ones in R^n. Since $0 \in \text{int} C$ one can always select $\alpha > 0$ small enough such that $-\alpha e \in C$.

Proposition 7. $C^* \subset S_1 = \{t \in R_-^n : -\sum_{j=1}^n t_j \leq \frac{1}{\alpha}\}$.

Proof. We have $(C^*)^* = C$ (see [6]). Hence $-\alpha e \in C$ implies $\langle -\alpha e, t \rangle \leq 1$ for all $t \in C^*$; i.e., $-\alpha \sum_{j=1}^n t_j \leq 1$ for all $t \in C^*$. □

Thus, starting with S_1, we can build a sequence of nested polytopes in R_-^n
$$S_1 \supset S_2 \supset \ldots \supset S_k \supset \ldots \supset C^*$$
such that, for each k, either $S_k \subset D^*$ (then the current best $\bar{t}$ is a global optimum), or at least one $s^k \in -S_k$ has $\langle t^o, s^k \rangle - \mu(s^k) > 1$. It may be that $t^k = h(x(s^k))$ is better than $\bar{t}$ (subproblem $(Q, \bar{t})$ has now been solved and we can proceed to $(\tilde{Q}, \bar{t}^1)$ with $\bar{t}^1$ equal to the just found t^k). Otherwise, t^k generates a cut which, adjoined to S_k, determines S_{k+1}. It turns out that if we always choose $s^k \in \text{argmax}_{s \in -V_k}\{\langle t^o, s \rangle - \mu(s)\}$ then convergence of the procedure is guaranteed.

In practice we combine $(\tilde{Q}, \bar{t}^o), (\tilde{Q}, \bar{t}^1), \ldots,$ into a unified process. Then we have the following:

Algorithm

Initialization. Let $\bar{t}$ be the best available feasible solution of problem $(\tilde{Q})$. Take a feasible solution t^o such that $f(t^o) < f(\bar{t})$ (preferably a t^o substantially worse than $\bar{t}$ because we want t^o to lie sufficiently far from the boundary of the set $\tilde{C}$). Set $\tilde{V}_1 = V_1$ (set of vertices of initial S_1) = $\{-\frac{1}{\alpha}e^i : i = 1, \ldots, n\}$. Set $k = 1$.

Iteration $k = 1, 2, \ldots$

Step 1. For each $s \in -V_k$ solve $R(s)$ to obtain the optimal value $\mu(s)$ and optimal solution $x(s)$.

Step 2. If $\max\{\langle t^o, s \rangle - \mu(s) : s \in -V_k\} \leq 1$, then terminate: t is the global optimal solution of $(\tilde{Q})$.

Step 3. Select $s^k \in \text{argmax}\{\langle t^o, s \rangle - \mu(s) : s \in -V_k\}$. Let $x^k = x(s^k)$, $t^k = h(x^k)$. If $f(t^k) > f(\bar{t})$ then reset $\bar{t} \leftarrow t^k$

(otherwise $\bar{t}$ is unchanged). Compute

$$\theta_k = \sup\{\lambda : f(t^o + \lambda(t^k - t^o)) \leq f(\bar{t})\}$$

and define

$$S_{k+1} = S_k \cap \{t : \langle t^k - t^o, t \rangle \leq \frac{1}{\theta_k}\}.$$

Step 4. Compute the vertex set V_{k+1} of S_{k+1} (knowing already the vertex set V_k of S_k). Set $\tilde{V}_{k+1} = V_{k+1} \setminus V_k$. Go to iteration $k+1$.

4 Convergence

Define a function $g : R^n \to [0, \infty]$ by setting for each $v \in R^n$.

$$g(v) = \sup\{\langle v, t \rangle : t \in D\}.$$

Obviously $g(v)$ is a convex function and $D^* = \{v : g(v) \leq 1\}$.

Lemma 1. *For every k, $t^k - t^o \in \partial g(-s^k)$.*

Proof. Let $v^k = -s^k$. Clearly

$$g(v^k) = \langle t^o, s^k \rangle - \mu(s^k) = \langle t^o, s^k \rangle - \langle t^k, s^k \rangle = \langle t^k - t^o, v^k \rangle. \quad (6)$$

Hence, $g(v) - g(v^k) = g(v) - \langle t^k - t^o, v^k \rangle \geq \langle v, t^k - t^o \rangle - \langle t^k - t^o, v^k \rangle = \langle t^k - t^o, v - v^k \rangle.$ □

Lemma 2. *Any cluster point of the sequence $v^k = -s^k$ belongs to D^*.*

Proof. Denote $\ell_k(v) = \langle t^k - t^o, v \rangle$. Now consider the set D^*, the sequence $\{v^k\}$, and the sequence of affine functions $\ell_k(v)$. We have (see (6))

$$\ell_k(v^k) = \langle t^k - t^o, v^k \rangle = g(v^k) > 1,$$

while

$$\ell_k(v) = g(v^k) + \langle t^k - t^o, v - v^k \rangle \leq g(v) \leq 1$$

for all $v \in D^*$. That is, each $\ell_k(\cdot)$ strictly separates v^k from D^*. Since the sequence $\{v^k\}$ is bounded (contained in S_1) and $t^k - t^o \in \partial g(v^k)$ by Lemma 1, it follows from well known results in outer approximation methods (see, e.g., [4]) that any cluster point $\hat{v}$ of $\{v^k\}$ will belong to D^*. □

Denote by $\bar{t}^k$ the current best solution in iteration k.

Theorem. *Either the algorithm terminates after finitely many iterations yielding a global optimal solution, or it generates* an infinite sequence $\{\bar{t}^k\}$ every cluster point of which is a global optimal solution.

Proof. Suppose the algorithm is infinite. We will show that in fact any cluster point $\hat{t}$ of sequence $\{\bar{t}^k\}$ is a global optimal solution. Let $\hat{t} = \lim_{\nu \to \infty} \bar{t}^{k_\nu}$. Without loss of generality, we can assume $v^{k_\nu} \to \hat{v}$. Since, by Lemma 2, $\hat{v} \in D^*$ it follows that $g(\hat{v}) \leq 1$. Let $\tilde{C}_k = \{t : f(t) \leq f(\bar{t}^k)\}$, $C_k = \tilde{C}_k - t^o$. Then

$$\begin{aligned} g(v^k) &= \max\{g(v) : v \in V_k\} = \max\{g(v) : v \in S_k\} \\ &\geq \max\{g(v) : v \in C_k^*\} \end{aligned}$$

since $S_k \supset C_k^*$ by Proposition 6. But $g(v^{k_\nu}) \leq g(\hat{v}) \leq 1$. Therefore,

$$\max\{g(v) : v \in \cap_{\nu=1}^\infty C_{k_\nu}^*\} \leq 1$$

which implies that $\cap_{\nu=1}^\infty C_{k_\nu}^* \subset D^*$, and consequently, $\cup_{\nu=1}^\infty C_{k_\nu} \supset D$. Thus, $\tilde{D} \subset \cup_{\nu=1}^\infty \tilde{C}_{k_\nu}$. That is, for any $t \in \tilde{D}$, there exists a k_ν such that $t \in \tilde{C}_{k_\nu}$; hence, $f(t) \leq f(\bar{t}^{k_\nu}) \leq f(\hat{t})$, proving the global optimality of $\hat{t}$. □

5 Discussion

If P_k denotes the polar of the polytope S_k then

$$P_1 \subset P_2 \subset \ldots \subset P_k \subset \ldots$$

That is, the method amounts to building a sequence of expanding polyhedra all contained in the convex set $C = \{t : f(t+t^o) \leq \gamma_{\text{opt}}\}$, where $\gamma_{\text{opt}} = \min f(D)$, and eventually covering all of D, whence the name *polyhedral annexation*.

Every subproblem $R(s)$ is an unconstrained convex minimization problem which can be solved by efficient standard methods. If the functions $h_j(x)$ are polyhedral (as assumed in [5]), then each $R(s)$ seeks to minimize a convex piecewise affine function over R^m. Recall that $m = 2$ for the practical problem considered in [5].

We assume that two feasible solutions t^o and $\bar{t}$ are available at the beginning with $f(t^o) < f(\bar{t})$. This assumption is innocuous here since for any x the vector $t = h(x)$ is a feasible solution.

Each polytope S_{k+1} is obtained from its predecessor S_k by adjoining a new linear constraint. Therefore, V_{k+1} can be derived from V_k using, e.g., the procedure of either Chen et al. [1] or Horst et al. [3]. If at some iteration k the set V_k becomes too large, then it is possible to restart, with the current best solution $\bar{t}^k$ as the starting $\bar{t}$. Thus, the growth of V_k may create difficulty for this method, as for similar methods of concave minimization. Based on the reported numerical experience in [1], [3], our method should be practical for values of n up to 10 or 15 on a microcomputer.

The problems $R(s)$ need not necessarily be solved to optimality. It is possible to develop an "approximate" variant of the Algorithm where each $R(s)$ can be solved to within some accuracy (but then the output of the Algorithm is an approximate global solution).

Finally, it is clear that the method can be applied to nonconvex functions h if the unconstrained global minimization of these functions can be done efficiently.

References

[1] P.-C. Chen, P. Hansen and B. Jaumard, "On-Line and Off-Line Vertex Enumeration by Adjacency Lists," *Operations Research Letters*, to appear.

[2] R. Durier and C. Michelot, "Geometrical Properties of the Fermat-Weber Problem," *European Journal of Operational Research* 20 (1985) 332-343.

[3] R. Horst, J. de Vries and N.V. Thoai, "On Finding New Vertices and Redundant Constraints in Cutting Plane Algorithms for Global Optimization," *Operations Research Letters* 7 (1988) 85-90.

[4] R. Horst and H. Tuy, *Global Optimization (Deterministic Approaches)*, Springer Verlag, Berlin, 1990.

[5] H. Idrissi, P. Loridan, and C. Michelot, "Approximation of Solutions for Location Problems," *Journal of Optimization Theory and Applications* 56 (1988) 127-143.

[6] R. T. Rockafeller, *Convex Analysis*, Princeton University Press, Princeton, 1970.

[7] J.-F. Thisse, J.E. Ward, and R.E. Wendell, "Some Properties of Location Problems with Block and Round Norms," *Operations Research* 32 (1984) 1309-1327.

[8] H. Tuy, "On Polyhedral Annexation Method for Concave Minimization," in *Functional Analysis Optimization and Mathematical Economics*, pp. 248-260, (Lev Leifman, ed.), Oxford University Press, New York, 1990.

A New Cutting Plane Algorithm for a Class of Reverse Convex 0-1 Integer Programs

Sihem BenSaad
AT&T Bell Laboratories
Crawfords Corner Road,
Holmdel, N.J. 07733

Abstract

A new algorithm is presented for a wide class of optimization problems where the objective function and the constraints are linear, with the exception of one nonlinear reverse convex constraint, and where the variables are possibly 0-1 integer. The algorithm is based on a cutting plane method. The cutting plane is constructed using the level sets of the functions involved as well as the edges of the polytope defining the linear constraints. It is a deeper cut than the Tuy-cut. We show that the algorithm converges finitely to a global optimum of this non-linear/0-1 integer class of optimization problems.

1 Introduction

In this paper, the following optimization problem is considered:

$$min_{x \in R^n} \quad c^T x$$
$$such\ that \quad Ax \leq b$$
$$g(x) \leq 0$$

where A is an $m \times n$ matrix, $m \geq n$, and b is an m-vector, and g is a concave function of R^n. All variables are real and all functions take on real values. The last constraint is called "reverse convex" following a paper by Meyer[9], and refers to the fact that the set $\{x \in R^n | g(x) \leq 0\}$ is the closure of the complement of a convex set. This problem is quite general since it encompasses the minimization of a concave function over a polytope, 0-1 linear optimization problems, some classes of non-linear 0-1 optimization problems. This is due to the fact that expressing that the variables x_i are either 0 or 1 is equivalent to $\sum_{i=1}^{n}(x_i - x_i^2) \le 0$ and $0 \le x_i \le 1$, for $i = 1, \ldots, n$. Also, the minimization of Problems of this form arise in the context of maximizing flow capacity in a network where the incremental capacity cost functions exhibit economies of scale (see Bansal and Jacobsen [2], Hillestad and Jacobsen[6]). Other examples are the optimal scheduling problem, the set covering and the set partitioning problems. Also, the minimization of a concave function subject to linear constraints is convertible to the above optimization problem by adding an extra variable, x_{n+1}, and minimizing x_{n+1} subject to the linear constraints and the additional reverse convex constraint $f(x) - x_{n+1} \le 0$. Because of the close relationship between reverse convex programs and concave minimization, the reader is referred to the recent survey article by Pardalos and Rosen [10]. For additional litterature on algorithms for reverse convex problems, the reader is referred to [3],[5],[7],[13],[4],[12], [8],[1],[11].

In this paper, we develop a cutting plane algorithm for the above optimization problem. Several cutting plane algorithms have been developed (see above literature) for reverse convex problems. They however had limited success in finding a global optimal solution relatively fast, even for small instances. Through this litterature search, we came to the conclusion that one of the main reasons for such a slow convergence is the fact that these cuts, most of which are based on the "Tuy-cut", are not deep enough, in the sense that they do not eliminate a large enough portion of the feasible region. From this observation, came the motivation to start with a "Tuy-cut" and refine it by moving along n edges of the polytope F_A defining the linear constraints. The hope is to eliminate n edges of this polytope, for most of the main iterations of the algorithm. In sections 3 and 4, we expand upon how these edges are constructed, and why they lead to a much improved cut.

In section 2, a set of definitions is presented. Section 3 contains a statement of the algorithm and a geometric example that demonstrates concretely the successive steps of the algorithm. Section 4 contains the arguments necessary to show that the cutting hyperplanes generated by the algorithm do not exclude any feasible point and that each major iteration does produce a strict improvement. These results allow us to show finite convergence to optimality.

2 Definitions

We briefly introduce the notation which will be used throughout the paper. We denote by F_A a bounded full-dimensional convex polytope in R^n defined by a system of linear inequalities; that is

$$F_A = \{x \in R^n | Ax \ge b\}$$

where A is an $m \times n$ matrix, $m \ge n$, and b is an m-vector. Let

$$G = \{x \in R^n | g(x) \leq 0\}$$

where g is a continuous concave function on R^n. We assume G and G^c are both non-empty. Let $F = F_A \cap G$. The problem to be considered in this paper is: $min\{c^T x | x \in F\}$.

Definition 1 *Let x^0 be on an edge of F_A and $x^0 \in G^c$. The function η_{x^0} is defined to be:*

$$\eta_{x^0} : R^n \longrightarrow R$$
$$z \longmapsto \eta_{x^0}(z) = \alpha_1$$

where $\alpha_1 = \min \{\alpha \geq 0 : g(x^0 + \alpha(z - x^0)) = 0\}$ if this minimum exists, and $\alpha_1 = $ some constant $\overline{\alpha}$ otherwise ($\overline{\alpha} > 1$).

Definition 2 *Let y be on an edge of F_A. Let e be a non-zero vector of R^n. The real $\eta_{y,e}$ is defined to be:*
$\eta_{y,e} = \min \{\alpha \geq 0 : g(y + \alpha e) = 0\}$ if this minimum exists, and $\eta_{y,e} = $ some constant $\overline{\alpha}$ otherwise ($\overline{\alpha} > 1$).

Definition 3 *Let $z^1, \ldots, z^n, x^0$ be $n+1$ vectors of R^n such that $z^1 - x^0, \ldots, z^n - x^0$ are linearly independent. $\mathcal{H}[z^1, \ldots, z^n]$ denotes the plane defined by $z^1, \ldots, z^n$ and $\mathcal{H}_+[z^1, \ldots, z^n]$ denotes the half-space not containing x^0, whose bounding hyperplane is $\mathcal{H}[z^1, \ldots, z^n]$.*

Definition 4 *Let $z^1, \ldots, z^n$ be n points of R^n. Then $H[z^1, \ldots, z^n]$ denotes the convex hull of these n points.*

Definition 5 *$C[v^1, \ldots v^n]$ denotes the cone originating at the origin and generated by the n linearly independent vectors $v^1, \ldots, v^n$ of R^n i.e., $C[v^1, \ldots, v^n] = \{z \in R^n \text{ s.t. } z = \alpha_1 v^1 + \ldots + \alpha_n v^n, \alpha_1 \geq 0, \ldots, \alpha_n \geq 0\}$*

At this point, we collect the major assumptions which are to hold throughout the remainder of the paper:

1. g is a continuous, concave function defined on R^n.

2. G^c is bounded. This assumption guarantees that certain line search problems, in the algorithm, will have solutions.

3. The solutions to the linear problems throughout are assumed to be non-degenerate vertices

3 Statement of the Algorithm

Step–I solve the linear program

$$\min\{c^T x | x \in F_A\}$$

and assume x^0 is a vertex optimal solution.

Step–II If $x^0 \in G$, let $xopt = x^0$ and go to step (V).
Otherwise,
Find n neighboring vertices of x_0. Denote them by $z_1, \ldots, z_n$.

Let $z_i := x_0 + \eta_{x_0}(z_i)(z_i - x_0)$
Let $k = 0$, $u_i^k := z_i$, $i = 1, \ldots, n$
and $\mathcal{H}^k := \mathcal{H}[u_1^k, \ldots, u_n^k]$

Step–III Find a vertex solution of

$$\min\{c^T x | x \in F_A \cap \mathcal{H}^k\}$$

Denote it by $y_1^{(k)}$.
If $g(y_1^{(k)}) \leq 0$, set $xopt = y_1^{(k)}$ and go to step (V).
Let $y_i^{(k)}, i = 2, \ldots, n$ be the $(n-1)$ neighboring vertices of $y_1^{(k)}$ with respect to $F_A \cap \mathcal{H}^k$, that belong to the plane $\mathcal{H}^k$.
Let $\tilde{y}_1^k$ be a neighboring vertex in $F_A \cap \mathcal{H}_+^k$ of y_1^k that does not belong to $\mathcal{H}^k$.

Step–IV Let $e_i^{(k)}$ be the direction along which the edge (unique by the nondegeneracy assumption) of F_A containing $y_i^{(k)}$ runs, oriented to point into $\mathcal{H}_+^k$, e.g. $e_i^{(k)} = b - a$, where $[a, b]$ is the edge in question with $b \in \mathcal{H}_+^k$.

Let $u_1^{(k+1)} = y_1^{(k)} + \eta_{y_1^{(k)}}(\tilde{y}_1^k)(\tilde{y}_1^k - y_1^k)$
and
Let $u_i^{(k+1)} = y_i^{(k)} + \eta_{y_i^{(k)}, e_i^{(k)}} e_i^{(k)}$ for $i = 2, \ldots, n$

Let $\mathcal{H}^{(k+1)} = \mathcal{H}[u_1^{(k+1)}, \ldots, u_n^{(k+1)}]$
Let $k = k + 1$
Return to step (III).

Step–V Return $xopt$ as an optimal vector. Exit

To illustrate the algorithm geometrically, we consider the two-dimensional example in figure 1. The feasible region is disconnected and is the union of the two shaded areas, together with the isolated point x^*.

The algorithm starts out with a vertex solution x^0 to the linear problem. It then constructs a preliminary cutting plane $\mathcal{H}^0[z_1, z_2]$, where z_1, z_2 are the extensions to the boundary of the reverse convex constraint of two edges defining x^0. This hyperplane is known in the literature as the Tuy cut. If we solve the linear problem over the original linear feasible region F_A intersected with this hyperplane $\mathcal{H}^0[z_1, z_2]$, we find y_1^0 to be a vertex optimal solution and y_2^0 to be its neighboring vertex in that additional hyperplane. These two vertices define at least two distinct edges of F_A on which they lie. Step (IV) of the algorithm identifies these edges and moves along them until the boundary of the reverse convex constraint is hit. This defines the two points u_1^1 and u_2^1. The key idea of this algorithm is that $\mathcal{H}[u_1^1, u_2^1]$ is a cutting plane that does not exclude any part of the feasible region and is "deeper" than the Tuy cut.

In this example, at the next main iteration of the algorithm, a linear program is solved over the intersection of F_A with the cut defined by the hyperplane $\mathcal{H}[u_1^1, u_2^1]$. y_1^1 is a vertex optimal solution and y_2^1 is a neighboring vertex in that plane. Again, the next step is to move on the two edges of F_A defined by these two points until the level set of the reverse convex constraint is reached, thus defining the points u_1^2 and u_2^2. The next linear program is solved over F_A intersected with this new cutting hyperplane $\mathcal{H}[u_1^2, u_2^2]$. It yields $u_1^2 = x^*$ as the optimal solution.

It is important to realize that these geometric arguments become much more difficult to envision in three or higher dimensions.

CUTTING PLANE ALGORITHM

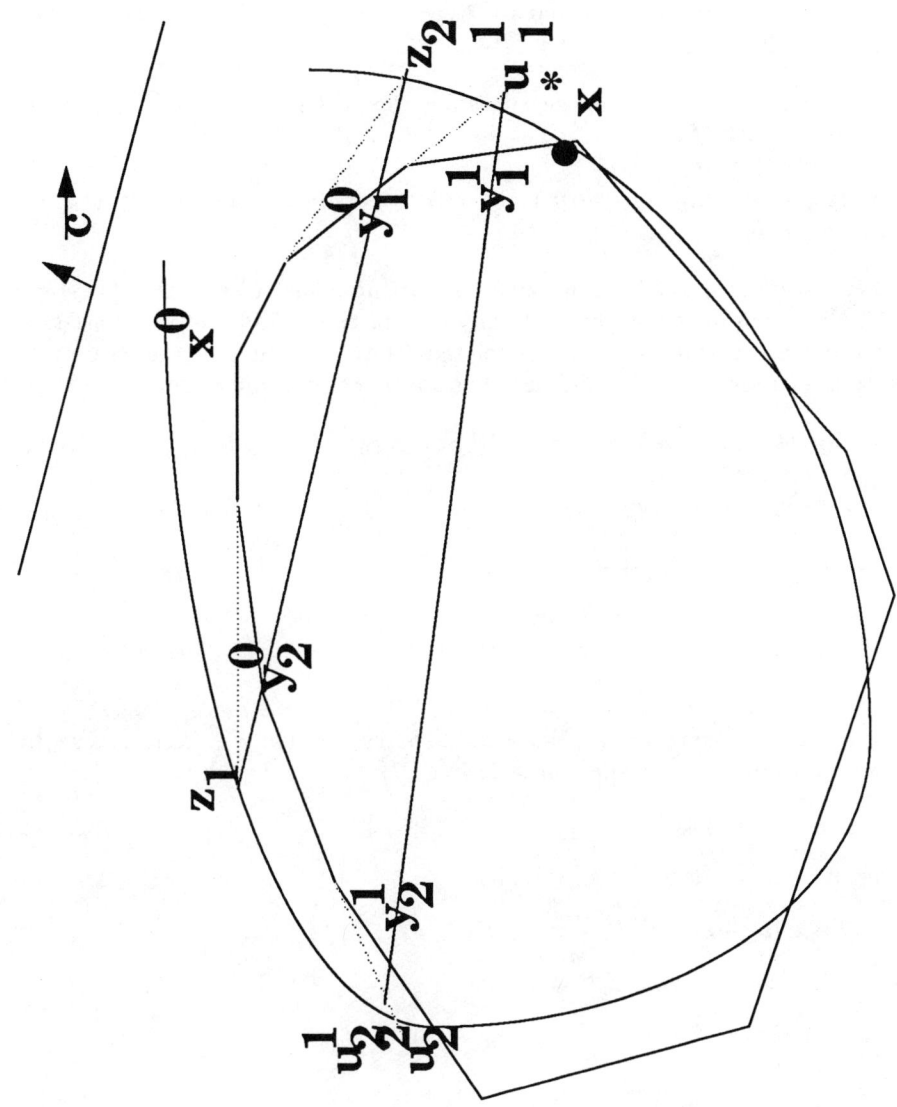

4 Convergence to a Global Optimum

The following proposition states that the hyperplane specified in step (II) of the algorithm is well defined.

Proposition 1 *Let x^0 be a non-degenerate vertex of F_A and let $e_1, \ldots, e_n$ be n neighboring vertices of x^0.*
Let $z_i = x_0 + \eta_{x_0}(e_i)(e_i - x_0)$
Then $z_1, \ldots, z_n$ are affinely independent with respect to x^0 and $\mathcal{H}[z_1, \ldots, z_n]$ is a hyperplane of R^n.

The following proposition shows that the cutting plane $\mathcal{H}^0[z_1, \ldots, z_n]$ in step II of the algorithm does not eliminate any part of the feasible region. Part (a) and its proof are a statement and proof that the "Tuy-cut", in the case of a reverse convex program, is a valid cut. For another proof, refer to [5].

Proposition 2 *Let x_0 be a vertex of F_A with $g(x_0) > 0$ and let $e_1, \ldots, e_n$ be n neighboring vertices of x_0.*
Let $z_i = x_0 + \eta_{x_0}(e_i)(e_i - x_0)$. Then:

$(a) F_A \cap \mathcal{H}_+[z_1, \ldots, z_n] \supset F$

$(b) x \in F_A \cap \mathcal{H}[z_1, \ldots, z_n] \Longrightarrow x \in H[z_1, \ldots, z_n]$

Proof:
(a)
Let x be in the feasible region F. We would like to show that $x \in \mathcal{H}_+[z_1, \ldots, z_n]$. It will be sufficient to show that $\exists \alpha \geq 1, \exists \tilde{x} \in H[z_1, \ldots, z_n]$ such that:

$$x - x_0 = \alpha(\tilde{x} - x_0)$$

However, if x_0 is nondegenerate, we have:

$$\{x_0\} + C[z_1 - x_0, \ldots, z_n - x_0] \supset F_A \supset F$$

Consequently, $\exists a_1, \ldots, a_n \in R^+$ such that:

$$x - x_0 = a_1(z_1 - x_0) + \ldots + a_n(z_n - x_0)$$

i.e.

$$x - x_0 = (\sum_{i=1}^{n} a_i)(\frac{a_1}{\sum_{i=1}^{n} a_i} z_1 + \ldots + \frac{a_n}{\sum_{i=1}^{n} a_i} z_n - x_0)$$

So, if we define $\tilde{x}$ to be:

$$\tilde{x} = (\frac{a_1}{\sum_{i=1}^{n} a_i} z_1 + \ldots + \frac{a_n}{\sum_{i=1}^{n} a_i} z_n),$$

CUTTING PLANE ALGORITHM

Then $\tilde{x} \in H[z_1,\ldots,z_n]$ and we only need to show that $(\sum_{i=1}^{n} a_i) \geq 1$. By contradiction, let us assume that $(\sum_{i=1}^{n} a_i) < 1$. We have

$$x = a_1 z_1 + \ldots + a_n z_n - (1 - \sum_{i=1}^{n} a_i) x_0.$$

Since g is concave and we are assuming that $(1 - \sum_{i=1}^{n} a_i) > 0$,

$$\begin{aligned} g(x) &\geq a_1 g(z_1) + \ldots + a_n g(z_n) + (1 - \sum_{i=1}^{n} a_i) g(x_0) \\ &\geq (a_1 + \ldots + a_n + (1 - \sum_{i=1}^{n} a_i)) max\{g(z_1),\ldots,g(z_n), g(x_0)\} \\ &\geq g(x_0) \\ &> 0. \end{aligned}$$

This contradicts the fact that $x \in F$. □

(b)
Again,

$$\{x_0\} + C[z_1 - x^0,\ldots,z_n - x^0] \supset F_A.$$

So, if $x \in F_A \cap \mathcal{H}[z_1,\ldots,z_n]$ then $x \in \mathcal{H}[z_1,\ldots,z_n] \cap (\{x_0\} + C[z_1 - x^0,\ldots,z_n - x_0])$. This means that $x \in H[z_1,\ldots,z_n]$.

Proposition 3 *At each iteration k of the algorithm, let $u_1^k,\ldots,u_n^k$ be the points as defined in step (IV) of the algorithm. Then, $\mathcal{H}[u_1^k,\ldots,u_n^k]$ defines a hyperplane of R^n and:*

(a) $F_A \cap \mathcal{H}_+[u_1^k,\ldots,u_n^k] \supset F$

(b) $x \in F_A \cap \mathcal{H}[u_1^k,\ldots,u_n^k] \Longrightarrow x \in H[u_1^k,\ldots,u_n^k]$

<u>Proof:</u> The proof is by induction over k:
<u>$k = 0$:</u>
This was shown in proposition 2.

So, by induction over k, let us assume that:
<u>for $k \geq 0$:</u>

(a) $F_A \cap \mathcal{H}_+[u_1^k,\ldots,u_n^k] \supset F$

and that

$(b) x \in F_A \cap \mathcal{H}[u_1^k, \ldots, u_n^k] \Longrightarrow x \in H[u_1^k, \ldots, u_n^k]$

<u>For $k+1$:</u>
<u>Part (a):</u>
By construction (step (III)):

$$y_1^k, \ldots, y_n^k \in F_A \cap \mathcal{H}[u_1^k, \ldots, u_n^k]$$

So, by induction hypothesis:

$$y_1^k, \ldots, y_n^k \in H[u_1^k, \ldots, u_n^k]$$

Consequently, $g(y_1^k) \geq 0, \ldots, g(y_n^k) \geq 0$.
y_1^k is a vertex of $F_A \cap \mathcal{H}_+[y_1^k, \ldots, y_n^k]$ with neighboring vertices $\tilde{y}_1^k, y_2^k, \ldots, y_n^k$.
So:

$$\begin{aligned}\{y_1^k\} + C[\tilde{y}_1^k - y_1^k, \ldots, y_2^k - y_1^k, \ldots, y_n^k - y_1^k] &\supseteq F_A \cap \mathcal{H}_+[y_1^k, \ldots, y_n^k] \\ &= F_A \cap \mathcal{H}_+^k \\ &\supseteq F\end{aligned}$$

The last inclusion is derived by induction hypothesis part(a). We also have that:

$$\{y_1^k\} + C[\tilde{y}_1^k - y_1^k, y_2^k - y_1^k, \ldots, y_n^k - y_1^k] = \{y_1^k\} + C[u_1^{k+1} - y_1^k, y_2^k - y_1^k, \\ \ldots, y_n^k - y_1^k]$$

and $g(u_1^{k+1}) \geq 0$, $g(y_2^k) \geq 0, \ldots, g(y_n^k) \geq 0$. So, by the same argument as in proposition 2:

$$F_A \cap \mathcal{H}_+[u_1^{k+1}, y_2^k, \ldots, y_n^k] \supseteq F$$

and

$$x \in F_A \cap \mathcal{H}[u_1^{k+1}, y_2^k, \ldots, y_n^k] \Longrightarrow x \in H[u_1^{k+1}, y_2^k, \ldots, y_n^k]$$

By induction, assume this is true up to index $i > 1$, i.e.

$$F_A \cap \mathcal{H}_+[u_1^{k+1}, u_2^{k+1}, \ldots, u_i^{k+1}, y_{i+1}^k, \ldots, y_n^k] \supseteq F$$

and

$$x \in F_A \cap \mathcal{H}[u_1^{k+1}, u_2^{k+1}, \ldots, u_i^{k+1}, y_{i+1}^k, \ldots, y_n^k] \Longrightarrow x \in H[u_1^{k+1}, u_2^{k+1}, \\ \ldots, u_i^{k+1}, y_{i+1}^k, \ldots, y_n^k]$$

Then, y_{i+1}^k is a vertex of $F_A \cap \mathcal{H}_+[u_1^{k+1}, u_2^{k+1}, \ldots, u_i^{k+1}, y_{i+1}^k, \ldots, y_n^k]$ with neighboring vertices $u_1^{k+1}, u_2^{k+1}, \ldots, u_i^{k+1}, u_{i+1}^{k+1}, y_{i+2}^k, \ldots, y_n^k$. ($u_{i+1}^{k+1}$ is actually the extension of a neighboring vertex to the boundary of the reverse convex constraint). So:

CUTTING PLANE ALGORITHM 161

$$\{y_{i+1}^k\} + C[u_1^{k+1} - y_{i+1}^k, u_2^{k+1} - y_{i+1}^k, \ldots, u_i^{k+1} - y_{i+1}^k,$$
$$u_{i+1}^k - y_1^k, y_{i+2}^k - y_{i+1}^k, \ldots, y_n^k - y_1^k] \supseteq F_A \cap \mathcal{H}_+[u_1^{k+1},$$
$$\ldots, u_i^{k+1}, y_{i+1}^k, \ldots, y_n^k]$$
$$\supseteq F.$$

And we also have that $g(u_1^{k+1}) \geq 0, g(u_2^{k+1}) \geq 0, \ldots = g(u_i^{k+1}) \geq 0, g(y_{i+1}^k) \geq 0, \ldots, g(y_n^k) \geq 0$. Consequently, by the same argument as in proposition 2,

$$F_A \cap \mathcal{H}_+[u_1^{k+1}, u_2^{k+1}, \ldots, u_{i+1}^{k+1}, y_{i+2}^k, \ldots, y_n^k] \supset F,$$

and

$$x \in F_A \cap \mathcal{H}[u_1^{k+1}, u_2^{k+1}, \ldots, u_{i+1}^{k+1}, y_{i+2}^k, \ldots, y_n^k] \implies x \in H[u_1^{k+1}, u_2^{k+1}, \ldots,$$
$$u_{i+1}^{k+1}, y_{i+2}^k, \ldots, y_n^k].$$

So, by induction over i:

$$F_A \cap \mathcal{H}_+[u_1^{k+1}, u_2^{k+1}, \ldots, u_{i+1}^{k+1}, \ldots, u_n^{k+1}] \supset F,$$

and

$$x \in F_A \cap \mathcal{H}[u_1^{k+1}, u_2^{k+1}, \ldots, u_{i+1}^{k+1}, \ldots, u_n^{k+1}] \implies x \in H[u_1^{k+1}, u_2^k,$$
$$\ldots, u_{i+1}^{k+1}, \ldots, u_n^{k+1}].$$

i.e. the proposition is true for $k + 1$. $\square$

Theorem 1 *The algorithm outlined above converges in a finite number of steps to a global optimum of this reverse convex class of optimization problems.*

Proof:
At each iteration of the algorithm, the following monotonicity property holds:

$$F_A \cap \mathcal{H}_+^{k+1} \subset F_A \cap \mathcal{H}_+^k$$

This is true since:

$$F_A \cap \mathcal{H}_+[u_1^{k+1}, \ldots, u_n^{k+1}] \subseteq F_A \cap (\cup_{i=1}^n \{y_i^k\} + C[y_1^k - y_i^k, \ldots, y_{i-1}^k - y_i^k,$$
$$u_i^k - y_i^k, y_{i+1}^k - y_i^k, \ldots, y_n^k - y_i^k])$$

and:

$$F_A \cap (\cup_{i=1}^n \{y_i^k\} + C[y_1^k - y_i^k, \ldots, y_{i-1}^k - y_i^k,$$
$$u_i^k - y_i^k, y_{i+1}^k - y_i^k, \ldots, y_n^k - y_i^k]) \subseteq F_A \cap \mathcal{H}_+[y_1^k, \ldots, y_n^k]$$

but,

$$\mathcal{H}_+[y_1^k, \ldots, y_n^k] = \mathcal{H}_+[u_1^k, \ldots, u_n^k]$$

Therefore:
$$F_A \cap \mathcal{H}_+^k \supset F_A \cap \mathcal{H}_+^{k+1}$$

This, together with the previous proposition shows that the algorithm is generating tighter and tighter supersets of the feasible region:
$$F_A \cap \mathcal{H}_+^k \supset F_A \cap \mathcal{H}_+^{k+1} \supset F$$

Consider the sequence $\{y_1^k\}_k$ generated by the algorithm. Let $E_k = [a_k, b_k]$ be the edge of F_A picked by the algorithm in step (IV) and such that:
$$y_1^k \in E_k$$

a_k, b_k denote the two end-points of E_k. By convention, if $g(a_k)g(b_k) \leq 0$, we will assume that $g(b_k) \geq 0$.

We would like to show that, as long as the algorithm has not converged, the newly generated point y_1^{k+1} cannot belong to the collection of edges of F_A on which the previous $y_1^i, i = 1, \ldots, k$ lie; i.e. we would like to show that:
$$y_1^{k+1} \notin \cup_{i=1}^k [a_i, b_i]$$

By contradiction, let us assume that $y_1^{k+1} \in \cup_{i=1}^k [a_i, b_i]$. More particularly, let us assume that for some $i \in \{1, \ldots, k\}$
$$y_1^{k+1} \in [a_i, b_i]$$

Since $i \leq k$, we can use the monotonicity property shown above and state that:
$$F_A \cap \mathcal{H}_+^{i+1} \supset F_A \cap \mathcal{H}_+^{k+1}$$
$$\Rightarrow y_1^{k+1} \in F_A \cap \mathcal{H}_+^{i+1}$$

Two cases can arise:

Case–a $u_1^i \notin [a_i, b_i]$

So $[a_i, b_i] \subseteq (\mathcal{H}_+^{i+1})^c$

$\Rightarrow y_1^{k+1} \notin [a_i, b_i]$

Case–b $u_1^i \in [a_i, b_i]$

So $[a_i, u_1^i] \subseteq \mathcal{H}_+^{i+1}$

And $[u_1^i, b_i] \subseteq (\mathcal{H}_+^{i+1})^c$

So $y_1^{k+1} \in [a_i, u_1^i]$

But we have that $g(b_i) \geq 0$ and $g(u_1^i) = 0$.

Therefore, by concavity of g, $g(a_i) \leq 0$ and $g(x) \leq 0$ for any $x \in [a_i, u_1^i]$. In particular, $g(y_1^{k+1}) \leq 0$. This contradicts the fact that the algorithm has not stopped.

□

5 Conclusion

The algorithm presented here covers a wide variety of nonlinear/0-1 integer optimization problems and is motivated by the observation that we can use the level sets of the reverse convex constraint and the geometry of the polytope involved to obtain a "deeper" cut than the Tuy-cut. This cut does not cut away any feasible point and is constructed in such a way that, for most iterations, it will eliminate n edges of the original polytope. This hints that this algorithm might be quite efficient in providing global optimal solutions for this NP-hard class of problems. Numerical computations are underway to test the algorithm and will be reported subsequently. Future work also includes generalizing this cutting plane algorithm to the degenerate case.

References

[1] Jr. A. F. Veinott. Minimum concave cost solution leontiev substitution models of multi-facility inventory systems. *Operations Research*, 17:262–291.

[2] P. P. Bansal and S. E. Jacobsen. An algorithm for optimizing network flow capacity under economies of scale. *Journal of Optimization Theory and Applications*, 15(5), 1975.

[3] S. BenSaad and S. E. Jacobsen. A level set algorithm for a class of reverse convex programs. *Annals of Operations Research*, 27, 1990.

[4] M. Hamami and S. E. Jacobsen. Exhaustive non-degenerate conical processes for concave minimization on convex polytopes. *Math. Oper. Res.*, 13(3):479–487, 1988.

[5] R. J. Hillestad and Jacobsen S. E. Reverse convex programming. *Applied Mathematics and Optimization*, 6:63–78, 1980.

[6] R. J. Hillestad and S. E. Jacobsen. Linear programs with an additional reverse convex constraint. *Appl. Math. Optim.*, 1980.

[7] R.J. Hillestad. Optimization problems subject to a budget constraint with economies of scale. *Operations Research*, 23(6):1091–1098, 1975.

[8] K. L. Hoffman. A method for globally minimizing concave functions over convex sets. *Math. Prog.*, 20:22–32, 1981.

[9] R. Meyer. The validity of a family of optimization methods. *SIAM J. Control*, 8:41–54, 1970.

[10] P. M. Pardalos and J. B. Rosen. Methods for global concave minimization: A bibliographic survey. *SIAM Review*, 1986.

[11] J. B. Rosen. Global minimization of a linearly constrained concave function by partition of feasible domain. *Mathematics of Operations Research*, 8(2):215–230, May 1983.

[12] S. Sen and A. Whiteson. A cone splitting algorithm for reverse convex programming. *Proc. IEEE Conference on Systems, Man, and Cybernetics*, 1985.

[13] N. V. Thoai and H. Tuy. Convergent algorithms for minimizing a concave function. *Math. of Oper. Res.*, 5(4):556–566, 1980.

Global Optimization of Problems with Polynomial Functions in One Variable

V. Visweswaran[1]   C.A. Floudas[1,2]

Abstract

In Floudas and Visweswaran (1990, 1991), a new global optimization algorithm (**GOP**) was proposed for solving constrained nonconvex problems involving quadratic and polynomial functions in the objective function and/or constraints. In this paper, the application of this algorithm to the special case of polynomial functions of one variable is discussed. The special nature of polynomial functions enables considerable simplification of the **GOP** algorithm. The primal problem is shown to reduce to a simple function evaluation, while the *relaxed dual* problem is equivalent to the simultaneous solution of two linear equations in two variables. In addition, the one-to-one correspondence between the x and y variables in the problem enables the iterative improvement of the bounds used in the relaxed dual problem. The simplified approach is illustrated through a simple example that shows the significant improvement in the underestimating function obtained from the application of the modified algorithm. The application of the algorithm to several unconstrained and constrained polynomial function problems is demonstrated. Comparisons of the algorithm with recent approaches for *d.c. programming* problems are also provided.

Keywords : Global Optimization, Polynomial functions, Unconstrained and Constrained Optimization, The GOP algorithm.

1 Introduction

Polynomial functions of one variable occur frequently in mathematical programming problems. Problems involving the unconstrained or constrained optimization of these functions are interesting not only because of the inherent simplicity of the problem structure, but also because these functions form the backbone of larger optimization problems involving more variables. Often, the solution of these larger problems becomes much easier if a few of the variables are fixed. Consequently, they can be viewed as parametric problems in one variable. The solution of optimization problems involving one (or a few) variable(s) can often provide significant insight into the nature of larger problems.

---

[1] Chemical Engineering Department, Princeton University, Princeton, N.J. 08544
[2] Author to whom all correspondence should be addressed.

The unconstrained minimization of Lipschitz continuous functions (of which polynomial functions are a subset) has been studied extensively in the past two decades. Algorithms for solving this problem have been proposed by Evtushenko (1971), Piyavskii (1972), and Timonov (1977), among others. Shen and Zhu (1987) proposed an interval version of Schubert's algorithm for univariate functions. Galperin (1987) and Pinter(1988) also considered the incorporation of constraints in the problem. Hansen (1979) proposed an algorithm for minimizing univariate functions using interval analysis. A comprehensive review of global optimization of univariate Lipschitz functions (including functions other than polynomials) is given in Hansen *et al* (1989a). They provide the necessary conditions for finite convergence of algorithms addressing this problem and the characteristic that a best possible algorithm should have. An extensive comparison of the computational aspects of these algorithms as well as new improved algorithms are provided in Hansen *et al* (1989b). Wingo (1985) proposed a method for locally approximating the polynomial function to enable the solution without evaluating derivatives. However, the algorithm fails to identify the global solution in some cases. Dixon (1990) proposed several methods for accelerating the search procedure using interval methods to locate the global solution for functions of one variable. Horst *et al* (1991a) and Horst *et al* (1991b) have proposed algorithms for solving *D. C. programming problems*.

Floudas and Visweswaran (1990, 1991) proposed a deterministic approach for global optimization of problems involving quadratic and polynomial functions in the objective function and/or constraints. They made use of primal-dual decomposition to solve the originally nonconvex problem through a series of primal and *relaxed dual* subproblems. The algorithm (**GOP**) was shown to have finite convergence to an ϵ-global minimum of the problem. The algorithm was applied to several classes of problems (Visweswaran and Floudas, 1990a) including polynomial function problems. Visweswaran and Floudas (1990b) presented new properties that exploit the structure of the Lagrange function and showed that they enhance the computational efficiency of the **GOP** algorithm when applied to problems with quadratic terms in the objective function and/or constraints.

In this paper, the application of the **GOP** algorithm to the special case of polynomial functions in one variable is discussed. Use is made of the new properties presented in Visweswaran and Floudas (1990b) as well as the structure of polynomial functions to reduce the **GOP** algorithm to an extremely simple form of application. The modified algorithm is also shown to be applicable to constrained optimization problems with polynomial functions in one variable. The improvements over the original **GOP** algorithm are illustrated both computationally and geometrically through the use of a simple example. In addition, several examples of unconstrained and constrained problems help to highlight the effectiveness of the proposed algorithm. Finally, two examples of *d.c. programming problems* are provided as a means of comparison of the modified **GOP** algorithm with an algorithm proposed by Horst *et al* (1991a).

2 Problem Statement

In this paper, the application of the **GOP** algorithm to optimization problems involving polynomial functions of one variable in the objective function and/or constraints is presented. These problems have the following form :

$$\min \quad F(y) = a_0 + a_1 y + a_2 y^2 + \ldots\ldots + a_N y^N$$

$$A_{j0} + A_{j1}y + A_{j2}y^2 + \ldots + A_{jN}y^N \leq 0 \quad \forall\, j = 1, 2, ..., J \quad (1)$$
$$B_{m0} + B_{m1}y + B_{m2}y^2 + \ldots + B_{mN}y^N = 0 \quad \forall\, m = 1, 2, ..., M$$
$$y^L \leq y \leq y^U ,$$

where y is a *single* variable and A_{ji}, A_{mi} are the coefficients of y^i in the jth inequality and mth equality constraint respectively. The nonconvexities in this problem arise due to the existence of polynomial terms in either the objective function or the set of constraints. It is assumed that the polynomials have nonconvex terms right up to the Nth degree term.

Consider the following transformations:

$$\begin{aligned}
x_0 &= 1 \\
x_1 &= y \\
x_2 &= y^2 = (x_1)y \\
x_3 &= y^3 = (x_2)y \\
&\vdots \\
x_N &= y^N = (x_{N-1})y.
\end{aligned}$$

Then, the problem can be written in the following equivalent form:

$$\min_{x,y} \sum_{i=0}^{N} a_i x_i$$

$$\sum_{i=0}^{N} A_{ji} x_i \leq 0 \quad j = 1, 2, ..., J \quad (2)$$

$$\sum_{i=0}^{N} B_{mi} x_i = 0 \quad m = 1, 2, ..., M$$

$$x_i - x_{i-1} y = 0 \quad i = 1, 2, ..., N$$

where $x_0 = 1$.

The case of unconstrained optimization problems is considered in the following section. The theoretical development is extended to the case of constrained problems in Section 7.

3 Unconstrained Problems

For the case of unconstrained optimization, problem (2) can be simplified to the following form :

$$\min_{x,y} \sum_{i=0}^{N} a_i x_i$$

$$x_i - x_{i-1} y = 0 \quad i = 1, 2, ..., N, \qquad (3)$$

where $x_0 = 1$. For a fixed $y = y^K$, the primal problem can be written as

$$\min_{x} \sum_{i=0}^{N} a_i x_i$$

$$\text{s.t.} \quad x_i - x_{i-1} y^K = 0 \quad i = 1, 2, ..., N \qquad (4)$$

with $x_0 = 1$.

Note that for any fixed value of $y = y^k$, all the x variables are uniquely determined. Therefore, this problem is simply one of function evaluation, with the solution being as follows :

$$\begin{aligned} x_0 &= 1 \\ x_1 &= y^K \\ x_2 &= (y^K)^2 \\ x_3 &= (y^K)^3 \\ &\vdots \\ x_N &= (y^K)^N \\ F(y^K) &= \sum_{i=0}^{N} a_i (y^K)^i. \end{aligned}$$

The KKT gradient conditions for the primal problem (4), for a fixed $y = y^K$, can be written as :

$$\nabla_{x_i} L(x, y^K, \nu^K) = a_i + \nu_i^K - \nu_{i+1}^K y^K = 0 \quad \forall \ i = 1, 2, ..., N$$

where ν_i^K are the Lagrange multipliers for the new equality constraints introduced, with $\nu_0^K = \nu_{N+1}^K = 0$. Here, the Lagrange multipliers can be found by backward substitution:

$$\begin{aligned}
\nu_N^K &= -a_N \\
\nu_{N-1}^K &= \nu_N^K y^K - a_{N-1} &= -a_N(y^K) - a_{N-1} \\
\nu_{N-2}^K &= \nu_{N-1}^K y^K - a_{N-2} &= -a_N(y^K)^2 - a_{N-1}(y^K) - a_{N-2} \\
&\vdots \\
\nu_1^K &= \nu_2^K y^K - a_1^K &= -a_N(y^K)^{N-1} \cdots - a_2(y^K) - a_1.
\end{aligned}$$

The Lagrange function formulated from the primal problem (4) can be written as

$$L(x, y, \nu^K) = \sum_{i=0}^{N} a_i x_i + \sum_{i=1}^{N} \nu_i^K (x_i - x_{i-1} y).$$

Separating the terms in x, this can be rewritten as

$$L(x, y, \nu^K) = \sum_{i=0}^{N} \left[a_i + \nu_i^K - \nu_{i+1}^K y \right] x_i. \tag{5}$$

Using the KKT gradient conditions, the Lagrange function becomes

$$L(x, y, \nu^K) = \sum_{i=0}^{N} \nu_{i+1}^K (y^K - y) x_i.$$

Thus, the *qualifying* constraint for each x_i is of the form

$$y^K - y \geq 0, \; or \; y^K - y \leq 0,$$

depending on whether ν_{i+1}^k is greater than or less than zero respectively. Therefore, two relaxed dual problems are solved at every iteration. These two problems are separately considered below.

First Relaxed Dual Problem

Consider the relaxed dual problem solved for the region $y^K - y \geq 0$. Before solving the relaxed dual problems, the Lagrange functions from previous iterations are selected. From the kth iteration ($k = 1, 2, ..., K-1$), the two Lagrange functions have *qualifying* constraints of the form $y^k - y \geq 0$ and $y^k - y \leq 0$. These Lagrange functions are selected on the basis of satisfaction of their *qualifying* constraints at $y = y^K$. Since $y^k - y^K$ must be either positive or negative, exactly one Lagrange function from every iteration will be present for the current relaxed dual problems.

Now, the previous iterations correspond to values of $y = y^1, \ldots, y^{K-1}$. Some of these fixed values of y will be less than y^K. Suppose that

$$y^L = \max_{j=1,\ldots,K-1} \{y^j : y^j < y^K\}.$$

Then, the Lagrange function formulated from the Lth iteration (for which the fixed value of y is y^L) has the *qualifying* constraint $y^L - y \leq 0$. Therefore, for the current relaxed dual problem, the lower bound on the y variable is y^L. At the same time, the *qualifying* constraint from the current iteration ensures that the upper bound on y is y^K. Thus, for the current relaxed dual problem, $y^L \leq y \leq y^K$.

Now, the x variables are related to y through the equivalence relations

$$\begin{aligned} x_0 &= 1 \\ x_1 &= y \\ x_2 &= y^2 = (x_1)y \\ &\vdots \\ x_N &= y^N = (x_{N-1})y. \end{aligned}$$

Therefore, the bounds on the x variables can be changed to suit the new bounds on y for the current relaxed dual problem. This is done in the following manner:

$$\begin{aligned} x_1^L &= y^L, & x_1^U &= y^K \\ x_2^L &= MIN[(y^L)^2, (y^K)^2], & x_2^U &= MAX[(y^L)^2, (y^K)^2] \\ x_3^L &= MIN[(y^L)^3, (y^K)^3], & x_3^U &= MAX[(y^L)^3, (y^K)^3] \\ &\vdots & &\vdots \\ x_N^L &= MIN[(y^L)^N, (y^K)^N], & x_N^U &= MAX[(y^L)^N, (y^K)^N]. \end{aligned}$$

If $y^L < 0$ and $y^K > 0$, then $x_2^L = x_4^L = x_6^L \ldots = 0$.

The reason for the use of these expressions is as follows. If $y^L \geq 0$, then the lower and upper bounds on $x_1, x_2, \ldots x_N$ are simply given by

$$\begin{aligned} x^L &= (y^L, (y^L)^2, \ldots (y^L)^N), \text{ and} \\ x^U &= (y^K, (y^K)^2, \ldots (y^K)^N). \end{aligned}$$

However, if $y^L < 0$ and $y^K > 0$, then the value of y can be either positive or negative. Therefore, any even power of y can be as low as zero. Hence, $x_2^L = x_4^L \ldots = 0$. Furthermore, $|y^L|$ can be either greater than or less than $|y^K|$. Therefore, for each of the powers of y, the minimum and maximum values are given by the minimum and maximum values of the corresponding terms in *both* y^L and y^K.

These new bounds for the x variables are then used in the Lagrange functions in the following form:

$$\begin{aligned} If \quad \nu_{i+1} &\geq 0, \quad then \quad x_i^B = x_i^L. \\ If \quad \nu_{i+1} &< 0, \quad then \quad x_i^B = x_i^U. \end{aligned}$$

It should be noted here that due to the nature of the transformation variables that are introduced into the problem, the Lagrange function is linear in x for every fixed y. Therefore, there is no need to linearize the Lagrange function with respect to x (Visweswaran and Floudas, 1990).

Now, there are $(K - 1)$ Lagrange functions from previous iterations that can be used for the current relaxed dual problem. However, it is sufficient to consider only one of these Lagrange functions, since the omission of the remaining constraints does not destroy the validity of the relaxed dual problem as a lower bound on the global solution. The obvious choice for this is to use the Lagrange function corresponding to the nearest point on the left side, that is, the Lagrange function corresponding to y^L, and ignoring all other Lagrange functions from previous iterations. The current relaxed dual problem then has the following form:

$$\min_{y, \mu_B} \mu_B$$

s.t.

$$\begin{aligned} \mu_B &\geq L(x^{B_L}, y, \nu^L) \\ y &\geq y^L \\ \mu_B &\geq L(x^{B_K}, y, \nu^K) \\ y &\leq y^K \end{aligned}$$

where B_L and B_K are two combinations of bounds of the x variables that are being used in the Lagrange functions from the Lth and Kth iterations, respectively.

Again, due to the nature of the transformations, the two constraints are linear in y. Therefore, it is clear that the solution to this problem will be either at the intersection of the two constraints, or at one of the two bounds før y, namely, y^L or y^K. Consider the three cases.

(i) The solution of the current relaxed dual problem lies at y^L. In this case, the value of the first Lagrange function (that is, the one formulated from y^L) must equal the value of the objective function at y^L. This arises as a result of the strong duality theorem. This means that due to the presence of the first constraint, the value of μ_B must equal the value of the objective function at y^L. Therefore, it is not necessary to consider this solution.

(i) The solution of the current relaxed dual problem lies at y^K. In this case, the value of the second Lagrange function (that is, the one formulated from y^K) must equal the value of the objective function at y^K. This arises as a result of the strong duality theorem. This means that due to the presence of the second constraint, the value of μ_B must equal the value of the objective function at y^K. Therefore, again it is not necessary to consider this solution.

(iii) The solution of the current relaxed dual problem lies at the intersection of the two constraints. In this case, the solution cannot be omitted.

Thus, the relaxed dual problem can be solved easily by considering only the third possibility and solving for the intersection of the two constraints in the problem. This is very easy to do since the constraints are linear in y.

Second Relaxed Dual Problem

For the case when $y^K - y \leq 0$, the nearest point to the right side of y^K becomes the upper bound for y, while the value of y^K becomes the lower bound for y. That is, for this case, $y^K \leq y \leq y^R$, where now y^R is found as

$$y^R = \min_{j=1,\ldots,K-1} \{y^j : y^j > y^K\}.$$

Similarly, the bounds for the x variables can be found as follows:

$$\begin{aligned}
x_1^L &= y^K, & x_1^U &= y^R \\
x_2^L &= MIN[(y^K)^2, (y^R)^2], & x_2^U &= MAX[(y^K)^2, (y^R)^2] \\
x_3^L &= MIN[(y^K)^3, (y^R)^3], & x_3^U &= MAX[(y^K)^3, (y^R)^3] \\
&\vdots & &\vdots \\
x_N^L &= MIN[(y^K)^N, (y^R)^N], & x_N^U &= MAX[(y^K)^N, (y^R)^N].
\end{aligned}$$

If $y^K < 0$, then $x_2^L = x_4^L = x_6^L \ldots = 0$.

In this case, the relaxed dual problem is solved by considering the current Lagrange function, that is, the Lagrange function formulated from the current iteration corresponding to $y^K - y \leq 0$, and the Lagrange function from the Rth iteration corresponding to $y^R - y \geq 0$. Again, the relaxed dual problem can be solved simply by considering the intersection of the two Lagrange functions and comparing the corresponding value of μ_B to the values of the objective function at y^K and y^R.

4 The Improved GOP Algorithm

The modified algorithm for minimizing unconstrained polynomial functions in one variable can be stated in the following steps:

STEP 0- *Initialization of parameters:*

Define the storage parameters $\mu_B^{stor^1}(K^{max})$, $\mu_B^{stor^2}(K^{max})$, $y^{stor^1}(K^{max})$, $y^{stor^2}(K^{max})$, and $y^k(K^{max})$ over the maximum expected number of iterations K^{max}. Define P^{UBD} and R^{LBD} as the upper and lower bounds obtained from the primal and relaxed dual problems respectively. Also define the parameters y^{LEFT} and y^{RIGHT} and initialize them to the original lower and upper bounds y^L, y^U on y. Set

$$\mu_B^{stor^1}(K^{max}) = U, \mu_B^{stor^2}(K^{max}) = U$$

$$P^{UBD} = U, \text{ and } R^{LBD} = L.$$

where U is a very large *positive* number and L is a very large *negative* number. Define the logical variables LRD and RRD (for the left and right relaxed dual problems, respectively). Select a starting point y^1 for the algorithm. Set the counter K equal to 1. Select a convergence tolerance parameter ϵ.

STEP 1- *Primal problem:*

Store the value of y^K. Calculate the solution of the problem as follows:

$$\begin{aligned} x_0 &= 1 \\ x_1 &= (y^K) \\ x_2 &= (y^K)^2 \\ x_3 &= (y^K)^3 \\ &\vdots \\ x_N &= (y^K)^N \\ F(y^K) &= \sum_{i=0}^{N} a_i (y^K)^i . \end{aligned}$$

Also find the Lagrange multipliers for the problem as follows:

$$\begin{aligned} \nu_N^K &= -a_N \\ \nu_{N-1}^K &= \nu_N^K y^K - a_{N-1} &= -a_N(y^K) - a_{N-1} \\ \nu_{N-2}^K &= \nu_{N-1}^K y^K - a_{N-2} &= -a_N(y^K)^2 - a_{N-1}(y^K) - a_{N-2} \\ &\vdots \\ \nu_1^K &= \nu_2^K y^K - a_1^K &= -a_N(y^K)^{N-1} \cdots - a_2(y^K) - a_1 . \end{aligned}$$

Store the lagrange multipliers ν^K. Update the upper bound so that

$$P^{UBD} = MIN(P^{UBD}, F(y^K)).$$

STEP 2- *Determination of nearest points from the previous values of y:*

Set $y^{LEFT} = y^L, y^{RIGHT} = y^U$. Set $LRD = YES$, $RRD = YES$. If $y^K = y^L$, then set $LRD = NO$. If $y^K = y^U$, then set $RRD = NO$.

(a) If $K = 1$, then set $y^{LEFT} = y^L$, $y^{RIGHT} = y^U$. Go to Step 3.

(b) If $K = 2$, then y^K is either the lower bound y^L or the upper bound y^U.

  (i) If $y^2 = y^L$, then set $y^{LEFT} = y^L$, $y^{RIGHT} = y^1$. Go to Step 3.
  (ii) If $y^2 = y^U$, then set $y^{LEFT} = y^1$, $y^{RIGHT} = y^U$. Set Go to Step 3.

(c) If $K > 2$, then then do the following steps for $k = 1, 2, \ldots, K - 1$.

  (i) If $y^k < y^K$, then set $y^{LEFT} = MAX(y^k, y^{LEFT})$, $L = k$
  (ii) If $y^k > y^K$, then set $y^{RIGHT} = MIN(y^k, y^{RIGHT})$, $U = k$

STEP 3– *First Relaxed dual problem (i.e. for $y^K - y \geq 0$)* :

Updating bounds on x:
 Reset the bounds on the x variables as follows :

$$\begin{array}{llll}
x_1^L &= y^{LEFT}, & x_1^U &= y^K \\
x_2^L &= (y^{LEFT})^2, & x_2^U &= (y^K)^2 \\
x_3^L &= (y^{LEFT})^3, & x_3^U &= (y^K)^3 \\
\vdots & \vdots & \vdots & \vdots \\
x_N^L &= (y^{LEFT})^N, & x_N^U &= (y^K)^N.
\end{array}$$

Formulating the Lagrange functions

(i) For the Lagrange function from the Lth iteration (i.e. the Lagrange function from the Lth constraint with the *qualifying* constraint $y^L - y \leq 0$) :

$$\left.\begin{array}{ll} \text{If} & v_{i+1}^L \geq 0, \text{ then } x_i^{B_L} = x_i^U \\ \text{If} & v_{i+1}^L < 0, \text{ then } x_i^{B_L} = x_i^L \end{array}\right\} \forall\, i = 1, 2, \ldots, N - 1.$$

(ii) For the Lagrange function from the Kth iteration (i.e. the Lagrange function from the current iteration with the *qualifying* constraint $y^K - y \geq 0$) :

$$\left.\begin{array}{ll} \text{If} & v_{i+1}^K \geq 0, \text{ then } x_i^{B_K} = x_i^L \\ \text{If} & v_{i+1}^K < 0, \text{ then } x_i^{B_K} = x_i^U \end{array}\right\} \forall\, i = 1, 2, \ldots, N - 1.$$

POLYNOMIAL FUNCTIONS IN ONE VARIABLE

Solving the Relaxed dual problem
Find the intersection of the following two constraints :

$$\mu_B = L(x^{B_L}, y, \nu^L), \text{ and}$$
$$\mu_B = L(x^{B_K}, y, \nu^K).$$

If $\mu_B < P^{UBD}$, then store the solution of the problem. That is, set

$$\mu_B^{stor^1}(K) = \mu_B^{int}, \quad \text{and} \quad y^{stor^1}(K) = y^{int},$$

where *int* denotes the value of the variable obtained by the intersection.

STEP 4– *Second Relaxed dual problem: (i.e. for $y^K - y \leq 0$) :*

Updating bounds on x:
Reset the bounds on the x variables as follows :

$$\begin{aligned}
x_1^L &= y^K, & x_1^U &= y^{RIGHT} \\
x_2^L &= (y^K)^2, & x_2^U &= (y^{RIGHT})^2 \\
x_3^L &= (y^K)^3, & x_3^U &= (y^{RIGHT})^3 \\
&\vdots & &\vdots \\
x_N^L &= (y^K)^N, & x_N^U &= (y^{RIGHT})^N.
\end{aligned}$$

Formulating the Lagrange functions

(i) For the Lagrange function from the Uth iteration :

$$\left. \begin{array}{l} If \;\; \nu_{i+1}^U \geq 0, \;\; then \;\; x_i^{B_U} = x_i^L. \\ If \;\; \nu_{i+1}^L < 0, \;\; then \;\; x_i^{B_U} = x_i^U. \end{array} \right\} \;\; \forall \; i = 1, 2, ..., N-1 \;.$$

(ii) For the Lagrange function from the Kth iteration :

$$\left. \begin{array}{l} If \;\; \nu_{i+1}^K \geq 0, \;\; then \;\; x_i^{B_K} = x_i^U. \\ If \;\; \nu_{i+1}^K < 0, \;\; then \;\; x_i^{B_K} = x_i^L. \end{array} \right\} \;\; \forall \; i = 1, 2, ..., N-1 \;.$$

Solving the Relaxed dual problem
Find the intersection of the following two constraints :

$$\mu_B = L(x^{B_U}, y, \nu^U), \text{ and}$$
$$\mu_B = L(x^{B_K}, y, \nu^K).$$

If $\mu_B int < P^{UBD}$, then store the solution of the problem. That is, set

$$\mu_B^{stor^2}(K) = \mu_B^{int}, \quad \text{and} \quad y^{stor^2}(K) = y^{int}.$$

where again *int* denotes the value of the variable obtained by the intersection.

STEP 5- *Selecting a new lower bound and* y^{K+1}:
From the stored sets $\mu_B^{stor^1}$ and $\mu_B^{stor^2}$, select the minimum μ_B^{min} (including the solutions from the current iteration). Also, select the corresponding stored value of y as y^{min}. Set $R^{LBD} = \mu_B^{min}$, and $y^{K+1} = y^{min}$. Delete μ_B^{min} and y^{min} from the stored set.

STEP 6- *Check for convergence:*
Check if $R^{LBD} > P^{UBD} - \epsilon$. IF yes, STOP. Else, set $K = K+1$ and return to step 1.

Remark 1 : Since the primal problem in Step 1 is solved for a fixed value of y, the problem is just a function evaluation at $y = y^K$.

Remark 2 : Each of the two relaxed dual problems solved (in Steps 3 and 4) are solved by calculation of the intersection of two linear equality constraints in two variables and the comparison of the resulting solution for μ_B with the values of the objective function at the two relevant bounds of y.

5 An Illustrating Example

Consider the following unconstrained optimization problem :

$$\min_y -y^3 + 4.5y^2 - 6y$$

$$\text{s.t.} \quad 0 \leq y \leq 3 .$$

Thus, $a_0 = 0, a_1 = -6, a_2 = 4.5$, and $a_3 = -1$. The nonconvexity in this problem arises due to the presence of the term $-y^3$ in the objective function.

Introduction of three transformation variables x_1, x_2 and x_3 and their equivalence relationships to y results in the following equivalent form of the problem :

$$\min -6x_1 + 4.5x_2 - x_3$$

$$\begin{aligned}
\text{s.t.} \quad x_1 - y &= 0 \\
x_2 - x_1 y &= 0 \\
x_3 - x_2 y &= 0 \\
0 \leq y &\leq 3
\end{aligned}$$

POLYNOMIAL FUNCTIONS IN ONE VARIABLE

$$0 \leq x_1 \leq 3$$
$$0 \leq x_2 \leq 9$$
$$0 \leq x_3 \leq 27 .$$

Consider a starting point of $y^1 = 1$ for the **GOP** algorithm. The solution of the primal problem can be found as follows:

$$
\begin{aligned}
x_0 &= 1 \\
x_1 &= (y^1) = 1 \\
x_2 &= (y^1)^2 = 1 \\
x_3 &= (y^1)^3 = 1 \\
F(y^1) &= \sum_{i=0}^{3} a_i(y^1)^i = -2.5 \\
\nu_3^1 &= -a_3 = 1 \\
\nu_2^1 &= \nu_3^1 y^1 - a_2 = -3.5 \\
\nu_1^1 &= \nu_2^1 y^1 - a_1 = 2.5 .
\end{aligned}
$$

Since $K = 1$, and $y^1 = 1$, set $y^{LEFT} = y^L = 0$ and $y^{RIGHT} = y^U = 3$. Also, set $LRD = YES$ and $RRD = YES$.

The Lagrange function for the problem can be formulated as

$$L(x, y, \nu^1) = -3.5x_1(1-y) + x_2(1-y) - 2.5y .$$

First Relaxed dual problem (Solved for $1 - y \geq 0$)

According to Step 3 of the modified algorithm, set:

$$
\begin{aligned}
x_1^L &= y^{LEFT} = 0, & x_1^U &= y^1 = 1 \\
x_2^L &= (y^{LEFT})^2 = 0, & x_2^U &= (y^1)^2 = 1 \\
x_3^L &= (y^{LEFT})^3 = 0, & x_3^U &= (y^1)^3 = 1 .
\end{aligned}
$$

The proper bounds to be used for the x variables in the Lagrange function are then determined as follows:

$$\text{Since } \nu_2^1 < 0, \ x_1^{B_1} = x_1^U = 1.$$
$$\text{Since } \nu_3^1 > 0, \ x_2^{B_1} = x_2^L = 0.$$

Using these bounds for the x variables, the current Lagrange function becomes

$$L(x, y, \nu^1) = -3.5(1-y) - 2.5y = y - 3.5.$$

Since this is the first iteration, there are no previous Lagrange functions. Therefore, the solution to this problem will lie at $y = 0$, since the coefficient of y in the Lagrange function is positive. The objective value, that is, the value of μ_B, is -3.5 .

Second Relaxed dual problem (Solved for $1 - y \leq 0$)
Set
$$x_1^L = y^1 = 1, \quad x_1^U = y^{RIGHT} = 1,$$
$$x_2^L = (y^1)^2 = 1, \quad x_2^U = (y^{RIGHT})^2 = 9,$$
$$x_3^L = (y^1)^3 = 1, \quad x_3^U = (y^{RIGHT})^3 = 27.$$

The proper bounds to be used for the x variables in the Lagrange function are then determined as follows:

$$\text{Since } \nu_2^1 < 0, \quad x_1^{B_1} = x_1^L = 1.$$
$$\text{Since } \nu_3^1 > 0, \quad x_2^{B_1} = x_2^U = 9.$$

Using these bounds for the x variables, the current Lagrange function becomes

$$L(x, y, \nu^1) = -3.5(1-y) + 9(1-y) - 2.5y = -8y + 5.5.$$

The solution can be found by inspection. In this case, since the coefficient of y is negative, the solution lies at the upper bound, that is, $y = 3, \mu_B = -18.5$.

Thus, at the end of the first iteration, the lower bound from the two relaxed dual problems is -18.5. The value of y for the next iteration is 3.

It is interesting to compare the two solutions found in the first iteration to the solutions that would be found using the original bounds for the x variables. If the bounds for x_1, x_2 and x_3 were fixed to be $[0, 3]$, $[0, 9]$ and $[0, 27]$ (which are the original bounds for the x variables), then the solutions found are $y = 0, \mu_B = -10.5$ and $y = 3, \mu_B = -25.5$. Thus, using the improved bounds for x results in a tighter lower bound from the relaxed dual problems.

According to Step 5, the next y is selected as the one that corresponds to the minimum of the two relaxed dual problems, that is, $y^2 = 3$.

Second Iteration

The primal problem is solved for $y^2 = 3$. The solution to this problem is:

$$x_0 = 1$$
$$x_1 = y^2 = 3$$
$$x_2 = (y^2)^2 = 9$$
$$x_3 = (y^2)^3 = 27$$
$$F(y^2) = \sum_{i=0}^{3} a_i (y^1)^i = -4.5$$
$$\nu_3^2 = -a_3 = 1$$
$$\nu_2^2 = \nu_3^2 y^2 - a_2 = -1.5$$
$$\nu_1^2 = \nu_2^2 y^2 - a_1 = 1.5.$$

POLYNOMIAL FUNCTIONS IN ONE VARIABLE

Since $K = 2$, and $y^1 = 1 < y^2$, set $y^{LEFT} = y^1 = 1$ and $y^{RIGHT} = y^U = 3$. Also, since $y^2 = y^U$, set $LRD = YES$ and $RRD = NO$.

The Lagrange function for the problem can be formulated as

$$L(x, y, \nu^2) = -1.5x_1(3-y) + x_2(3-y) - 1.5y.$$

Since $RRD = NO$, only the relaxed dual problem for $y \leq 3$ needs to be solved.

<u>Relaxed dual problem</u> (Solved for $3 - y \geq 0$)

Set

$$\begin{aligned} x_1^L &= y^{LEFT} = 1, & x_1^U &= y^1 = 3, \\ x_2^L &= (y^{LEFT})^2 = 1, & x_2^U &= (y^1)^2 = 9, \\ x_3^L &= (y^{LEFT})^3 = 1, & x_3^U &= (y^1)^3 = 27. \end{aligned}$$

The proper bounds to be used for the x variables in the Lagrange function from the current iteration are then determined as follows:

$$\text{Since} \quad \nu_2^2 < 0, \quad x_1^{B_2} = x_1^U = 3.$$
$$\text{Since} \quad \nu_3^2 > 0, \quad x_2^{B_2} = x_2^L = 1.$$

Using these bounds for the x variables, the current Lagrange function becomes

$$L(x, y, \nu^2) = -4.5(3-y) + 1(3-y) - 1.5y = 2y - 10.5.$$

For the previous iteration, the Lagrange function has the following bounds for x:

$$\text{Since} \quad \nu_3^2 > 0, \quad x_2^{B_1} = x_2^U = 9.$$
$$\text{Since} \quad \nu_2^2 < 0, \quad x_1^{B_1} = x_1^L = 1.$$

From this, the Lagrange function from the first iteration becomes

$$L(x, y, \nu^1) = -3.5(1-y) + 9(1-y) - 2.5y = -8y + 5.5.$$

The solution to the current relaxed dual problem lies at the intersection of the two Lagrange functions, or at one of the two bounds for y in the current problem, that is, at either $y = 1$ or $y = 3$. If the solution lies at either of these two bounds, then it need not be considered. Therefore, the only case to be considered is when the solution lies at the intersection of the Lagrange functions. Here, the intersection lies at $y = 1.6$, $\mu_B = -7.3$. This is less than the objective function values at $y = 1$ or $y = 3$, so this is the solution of the current relaxed dual problem. Therefore, $R^{LBD} = -7.3$ and $y^3 = 1.6$.

Third Iteration

The primal problem is solved for $y^3 = 1.6$. The solution to this problem is:

$$x_0 = 1$$
$$x_1 = y^3 = 1.6$$
$$x_2 = (y^3)^2 = 2.56$$
$$x_3 = (y^3)^3 = 4.096$$
$$F(y^3) = \sum_{i=0}^{3} a_i (y^3)^i = -2.176$$
$$\nu_3^3 = -a_3 = 1$$
$$\nu_2^3 = \nu_3^3 y^3 - a_2 = -2.9$$
$$\nu_1^3 = \nu_2^3 y^3 - a_1 = 1.36.$$

Since $y^1 = 1 < y^3$, and $y^2 = 3 > y^3$, set $y^{LEFT} = y^1 = 1$ and $y^{RIGHT} = y^2 = 3$. Also, set $LRD = YES$ and $RRD = YES$.

The Lagrange function for the problem becomes

$$L(x, y, \nu^3) = -2.9x_1(1.6 - y) + x_2(1.6 - y) - 1.36y.$$

First Relaxed dual problem (Solved for $1.6 - y \geq 0$)
Set

$$x_1^L = y^{LEFT} = 1, \quad x_1^U = y^3 = 1.6,$$
$$x_2^L = (y^{LEFT})^2 = 1, \quad x_2^U = (y^3)^2 = 2.56,$$
$$x_3^L = (y^{LEFT})^3 = 1, \quad x_3^U = (y^3)^3 = 4.096.$$

The proper bounds to be used for the x variables in the Lagrange function are then determined as follows:

$$\text{Since } \nu_3^3 > 0, \quad x_2^{B_3} = x_2^L = 1.$$
$$\text{Since } \nu_2^3 < 0, \quad x_1^{B_3} = x_1^U = 1.6.$$

Using these bounds for the x variables, the current Lagrange function becomes

$$L(x, y, \nu^3) = 2.28y - 5.824.$$

Since $y^1 = 1 < y^3$ and the current relaxed dual problem is being solved for $y \leq 1.6$, the Lagrange function from the first iteration for $y \geq 1$ will be present. With the above bounds for x_1 and x_2, this Lagrange function becomes

$$L(x, y, \nu^1) = -3.5(1 - y) + 2.56(1 - y) - 2.5y = -1.56y - 0.94.$$

The intersection of the two Lagrange functions lies at $y = 1.2718$ and $\mu_B = -2.924$. Since the value of μ_B is greater than the best upper bound from the primal problems (-4.5), this solution need not be considered for future iterations.

Second Relaxed dual problem (Solved for $1.6 - y \leq 0$)
Set

$$\begin{aligned} x_1^L &= y^3 = 1.6, & x_1^U &= y^{RIGHT} = 3, \\ x_2^L &= (y^3)^2 = 2.56, & x_2^U &= (y^{RIGHT})^2 = 9, \\ x_3^L &= (y^3)^3 = 4.096, & x_3^U &= (y^{RIGHT})^3 = 27. \end{aligned}$$

The proper bounds to be used for the x variables in the Lagrange function are then determined as follows :

$$\begin{aligned} \text{Since} \quad \nu_2^3 &< 0, \quad x_1^{B_3} = x_1^L = 1.6. \\ \text{Since} \quad \nu_3^3 &> 0, \quad x_2^{B_3} = x_2^U = 9. \end{aligned}$$

Using these bounds for the x variables, the current Lagrange function becomes

$$L(x, y, \nu^3) = 6.976 - 5.72y.$$

Since $y^2 = 3 > y^3$ and the current relaxed dual problem is being solved for $y \geq 1.6$, the Lagrange function from the second iteration for $y \leq 3$ will be present. With the above bounds for x_1 and x_2, this Lagrange function becomes

$$L(x, y, \nu^2) = -4.5(3 - y) + 2.56(3 - y) - 1.5y = 0.44y - 5.82.$$

The solution to this relaxed dual problem lies at the intersection of the two Lagrange functions, which occurs at $y = 2.0772$ and $\mu_B = -4.9059$.

At the fourth iteration, the primal problem has a solution of -2.009. When the two relaxed dual problems are solved, they both have solutions greater than -4.5. Therefore, the algorithm converges at the end of the fourth iteration. In comparison to this, the original **GOP** algorithm takes 17 iterations to converge to the global solution.

6 Geometrical Interpretation

The application of the modified **GOP** algorithm to the example in the previous section can also be illustrated geometrically. Figure 1(a) shows the plot of the objective function $F(y)$ as a function of y. Since the problem is one of unconstrained minimization, is also the plot of the solutions of the primal problem as a function of y.

For a starting point of $y^1 = 1$, the sequence of points generated by the algorithm is graphically illustrated in Figures 1(b)-1(e). For the first iteration (Figure 1(b)), with an optimal value of -2.5 for the primal problem, the relaxed dual problems are solved for $y \leq 1$ and for $y \geq 1$. For the relaxed dual problem corresponding to $y \leq 1$, the bounds for x_1 and x_2 (which originally were [0,3] and [0,9] respectively) can be improved to [0,1] and [0,1] respectively. L_1^1 is the Lagrange function that results from using the improved bounds for the x variables, while P_1^1 is the Lagrange function obtained from using the original bounds for x_1 and x_2. Thus, the use of the improved bounds results in a tighter underestimator for the relaxed dual problem. Similarly, for the relaxed dual problem corresponding to $y \geq 1$, the bounds for x_1 and x_2 can be improved to [1,3] and [1,9] (instead of [0,3] and [0,9]). This results in a tighter Lagrange function (L_2^1) when the modified bounds are used as compared to the Lagrange function (P_2^1) that results from the use of the original bounds.

For the second iteration, $y^2 = 3$ for the primal problem, and the optimal solution is -4.5. For this iteration, only one relaxed dual problem needs to be solved, namely for $y \leq 3$. For this problem, the Lagrange function from the first iteration corresponding to $y \geq 1$ is present. Therefore, the bounds on x_1 and x_2 can be modified to [1,3] and [1,9] respectively. The Lagrange function that results from the use of the tighter bounds on the x variables, namely L_1^2, is a tighter underestimator of the objective function than the Lagrange function P_1^2 obtained from using the original bounds for x_1 and x_2. The solution of this relaxed dual problem is shown by point B on Figure 1(c). In contrast, the solution obtained by the original **GOP** algorithm at a similar juncture is shown by point A. As can be seen, the lower bound obtained from the modified algorithm is tighter than the lower bound from the original **GOP** algorithm. It should be noted that the original **GOP** algorithm and the modified **GOP** algorithm differ in the subsequent selections of y for the third and further iterations.

Figure 1(d) shows the relaxed dual problems solved at the third iteration, for which the corresponding primal problem has been soved for $y = 1.6$. For this iteration, the nearest points from previous iterations on the left and right sides are $y = 1$ and $y = 3$. Consider the relaxed dual problem solved for $y \leq 1.6$. The bounds on x_1 and x_2 can be improved to [1,1.6] and [1,2.56] respectively. This results in the Lagrange function L_1^3 formulated from the current iteration. It is interesting to note that due to the improvement of the bounds on the x variables results in an even tighter form of the Lagrange function from the first iteration. Originally, at the second relaxed dual problem in the first iteration, this Lagrange function had been formulated using the bounds of [1,3] and [1,9] for x_1 and x_2 respectively, and is represented by L_2^1 in Figure 1(d). Now, how-

POLYNOMIAL FUNCTIONS IN ONE VARIABLE

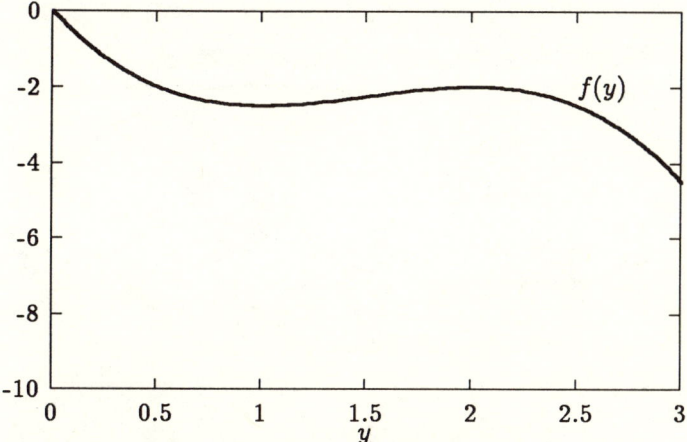

Figure 1(a) : Objective Function

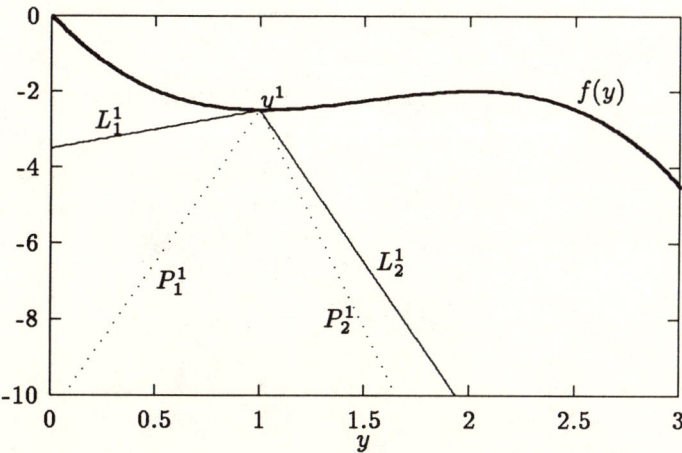

Figure 1(b) : Iteration 1 of the modified **GOP** algorithm

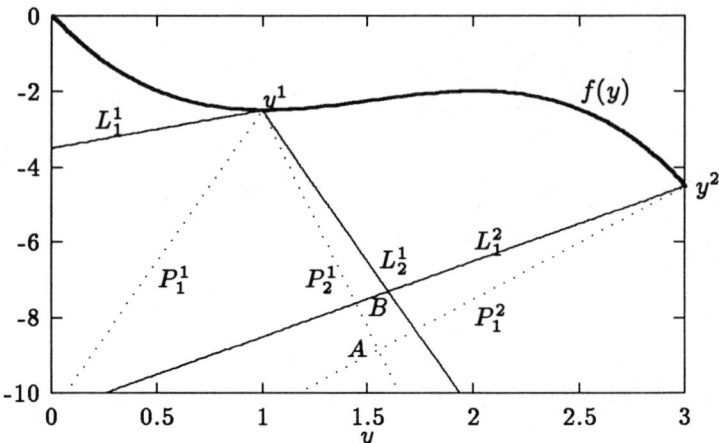

Figure 1(c) : Iteration 2 of the modified **GOP** algorithm

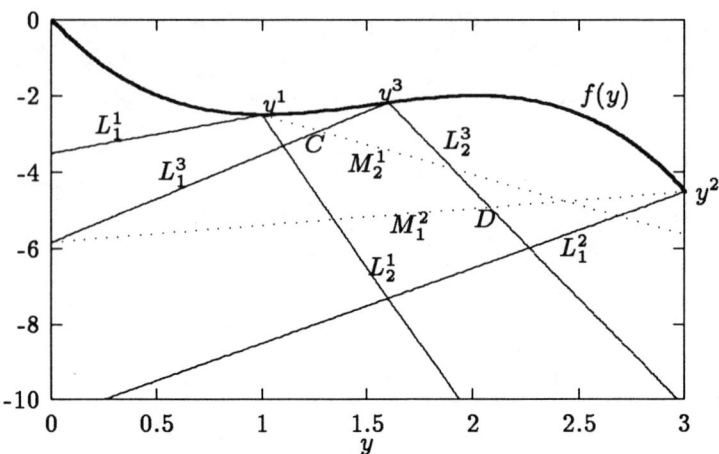

Figure 1(d) : Iteration 3 of the modified **GOP** algorithm

POLYNOMIAL FUNCTIONS IN ONE VARIABLE

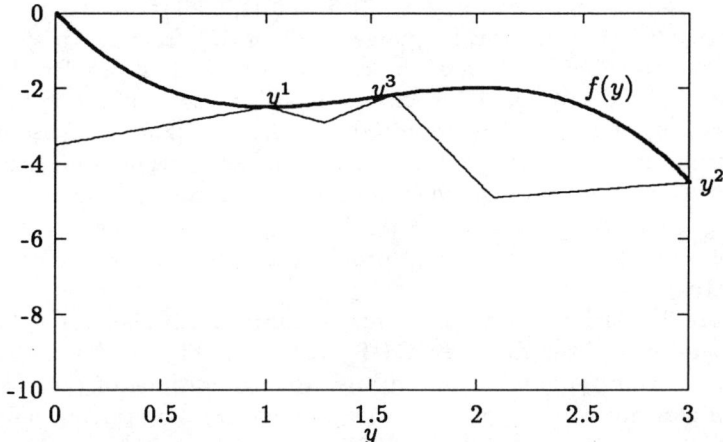

Figure 1(e) : Underestimating Function After 3 Iterations

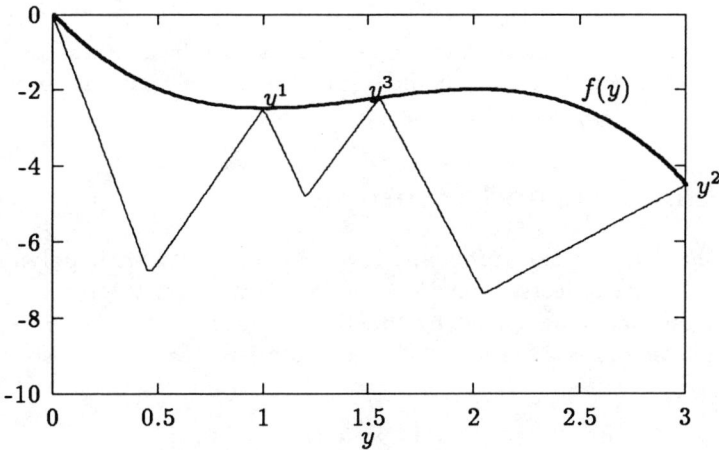

Figure 1(f) : Underestimating Function after 3 Iterations of the Original **GOP** Algorithm

ever, the bounds on x_1 and x_2 are [1,1.6] and [1,2.56] respectively. Using these bounds results in the Lagrange function M_2^1 from the first iteration. Therefore, the solution of the current relaxed dual problem lies at the point C. Similarly, for the relaxed dual problem solved for $y \geq 1.6$, the bounds on x_1 and x_2 can be improved to [1.6,3] and [2.56,9] respectively. This results in the Lagrange function L_2^3 from the current iteration. At the same time, the Lagrange function from the second iteration moves up from L_1^2 to M_1^2 due to the use of the tighter bounds on the x variables. The solution of this relaxed dual problem is shown by the point D on Figure 1(d).

Figure 1(e) shows the underestimating function that is obtained after three iterations of the modified **GOP** algorithm. Figure 1(f) shows the underestimating function obtained after three iterations of the original **GOP** algorithm when applied to this problem from a starting point of $y^1 = 1$. As can be seen, the modified algorithm provides a tighter underestimator as compared to the the one that is obtained by the original algorithm. Moreover, the modified algorithm eliminates whole regions of the problem for future iterations due to the tightness of the underestimator.

At the fourth iteration, the Lagrange functions from the second and third iteration move up so that the solution of the relaxed dual problems is -4.5, and the algorithm terminates.

7 Constrained Problems

When the optimization problems involve constraints with polynomial functions in the objective function and/or constraints, the primal problem can no longer be solved by function evaluation. In this case, the primal problem, for a fixed $y = y^K$, can be written as

$$\min_{x} \sum_{i=0}^{N} a_i x_i$$

$$\sum_{i=0}^{N} A_{ji} x_i \leq 0 \quad j = 1, 2, ..., J$$

$$\sum_{i=0}^{N} B_{mi} x_i = 0 \quad m = 1, 2, ..., M$$

$$x_i - x_{i-1} y^K = 0 \quad i = 1, 2, ..., N,$$

where $x_0 = 1$.

The solution of this problem is obtained by solving a linear programming problem in the x variables, and provides the multipliers used in formulating the Lagrange function as well as an upper bound on the global solution.

The KKT gradient conditions for this problem are

$$\nabla_{x_i} L(x, y^K, \lambda^K, \mu^K, \nu^K) = a_i + \sum_{j=1}^{J} \mu_j^K A_{ji} + \sum_{m=1}^{M} \lambda_m^K B_{mi}$$
$$+ \nu_i^K - \nu_{i+1}^K y^K$$
$$= 0$$

where λ^K and μ^K correspond to the original equality and inequality constraints, and ν^K corresponds to the new equality constraints introduced, with $\nu_0^K = \nu_{N+1}^K = 0$.

The Lagrange function for this problem is given by

$$L(x, y, \lambda^K, \mu^K, \nu^K) = \sum_{i=0}^{N} [a_i x_i + \sum_{j=1}^{J} \mu_j^K A_{ji} x_i$$
$$+ \sum_{m=1}^{M} \lambda_m^K B_{mi} x_i + \nu_i^K (x_i - x_{i-1} y)].$$

Separating the terms in x, this can also be written as

$$L(x, y, \lambda^K, \mu^K, \nu^K) = \sum_{i=0}^{N} \left[a_i + \sum_{j=1}^{J} A_{ji} \mu_j^K + \sum_{m=1}^{M} B_{mi} \lambda_m^K \right.$$
$$\left. + \nu_i^K - \nu_{i+1}^K y \right] x_i.$$

Using the KKT conditions, the Lagrange function can be written as

$$L(x, y, \lambda^K, \mu^K, \nu^K) = \sum_{i=0}^{N} \nu_{i+1}^K (y^K - y) x_i.$$

Thus, it can be seen that the Lagrange function formulated from the primal problem is identical to the one formulated in the case of unconstrained optimization problems. Therefore, Steps 2-6 of the improved **GOP** algorithm presented in Section 4 remain the same as for unconstrained optimization problems.

It is possible that for some values of $y = y^K$, the primal problem is infeasible. In this case, it is necessary to solve a relaxed primal problem involving the minimization of the sum of infeasibilities. This problem is shown below:

$$\min_x \sum_{j=1}^{J} \alpha_j + \sum_{m=1}^{M} (\beta_m + \gamma_m)$$

$$\sum_{i=0}^{N} A_{ji} x_i \leq \alpha_j \quad j = 1, 2, ..., J$$

$$\sum_{i=0}^{N} B_{mi} x_i = \beta_m - \gamma_m \quad m = 1, 2, ..., M$$

$$x_i - x_{i-1} y^K = 0 \quad i = 1, 2, ..., N.$$

The Lagrange function for this problem is given by

$$L(x, y, \lambda^K, \mu^K, \nu^K) = \sum_{j=1}^{J} \alpha_j + \sum_{m=1}^{M} (\beta_m + \gamma_m) + \sum_{j=1}^{J} \mu_j^K \sum_{i=1}^{N} A_{ji} x_i - \alpha_j$$
$$+ \sum_{m=1}^{M} \lambda_m^K (\sum_{i=1}^{N} B_{mi} x_i - \beta_m + \gamma_m)$$
$$+ \sum_{i=1}^{N} \nu_i^K (x_i - x_{i-1} y).$$

Using the KKT gradient conditions for the relaxed primal problem and separating the terms in x, this again be reduces to

$$L(x, y, \lambda^K, \mu^K, \nu^K) = \sum_{i=0}^{N} \nu_{i+1}^K (y^K - y) x_i.$$

In this case, however, the Lagrange function is added to the relaxed dual problem in the following form:

$$0 \geq L(x, y, \lambda^K, \mu^K, \nu^K).$$

This has the same form as the regular Lagrange function, except that μ_B has been replaced by 0. Steps 2-6 can again be applied towards solving relaxed dual problems in these iterations, with the above replacement being the only change.

8 Computational Experience

8.1 Unconstrained Problems

Example 1 :

This example is taken from Wingo (1985).

$$\min_y \; y^6 - \frac{52}{25}y^5 + \frac{39}{80}y^4 + \frac{71}{10}y^3 - \frac{79}{20}y^2 - y + \frac{1}{10}$$

$$-2 \le y \le 11$$

This function has a local minimum at 0, with a value of $\frac{1}{10}$. The best solution reported by Wingo (1985) is -23627.1758, occurring at $y = 11$. However, the global minimum of the function occurs at $y = 10$, with an objective value of -29763.233.

For this problem,

$$N = 6, \quad a = (\frac{1}{10}, -1, -\frac{79}{20}, \frac{71}{10}, \frac{39}{80}, -\frac{52}{25}, 1), \quad M = J = 0.$$

When the **(GOP)** was applied to this problem, it converged to the global solution of -29763.233 taking around 175 iterations from different starting points. When the improved **GOP** algorithm was applied, however, the global optimum was identified from all starting points in less than 24 iterations. For a relative tolerance of 10^{-3}, the algorithm takes 11 iterations, while the number of iterations required for relative error between the upper and lower bounds to be less than 10^{-8} is 23. After 26 iterations, the algorithm converges exactly (with an absolute error of 0) to the global solution.

Example 2 :

This example is taken from Moore (1979). It involves the minimization of a 50th degree polynomial in one variable.

$$\min_y \; \sum_{i=1}^{50} a_i y^i$$

$$1 \le y \le 2$$

where

$a = \{-500.00000, 2.5000000, 1.666666666, 1.2500000, 1.000000,$
$\phantom{a = \{}0.8333333, 0.714285714, 0.625000000, 0.555555555, 1.0000000,$
$\phantom{a = \{}-43.6363636, 0.41666666, 0.384615384, 0.357142857, 0.3333333,$
$\phantom{a = \{}0.312500000, 0.294117647, 0.277777777, 0.263157894, 0.2500000,$
$\phantom{a = \{}0.238095238, 0.227272727, 0.217391304, 0.208333333, 0.2000000,$
$\phantom{a = \{}0.192307692, 0.185185185, 0.178571428, 0.344827586, 0.6666666,$
$\phantom{a = \{}-15.483870970, 0.15625000, 0.1515151, 0.14705882, 0.14285712,$

0.138888888, 0.135135135, 0.131578947, 0.128205128, 0.1250000,
0.121951219, 0.119047619, 0.116279069, 0.113636363, 0.1111111,
0.108695652, 0.106382978, 0.208333333, 0.408163265, 0.8000000}.

This function has the global minimum at $y = 1.0911$, with a value of -663.5. The improved version of the **GOP** algorithm takes 45 iterations to find the global solution from a starting point of $y = 1$.

Example 3 :

This example is taken from Wilkinson (1963).

$$\min_{y} \ 0.000089248y - 0.0218343y^2 + 0.998266y^3 - 1.6995y^4 + 0.2y^5$$

$$0 \leq y \leq 10.$$

This problem has local minima at $y = 6.325, f = -443.67$, $y = 0.4573$, $f = -0.02062$, $y = 0.01256, f = 0.0$ and $y = 0.00246, f = 0$ among others. The improved **GOP** algorithm identifes the global solution from different starting points within 25 iterations.

Example 4 :

This example is taken from Dixon and Szego (1975).

$$\min_{y} \ 4y^2 - 4y^3 + y^4$$

$$-5 \leq y \leq 5.$$

This problem has two global minima at $y = 0, f = 0$, and $y = 2, f = 0$. There is a local maximum at $y = 1$. When the original **GOP** algorithm was applied to this problem, the global solutions were identified after around 150 iterations. However, the improved **GOP** algorithm identifes the global solution from different starting points within 50 iterations.

Example 5 (Three-hump camel-back function):

This example is taken from Dixon and Szego (1975).

$$\min \ F(y) = 2y_1^2 - 1.05y_1^4 + \frac{1}{6}y_1^6 - y_1 y_2 + y_2^2$$

$$-5 \leq y_1, y_2 \leq 5.$$

The nonconvexities in the problem are due to the $-1.05y_1^4$ and $-y_1 y_2$ terms in the function. It can be seen that at the optimal solution, the value of y_2^* must be either -5, 5 or $\frac{y_1^*}{2}$. Assuming that the solution does not lie at −5 or 5, y_2 can be replaced by $\frac{y_1}{2}$. Then, the problem can be converted to the following unconstrained optimization problem :

$$\min\ F(y) = 1.75y_1^2 - 1.05y_1^4 + \frac{1}{6}y_1^6$$

$$-5 \leq y_1 \leq 5.$$

The modified **GOP** algorithm was applied to the problem in this form. The solution of $y_1 = 0, F(y) = 0$ was identified in 31 iterations for an absolute error of 10^{-5} between the upper and lower bounds on the global solution.

Example 6 :

This example is taken from Goldstein and Price (1971).

$$\min\ y^6 - 15y^4 + 27y^2 + 250$$

$$-5 \leq y \leq 5.$$

This function has local minima at (0,250), (3,7) and (-3,7). When the **GOP** algorithm was applied to the problem, the two global solutions were simultaneously identified in 68 iterations for a relative tolerance of 10^{-3}.

Example 7 :

This example is taken from Dixon (1990).

$$min\ y^4 - 3y^3 - 1.5y^2 + 10y$$

$$-5 \leq y \leq 5.$$

This function has a global solution of -7.5 at $y = -1$. When the modified **GOP** algorithm was applied to the problem, the global solution was identified in 24 iterations for a relative tolerance of 10^{-3}.

Remarks :

The number of iterations taken by the above problems for different orders of accuracy in the convergence of the upper and lower bounds is given in Table 1. Some interesting points to note about the computational results of the modified algorithm are given below :

(a) For all the problems for which the algorithm was applied, the number of iterations required for convergence increases almost linearly with the accuracy desired. For Example 1, the number of iterations required for an accuracy of 10^{-3} is 11, while the number of iterations required for the bounds to be within 10^{-9} is only 26. This is in direct contrast to most algorithms that require an increasingly larger number of iterations as the accuracy is increased.

(b) For three of the problems considered (Examples 1,2 and 3), the algorithm terminates exactly after a certain number of iterations. This is because after some iterations, the use of the improved bounds for the x variable results in tighter underestimating functions. Eventually, at some iteration, all the points in the stored sets are used up, and the Lagrange functions formulated for that iteration are such that no new points are generated that can improve the solution. Thus, for these problems, the algorithm converges to the global solution exactly.

(c) For Examples 4 and 5, the convergence is in terms of the absolute difference between the upper and lower bounds. This is because the global solution of these problems is 0.

(d) At every iteration of the algorithm, there is one function evaluation associated with the solution of the primal problem, and two problems involving the solution of the simultaneous solution of two linear equations in two variables (which correspond to the solution of the two relaxed dual problems).

Problem	*Relative Tolerance for Convergence*			
	10^{-3}	10^{-5}	10^{-7}	10^{-8}
Example 1	11	15	19	23
Example 2	34	39	45	45
Example 3	13	18	28	33
Example 4	30	40	54	56
Example 5	27	31	34	36
Example 6	68	161	-	-
Example 7	24	77	216	491

Table 1: Number of iterations of the modified **GOP** algorithm

8.2 Constrained Problems

Example 8 :
This example is taken from Soland (1971).

$$\min_{y} \; -12y_1 - 7y_2 + y_2^2$$

$$\text{subject to} \quad -2y_1^4 + 2 - y_2 = 0$$
$$0 \le y_1 \le 2$$
$$0 \le y_2 \le 3.$$

The nonconvexity in this problem comes from the presence of the polynomial term $-2y_1^4$ in the first constraint.

When the original **GOP** algorithm was applied to the problem in this form, from a starting point of 0 for y_1, the algorithm converged to the global solution of -16.73889 at $y = (0.7175, 1.47)$ in 89 iterations, solving 3 subproblems at every iteration. When the improved **GOP** algorithm was applied from the same starting point, the global solution was identified in 14 iterations.

It should be noted that the constraint can be written in the following form :

$$y_2 = 2 - 2y_1^4.$$

Then, using this constraint to substitute for y_2 and utilizing the bounds on y_2, the problem can be converted into the following unconstrained optimization problem :

$$min \ 4y_1^8 + 6y_1^4 - 12y_1 - 10$$
$$0 \le y_1 \le 1.$$

When the modified **GOP** algorithm was applied to the problem in this form, the algorithm converges exactly to the global solution in 14 iterations. It should also be noted that in this form, the problem is convex, and therefore any conventional solver should be able to identify the global solution.

Example 9:

This is a test example that has a feasible region consisting of two disconnected sub-regions.

$$\min_{y} \quad -y_1 - y_2$$
$$y_2 \le 2 + 2y_1^4 - 8y_1^3 + 8y_1^2$$
$$y_2 \le 4y_1^4 - 32y_1^3 + 88y_1^2 - 96y_1 + 36$$
$$0 \le y_1 \le 3$$
$$0 \le y_2 \le 4.$$

The constraint region for this problem is given in Figure 2(a). As can be seen, there are two distinct regions where the problem is feasible. Because of this reason, if a conventional **NLP** solver were applied to this problem, it is highly unlikely that the solver would converge to the global solution at point C. Depending on the starting point, the solution will be one of the points A, B, or C.

From a starting point of 0 for y_1, the original **GOP** algorithm takes 210 iterations to converge to the global solution of -5.50796 (occurring at $y_1 = 2.3295$). When the improved **GOP** algorithm was applied to it, however, the algorithm took 24 iterations to converge to the global solution.

8.3 D.C. Programming Problems

These problems consist of an objective function that can be represented as the difference of two convex functions, and convex constraints. Many polynomial problems with convex constraints can be represented in this form.

Example 10
This problem is taken from Horst *et al* (1991a).

$$\min_{y_1,y_2}\; 4y_1^4 + 2y_2^2 - 4y_1^2$$

$$y_1^2 - 2y_1 - 2y_2 - 1 \le 0$$
$$-1 \le y_1 \le 1$$
$$-1 \le y_2 \le 1$$

The problem has an optimal solution of (0.707,0.000,-1.000).

To apply the **GOP** algorithm, the problem was converted to the appropriate form by the introduction of a new variable x_1 and a new constraint, as follows :

$$\min_{x,y}\; 4y_1^4 + 2y_2^2 - 4x_1y_1$$

$$x_1 - y_1 = 0$$
$$y_1^2 - 2y_1 - 2y_2 - 1 \le 0$$
$$-1 \le y_1 \le 1$$
$$-1 \le y_2 \le 1$$

By projecting on the y variables, the primal problem is solved in the space of x_1 while the *relaxed dual* problem is solved in the space of the y variables.

Iteration	Upper bound	Lower Bound
1	0.0000000	-1.8898815
2	-0.9574405	-1.7531346
3	-0.9574405	-1.1054029
4	-0.9901436	-1.0205925
5	-0.9901436	-1.0111922
6	-0.9995872	-1.0012325
7	-0.9998384	-1.0001769
8	-0.9998384	-1.0001318
9	-0.9999983	-1.0000135
10	-0.9999983	-1.0000064

Table 2: Progress of bounds for the GOP algorithm

Since $x_1 = y_1$, there is a one-to-one correspondence between the two variables, and therefore the bounds on the x variables can be updated from one iteration to another depending on the particular region where the *relaxed dual* problem is being solved. There are two *relaxed dual* problems being solved at every iteration.

The starting point for the y variables is (0,0). The algorithm takes **10 iterations** to converge to the optimal solution with **a tolerance of** 10^{-6}. The sequence of lower and upper bounds obtained by the algorithm in various iterations is given in Table 2. From other starting points, the algorithm performs similarly, converging in an average of 8-9 iterations. As a comparison, the algorithm proposed by Horst *et al* (1991a) takes 34 iterations to terminate with a tolerance of 0.05.

Example 11

This problem contains three variables and two constraints. It is taken from Horst *et al* (1991a).

$$\min_{y} \ (y_1^4 + y_2 + y_3) - (y_1 + y_2^2 - y_3)$$

$$(y_1 - y_2 - 1.2)^2 + y_2 \leq 4.4$$
$$y_1 + y_2 + y_3 \leq 6.5$$
$$y_1 \geq 1.4$$
$$y_2 \geq 1.6$$
$$y_3 \geq 1.8$$

The problem has an optimal solution of (1.400, 1.809502, 1.8000) with an objective value of 4.5768.

As in the previous example, the nonconvexity arises due to a single term in the objective term $(-y_2^2)$. By introducing a new variable x_2, the problem is converted to a bilinear form as follows:

$$\min_{x,y} \quad y_1^4 - y_1 + y_2 + 2y_3 - x_2 y_2$$

$$(y_1 - y_2 - 1.2)^2 + y_2 \leq 4.4$$
$$y_1 + y_2 + y_3 \leq 6.5$$
$$x_2 - y_2 = 0$$
$$y_1 \geq 1.4$$
$$y_2 \geq 1.6$$
$$y_3 \geq 1.8$$
$$x_2 \geq 1.6$$

The primal problem is solved by projecting on all the y variables, in effect making it a function evaluation. This enables the constraints from the original problem to be used in the *relaxed dual* problems. Again, the equivalence between x_2 and y_2 enables the update of the bounds of x_2 from one iteration to another.

The **GOP** algorithm was applied to the problem in this form. From all considered starting points (For example, starting from $y_1 = y_2 = y_3 = 2$), the algorithm converged **absolutely** to the optimal solution in **2 iterations**, in contrast to the algorithm proposed by Horst *et al* (1991) for *d. c.* programming problems, which requires 18 iterations to converge to within a tolerance of 0.01.

Conclusions

In this paper, the application of the **GOP** algorithm (Floudas and Visweswaran, 1990, 1991) to problems involving polynomial functions in one variable is considered. The problem is solved by decomposition into a series of primal and *relaxed dual* problems. The solution of the primal problem can be obtained by simple function evaluations. The relaxed dual problem can be solved through two subproblems, each of which is shown to reduce to a problem of finding the intersection of two linear constraints in two variables. The simplfied primal and relaxed dual problems result in a modified algorithm that is computationally very efficient. The application of the modified algorithm is shown through an illustrating example that details the improvement of the modified algorithm over the original **GOP** algorithm both numerically and geometrically. Several examples of unconstrained and constrained polynomial function problems are presented to highlight the efficiency of the new algorithm. A comparison with an algorithm proposed by Horst *et al* (1991a) for *d. c. programming problems* serves to illustrate the applicability of the proposed algorithm to these problems.

Acknowledgement

The authors gratefully acknowledge financial support from the National Science Foundation under Grant CBT-8857013, as well as support from Amoco Chemical Co., Tennessee Eastman Co., Shell Development Co, and Mobil Research and Development Co.

References

[1] Dixon, L. C. W., On Finding the Global Minimum of a Function of One Variable, Presented at the SIAM National Meeting, Chicago (1990).

[2] Dixon, L. C. W., and Szegö, G. P., *Towards Global Optimization*, North Holland, Amsterdam (1975).

[3] Evtushenko, Y. G., Numerical Methods for Finding Global Extrema (Case of Nonuniform mesh), *USSR Computational Mathematics and Mathematical Physics*, 11, 1390 (1971).

[4] Floudas, C.A. and Visweswaran, V., A Global Optimization Algorithm (GOP) for Certain Classes of Nonconvex NLPs -I. Theory, *Computers and Chemical Engineering*, 14, 1397 (1990).

[5] Floudas, C.A., and Visweswaran, V., A Primal-Relaxed Dual Global Optimization Approach, *Journal of Optimization Theory and Applications*, Accepted for Publication (1991).

[6] Galperin, E.A., The Cubic Algorithm, *Journal of Mathematical Analysis and Applications*, 112, 635 (1985).

[7] Galperin, E.A., The Beta-Algorithm, *Journal of Mathematical Analysis and Applications*, 126, 455 (1987).

[8] Goldstein, A. A., and Price, J. F., On Descent From Local Minima, *Mathematics of Computation*, 25 , 569 (1971).

[9] Hansen, E.R., Global Optimization using Interval Analysis: The One-Dimensional Case, *Journal of Optimization Theory and Applications*, **29**, 331 (1979).

[10] Hansen, P., Jaumard, B., and Lu, S-H., Global Optimization of Univariate Lipschitz functions : I. Survey and Properties, *To appear in Mathematical Programming* (1991).

[11] Hansen, P., Jaumard, B., and Lu, S-H., Global Optimization of Univariate Lipschitz functions : II. New Algorithms and Computational Comparisons, *To appear in Mathematical Programming* (1991).

[12] Hansen, P., Lu, S-H., and Jaumard, B., Global Minimization of Univariate Functions by Sequential Polynomial Approximation, *International Journal of Computer Mathematics*, **28**, 183 (1989).

[13] Horst, R., Phong, T. Q., Thoai, Ng. V., and Vries, J. de., On Solving a D. C. Programming Problem by a Sequence of Linear Programs, *Journal of Global Optimization*, **1(2)**, 183-203 (1991a).

[14] Horst, R., Thoai, Ng. V., and Benson, H. P., Concave Minimization via Conical Partitions and Polyhedral Outer Approximations, *Mathematical Programming*, To appear (1991b).

[15] Moore, R., Methods and Applications in Interval Analysis, *SIAM Studies in Applied Mathematics*, SIAM, Philadelphia (1979).

[16] Piyavskii, S.A., An Algorithm for finding the Absolute Extremum of a Function, *USSR Comput. Math. Phys.*, **12**, 57 (1972).

[17] Ratschek, H., and Rokne, J., *New Computer Methods for Global Optimization*, Halsted Press (1988).

[18] Shen, Z., and Zhu, Y., An Interval Version of Schubert's Iterative Method for the Localization of the Global Maximum, *Computing*, **38**, 275 (1987).

[19] Shirov, V. S., Search for the Global Extremum of a Polynomial on a Parallelopiped, *USSR Comput. Maths. Math. Phys.*, **25**, 105 (1985).

[20] Timonov, L.N., An Algorithm for Search of A Global Extremum, *Engineering Cybernetics*, **15**, 38 (1977).

[21] Visweswaran, V. and Floudas, C.A., A Global Optimization Algorithm (GOP) for Certain Classes of Nonconvex NLPs -II. Application of Theory and Test Problems, *Computers and Chemical Engineering*, **14**, 1419 (1990a).

[22] Visweswaran, V. and Floudas, C.A., New Properties and Computational Improvement of the GOP Algorithm For Problems With Quadratic Objective Function and Constraints, *Submitted for Publication* (1990b).

[23] Wilkinson, J. H., *Rounding Errors in Algebraic Processes*, Prentice-Hall, Engelwood Cliffs, N.J. (1963).

[24] Wingo, D.R., Globally Minimizing Polynomials Without Evaluating Derivatives, *International Journal of Computer Mathematics*, **17**, 287 (1985).

One Dimensional Global Optimization Using Linear Lower Bounds*

Matthew Bromberg[†]   Tsu-Shuan Chang[‡]

Abstract

A new technique utilizing linear functions as lower bounds to the original cost function is developed. Since linear functions have level sets that are easy to characterize, it is possible to remove regions where the global minimum can not occur. Starting with functions that can be represented as the difference of two convex functions, it is demonstrated that linear lower bounds can be found for virtually any cost function that can be written in closed form. Based upon this technique, a Linear Lower Bound algorithm has been developed. The algorithm is guaranteed to converge to a global minimum within a specified tolerance in a finite number of steps. Numerical testing on certain standard test problems suggests that our algorithm is highly competitive with some existing algorithms.

1. Introduction

Because so many problems can be formulated as optimization problems, much effort has been spent searching for good optimization techniques. Unfortunately most algorithms can only find local

---

* The work was supported in part by NSF grant ECS-87-17235.

[†] was a graduate student in the Department of Electrical Engineering and Computer Science, University of California, Davis, and is now with Argosystems, Sunnyvale, CA 94086.

[‡] Department of Electrical Engineering and Computer Science, University of California, Davis, CA 95616.

minima, and once a local minimum is reached, a vanishing gradient can not distinguish a local minimum from a global one. In recent years there has been some progress in finding solution techniques for this difficult problem.

One concept frequently used in global optimization is the covering method. The basic idea is to detect and throw away subregions not containing the global minimum until the remaining set is small and is known to contain it. Many such methods rely on an estimate of Lipschitz constant (e.g. [11,14]). For example, the Pijavskij-Shubert algorithm [11] uses the Lipschitz constant to approximate the function in a piecewise linear fashion. Segments not containing the global minimum are then eliminated sequentially. Another way is to make use of interval analysis (e.g., [6,7,12]). Recently, a general theory for branch and bound methods has been developed. This theory has been successful in analyzing the convergence of a wide class of algorithms [9].

In monotonicity analysis (e.g., [8,17]), the basic monotonicity principle is used to determine global minima. For example, branch and bound techniques can be used by dividing a feasible region into smaller regions to make a non-monotonic function become monotonic in those smaller regions.

Geometric programming [1] works for a class of problems whose non-convex cost functions can be converted into convex ones through nonlinear transformations. This allows the function to be minimized in a fashion similar to ordinary convex functions.

In [10], convex functions are constructed iteratively to underestimate tightly a given factorable cost function. The global optimum of these functions can be found in a fashion similar to finding the global optimum of their convex envelopes.

In spite of the recent progress in global optimization, the need for a general theory or for efficient algorithms has not been mitigated. Since some algorithms are based on heuristics, it is desirable to find algorithms that are guaranteed to converge to global minima. It is even more desirable if the algorithm converges in a finite number of iterations.

In an attempt to develop a new conceptual framework for global optimization, we have recently introduced a method called the Function Imbedding Technique (FIT) [4]. This method proposes to convert nonconvex problems into convex ones by imbedding the cost function into a higher dimensional cost function. A deterministic algorithm based upon FIT has been developed for one-dimensional global optimization problems [5].

In order to develop more efficient algorithms for global optimization problems, we introduce in this paper the concept of a Linear Lower Bound (LLB), and use it to develop a new algorithm called the Linear Lower Bound Algorithm. The basic idea is to generate linear lower bounds for nonconvex cost functions. The linear lower bounds can then be used to eliminate regions which can not contain global minima. The algorithm finds the globally optimal solution within a specified tolerance for a large class of global optimization problems in a finite number of iterations. In addition numerical tests indicate that the algorithm is highly competitive with existing methods.

The paper is organized as follows. In section 2, the general framework of the Linear Lower Bound technique is presented. In section 3, the Linear Lower Bound algorithm and a convergence proof is given. In section 4, methods for finding linear lower bounds for a large class of cost functions are presented. In section 5, comparison tests of the numerical performance of the algorithm is presented. The paper is concluded in section 6.

2. The Linear Lower Bound Concept

Let us consider the following optimization problem.

$$\min_{x \in K \subset R} f(x), \qquad (2.1)$$

where K is a compact region in R and f(x) is a continuous function. Let A be a compact subset of K, then a linear lower bound l(x) on a set A is a linear function which satisfies the following conditions.

Linear Lower Bound

A function f(x) is said to have a linear lower bound on A bounded by G, if there exists $p \, \varepsilon \, A \subset R^n$, $\beta \, \varepsilon \, R$, $\gamma \, \varepsilon \, R^n$ such that

$$f(x) \geq \gamma \cdot x + \beta, \quad \text{for all } x \, \varepsilon \, A, \qquad (2.2a)$$

$$f(p) = l(p) = \gamma \cdot p + \beta \qquad (2.2b)$$

and γ is bounded, i.e.,

$$|\gamma| \leq G, \qquad (2.2c)$$

where G is a finite number.

To illustrate how a linear lower bound can be used to obtain a global minimum within ε, i.e. a ε-global minimum, let us consider the one dimensional global optimization problem depicted in Figure

1. Note that the original compact interval K can be divided into two sub-intervals determined by the local minimizer m; the local minimizer m can be obtained by using any standard local minimization algorithm. Thus we can assume the current region of interest is $A^0 \equiv [m, M]$.

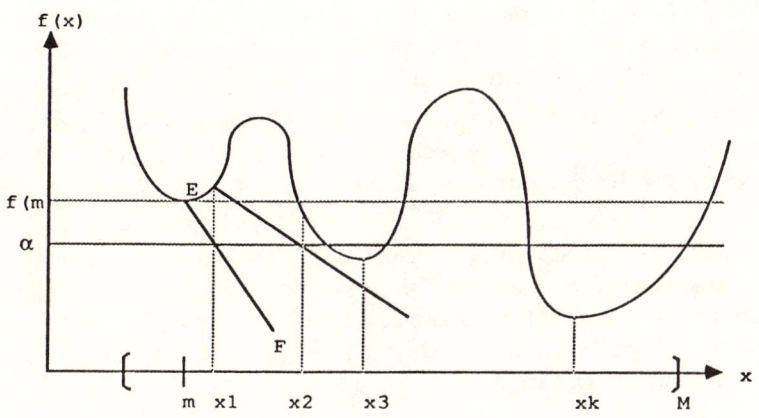

Fig. 1 A Linear Lower Bound Algorithm

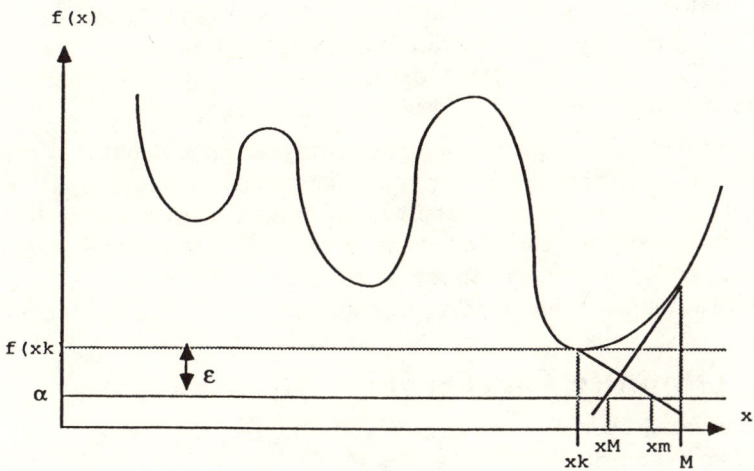

Fig. 2 A Stopping Criterion

Assuming that linear lower bounds can be constructed, one way to use them to obtain an ε-global minimum is demonstrated as follows. Let us consider p to be an end point of A^0, say $p = m$. Graphically we can draw the linear lower bound as a straight line EF which lies below f(x) for all $x \in A^0 = [m, M]$. Let

$$\alpha \equiv f(m) - \varepsilon, \quad \varepsilon > 0. \qquad (2.3)$$

We can then get the intersection point at $x = x^1$. Since $f(x^1) > f(m)$, we know that the ε-global minimum can not occur at the interval $[m, x^1]$. In other words, the new region of interest is $I^1 \equiv [x^1, M]$. The procedure is then applied to I^1.

By repeating this process we can eventually find a point x^2 such that $f(x^2) < f(m)$. A local minimization is then performed to find the new local minimum x^3 and to obtain the new region $I^3 \equiv [x^3, M]$.

Note that the size of the region of interest is strictly decreasing. Repeating the aforementioned process, we will reach the global minimum x^k. The question is then, how do we know that this point is indeed a global minimum within a tolerance of ε? To answer this question, consider the scenario in Figure 2.

Suppose the assumption is made that there exists a global minimum at least ε smaller than the current minimizer found. This assumption can be tested as follows. For the given interval $I^k \equiv [x^k, M]$, we can draw the linear lower bounds from both end points. From the linear lower bound starting at x^k, we can see that the ε-global minimizer must lie in $[x_m, M]$. On the other hand, starting from M, one must also have that the ε-global minimizer lies in $[x^k, x_M]$. This leads to a contradiction. Thus the global minimum must have a value equal to $f(x^k)$ within ε, as depicted.

Intuitively this algorithm must converge to an ε-global minimizer, since we can always use the linear lower bound to cut away a region which does not contain the global minimizer. In addition the sub-regions which are cut away have a lower bound on their size. Since the original region of interest is compact and has a finite size, the algorithm must converge in a finite number of iterations.

3. A Globally Convergent Algorithm

Motivated by the previous discussion, let us present the Linear Lower Bound Algorithm.

The Linear Lower Bound Algorithm

Let P be a collection of subsets of K and its elements are arranged according to the order of their creation. Initially P is any finite partition of K.

(1) From the current point p, use a local descent method to obtain p' and set $\alpha = f(p') - \varepsilon$. Save p'.

(2) Choose a point p from the most recently created element A in P according to (2.2), and obtain a linear lower bound parametrized by γ and β.

(3) Set

$$A' = A \cap \{x : \gamma \cdot x + \beta \leq \alpha\}. \qquad (3.1)$$

If A' is empty, remove A; otherwise replace A by A'.

(4) If A' is empty, assign the new A to be the previous created element. After this, assign new A to be the most recent region added. If P is empty, then STOP.

(5) Choose a point $p \, \varepsilon \, A$ for testing if a nonempty A is available. Otherwise if $f(p) < f(p')$, then go to step (1) else go to step (2).

Theorem 3.1 : Convergence to an ε-Solution in a Finite Number of Steps

For a continuous cost function f(x), if the linear lower bounds in (2.2) can be obtained for any partition element A such that each linear lower bound is bounded by the same fixed G, then the LLB algorithm finds a point x^* such that

$$f(x^*) < \min_{x \, \varepsilon \, K} f(x) + \varepsilon. \qquad (3.2)$$

Such a point can always be found in a finite number of iterations.

Proof: Let P^i be the collection of subsets after the ith iteration. Let z_i be the point chosen in step (2) after the ith iteration.

This theorem shall be proved in two parts. First one notes that either the stopping criteria in step (4) is satisfied, and the algorithm terminates in a finite number of iterations, or the stopping criteria is never satisfied and the algorithm never terminates. Suppose that stopping criteria is satisfied, then every point in K was eliminated at some time during step (3) of the algorithm. This means from equations (3.1) and (2.2), that for every x eliminated from K, there exists

γ_x and β_x such that,

$$f(x) \geq \gamma_x \cdot x + \beta_x > \alpha \geq \bar{\alpha} \qquad (3.3)$$

where $\bar{\alpha}$ is the terminal α. From step (1) one also has a terminal $\bar{p}'$ satisfying

$$\bar{\alpha} = f(\bar{p}') - \varepsilon. \qquad (3.4)$$

But from equation (3.3) this means that for all x in K

$$f(\bar{p}') \leq f(x) + \varepsilon. \qquad (3.5)$$

This implies that $\bar{p}'$ is an ε solution as desired.

On the other hand, suppose that the algorithm never satisfies its stopping criterion and therefore never terminates. We shall show that this leads to a contradiction. Let α_j and p_j' be the current α and p' at iteration j. Let x_* be the minimizer of f(x) on K. Note that α_j is a non-increasing sequence bounded below by $f(x_*) - \varepsilon$. Therefore α_j has a limit which is also its greatest lower bound.

$$\lim_{j \to \infty} \alpha_j \equiv \alpha \text{ and } \alpha_j \geq \alpha. \qquad (3.6)$$

Let us pick a subsequence from z_i as follows. Since the sequence is infinite and the partition initially contains a finite number of subsets of K, one of these subsets contains an infinite subsequence of the z_i. Call this subset A_0 and remove all points of z_i not in A_0. In addition relabel the sequence so that the first element of the sequence chosen in step 5 which lies in A_0 is called p_0. Since A_0 is not removed in step (3), it is reduced in step (3) to A_0' and then partitioned into a finite number of subsets. One of these subsets will contain an infinite number of the z_i. Choose one of them and call it A_1. Let p_1 be the first point in z_i which is chosen in A_1.

By induction generate the sequence of subsets A_i and p_i as above. Choose A_{i+1} to be a subset of A_i' containing an infinite number of the z_i and p_{i+1} is the first element of z_i which is found in A_{i+1}.

This determines a nested sequence of compact sets. That is

$$A_0 \supset A_1 \supset A_2 \supset A_3 \cdots \text{ etc.} \qquad (3.7)$$

Notice by step (2) of the algorithm, equation (2.2) and equation (3.6) one has,

$$f(p_i) = \gamma \cdot p_i + \beta \geq f(p_i') = \alpha_i + \varepsilon \geq \alpha. \qquad (3.8)$$

Since the p_j are in the compact set K there is a convergent subsequence which converges to a point p_*. It is clear that $p_* \, \varepsilon \, A_i$ for all i since the A_i are closed and only a finite number of p_j are not contained in A_i. Therefore

$$p_* \, \varepsilon \, A_\infty \equiv \bigcap_{i=0}^{\infty} A_i. \tag{3.9}$$

Throw out the points not in the convergent subsequence and relabel p_i and A_i. If the order of the sequence is preserved after relabeling, then equation (3.7) is still valid.

The question can now be asked, how much was eliminated from A_i in step (3) of the LLB algorithm? Recall that

$$f(p_i) = \gamma \cdot p_i + \beta \geq f(p_i') > \alpha_i. \tag{3.10}$$

Thus p_i is not in A_i'. We will show that at least some small fixed ball about p_i is eliminated, preventing p_* from being in that ball. This will be a contradiction. One is therefore interested in,

$$\min_{x \, \varepsilon \, A_i'} |x - p_i| \geq \min_{\gamma \cdot x + \beta \leq \alpha_i} |x - p_i| \tag{3.11}$$

$$= \min_{\gamma \cdot (x - p_i) \leq \alpha_i - f(p_i)} |x - p_i| \tag{3.12}$$

$$= \frac{f(p_i) - \alpha_i}{|\gamma|} \geq \frac{f(p_i) - \alpha_i}{G} \geq \frac{\varepsilon}{G}. \tag{3.13}$$

In summary,

$$\min_{x \, \varepsilon \, A_i'} |x - p_i| \geq \frac{\varepsilon}{G}. \tag{3.14}$$

These results are derived by noting the inequalities in equation (3.8) and the bound on γ from the assumed bounded gradients of the linear lower bounds. Note that it is true for all i. However p_i converges to p_*, therefore there exists an N such that if $i > N$ then

$$|p_* - p_i| < \frac{\varepsilon}{G}. \tag{3.15}$$

By the choice of p_i, however, $p_i \, \varepsilon \, A_i$ and A_i is never entirely eliminated. Since A_i is closed one must have $p_* \, \varepsilon \, A_i$ and $p_* \, \varepsilon \, A_i'$ for all i. Thus by equation (3.14)

$$|p_* - p_i| \geq \frac{\varepsilon}{G}, \tag{3.16}$$

contradicting equation (3.15).

This means that the assumption that the algorithm fails to terminate is false. Thus after executing steps (1) through (5) of the LLB algorithm a finite number of times the stopping criteria must be satisfied and the algorithm will stop. We have already shown however, that if the algorithm stops it must stop at an ε solution. This proves the theorem.

With this theorem, in order to solve the global minimization problem, we need only know how to find the linear lower bounds l(x) for a given function f(x) on a set A. Furthermore, note that there is no need to obtain exact local minima in step (1). If the local search is restricted to a finite number of steps, the algorithm will also converge in a finite number of steps. The reason this is true is because an examination of the above proof reveals that in theory no local minimization is necessary at all for finite convergence. The proof of finite convergence would remain exactly the same. The local minimization only serves to decrease α more rapidly than if no local minimization is performed.

4. Finding Linear Lower Bounds

The basic idea of our method is to decompose a given function in closed form into generic terms. Once linear lower bounds for the generic terms are found, they are put together to obtain a linear lower bound for the original function. We shall illustrate how linear lower bounds can be recursively obtained for a wide class of functions defined on a given interval A.

$$A = [x_0, x_1], \quad x_0 < x_1. \tag{4.1}$$

4.1 The Difference of Convex Functions

To understand the essential idea, let us first consider the case when the functional form of f(x) is the difference of two convex functions.

$$f(x) = c_1(x) - c_2(x). \tag{4.2}$$

If $c_1(x) \geq l_-^1(x)$ and $c_2(x) \leq l_+^2(x)$, where $l_-^1(x)$ and $l_+^2(x)$ are linear lower and upper bounds, then

$$f(x) \geq l_-^1(x) - l_+^2(x), \tag{4.3}$$

so that $l_-^1(x) - l_+^2(x)$ is a linear lower bound for f(x) on A.

Given a continuously differentiable convex function C(x): A = $[x_0, x_1] \to$ R, we have

ONE DIMENSIONAL GLOBAL OPTIMIZATION

$$\gamma_- \cdot (x - x_0) + C(x_0) \leq C(x) \leq \gamma_+ \cdot (x - x_0) + C(x_0) \quad (4.4)$$

for all $x \in A$, where

$$\gamma_- = \left.\frac{dC}{dx}\right|_{x_0}, \quad (4.5)$$

$$\gamma_+ = \frac{\Delta C}{\Delta x}, \quad (4.6)$$

$$\Delta C \equiv \left[C(x_1) - C(x_0)\right], \quad (4.7)$$

$$\Delta x \equiv \left[x_1 - x_0\right]. \quad (4.8)$$

Note that the left hand side of equation (4.4) is the usual subgradient inequality [13], and the right hand side is just the line passing through the point $(x_0, C(x_0))$ and $(x_1, C(x_1))$. Therefore from (4.3) and (4.4), it is always possible to obtain a linear lower bound for a function which is a difference of two convex functions.

One can also observe that the $\gamma_\pm$ for the linear bound for these two cases can always be bounded. In particular γ_- is bounded above by the maximum of $\left|\frac{dC(x)}{dx}\right|$ over $x \in A$ which is always finite if $C(x)$ has a continuous derivative. Also by the mean value property, we have

$$\frac{\Delta C}{\Delta x} = \left.\frac{dC}{dx}\right|_{\bar{x}} \quad (4.9)$$

for some $\bar{x}$ between x_0 and x_1. Since the right hand side of (4.9) is bounded on a compact set, $\frac{\Delta C}{\Delta x}$ will be bounded. One may object, however, to calculate this when Δx becomes small. To avoid this problem, at this time the following observation is made: If Δx is not too small one does not worry about the calculation of γ_+. However if γ_+ exceeds some predetermined bound, one can simply set $\gamma_+ = 0$ provided that $C(x_0) \geq C(x_1)$. This still keeps the desired result of equation (4.5a) intact and guarantees that γ_- and γ_+ are bounded.

Note that if they are bounded for each term, then the γ associated for the linear lower bound in equation (4.3) will also be bounded. Thus by restricting the regions of interest to be intervals, and with only the possible added inconvenience of choosing the end point of A which maximizes $c_2(x)$, in step (2) of the LLB algorithm, the linear lower bounds generated in this way will satisfy (3.2) for f(x) expressed as the difference of two convex functions.

One should also be aware that the class of functions which can be expressed as the difference of two convex functions is quite large. In fact it is not hard to see that any function defined on a compact set that has continuous second derivatives can be expressed in this way. The idea of the proof behind this is to express the function as

$$f(x) = \left[f(x) + \frac{\lambda}{2}x^2\right] - \frac{\lambda}{2}x^2. \qquad (4.10)$$

Here λ is chosen large enough to make $f(x) + \frac{\lambda}{2}x^2$ convex. It is desirable, however, to find more efficient decompositions than (4.10).

4.2 A Decomposition Using Monotonic Convex Functions

Let C(.) always denote convex functions. To indicate a monotonically non-decreasing function, the superscript + will be used and correspondingly the superscript - will be used for monotonically non-increasing functions. For example one writes

$$C^+(x) \equiv x, \qquad (4.11)$$

$$\text{and} \quad C^-(x) \equiv -x, \qquad (4.12)$$

to indicate that the first function is monotonically non-decreasing and the second function is monotonically non-increasing.

Given an arbitrary function Y(x) and a monotonic convex function C^s, s= + or -, $C^s(Y(x))$ is then called a composition of monotonic convex function. If one has linear upper and lower bounds for $Y(x)$,

$$l_-(x) \leq Y(x) \leq l_+(x), \quad \text{where} \quad l_\pm(x_0) = Y(x_0), \qquad (4.13)$$

one can obtain from (4.15) linear lower and upper bounds for

$$f(x) \equiv C^s(Y(x)), \qquad (4.14)$$

where both the upper and lower linear bounds match f(x) at $x = x_0$.

$$l_l(x) \leq C^s(l_{-s}(x)) \leq C^s(Y(x)) \leq C^s(l_s(x)) \leq l_u(x), \qquad (4.15)$$

$$\text{where} \quad l_{-s} \equiv \begin{cases} l_- & \text{if } s=+, \\ l_+ & \text{if } s=-. \end{cases} \qquad (4.16)$$

The inner two inequalities are due to the monotonicity. Since a convex function of a linear function is also convex, we can use (4.4) to obtain an upper linear bound for

$$C^s(l_s(x)) \le l_u(x), \tag{4.17}$$

and a lower linear bound for

$$C^s(l_{-s}(x)) \ge l_l(x), \tag{4.18}$$

which agree with $C^s(Y(x))$ at x_0, since all inequalities become equalities when $x = x_0$.

On the other hand, if one has linear upper and lower bounds for an arbitrary function $T_i(x)$,

$$l_-^i(x) \le T_i(x) \le l_+^i(x), \tag{4.19}$$

such that $l_\pm^i(x_0) = T_i(x_0)$, then one can obtain linear upper and lower bounds for

$$f(x) = \sum_i \alpha_i C_i(x) \tag{4.20a}$$

from (4.20b) below.

$$\sum_i \alpha_i l_{-s(\alpha_i)}^i(x) \le f(x) \le \sum_i \alpha_i l_{s(\alpha_i)}^i(x), \tag{4.20b}$$

$$\text{where} \quad s(z) \equiv \begin{cases} - & \text{if } z < 0 \\ + & \text{otherwise} \end{cases}, \tag{4.21}$$

and both the upper and lower linear bounds match f(x) at $x = x_0$, since these inequalities also become equalities at $x = x_0$.

Note that from (4.14) and (4.20a), we can construct a lot of functions recursively. From (4.15) and (4.20b), we can then find their linear lower bounds. The result is summarized in Theorem 4.1.

Theorem 4.1 : Construction of Linear Lower Bounds

If f(x) can be expressed recursively as a composition of monotonic convex functions and any arbitrary number of repeated linear combinations of such functions, then linear lower and upper bounds can be found for f(x) on A= $[x_0,x_1]$ such that the bounds match the value of f(x) at x= x_0.

4.3 Product and Division of Two Functions

We shall now demonstrate how to obtain linear lower bounds using the above approach by decomposing some common generic functions. Let us consider

$$f(x) = (y(x))^2 \quad \text{for } x \in [x_0, x_1], \tag{4.22}$$

and suppose

$$l_-(x) \leq y(x) \leq l_+(x), \text{ and } l_-(x_0)=l_+(x_0)=y(x_0). \quad (4.23)$$

Define
$$[z]_+ \equiv \max\{z,0\}, \quad (4.24)$$

$$r^+(y) \equiv [y]_+ y = \begin{cases} y^2 & \text{if } y \geq 0, \\ 0 & \text{otherwise.} \end{cases} \quad (4.25)$$

$$r^-(y) \equiv -[-y]_+ y = \begin{cases} y^2 & \text{if } y \leq 0, \\ 0 & \text{otherwise.} \end{cases} \quad (4.26)$$

Note $r^+(y)$ is non-decreasing, C^1 and convex, while $r^-(y)$ is non-increasing, C^1 and convex. In addition

$$f(x) = (y(x))^2 = r^+(y(x)) + r^-(y(x)). \quad (4.27)$$

Thus we have

$$2y(x_0)(y(x)-y(x_0)) + y(x_0)^2 \leq f(x) \leq r^+(l_+(x)) + r^-(l_-(x)). \quad (4.28)$$

The right inequality is due to (4.20) and the left one is just the tangent line of the function y^2 at $y(x_0)$. Using (4.20) on the left side, we have

$$2y(x_0) \cdot l_{-s(y(x_0))}(x) - y(x_0)^2 \leq (y(x))^2. \quad (4.29)$$

Applying (4.4) on the right side of (4.28), we get, if the right hand of (4.30) exists,

$$y(x)^2 \leq \frac{\Delta r^+}{\Delta l_+}(l_+(x)-l_+(x_0)) + \frac{\Delta r^-}{\Delta l_-}(l_-(x)-l_-(x_0)) + y(x_0)^2. \quad (4.30)$$

This achieves the linear lower and upper bounds for $f(x) = (y(x))^2$.

This result can be utilized when it is desired to find linear lower and upper bounds for the product of two function for which one already has linear lower and upper bounds. This is because,

$$y_1 \cdot y_2 = \frac{1}{4}\left[(y_1 + y_2)^2 - (y_1 - y_2)^2\right]. \quad (4.31)$$

The linear lower and upper bounds for $y_1(x) + y_2(x)$ and $y_1(x) - y_2(x)$ are easily obtained once we have them for $y_1(x)$ and $y_2(x)$. One then applies equations (4.29) and (4.30) to the two square terms in equation (4.31) to obtain the desired bounds. Although this is not always the best way to find lower bounds for the products of functions it demonstrates the wide scope of the Linear Lower Bound concept.

Once this is obtained it is easy to see how to achieve linear lower bounds for the division of two functions. If $f(x) = \dfrac{y_1(x)}{y_2(x)}$ then define

$$\bar{y}_2(x) \equiv \frac{1}{y_2(x)}. \qquad (4.32)$$

For convenience assume that $y_2(x) > 0$. With this assumption it is clear that $c(y_2) \equiv \dfrac{1}{y_2}$ is a decreasing convex function. Therefore by Theorem 4.1 lower and upper linear bounds can be obtained for $\bar{y}_2(x) = c(y_2(x))$. However this means that

$$f(x) = y_1(x) \cdot \bar{y}_2(x) \qquad (4.33)$$

and both $y_1(x)$ and $\bar{y}_2(x)$ have linear lower bounds. From equation (4.31) and the comments afterwards, it follows that linear lower and upper bounds can be obtained for f(x) expressed as the division of two functions with known bounds.

From Theorem 4.1, linear lower and upper bounds can be obtained for any linear combination of functions with known linear bounds. Since sums, products and divisions form no obstacle to the construction of linear bounds, it thus seems possible to build up recursively a library of linear bounds for a large class of functions that can be expressed in closed form.

4.4 Arbitrary One Dimensional Functions

In order to find linear lower bounds for arbitrary one dimensional functions, it is useful to recall the fact that a necessary and sufficient condition for smooth function to be convex is that their second derivatives be non-negative. Suppose f(x) is C^2 so that a decomposition into the difference of two convex functions exists.

$$f(x) \equiv c_1(x) - c_2(x). \qquad (4.34)$$

We thus have

$$c_1(x) = f(x) + c_2(x). \qquad (4.35)$$

The decomposition is clearly not unique, therefore it is desirable, to somehow find the "best" such decomposition. Qualitatively the best solution is one which minimizes the curvature of $c_2(x)$ subject to the conditions that

$$c_2''(x) \geq 0 \qquad (4.36)$$

and $\quad f''(x) + c_2''(x) \geq 0. \qquad (4.37)$

This is because when $c_2''(x)$ is minimized, the curvature of the convex function $c_1(x) = f(x) + c_2(x)$ is as close to the original function f(x) as possible which means that the lower bounds obtained contain as much information from f(x) as is possible. The solution of this minimization problem is simply to let

$$c_2''(x) = max(-f''(x), 0) = [-f''(x)]_+ \qquad (4.38)$$

and $\qquad c_1''(x) = max(f''(x), 0) = [f''(x)]_+. \qquad (4.39)$

Therefore for one dimensional C^2 functions the "best" decomposition is

$$f(x) = \int_{x_0}^{x}\int_{x_0}^{y} [f''(z)]_+ \, dz \, dy - \left[\int_{x_0}^{x}\int_{x_0}^{y} [f''(z)]_+ \, dz \, dy - f(x) \right], (4.40)$$

where the first term in the right hand side difference is $c_1(x)$ and the second term in paranthesis is $c_2(x)$.

Unfortunately it may not always be possible to easily evaluate $\int_{x_0}^{x}\int_{x_0}^{y} [f''(z)]_+ \, dz \, dy$ from the decomposition of equation (4.40). In this case it is necessary to try to decompose f(x) into simpler functions using the techniques previously developed.

In the case of analytic functions, the power series may be known, in which case it is useful to have formulas for the linear lower bounds of polynomial functions. Since the even power terms of the form x^2 are already convex and linear combinations of these can be handled using Theorem 4.1, one need only concentrate on the odd power terms. Now the odd power terms could easily be decomposed using the decomposition in equation (4.40). However it is possible to improve upon that. If the bounds are given for x as $x \in [a,b]$, it is possible to find the greatest linear lower bound for $\pm x^{2n-1}$ at either the point a or the point b directly. That is one can find the linear lower or upper bound which touches x^{2n-1} at either a or b and at one other point.

Perhaps more powerful approach is to find the optimal decomposition for two terms at once. That is if

$$f_i(x) = \alpha x^{2i} + \beta x^{2i+1}, \qquad (4.41)$$

then from equation (4.39) one obtains

$$c_1''(x) = 2ix^{2i-2}\Big[\alpha(2i - 1) + \beta(2i + 1)x\Big]_+. \qquad (4.42)$$

This can be integrated fairly easily once the values of α and β are known. Thus $f_i(x)$ can be decomposed into the difference of convex functions. In general a polynomial will be a linear combination of $f_i(x)$, using possibly different α and β for different i. A representation of any polynomial f(x) as the difference of two convex functions follows in a straightforward manner from the decomposition of each $f_i(x)$.

As mentioned earlier for analytic functions, if they can not easily be decomposed using equation (4.40), they may be written in a power series preferably about the point x_0. By then using the techniques described above upon the polynomial power series, it is possible to obtain linear lower bounds or a direct representation as the difference of convex functions for the given analytic function.

5. Numerical Examples

Based upon the LLB algorithm presented in section 4, we have implemented our algorithm for some standard test problems given in [16]. Our results indicate that in many cases, the LLB algorithm performs more efficiently than existing algorithms. Below is a summary of our test results.

The following one dimensional multi-modal cost functions were tested.

$$f_1(x) = \sin(x) + \sin(\frac{10x}{3}) + \ln(x) - 0.84x, \quad 2.7 \leq x \leq 7.5, \quad (5.1)$$

$$f_2(x) = \sin(x) + \sin(\frac{2x}{3}), \quad 3.1 \leq x \leq 20.4 \quad (5.2)$$

$$f_3(x) = -\sum_{i=1}^{5} \sin((i+1)x + i), \quad -10 \leq x \leq 10 \quad (5.3)$$

$$f_4(x) = (x + \sin(x))e^{-x^2}, \quad -10 \leq x \leq 10. \quad (5.4)$$

$$f_5(x) = -\sum_{i=1}^{10} \frac{1}{(k_i(x - a_i))^2 + c_i}, \quad 0 \leq x \leq 10. \quad (5.5)$$

The k_i, a_i and c_i are parameters chosen and are varied to create different problems. $f_6(x)$ has the same functional form as $f_5(x)$. The parameters used in $f_5(x)$ and $f_6(x)$ are the same as those in [16].

The performance of the the LLB algorithm can now be compared with other one dimensional algorithms as reported in [16]. To be consistent with the performance figures reported, the stepsize $\varepsilon = 10^{-6}$ was used. In [16] this same ε of 10^{-6} was used in the stopping criteria of the algorithms reported there. Table 1 compares function evaluations among competing algorithms. The number of function evaluations for the LLB algorithm is under the column heading LLB. The row heading $f_7(x)$ actually refers to an average of 100 trials for a function of the form of $f_5(x)$ in (5.5). One chooses the coefficients at random, using the uniform distribution, from $a_i \in [0,10]$, $k_i \in [1,3]$ and $c_i \in [0.1,0.3]$.

Table 1: Comparison with Other Algorithms

function	LLB	Zil 1	Zil 2	Strong	Pijav	Brent	Batish
f_1	7	33	29	45	462	25	120
f_2	7	37	38	442	448	45	158
f_3	15	125	165	150	3817	161	816
f_4	7	35	34	98	376	229	83
f_5	29	42	41	102	280	294	484
f_6	24	45	44	69	624	492	325
f_7	18	32	44	94	360	376	422

The column headings correspond to the following algorithms:
(1) LLB: The LLB algorithm presented in this paper.
(2) Zil1: The P^* algorithm of Zilinskas [18].
(3) ZIL2: The algorithm of Zilinskas in [19].
(4) Strong: The algorithm of Strongin in [15].
(5) Pijav: The algorithm of Pijavskij and Shubert in [14].
(6) Brent: The algorithm of Brent in [3].
(7) Batish: The algorithm of Batishchev in [2].

If the number of function evaluations is the sole performance criterion, then the LLB algorithm certainly excels all the other algorithms in Table 1. However this is not a fair comparison since the calculation of the linear lower bound about a given point also yields the function value at that point, so that additional function evaluations are not necessary. In Table 2, the number of function evaluations, the number of derivative evaluations and the number of linear lower bound evaluations for the LLB algorithm are listed. For $f_7(x)$ the numbers are averaged from 100 different random trials of functions of the form of $f_5(x)$. The summation of these three numbers is also given. In this case, the worst scenario of our algorithm is compared with other algorithms, since there is not much overhead beyond those evaluations. An examination of Table 2 reveals that the LLB algorithm is still competitive with existing algorithms. For extra information, the CPU time using the "time" command on VAX 11/780 is negligible for f_1, f_2, f_4, 0.4 second for f_3, f_5, 0.3 second for f_6, and 33.2 seconds for f_7.

Table 2: Performance of the LLB Algorithm

function	No. Func.	No. Deriv.	No. Low. Bnd.s	Total No.	minimum
f_1	7	15	24	46	-4.601308
f_2	7	10	12	29	-1.905961
f_3	15	57	167	239	-3.372898
f_4	7	12	26	45	-0.824239
f_5	29	64	103	196	-14.163855
f_6	24	55	85	164	-13.922345
f_7	18	35	40	93	**

In Table 3, the performance of the LLB algorithm is presented for $\varepsilon = 10^{-3}$ while keeping all the other parameters the same. It is not surprised to see that global minima are also achieved in this case, since the test problems have no local minimum different from the global one by $\varepsilon = 10^{-3}$. The global minimum is typically found pretty early. Most of iterations are used to verify that the solution is indeed an ε-global minimum. When ε is smaller, more time is used for the verification. As anticipated, we can see that the total number of evaluations in Table 3 is much less than those in Table 2, particularly true when more numbers of evaluations are needed.

Table 3: Performance of the LLB Algorithm with $\varepsilon = 10^{-3}$

function	No. Func.	No. Deriv.	No. Low. Bnd.s	Total No.	minimum
f_1	7	13	18	38	-4.601308
f_2	7	10	12	29	-1.905961
f_3	8	31	116	145	-3.372898
f_4	7	11	23	41	-0.824239
f_5	29	50	66	145	-14.163855
f_6	24	47	60	131	-13.922345
f_7	18	31	31	80	**

To conclude this section, a few words regarding the implementation of the LLB algorithm is in order here. Although the LLB algorithm has been presented for general problems, each problem still has its own special structure. We can use this special structure to enhance the algorithm's performance. For example by following the methods presented in section 4, a given cost function may have many ways to obtain a linear lower bound. The structure of the cost function however may suggest one way which is superior to other methods of obtaining the linear lower bound. An example of this occurs in the decomposition of polynomial functions mentioned before equation (4.42).

Heuristics can also be used to improve algorithm performance. For example, an interval is bisected when it is found that more than a certain fraction of the interval remains after the interval is reduced during step (3) of the Linear Lower Bound Algorithm. Also if a local minimum is known to exist in the interior of a given interval, it might be a good idea to bisect the interval about that local minimum point. Finally, dividing the initial interval into several subintervals at the beginning of the algorithm can improve algorithm performance as measured by the number of function evaluations required to converge.

6. Conclusion

This paper has demonstrated that finding the linear lower bounds to known cost functions can be a very effective means of obtaining global minimizers. The level sets of the linear lower bounds are utilized to eliminate unpromising regions. It was shown in fact that this procedure leads to an algorithm which converges to an ε solution in a finite number of iterations. Although the algorithm requires the development of linear lower bounds for the cost function, this paper has demonstrated that for virtually any function that one can express algebraically, one can also express a formula for its linear lower bound. Thus a large class of cost functions can be minimized by the LLB method. Furthermore, the numerical testing of the one dimensional algorithm demonstrates that this algorithm is highly competitive with existing algorithms.

Regarding the numerical testing, the best way to compare algorithms is to find their CPU time needed since they have to be run on computers to solve real problems. However it is known that a fair comparison is typically difficult unless efficient codes are written by the same group of people and run on the same computer. As mentioned in [16], the number of function evaluations reported in Table 1 is most likely close to the lowest numbr the authors can get. In our case, the number presented is also a better result. In general the algorithm could be sensitive to certain parameters and heuristics used, such as the number of initial intervals created. Even though our algorithm works for a range of parameters, it is still better to have a more robust algorithm. Also the algorithm performance depends upon the linear lower bounds generated. Better linear lower bounds should be used when feasible. Recently we have performed research along both directions. Preliminary numerical testing indicates our new algorithm is faster and more robust. Once the new approach is thoroughly studied, the results will be reported.

A careful examination of the LLB concepts offered above, show that almost all of the material presented can be extended to the multi-dimensional case. Instead of finding linear lower bounds on intervals, simplices are used instead. A conceptual difficutly arises when a simplex is truncated by intersecting it with a level set of a linear lower bound as in step 3 of the LLB algorithm. In order to deal with this problem it is necessary to develop a theory which allows us to decompose truncated simplices into several simplices. Preliminary numerical testing of the n-dimensional algorithm seems to indicate that it will also be competitive with existing algorithms. Thus the extension of the LLB algorithm will be reported in a forthcoming paper.

References

[1] M. Avriel, "Nonlinear Programming: Analysis and Methods," Prentice Hall, 1976.

[2] D.I. Batishchev, "Search Methods of Optimal Design", Sovetskoje Radio, 1975 216 p (in Russian).

[3] R.P. Brent, "Algorithms for minimization without Derivatives", Prentice-Hall New Jersey 1973 195 p.

[4] M. Bromberg, T. S. Chang, P. B. Luh, "A Concept for Global Optimization Using the Function Imbedding Technique", Proceedings of the 1989 American Control Conference, June 1989, pp 786-793.

[5] M. Bromberg, T.S. Chang, "A Function Imbedding Technique for a Class of Global Optimization Problems: One Dimensional Global Optimization", Proceedings of the 28th Conference on Decision and Control Tampa, Florida, Dec. 1989, pp 2451-2456.

[6] E.R. Hansen, "Global Optimization Using Interval Analysis: The One-Dimensional Case", Journal of Optimization Theory and

Applications Vol. 29, No. 3, Nov. 1979, pp. 331-344.

[7] E. Hansen, "Global Optimization Using Interval Analysis - The Multi-Dimensional Case," Numerische Mathematik, 34, 1980, pp. 247-270.

[8] P. Hansen, B. Jaumard, S.H. Lu, "Some Further Results on Monotonicity in Globally Optimal Design", submitted to ASME Journal on Mech. Trans. & Auto. in Des..

[9] R. Horst and H. Tuy, "On the Convergence of Global Methods in Multiextremal Optimization", Journal of Optimization Theory and Applications: Vol. 54, No. 2, August 1987, pp. 253-271.

[10] G.P. McCormick, "Computability of Global Solutions to Factorable Nonconvex Programs : Part I - Convex Underestimating Problems." Mathematical Programming, Vol. 10, No. 2 April 1976, pp. 147-175.

[11] C.C. Meewella and D.Q. Mayne, "Efficient Domain Partitioning Algorithms for Global Optimization of Rational and Lipschitz Continuous Functions", Journal of Optimization Theory and Applications, Vol. 61, No 2, May 1989, pp. 247-271.

[12] R.E. Moore, "Interval Analysis", Prentice-Hall, Englewood Cliffs, New Jersey, 1966.

[13] R. T. Rockafellar, "Convex Analysis", Princeton University Press, 1970.

[14] Shubert, B.O."A sequential Method for Seeking a Global Minimum of a Function", SIAM, J. of Numerical Analysis, 9:3, 1972, pp.379-388.

[15] R.G. Strongin, "Numerical methods of Multiextremal Optimization", Nauka Moscow 239 p (in Russian).

[16] Aimo Torn, Antanas Zilinskas, "Global Optimization", Springer Verlag 1989.

[17] D. J. Wilde, "Globally Optimal Design" , John Wiley & Sons, Inc., 1978.

[18] A. Zilinskas, "Two algorithms for One-dimensional Multimodal Minimization", 1981 Math. Operat. Stat. 12, Ser. Optimization, 53-63.

[19] A. Zilinskas, "Optimization of One-dimensional Multimodal Functions", 1978 Algorithm AS 133, Applied Statistics 23, 367-375.

Optimizing the Sum of Linear Fractional Functions

James E. Falk[†]   Susan W. Palocsay[§]

Abstract

In this paper, we present a new algorithm that will yield a global solution to the problem of maximizing a sum of ratios of linear functions over linear polyhedra. First, we describe an applied problem related to this algorithm and discuss the incorrect results obtained by Almogy and Levin (1971) for problems with this structure. Then we present the details of our algorithm, including the necessary theoretical foundations, for the case of two ratios and solve several example problems. We also describe a computer implementation of the algorithm and discuss some computational considerations. Finally, we indicate how the algorithm can be extended to problems involving the sum of more than two ratios.

1. Introduction

The term "fractional program" refers to a nonlinear program which involves the optimization of a ratio of functions. These ratio problems appeared in the literature as early as 1956 when Isbell and Marlow considered the strategic military problem of deciding on a distribution of fire over enemy targets of several types. The formulation of this minimax problem led them to develop an iterative process for optimizing a linear fractional function subject to linear constraints. In 1962, the classical paper in the field by Charnes and Cooper appeared. They showed that any linear fractional programming problem can be replaced with at

---

[†]Department of Operations Research, George Washington University, Washington, D.C. 20052.
[§]Department of Information and Decision Sciences, James Madison University, Harrisonburg, VA 22807.
This work was supported in part by Office of Naval Research contract N00014-90-J-1537.

most two linear programming problems that differ only slightly from each other depending on the sign of the denominator at optimality. We use their result to solve the linear fractional subproblems of our algorithm to maximize a sum of ratios.

Schaible (1981) summarizes the considerable amount of work which has been done since 1962 on the problem of minimizing a single ratio of linear functions. Of particular note is the paper by Dinkelbach (1967) who introduced an innovative algorithm to solve this problem. However, none of the methods heretofore introduced extend to the general case where the objective function consists of a sum of ratios. A search of the literature indicates that there have been very few results that address this problem except for special cases (Schaible, 1981a). Ritter (1967) solved problems with different types of objective functions including the sum of a linear function and a linear fractional function using a parametric quadratic programming method. Schaible (1977) also treated this special case and showed that in general the objective function is neither quasi-convex nor quasi-concave. In another approach to the problem, Cambini, Martein, and Schaible (1989) developed an algorithm for the case of two ratios based on the concept of "optimal level solutions", by which they generate a sequence of ever improving local solutions which must, in a finite number of steps, terminate at the desired globally optimal point.

Almogy and Levin (1971) presented theoretical results for problems where the objective function is a sum of ratios of linear or quasi-concave functions and algorithmic results for problems involving up to three separable linear ratios. However, they use a parametric approach based on Dinkelbach's work that we show is in error in Section 3. The development of this approach was motivated by the need to solve a real-world decision problem involving optimal shipping schedules. We describe this problem in Section 2 as an example of a fractional program whose objective function is a sum of linear ratios.

The general problem which we address in this paper has the form

$$\mathcal{P}: \quad \underset{x \in S}{\text{maximize}} \; f(x) = \sum_{i=1}^{m} \left(\frac{n_i(x)}{d_i(x)} \right)$$

where $S = \{x \mid Ax \leq b, \; x \geq 0\}$. The numerator n_i and denominator d_i of each ratio are the linear functions $c_i^T x + \gamma_i$ and $h_i^T x + \delta_i$, respectively, where c_i and h_i are n-component vectors and γ_i and δ_i are constants. We assume that the denominator of each ratio is positive for all feasible values of x and that S is compact.

The new algorithm presented here takes advantage of the special structure of the sum of ratios problem by transforming the problem into a "state space" where each ratio corresponds to a single state variable. The procedure begins by placing bounds on the feasible region (in the state space) to identify a subset of the feasible region that contains the solution. This subset is defined by an upper bound on each ratio and a linear isovalue contour through the current best feasible lower bound.

The bounds on the feasible subset are tightened iteratively to reduce the size of the search space until either an optimal solution is obtained or the procedure cannot improve the current bounds. If the latter occurs, subproblems defined at the current iteration are solved to restart the algorithm and to locate the solution. Both the initialization and the iterative parts of the procedure involve the solution of linear fractional programs which is accomplished using the Charnes and Cooper approach.

We present the algorithm for a sum of two ratios and related proofs in Section 4. A computer implementation of the algorithm is described in Section 5 and an extension of the algorithm for sums of more than two ratios is suggested in Section 6.

2. An Applied Problem for the Algorithm

We begin this paper by presenting a decision problem from the literature that demonstrates the need for an algorithm to maximize a sum of linear ratios. The formulation of this problem is based on a shipping company's problem that was studied by Almogy and Levin (1969).

In their paper, Almogy and Levin discuss the problem confronting a shipping company of determining the optimal amount of cargo to ship from port to port in order to maximize profit per unit time. According to the problem statement, a ship must make an ordered sequence of visits to N ports where each port has cargo available for shipping to the remaining ports to be visited. Almogy and Levin treat the shipping problem as an M-stage decision problem by grouping the N ports into M disjoint sets with the sequence maintained and M less than or equal to N. Under the assumption that the amounts of available cargo for each set of ports are independent discrete random variables, the problem becomes a stochastic multi-stage decision problem with the objective in each stage of maximizing expected profit per unit time in the subsequent stages.

The formulation of the shipping problem is developed by first defining a fractional function f_m for $m = 1, \ldots, M$, which gives the net profit per unit time in each stage. Net profit in a particular stage is defined as the difference in profit per unit time between sailing the subroute in stage m and staying at the first port in this set. Since it is assumed that all direct costs and loading and unloading times are linear functions of the cargo amounts, the function f_m is a ratio of linear functions. The objective function for the problem has the general form

$$\underset{\overline{Q}^1}{\text{maximize}} \{f_1 + \underset{\overline{Q}^2|\overline{Q}^1}{E\{\text{maximize}} \{f_2 + \cdots + \underset{\overline{Q}^m|(\overline{Q}^1,\ldots,\overline{Q}^{m-1})}{E\{\text{maximize}} \quad f_m\}\ldots\}$$

where $\overline{Q}^m$ is the decision vector in stage m for cargo amounts to be carried between ports. It is maximized subject to linear constraints that guarantee that a decision is not made to ship more cargo than is available or to overflow the capacity of the ship.

Almogy and Levin present the two-stage problem in detail before extending their discussion to the general multi-stage problem. In the two-stage problem, they assume for explanatory purposes that the first stage contains only the first port and the remaining ports are contained in the second stage. The resulting objective function is expressed as

$$\underset{Q}{\text{maximize}} \quad \{f_o + \sum_{p=1}^{k} f_p \cdot \delta_p\}$$

where f_o and $\{f_p, p = 1, \ldots, k\}$ are the fractional functions representing profit per unit time in stages one and two, respectively, and $\{\delta_p, p = 1, \ldots, k\}$ is the set of probabilities assigned to the discrete random variable for the amounts of available cargo in stage two.

Thus, the multi-stage shipping decision problem involves maximizing a sum of linear ratios subject to linear inequality constraints. The need to solve this particular problem led Almogy and Levin to develop a parametric algorithm based on an approach that is discussed in the next section.

3. Almogy and Levin's Parametric Approach

Before we begin the presentation of our algorithm, it is necessary (for background purposes) to discuss the results developed in the paper by Almogy and Levin (1971) on maximizing sums of ratios. In their paper, they parameterize the sum of linear ratios problem by using Dinkelbach's basic approach to reformulate the objective function as a linear function. Applied to problem $\mathcal{P}$, their parametric linear programming problem is

$$H(r) = \underset{x \, \varepsilon \, S}{\text{maximize}} \; D(x, r) = \{\sum_{i=1}^{m} [n_i(x) - r_i \cdot d_i(x)]\}$$

where r is the m-component vector of parameters r_i and n_i and d_i are linear.

Almogy and Levin attempted to generalize Dinkelbach's results for this special case by claiming to prove that x^* is the optimal solution of problem $\mathcal{P}$ if and only if x^* solves the equation $H(r) = 0$, and $r_i^* = n_i(x^*)/d_i(x^*)$. If this were in fact true, it would be a very powerful result. It states that Dinkelbach's results extend directly to the case of sums of linear ratios and we need only find the values of the parameters r_i that meet this criteria to solve problem $\mathcal{P}$.

Unfortunately, this theorem is false in general as we demonstrate with the following counterexample. Consider the problem

$$\underset{x \, \varepsilon \, S}{\text{maximize}} \; \left\{ \frac{37x_1 + 73x_2 + 13}{13x_1 + 13x_2 + 13} + \frac{63x_1 - 18x_2 + 39}{13x_1 + 26x_2 + 13} \right\}$$

subject to
$$5x_1 - 3x_2 = 3$$
$$1.5 \leq x_1 \leq 3.$$

This problem has its solution at $x^* = (3, 4)$ with $r^* = (4, 1)$, as determined by graphically solving the problem and verifying that both necessary and sufficient optimality conditions hold at x^*. The corresponding parametric problem, $H(r)$, is

$$H(r) = \underset{x \,\varepsilon\, S}{\text{maximize}} \; \{(37x_1+73x_2+13) - r_1\cdot(13x_1+13x_2+13) + (63x_1-18x_2+39) - r_2\cdot(13x_1+26x_2+13)\}$$
$$= \text{maximum} \; \{572-104r_1-156r_2, \; 284.5-52r_1-71.5r_2\}$$

since the feasible region S for this problem has only two extreme points, (3, 4) and (1.5, 1.5), as shown in Figure 1.

Evaluating $H(r)$ at r^* gives $H(4, 1) = $ maximum $\{0, 5\} = 5$ with maximizing point (1.5, 1.5). Thus, the Dinkelbach objective function $D(x, r)$ is not maximized at x^* for $r = r^*$ and the Almogy and Levin theorem fails. To see why this happens, we look at the gradients with respect to x of f and $D(x, r^*)$ at x^*. Their graphs are sketched in Figure 2. We see that the gradient of $D(x, r^*)$ is very close to the gradient of f but falls just outside of the cone of binding constraints at x^*.

Although Dinkelbach's results do not extend directly, the function $H(r)$ has some useful properties in the context of our algorithm. These properties will be discussed as we develop the algorithm in the next section.

4. Solution Algorithm

In this section, we present an algorithm for locating a global solution to problem $\mathcal{P}$, the problem of maximizing a sum of linear ratios over linear polyhedra. We assume that m is equal to two throughout this section to simplify the presentation of the algorithm. The extension of the algorithm for m greater than two is straightforward and covered in Section 6.

In our algorithm, problem $\mathcal{P}$ is transformed from "X-space" into the state space we call "R-space" by mapping each ratio in the objective function into a dimension of the new space. The feasible region in R-space is defined as the generally nonconvex set

$$T = \{(r_1, r_2) \mid r_1 = n_1(x)/d_1(x), \; r_2 = n_2(x)/d_2(x); \text{ for some } x \,\varepsilon\, S\}.$$

Thus, each point $x = (x_1, \ldots, x_n)$ in S maps to one point $r = (r_1, r_2)$ in T. However, this mapping is not one-to-one since it

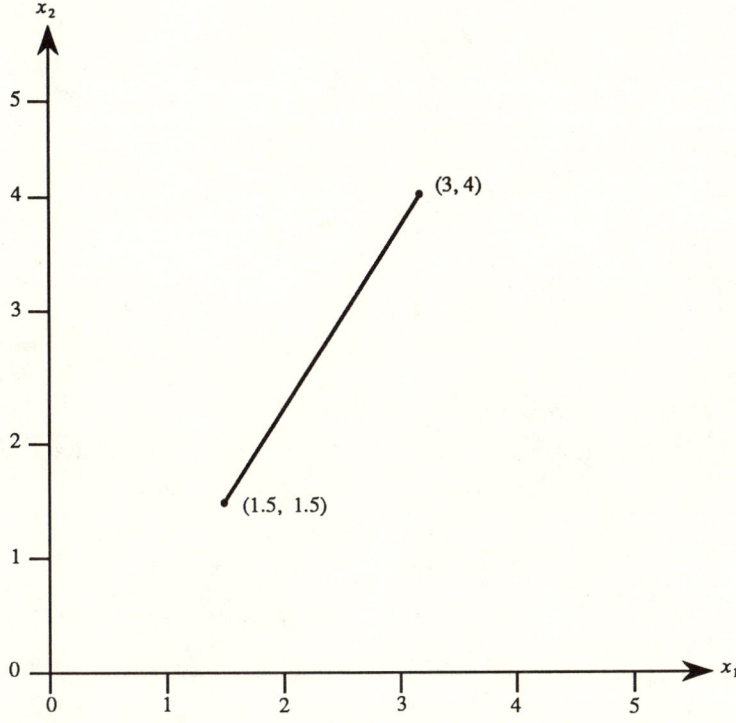

Figure 1
Feasible Region for Counterexample to
Almogy and Levin's Parametric Approach

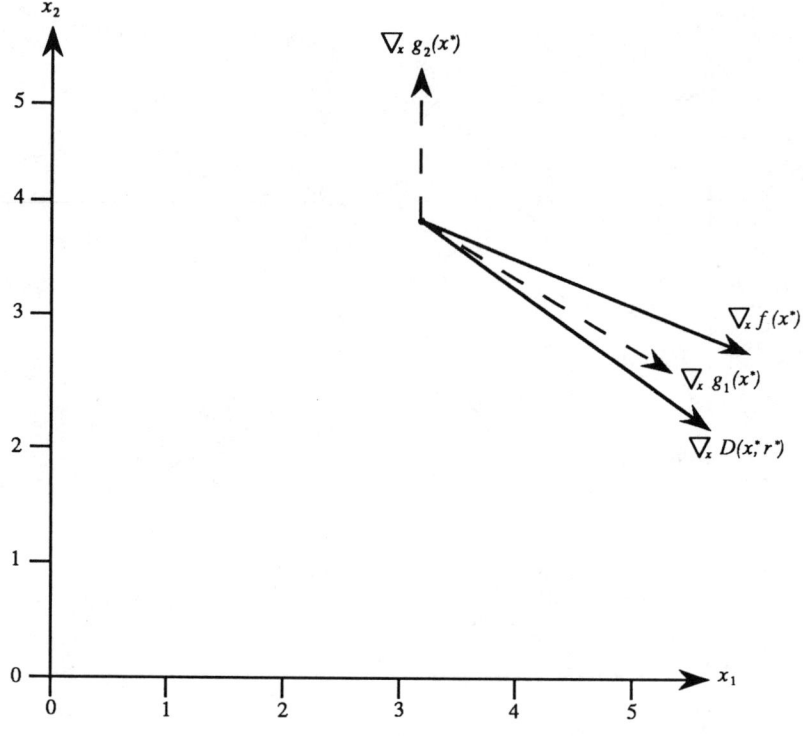

Figure 2
Gradients of f and D at x^* for
Counterexample to Almogy and Levin's Parametric Approach

is possible for more than one point x to map to the same point r. This mapping concept is illustrated in Figure 3.

When the problem is transformed into R-space, it becomes one of maximizing $r_1 + r_2$ over T. We observe that the new objective function is linear and therefore has linear isovalue contours. This observation is used to get initial bounds on a subset of R-space in which the solution is guaranteed to lie. Mathematically, we determine

$$u_1^0 = \operatorname*{maximize}_{x \,\varepsilon\, S} \frac{n_1(x)}{d_1(x)} \quad \text{and} \quad u_2^0 = \operatorname*{maximize}_{x \,\varepsilon\, S} \frac{n_2(x)}{d_2(x)}$$

to place initial upper bounds on r_1 and r_2, respectively. An upper bound on the subset of R-space containing the solution is $u^0 = (u_1^0, u_2^0)$ as shown in Figure 4. We use the notation $x^{1,0}$ and $x^{2,0}$ for solutions to the two optimization problems and note that these problems are linear fractional programs which can easily be solved, e.g., by using the Charnes and Cooper variable transformation. In this approach, solving the linear fractional problem

$$\begin{aligned}\text{maximize} \quad & \frac{c_i^T x + \gamma_i}{h_i^T x + \delta_i} \\ \text{subject to} \quad & Ax \leq b \\ & x \geq 0\end{aligned}$$

is equivalent to solving the linear programming problem

$$\begin{aligned}\text{maximize} \quad & \{c_i^T y + \gamma_i \cdot t\} \\ \text{subject to} \quad & Ay \leq b \cdot t \\ & h_i^T y + \delta_i \cdot t = 1 \\ & y, t \geq 0\end{aligned}$$

where the variable transformation $y = t \cdot x$ has been applied under the assumption that the denominator of the ratio is positive.

The solution points $x^{1,0}$ and $x^{2,0}$ provide the feasible points $r(x^{1,0})$ and $r(x^{2,0})$ in R-space where

$$r(x^{1,0}) = (r_1(x^{1,0}), r_2(x^{1,0})) = \left(\frac{n_1(x^{1,0})}{d_1(x^{1,0})}, \frac{n_2(x^{1,0})}{d_2(x^{1,0})}\right)$$

and

$$r(x^{2,0}) = (r_1(x^{2,0}), r_2(x^{2,0})) = \left(\frac{n_1(x^{2,0})}{d_1(x^{2,0})}, \frac{n_2(x^{2,0})}{d_2(x^{2,0})}\right).$$

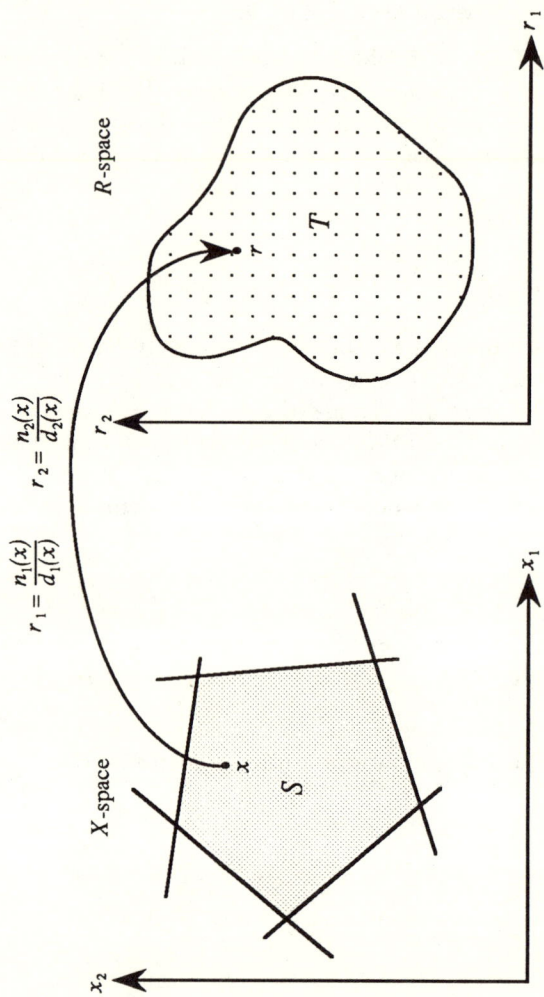

Figure 3
"X-space" to "R-space" Mapping Concept

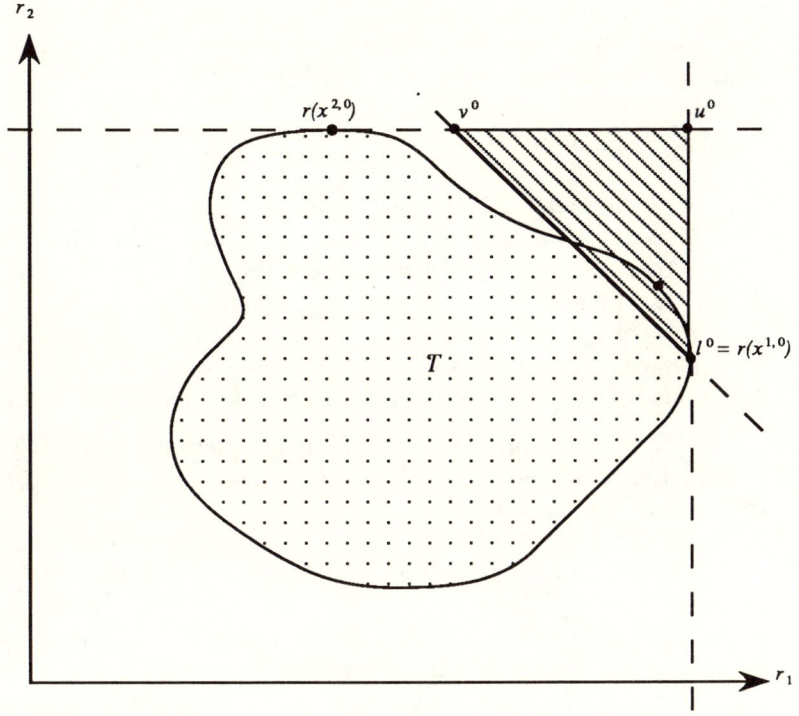

Figure 4
Initial Bounds on Optimal Solution in R-space

Both of these points are feasible and we select the one which provides the best lower bound for the feasible subset containing the solution, i.e., we choose the point which gives us

$$f_l = \text{maximum } \{(r_1(x^{1,0}) + r_2(x^{1,0})), (r_1(x^{2,0}) + r_2(x^{2,0}))\}.$$

Then we construct the isovalue contour $r_1 + r_2 = f_l$ and use it to determine an initial triangular subset of R-space that will contain the solution. We denote the second and third points of the triangle shown in Figure 4 as l^0 and v^0 where $l^0 = (l_1^0, l_2^0) = (u_1^0, f_l - u_1^0)$ and $v^0 = (v_1^0, v_2^0) = (f_l - u_2^0, u_2^0)$. Thus, we have reduced the size of the region in R-space to be searched and determined upper and lower bounds on the optimal solution value.

Now we return to the parametric function $H(r)$ which was discussed in the previous section on Almogy and Levin's results. Recall that $H(r)$ is defined as

$$H(r) = \underset{x \;\varepsilon\; S}{\text{maximize}} \;\; D(x, r) = \{\sum_{i=1}^{m} [n_i(x) - r_i \cdot d_i(x)]\}$$

in the case of m ratios and that we showed $H(r)$ is not required to be zero at the values of the parameters r_1^* and r_2^*. However, it turns out that $H(r)$ has several properties which can be applied to provide us a way of immediately determining, in some cases, if we have an optimal solution at this point in the algorithm. These properties are illustrated in Figure 5.

Theorem 1. $H(r) \geq 0$ for every $r \;\varepsilon\; T$.

Proof. Let $\tilde{x} \;\varepsilon\; S$ and $\tilde{r}_i = \dfrac{n_i(\tilde{x})}{d_i(\tilde{x})}$ so that $\tilde{r} \;\varepsilon\; T$. Then

$$H(\tilde{r}) = \underset{x \;\varepsilon\; S}{\text{maximize}} \;\; \{\sum_{i=1}^{m} [n_i(x) - \tilde{r}_i \cdot d_i(x)]\}$$

$$\geq \sum_{i=1}^{m} [n_i(\tilde{x}) - \tilde{r}_i \cdot d_i(\tilde{x})] = 0$$

and we have $H(\tilde{r}) \geq 0$ as desired. $\square$

Theorem 2. $H(r) > 0$ for every r such that $r_i < \dfrac{n_i(x^*)}{d_i(x^*)}$ for all i at x^*, the solution to problem $\mathcal{P}$.

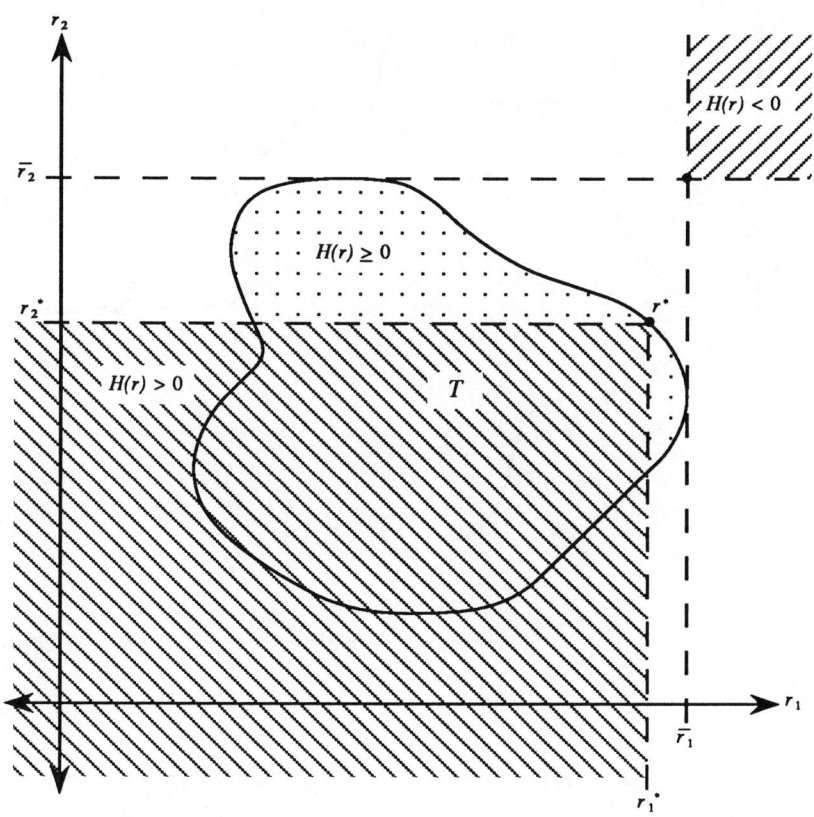

Figure 5
Properties of the Function $H(r)$

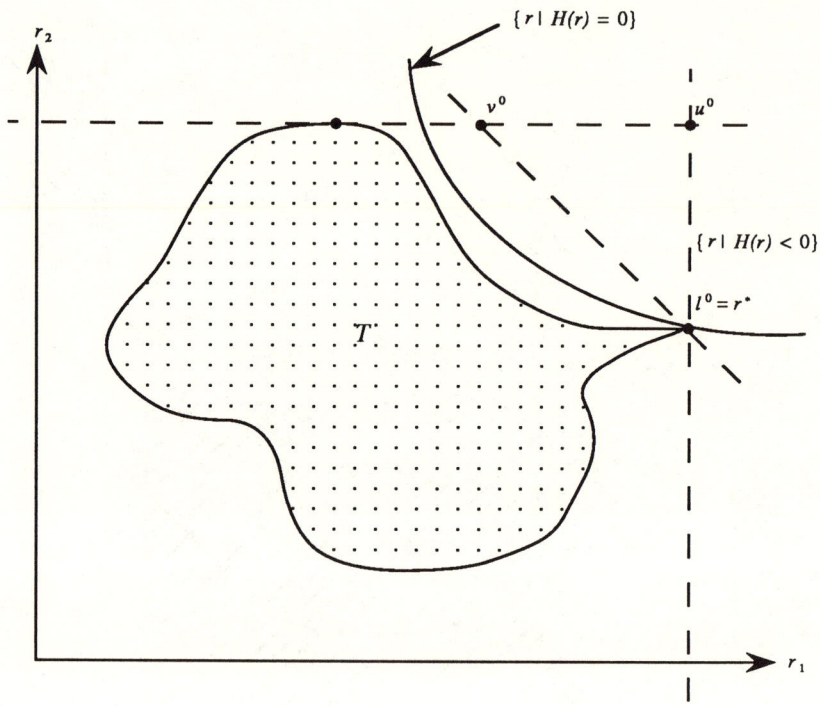

Figure 6
Determination of the Optimal Ratio During Initialization

Proof. Let x^* be optimal for problem $\mathcal{P}$ and set $\tilde{r}_i < r_i^*$ where $r_i^* = \frac{n_i(x^*)}{d_i(x^*)}$. Then

$$D(x^*, \tilde{r}) = \sum_{i=1}^{m} [n_i(x^*) - \tilde{r}_i \cdot d_i(x^*)] > 0$$

and therefore $H(\tilde{r})$, the maximum of $D(x, \tilde{r})$ over $x \in S$, is also strictly greater than zero. □

Theorem 3. $H(r) < 0$ for every r such that $r_i > \underset{x \in S}{\text{maximize}} \; \frac{n_i(x)}{d_i(x)}$ for every $i = 1, \ldots, m$.

Proof. Let $\overline{r}_i = \underset{x \in S}{\text{maximize}} \; \frac{n_i(x)}{d_i(x)}$ and $\tilde{r}_i > \overline{r}_i$. Then $n_i(x) - \tilde{r}_i \cdot d_i(x) < 0$ for every $x \in S$ which implies that

$$D(x, \tilde{r}) = \sum_{i=1}^{m} [n_i(x) - \tilde{r}_i \cdot d_i(x)] < 0$$

for every $x \in S$. Therefore $H(\tilde{r})$, the maximum of $D(x, \tilde{r})$ over $x \in S$, is also strictly less than zero. □

The convexity property of $H(r)$, which was proven by Dinkelbach for the single ratio case, also holds in the case of multiple ratios. This follows directly since H is the pointwise maximum of a linear (convex) function (Rockafellar, 1970).

Theorem 4. The function $H(r) = \underset{x \in S}{\text{maximize}} \; \{\sum_{i=1}^{m} [n_i(x) - r_i \cdot d_i(x)]\}$ is convex over E^m.

These properties allow us to determine in some cases that we have obtained an optimal solution during the initialization phase of the algorithm. This result is presented in the following theorem and illustrated in Figure 6.

Theorem 5. Let $r^1 \in T$ and $r^2, r^3 \in E^2$ such that $r^1, r^2,$ and r^3 are the extreme points of a triangular region in R-space. If $H(r^1) = 0$, $H(r^2) < 0$, and $H(r^3) < 0$, then $r^* = r^1$.

Proof. By the convexity of H in Theorem 4 and the conditions placed on $H(r^i)$ in the statement of this theorem, we have

$$H(\alpha_1 r^1 + \alpha_2 r^2 + \alpha_3 r^3) \leq \alpha_1 H(r^1) + \alpha_2 H(r^2) + \alpha_3 H(r^3) < 0$$

for $\sum_{i=1}^{3} \alpha_i = 1$, $0 \leq \alpha_i \leq 1$, and $\alpha_1 \neq 1$.

Since we know $H(r) \geq 0$ for every $r \ \varepsilon \ T$ by Theorem 2, there exists no point in the triangular region defined by r^1, r^2, and r^3 that can be a feasible solution except r^1. Therefore, r^* must be r^1. □

The three points u^0, l^0, and v^0 are located during the initialization phase of the algorithm in order to define an initial triangular subset of R-space containing the solution. While we know that $H(r^*)$ is not required to be zero, we can use the result in Theorem 5 to check for optimality when H is zero at either l^0 or v^0. For example, if $H(l^0)$ is zero, then if $H(v^0)$ and $H(u^0)$ are both negative, we know that r^* is the point l^0.

If H is not zero at either of these points or if the optimality check fails, then we are interested in efficiently searching for the optimal solution in the triangular region determined during the initialization phase of the algorithm. The iterative method we use is one of alternatively solving two linear fractional programs which maximize r_1 and r_2, respectively, in the defined region. In each of these fractional programs, we have replaced one of the ratios in the objective function of problem $\mathcal{P}$ with a parameter t and added a constraint on t based on the replaced ratio to obtain the problems

$$\text{maximize } \frac{n_1(x)}{d_1(x)} + t_2$$
$$\text{subject to}$$
$$x \ \varepsilon \ S$$
$$t_2 \leq \frac{n_2(x)}{d_2(x)} \quad \text{for fixed } t_2$$

and

$$\text{maximize } t_1 + \frac{n_2(x)}{d_2(x)}$$
$$\text{subject to}$$
$$x \ \varepsilon \ S$$
$$t_1 \leq \frac{n_1(x)}{d_1(x)} \quad \text{for fixed } t_1$$

where the parameters t_1 and t_2 are chosen at each iteration so that r_1 and r_2 remain within the limits of the rectangular region defined by the points u, v, and l of the current triangular region and the unlabeled point (v_1, l_2). The solutions of these problems are used to slice off pieces of the triangular region both vertically and horizontally and thereby reduce the size of the search space. Since we use the Charnes and Cooper tranformation for linear fractional programs, we are solving a sequence of linear programs.

As these iterations continue the best upper bound does not increase and the best lower bound does not decrease. If the best upper and lower bounds converge to a common value, it must be optimal.

However, in general, the procedure could "stall" and not be able to further reduce the triangular search region. This situation is illustrated in Figure 7. Note that stalling occurs at iteration k only when the two points v^k and l^k are feasible and on the same isovalue contour.

If the procedure stalls, we then (arbitrarily) divide the rectangle $[(v_1^k, l_2^k), l^k, u^k, v^k]$ into two rectangles

$$[(v_1^k, l_2^k), \tfrac{1}{2}((v_1^k, l_2^k) + l^k), \tfrac{1}{2}((v^k + l^k), v^k]$$

and

$$[\tfrac{1}{2}((v_1^k, l_2^k) + l^k), l^k, u^k \, \tfrac{1}{2}((v^k + l^k)]$$

as shown in Figure 8. We then consider two subproblems defined by the bounds on the two newly created rectangles. Note that if the best solution of either of these subproblems is better than the feasible points v^k and l^k, the basic procedure will continue with this better point defining a single new triangle which is known to contain the optimal solution.

Note that also if the solutions of the new subproblems are both worse than v^k and l^k, then the basic procedure applied to each subproblem will continue.

The only difficulty that remains is in the (unlikely) case that the best solution of one or both of the subproblems gives a value to $r_1 + r_2$ identical to v^k and l^k. In such a case, we would continue to split the rectangle(s) until either one happens on a point which gives a different value to $r_1 + r_2$ than the common value of v^k and l^k, or the area of the region known to contain an optimal ratio is below some given tolerance.

Note that the "area of uncertainty" is, in fact, halved in the worse case scenario where the new split fails to restart a stalled problem (see Figure 9).

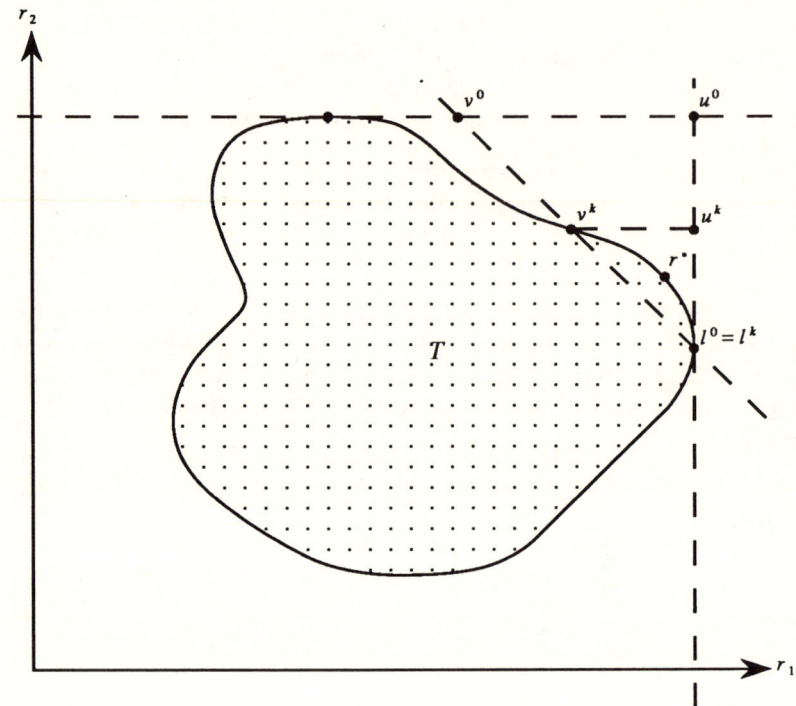

Figure 7
Stalling in the Iterative Procedure

SUM OF LINEAR FRACTIONAL FUNCTIONS

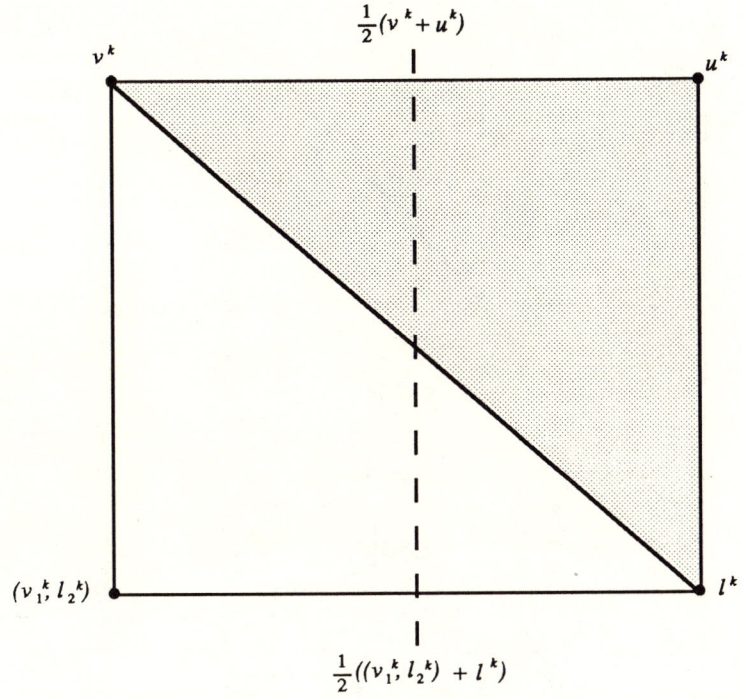

Figure 8
Restarting a Stalled Problem

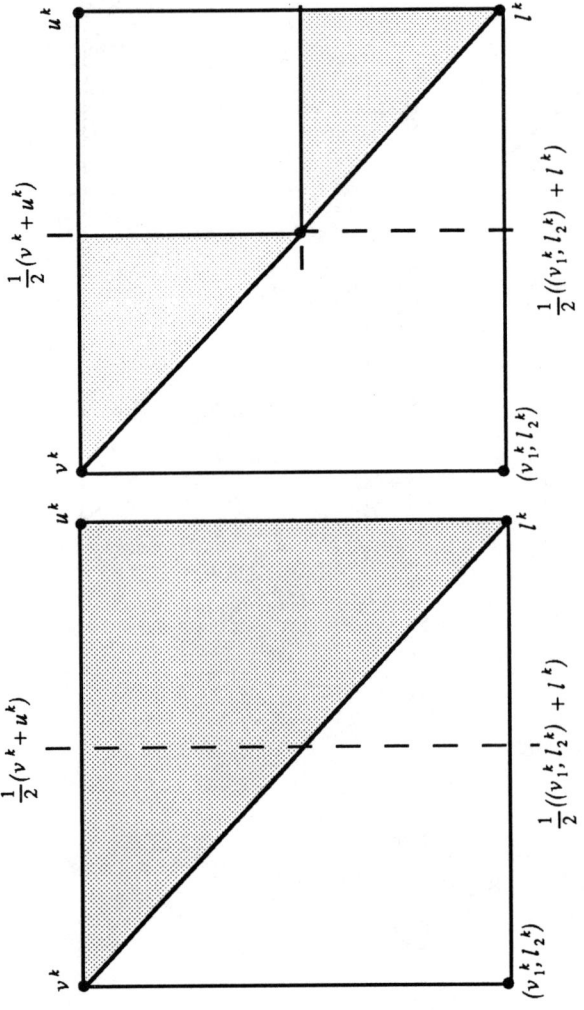

Figure 9
Region Containing the Optimal Ratio After a Split

Now we summarize the steps of the algorithm.

Step 1. Set $k = 0$ and determine an initial upper bound $u^0 = (u_1^0, u_2^0)$ on the optimal solution where

$$u_1^0 = \underset{x \in S}{\text{maximize}} \; \frac{n_1(x)}{d_1(x)} \quad \text{and} \quad u_2^0 = \underset{x \in S}{\text{maximize}} \; \frac{n_2(x)}{d_2(x)}.$$

Denote the solutions by $x^{1,0}$ and $x^{2,0}$.

Step 2. Compute the feasible points $r(x^{1,0})$ and $r(x^{2,0})$ where

$$r(x^{1,0}) = (r_1(x^{1,0}), r_2(x^{1,0})) = \left(\frac{n_1(x^{1,0})}{d_1(x^{1,0})}, \frac{n_2(x^{1,0})}{d_2(x^{1,0})} \right)$$

and

$$r(x^{2,0}) = (r_1(x^{2,0}), r_2(x^{2,0})) = \left(\frac{n_1(x^{2,0})}{d_1(x^{2,0})}, \frac{n_2(x^{2,0})}{d_2(x^{2,0})} \right).$$

Then compute f_l where

$$f_l = \text{maximum} \; \{(r_1(x^{1,0}) + r_2(x^{1,0}), (r_1(x^{2,0}) + r_2(x^{2,0}))\}.$$

Step 3. Determine the remaining points of the initial triangle:

$$l^0 = (l_1^0, l_2^0) = (u_1^0, f_l - u_1^0)$$

and

$$v^0 = (v_1^0, v_2^0) = (f_l - u_2^0, u_2^0).$$

Step 4. Compute $H(u^0)$, $H(l^0)$, and $H(v^0)$. If $H(l^0) = 0$, $H(v^0) < 0$, and $H(u_0) < 0$, then stop with $r^* = l^0$. If $H(v^0) = 0$, $H(l^0) < 0$, and $H(u_0) < 0$, then stop with $r^* = v^0$. The optimal value of the objective function of problem $\mathcal{P}$ is $r_1^* + r_2^*$ and an optimal global solution is x^* such that

$$\frac{n_1(x^*)}{d_1(x^*)} = r_1^* \quad \text{and} \quad \frac{n_2(x^*)}{d_2(x^*)} = r_2^*.$$

Step 5. Set $k = k + 1$. With l^{k-1} given, solve the problem

$$\text{maximize } \frac{n_1(x)}{d_1(x)} + l_2^{k-1}$$
$$\text{subject to}$$
$$x \in S$$
$$l_2^{k-1} \leq \frac{n_2(x)}{d_2(x)}.$$

Denote a solution by $x^{1,k}$ and let $u_1^k = \frac{n_1(x^{1,k})}{d_1(x^{1,k})}$. Then with v^{k-1} given, solve the problem

$$\text{maximize } v_1^{k-1} + \frac{n_2(x)}{d_2(x)}$$
$$\text{subject to}$$
$$x \in S$$
$$v_1^{k-1} \leq \frac{n_1(x)}{d_1(x)}.$$

Denote the solution by $x^{2,k}$ and let $u_2^k = \frac{n_2(x^{2,k})}{d_2(x^{2,k})}$.

Note that at least one of these two problems is already solved since one of the two points l^k or v^k is both feasible and a lower bound defined by u_1^0 or u_2^0, respectively.

Step 6. Determine the remaining points of the new triangle based on the new upper bound:

$$l^k = (l_1^k, l_2^k) = (u_1^k, f_l - u_1^k)$$

and

$$v^k = (v_1^k, v_2^k) = (f_l - u_2^k, u_2^k).$$

Step 7. If $u^k \doteq l^k$, then stop. The optimal value of the objective function of problem $\mathcal{P}$ is $l_1^k + l_2^k$ and an optimal global solution is x^* such that

$$\frac{n_1(x^*)}{d_1(x^*)} = l_1^k \quad \text{and} \quad \frac{n_2(x^*)}{d_2(x^*)} = l_2^k.$$

Step 8. If $u^k \neq u^{k-1}$, then return to Step 5.

Step 9. Otherwise $u^k \doteq u^{k-1}$ and the iterative procedure has stalled. Relabel f_l as $\hat{f}_l$. Repeat Step 1 adding the constraint

$$v_1^k \le \frac{n_1(x)}{d_1(x)} \le \tfrac{1}{2}(v_1^k + l_1^k)$$

to both maximization problems. Then repeat Step 2 relabeling f_l as $\overline{f}_l$. Repeat Steps 1 and 2 again with the added constraint

$$\tfrac{1}{2}(v_1^k + l_1^k) \le \frac{n_1(x)}{d_1(x)} \le l_1^k$$

in Step 1 and f_l relabed $\widetilde{f}_l$ in Step 2.

Select $f_l = $ maximum $\{\overline{f}_l, \widetilde{f}_l\}$. If $f_l > \hat{f}_l$, then set $u^0 = u^k$ and return to Step 3. Otherwise $f_l \le \hat{f}_l$ so execute Steps 3 through 9 of the algorithm in each of the two regions defined by the two constraints above on $n_1(x)/d_1(x)$. □

We conclude our presentation of the algorithm in this section with a description of a heuristic step that can sometimes be used to "speed up" the algorithm. It was added to the procedure after experimentation with several example problems indicated that it is sometimes possible to use information obtained when computing $H(r)$ at the three points of the initial triangular region in Step 4 to tighten the bounds on the search space before proceeding with the iterative part of the algorithm.

Specifically, when we compute $H(u^0)$, $H(l^0)$, and $H(v^0)$ we identify the points in S that maximize $D(x,r)$ for each of these three points. Since H is a linear program, we know that these solution points will be extreme points of S. Theoretically, it is possible to determine the areas of R-space where each extreme point of S solves $H(r)$. Let $\{\widetilde{x}^i\}$ denote the set of extreme points of S. Then this process consists of finding all the values of r for each extreme point $\widetilde{x}^j$ such that $D(\widetilde{x}^j, r) \ge D(\widetilde{x}^i, r)$, for all $i \ne j$. A subset of the lines $D(\widetilde{x}^j, r) = D(\widetilde{x}^i, r)$ will define these areas of the state space as shown in Figure 10. We are interested in the lines along which $H(r)$ equals zero in each of these areas.

In the algorithm, we take each unique pair of extreme points $\widetilde{x}^j$ and $\widetilde{x}^i$ obtained in the computation of $H(u^0)$, $H(l^0)$, and $H(v^0)$ and locate the intersecting points of the lines $D(\widetilde{x}^j, r) = 0$ and $D(\widetilde{x}^i, r) = 0$. If the value of $r_1 + r_2$ is greater than the value f_l associated with the initial isovalue contour, and a corresponding

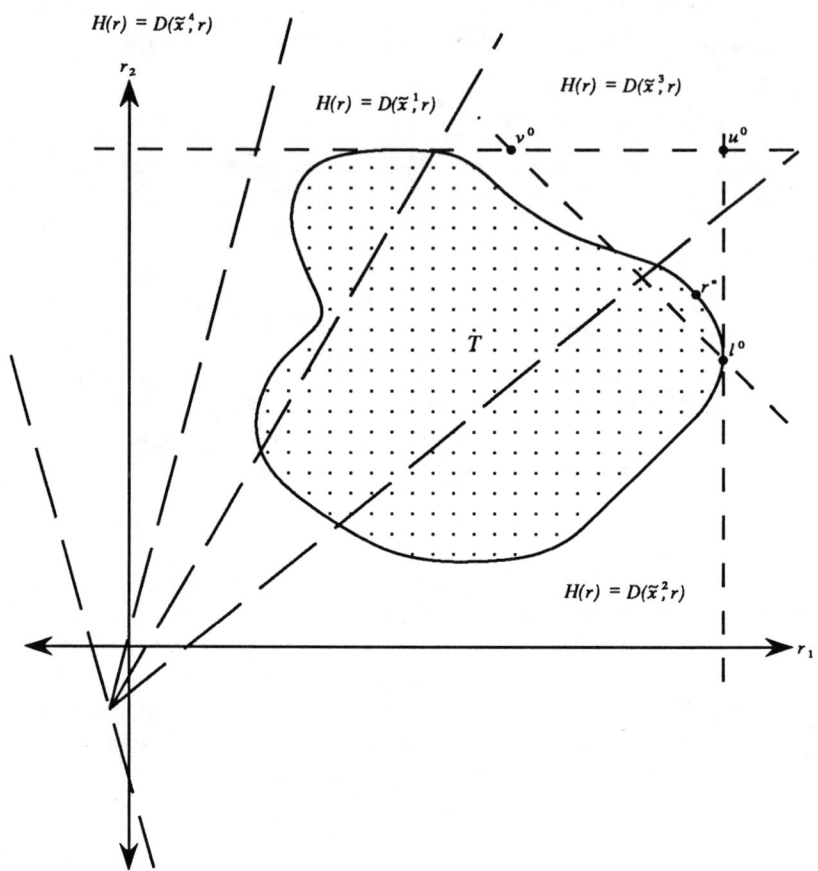

Figure 10
Extreme Point Solutions of $H(r)$

point in X-space is feasible, then we construct a new isovalue contour and revise l^0 and v^0 accordingly.

This step, which is outlined below, would occur after Step 4 of the algorithm to use our knowledge of the vertices associated with $H(u^0)$, $H(l^0)$, and $H(v^0)$ to look for a better isovalue contour value f_l before iterating in Step 5.

Heuristic Step. Denote the solutions of $H(u^0)$, $H(l^0)$, and $H(v^0)$ as $\tilde{x}^1$, $\tilde{x}^2$, and $\tilde{x}^3$, respectively. For each unique pair of extreme points $\tilde{x}^j$ and $\tilde{x}^i$, find the intersecting point of the lines $D(\tilde{x}^j, r) = 0$ and $D(\tilde{x}^i, r) = 0$ and denote it $\tilde{r}$. If $\tilde{r}_1 + \tilde{r}_2 \geq f_l$ and $\tilde{x}$ such that

$$\frac{n_1(\tilde{x})}{d_1(\tilde{x})} = \tilde{r}_1 \quad \text{and} \quad \frac{n_2(\tilde{x})}{d_2(\tilde{x})} = \tilde{r}_2$$

is feasible ($\tilde{x} \in S$), then let $f_l = \tilde{r}_1 + \tilde{r}_2$ and reset

$$l^0 = (l_1^0, l_2^0) = (u_1^0, f_l - u_1^0)$$

and

$$v^0 = (v_1^0, v_2^0) = (f_l - u_2^0, u_2^0).$$

5. Computer Implementation

A computer implementation of the algorithm presented in Section 4 was developed in a FORTRAN program called RATIO which interfaces with the eXperimental Mathematical Programming (XMP) package. The XMP package, which was developed by Dr. Roy E. Marsten (1981), is a collection of FORTRAN subroutines for optimization that can solve linear, linear mixed-integer, convex piecewise linear, and nonlinear programs. All of the routines for solving these different types of problems use the XMP code for linear programming which will solve a linear program using either the primal or the dual simplex method and will perform both sensitivity and parametric analyses.

The program RATIO uses the XMP package to solve the linear programming problems associated with the linear fractional programming problems in the initialization and iterative steps of the algorithm. Specifically, the following subset of XMP files is required to support the program RATIO: LPDEMO, XMAPS, MPSIN, MPSOUT, HPSORT, XLP1, XLP2, XLP3, XLP4,

XDATA, XRANG, LA05, and XDUALSM. These files contain the FORTRAN source code for the XMP subroutines involved in solving linear programming problems using the primal simplex method. They are treated as "black box" routines by RATIO with the exception of the two files LPDEMO and MPSOUT which were modified to manage file input and output operations. The XMP Technical Reference Manual (Marsten, 1987) contains additional detailed information on the XMP routines, the hierarchy structure of the XMP routines, and the XMP data structures.

In addition to XMP routines, the program RATIO contains subroutines for performing various functions involved in the algorithm. These routines are organized into three files: MAINPR which contains the main program, INIT which contains the initialization routines (for Steps 1 through 4), and ITRATE which provides the routines for the algorithmic iterations (in Steps 5 through 8). Because the program was written for the purpose of experimenting with the basic algorithm during research, the step to restart a stalled procedure (Step 9) and the heuristic step of the algorithm are not currently implemented.

The program RATIO is designed to solve problems of maximizing the sum of two linear ratios over a linear polyhedron. It can presently handle up to 20 variables and up to 20 constraints as defined by the dimension statements in the program. These statements may be changed if it is necessary to solve larger problems. The program requires that the problem data be provided in a slightly modified Mathematical Programming System (MPS) format to simplify the generation of MPS input files for the XMP package.

Now we use the program RATIO with several example problems to illustrate the results of the algorithm for solving sum of ratios maximization problems. The first problem to be discussed is the counterexample to Almogy and Levin's work which was discussed in Section 3. The problem is restated here for convenience:

$$\text{maximize} \left\{ \frac{37x_1 + 73x_2 + 13}{13x_1 + 13x_2 + 13} + \frac{63x_1 - 18x_2 + 39}{13x_1 + 26x_2 + 13} \right\}$$

subject to
$$5x_1 - 3x_2 = 3$$
$$1.5 \leq x_1 \leq 3.$$

Figure 11 shows the representation of this problem in R-space. The initial upper bounds placed on r_1 and r_2 in Step 1 of the algorithm are $u_1^0 = 4.00$ and $u_2^0 = 1.49$, respectively. Using the solution points $x^{1,0} = (3.00, 4.00)$ and $x^{2,0} = (1.50, 1.50)$ obtained from computing these bounds, we identify the feasible points $r(x^{1,0}) = (4.00, 1.00)$ and $r(x^{2,0}) = (3.42, 1.49)$.

Thus the initial isovalue contour is determined to be $f_l = 5.0$ which is the maximum of $\{5.00, 4.91\}$. The initial triangular region, based on an initial isovalue contour value of 5.00, is defined in Step 3 of the algorithm by the points $u^0 = (4.00, 1.49)$, $l^0 = (4.00, 1.00)$, and $v^0 = (3.51, 1.49)$ as shown in Figure 11.

Computing $H(r)$ at each of these three points in Step 4 yields $H(u^0) = -30.00$, $H(l^0) = 5.00$, and $H(v^0) = -4.55$. Although we know from Section 3 that the optimal solution in R-space occurs at $r^* = l^0$, we cannot determine that we have obtained the optimal solution at this point in the algorithm because $H(l^0)$ is not equal to zero (see Theorem 6).

Therefore, we proceed by executing the iterative steps of the algorithm (Steps 5 through 8) to solve this problem. The upper bound u converged to the value of the lower bound l^0 and identified l^0 as an optimal solution in R-space after 20 iterations. The corresponding solution point in X-space is $(3.00, 4.00)$ with an optimal objective function value of 5.00.

The second example problem considered is a problem whose constraints were presented in a problem solved by Falk and Hoffman (1976). It is of the form

$$\text{maximize} \left\{ \frac{3x_1 + x_2 - 2x_3 + 0.8}{2x_1 - x_2 + x_3} + \frac{4x_1 - 2x_2 + x_3}{7x_1 + 3x_2 - x_3} \right\}$$
subject to
$$x_1 + x_2 - x_3 \leq 1$$
$$-x_1 + x_2 - x_3 \leq -1$$
$$12x_1 + 5x_2 + 12x_3 \leq 34.8$$
$$12x_1 + 12x_2 + 7x_3 \leq 29.1$$
$$-6x_1 + x_2 + x_3 \leq -4.1$$
$$x_1 \geq 0, \ x_2 \geq 0, \ x_3 \geq 0.$$

The graph of the feasible region in R-space for this problem is illustrated in Figure 12. The initial upper bound on (r_1, r_2) is determined to be $u^0 = (1.90, 1.16)$ with corresponding solutions

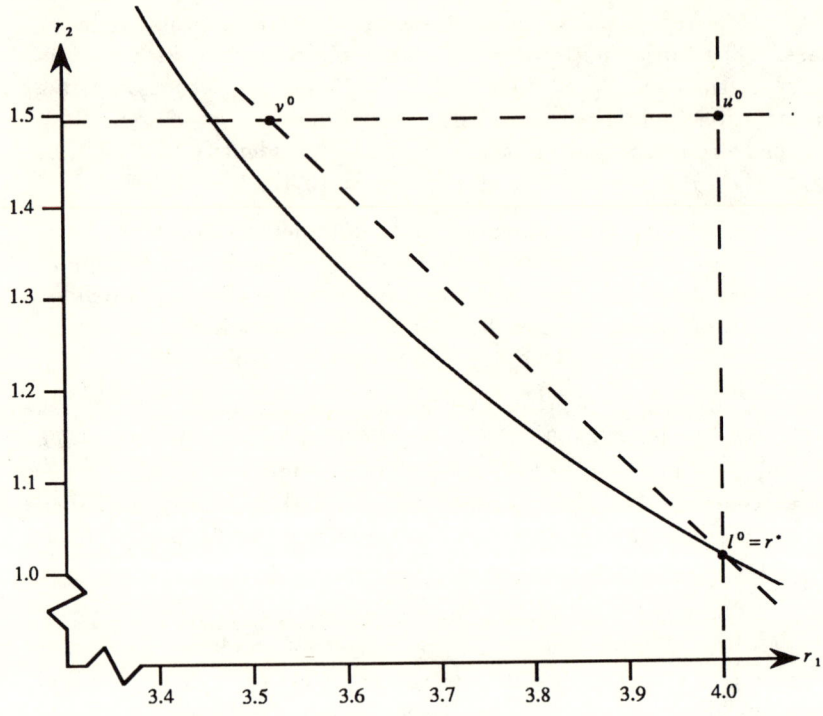

Figure 11
Graphical Representation of First Example Problem

$x^{1,0} = (1.00, 0.00, 0.00)$ and $x^{2,0} = (1.00, 0.00, 1.90)$. The feasible points $r(x^{1,0}) = (1.90, 0.57)$ and $r(x^{2,0}) = (0.00, 1.16)$ are computed from these solutions and the initial isovalue contour of $r_1 + r_2$ is located at $f_l = 2.47$. Therefore the initial triangular region determined in the initialization steps of the algorithm is defined by $u^0 = (1.90, 1.16)$, $l^0 = (1.90, 0.57)$, and $v^0 = (1.31, 1.16)$ as shown in Figure 12.

Computing $H(r)$ at u^0, l^0, and v^0 provides $H(u^0) = -2.73$, $H(l^0) = 0.00$, $H(v^0) = -1.56$. Since $H(l^0)$ is equal to zero and both $H(u^0)$ and $H(v^0)$ are negative, the algorithm immediately identifies l^0 as an optimal solution point in R-space using the results in Theorem 6. Thus, the corresponding optimal solution in X-space is (1.00, 0.00, 0.00) with an objective function value of 2.47.

The last example problem we consider is a minimization problem which has the form:

$$\text{minimize} \left\{ \frac{-x_1 + 2x_2 + 2}{3x_1 - 4x_2 + 5} + \frac{4x_1 - 3x_2 + 4}{-2x_1 + x_2 + 3} \right\}$$

subject to

$$x_1 + x_2 \leq 1.5$$
$$x_1 \leq x_2$$
$$0 \leq x_1 \leq 1$$
$$0 \leq x_2 \leq 1.$$

Figure 13 illustrates the feasible region in R-space for this problem. Since our objective in this problem is to minimize, we transformed the objective function into its equivalent maximization form to use program RATIO. The initial lower bound on (r_1, r_2) is $u^0 = (0.4, 0.25)$ with corresponding solutions $x^{1,0} = (0.00, 0.00)$ and $x^{2,0} = (0.00, 1.00)$. The feasible points $r(x^{1,0})$ and $r(x^{2,0})$ are computed as (0.40, 1.33) and (4.00, 0.25), respectively. The initial triangular region, shown in Figure 13, is defined by $u^0 = (0.4, 0.25)$, $l^0 = (0.40, 1.33)$, and $v^0 = (1.48, 0.25)$ with isovalue contour $r_1 + r_2$ equal to 1.73, which is the minimum of $\{1.73, 4.25\}$.

The results of Step 4 of the algorithm are $H(u^0) = 3.25$, $H(l^0) = -0.73$, $H(v^0) = -2.17$. Since the optimality check fails for this problem, we begin executing Steps 5 through 8 of the algorithm in order to reduce the size of the bounded region containing the optimal solution. However, in this case, the

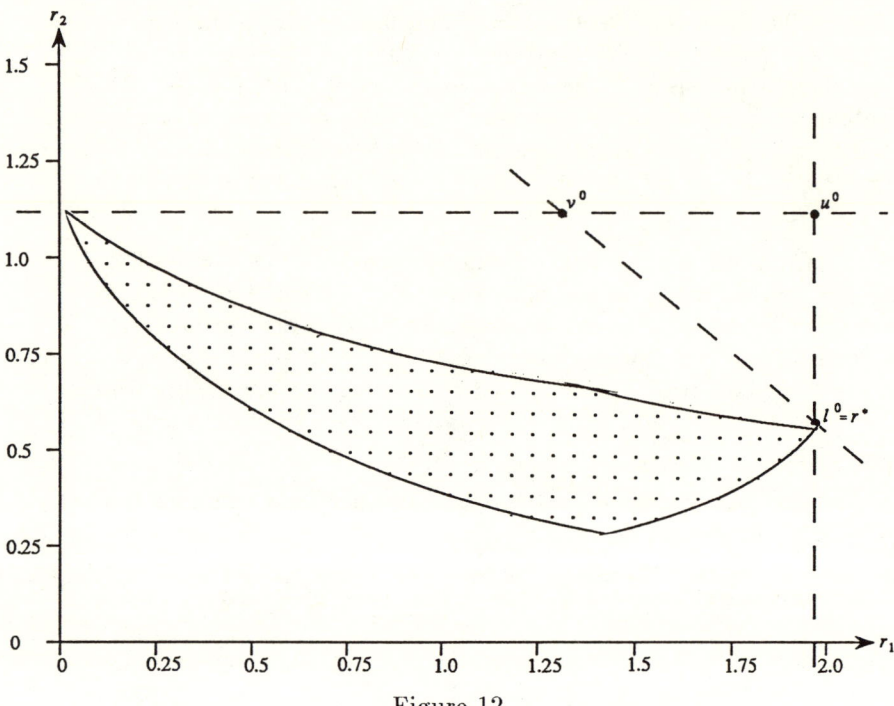

Figure 12
Graphical Representation of Second Example Problem

algorithm stalls after 12 iterations with $u^{12}=$ (0.40, 0.69), $l^{12}=$ (0.40, 1.33), and $v^{12}=$ (1.04, 0.69).

We proceed by following the approach outlined in Step 9. The rectangular region defined by u^{12}, l^{12}, v^{12}, and the point (v_1^{12}, l_2^{12}) is divided in half along the r_1 axis at $r_1 = 0.7195$ and Steps 1 and 2 are applied in each half of the region. The result is identification of a new isovalue contour for both regions that passes through the point (0.7195, 0.9068) and therefore has a value of 1.6263 which is less than 1.73, the value of the current isovalue contour.

The new triangular region determined by the new isovalue contour is defined by $u^0=$ (0.40, 0.6940), $l^0=$ (0.40, 1.2267), and $v^0=$ (0.9327, 0.6940). Now we reapply the procedure in Steps 5 through 8 to reduce the search space to the region defined by $u^0=$ (0.4580, 0.8948), $l^0=$ (0.4580, 1.1684), and $v^0=$ (0.7316, 0.8948). At this point, the algorithm stalls again and we repeat Step 9 to identify the optimal solution as approximately $r^*=$ (0.6654, 0.9586) at $x^*=$ (0.0, 0.2839).

6. Extension of the Algorithm to m Ratios

The algorithm discussed in this paper has proven to be useful in solving problems in which the objective function is a sum of two linear ratios. One extension of the algorithm is to apply the same approach to the problem of maximizing a sum of m linear ratios. By removing the limitation of two ratios, we substantially broaden the applicability of the algorithm.

The extension of the results in Section 4 is straightforward although it becomes more difficult to visualize. We are now transforming the problem into the m-dimensional space where each of the ratios in the objective function is mapped into one of the m dimensions. The feasible region in the new space is redefined more generally as

$$T = \{(r_1, \ldots, r_m) \mid r_i = n_i(x)/d_i(x),\ i = 1, \ldots, m;\\ \text{for some } x\ \varepsilon\ S\}.$$

The transformed problem consists of maximizing the sum of the r_i over T so that the isovalue contours of the objective function are hyperplanes.

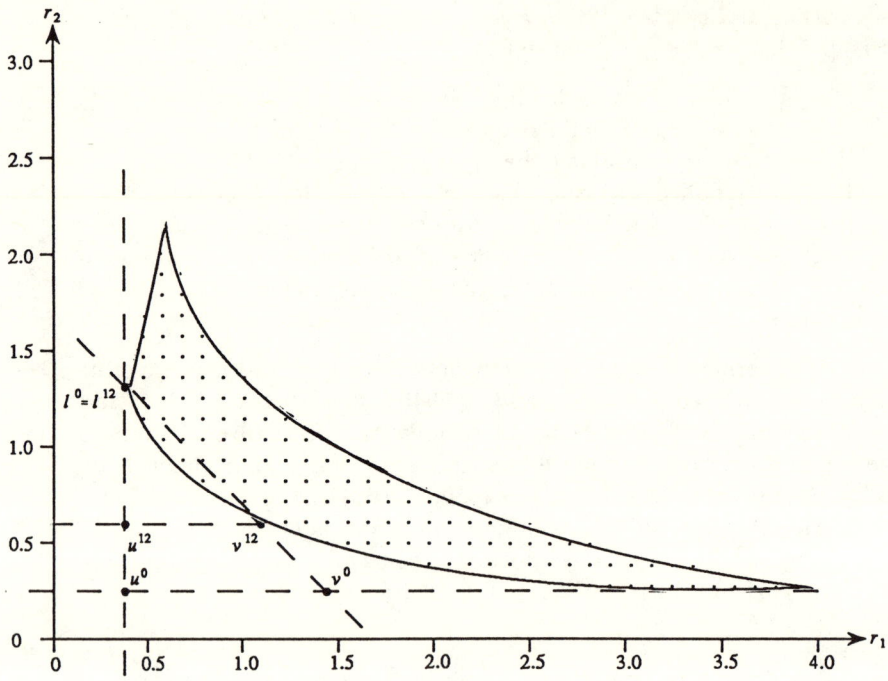

Figure 13
Graphical Representation of Third Example Problem

In the initialization part of the algorithm, we solve

$$\underset{x \in S}{\text{maximize}} \ \frac{n_i(x)}{d_i(x)}$$

for each of the m ratios to get upper bounds on each r_i and, therefore, an upper bound on the optimal solution. Then we use the solution points $\{x^{i,0}, i = 1, \ldots, m\}$ to construct the first isovalue contour which has the form $r_1 + r_2 + \cdots + r_m = f_l$ where the definition of f_l is extended as

$$f_l = \text{maximum} \ \{\sum_{i=1}^{m} r_i(x^{j,0}), j = 1, \ldots, m\}.$$

The points where this isovalue contour intersects the upper bounds, together with the upper bound point, define an m-simplex subset of the state space which contains the optimal solution.

The properties of $H(r)$ in Theorems 1, 2, 3, and 4 are already proven for m ratios. Also Theorem 5, which gives us a criteria for optimality, immediately extends to the case of m ratios by the convexity of H.

We extend the iterative steps in the algorithm by replacing all of the ratios in the objective function of $\mathcal{P}$ except one with the parameters t_i and adding the appropriate constraints on these parameters. Thus, each of these problems remains a linear fractional program which is suitable for the Charnes and Cooper transformation to a linear program. Convergence of the algorithm occurs in the same manner it did for the case of two ratios.

If the algorithm stalls, we apply the same approach described in Section 4. In the extension to the case of m ratios, we use a hyperplane to divide the m-simplex subset into two equal-sized subregions. Then we identify an isovalue contour in each subregion using the initialization steps and choose one of them to restart the iterations if it has a better isovalue contour value than the current one. If not, we continue by executing the algorithm separately in each of the two subregions until we obtain a global optimal solution in each subregion or we find a new improved isovalue contour of $r_1 + r_2$ for continued iterations.

Now we outline the steps of the extended algorithm.

Step 1. Set $k = 0$ and determine an initial upper bound $u^0 = (u_1^0, \ldots, u_m^0)$ on the optimal solution where

$$u_i^0 = \underset{x \, \varepsilon \, S}{\text{maximize}} \; \frac{n_i(x)}{d_i(x)}$$

with solutions $x^{i,0}$ for $i = 1, \ldots, m$.

Step 2. Compute the feasible points $r(x^{i,0})$ where

$$r(x^{i,0}) = (r_1(x^{i,0}), \ldots, r_m(x^{i,0})) = \left(\frac{n_1(x^{i,0})}{d_1(x^{i,0})}, \ldots, \frac{n_m(x^{i,0})}{d_m(x^{i,0})} \right)$$

for $i = 1, \ldots, m$. Then compute f_l where

$$f_l = \text{maximum} \left\{ \sum_{i=1}^{m} r_i(x^{j,0}), j = 1, \ldots, m \right\}.$$

Step 3. Determine the remaining points of the initial m-simplex:

$$l^{1,0} = (l_1^{1,0}, \ldots, l_m^{1,0}) = (f_l - \sum_{i=2}^{m} u_i^0, u_2^0, \ldots, u_m^0)$$

$$\vdots$$

$$l^{j,0} = (l_1^{j,0}, \ldots, l_m^{j,0}) = (u_1^0, \ldots, u_{j-1}^0, f_l - \sum_{i \neq j} u_i^0, u_{j+1}^0, \ldots, u_m^0)$$

$$\vdots$$

$$l^{m,0} = (l_1^{m,0}, \ldots, l_m^{m,0}) = (u_1^0, \ldots, u_{m-1}^0, f_l - \sum_{i=1}^{m-1} u_i^0).$$

Step 4. Compute $H(u^0)$ and $\{H(l^{i,0}), i = 1, \ldots, m\}$. If $H(l^{i,0}) = 0$ for some i, $H(u^0) < 0$, and $H(l^{j,0}) < 0$ for all $j = 1, \ldots, m$, $j \neq i$, then stop with $r^* = l^{i,0}$. The optimal value of the objective function of problem $\mathcal{P}$ is $\sum_{i=1}^{m} r_i^*$ and an optimal global solution is x^* such that

$$\frac{n_i(x^*)}{d_i(x^*)} = r_i^* \quad \text{for all } i = 1, \ldots, m.$$

SUM OF LINEAR FRACTIONAL FUNCTIONS

Step 5. Let $k = k + 1$. With $l^{i,k-1}$ given, solve the m problems

$$\text{maximize } \frac{n_1(x)}{d_1(x)} + t_2 + \cdots + t_m$$
$$\text{subject to } x \in S$$
$$t_i \leq \frac{n_i(x)}{d_i(x)}, \quad i = 2, \ldots, m$$

$$\vdots$$

$$\text{maximize } t_1 + \cdots + t_{m-1} + \frac{n_m(x)}{d_m(x)}$$
$$\text{subject to } x \in S$$
$$t_i \leq \frac{n_i(x)}{d_i(x)}, \quad i = 1, \ldots, m-1$$

where $t_i = l_i^{i,k-1}$. Denote the solutions by $x^{i,k}$ and let

$$u_i^k = \frac{n_i(x^{i,k})}{d_i(x^{i,k})} \text{ for } i = 1, \ldots, m.$$

Step 6. Determine the new m-simplex based on the new upper bound:

$$l^{1,k} = (l_1^{1,k}, \ldots, l_m^{1,0}) = (f_l - \sum_{i=2}^{m} u_i^k, u_2^k, \ldots, u_m^k)$$

$$\vdots$$

$$l^{j,k} = (l_1^{j,k}, \ldots, l_m^{j,k}) = (u_1^k, \ldots, u_{j-1}^k, f_l - \sum_{i \neq j} u_i^k, u_{j+1}^k, \ldots, u_m^k)$$

$$\vdots$$

$$l^{m,k} = (l_1^{m,k}, \ldots, l_m^{m,k}) = (u_1^k, \ldots, u_{m-1}^k, f_l - \sum_{i=1}^{m-1} u_i^k).$$

Step 7. If $u^k \doteq l^{j,k}$ for any j, then stop. The optimal value of the objective function of problem $\mathcal{P}$ is $\sum_{i=1}^{m} l_i^{j,k}$ and the optimal global solution is x^* such that

$$\frac{n_i(x^*)}{d_i(x^*)} = l_i^{j,k} \text{ for } i = 1, \ldots, m.$$

Step 8. If $u^k \neq u^{k-1}$, then return to Step 5.

Step 9. Otherwise, $u^k \doteq u^{k-1}$ and the iterative procedure has stalled. Use a hyperplane to divide the m-simplex containing the solution in half. Relabel f_l as $\hat{f}_l$. Repeat Steps 1 and 2 once for each of the two subregions. In Step 1, add the appropriate constraints on $m-1$ of the ratios to all m maximization problems to restrict the search to the particular subregion of the m-simplex. In Step 2, relabel f_l as $\overline{f}_l$ for the first subregion and $\widetilde{f}_l$ for the second subregion. Then select $f_l = \text{maximum}\{\overline{f}_l, \widetilde{f}_l\}$. If $f_l > \hat{f}_l$, then set $u^0 = u^k$ and return to Step 3. Otherwise $f_l \le \hat{f}_l$ so execute Steps 3 through 9 of the algorithm in each of the two subregions defined by the dividing hyperplane. □

The heuristic described in Section 4 for locating a better isovalue contour value f_l also generalizes when we have m unique extreme points to use in finding the intersection of the m hyperplanes $D(\widetilde{x}^i, r) = 0$. This step would occur between Steps 4 and 5 as follows:

Heuristic Step. Denote the solutions of $H(u^0)$ and $\{H(l^{i,0}), i = 1, \ldots, m\}$ as $\widetilde{x}^1, \widetilde{x}^2, \ldots, \widetilde{x}^{m+1}$, respectively. For each unique set of m extreme points, find the intersecting point of the hyperplanes $D(\widetilde{x}^j, r) = 0$ and denote it $\widetilde{r}$.

If $\sum_{i=1}^{m} \widetilde{r}_i \ge f_l$ and $\widetilde{x}$ such that

$$\frac{n_i(\widetilde{x})}{d_i(\widetilde{x})} = \widetilde{r}_i \quad \text{for all } i = 1, \ldots, m$$

is feasible, then let $f_l = \sum_{i=1}^{m} \widetilde{r}_i$ and reset the values of $\{l^{i,0}, i = 1, \ldots, m\}$ as in Step 3.

REFERENCES

Almogy, Y. and O. Levin (1969). Parametric Analysis of a Multi-Stage Stochastic Shipping Problem, *Proc. of the Fifth IFORS Conference*, Venice, 359-370.

Almogy, Y. and O. Levin (1971). A Class of Fractional Programming Problems, *Operations Research*, 19, 57-67.

Cambini, A., L. Martein, and S. Schaible (1989). On Maximizing a Sum of Ratios, *Journal of Information and Optimization Sciences*, 10, 65-79.

Charnes, A. and W. W. Cooper (1962). Programming with Linear Fractional Functionals, *Naval Research Logistics Quarterly*, 9, 181-186.

Dinkelbach, W. (1967). On Nonlinear Fractional Programming, *Management Science*, 13, 492-498.

Falk, James E. and Karla R. Hoffman (1976). A Successive Underestimation Method for Concave Minimization Problems, *Mathematics of Operations Research*, 1, 251-259.

Isbell, J. R. and W. H. Marlow (1956). Attrition Games, *Naval Research Logistics Quarterly*, 3, 71-93.

Marsten, Roy E. (1981). The Design of the XMP Linear Programming Library, *ACM Transactions on Mathematical Software*, 7, 481-497.

Marsten, Roy E. (1987). *XMP Technical Reference Manual*, University of Arizona, Tuscon, Arizona.

Ritter, K. (1967). A Parametric Method for Solving Certain Nonconcave Maximization Problems, *Journal of Computer and System Sciences*, 1, 44-54.

Rockafellar, R. (1970). *Convex Analysis*, Princeton University Press: Princeton, New Jersey.

Schaible, S. (1977). On the Sum of a Linear and Linear-Fractional Function, *Naval Research Logistics Quarterly*, 24, 691-693.

Schaible, S. (1981). A Survey of Fractional Programming, *Generalized Concavity in Optimization and Economics*, Academic Press, Inc.: New York, 417-440.

Minimizing and Maximizing the Product of Linear Fractional Functions *

Hiroshi Konno †   Yasutoshi Yajima ‡

October 1990

Abstract

Two algorithms for obtaining a global minimum and a global maximum of a product of linear fractional functions are proposed. The product of linear fractional functions are neither quasi-convex nor quasi-concave, and thus can have multiple local minima and maxima. We show that a parametric simplex algorithm and a parametric successive underestimation algorithm can be applied to this class of problems. Results of numerical experiments show that large scale problems can be successfully solved by each one of these algorithms.

1 Introduction

In a series of articles [8, 9, 10, 16], we proposed parametric simplex algorithms for globally minimizing a class of quasi-linear nonconvex functions over a polytope. This class of problems includes, among others the following problems:

i) linear multiplicative programming problems:

$$(LMP) \text{ minimize} \{c_0^t x + c_1^t x \cdot c_2^t x \mid x \in X\},$$

---

*This research was supported in part by Grant-in-Aid for Scientific Research of the Ministry of Education, Science and Culture, Grant No. 63490010.

†Institute of Human and Social Sciences, Tokyo Institute of Technology, 2-12-1 Meguro-ku, Tokyo 152, Japan

‡Department of Industrial Engineering and Management, Tokyo Institute of Technology

ii) minimization of a sum of two linear fractional functions:

$$(LFP)\ \text{minimize}\{\frac{d_1^t x + d_{10}}{c_1^t x + c_{10}} + \frac{d_2^t x + d_{20}}{c_2^t x + c_{20}}\ |\ x \in X\},$$

iii) bilinear programming problems:

$$(BLP)\ \text{minimize}\{c_0^t x + d_0^t y + \sum_{j=1}^{k} c_j^t x \cdot d_j^t y\ |\ x \in X,\ y \in Y\},$$

where $X \subset R^n, Y \subset R^{n'}$ are polytopes.

We demonstrated through numerical experiments on a number of randomly generated problems that certain variants of our parametric simplex algorithm work very well for these class of problems. In particular, we showed that

(a) a global minimum of (LMP) and (LFP) can be generated in less than 10% more computational work as needed to minimize $c_1^t x$ over X,

(b) a global minimum of a rank two bilinear programming problem (i.e., a bilinear function in which $k \leq 2$) is obtained in less than twice as much computation time as needed to minimize $c_1^t x$ over X and $d_1^t y$ over Y.

It is well known [5] that minimization of a concave quadratic function [13]

$$(CQF)\quad \text{minimize}\{c^t x + \frac{1}{2}x^t Q x\ |\ x \in X\}$$

can be converted to an equivalent bilinear program:

$$\text{minimize}\{cu + cv + u^t Q v\ |\ u \in X,\ v \in X\}.$$

Thus a global minimum of rank two concave quadratic programming problems can be obtained by applying the algorithm for solving rank two bilinear programs.

In another article [11], the authors proposed an alternative parametric programming approach for convex multiplicative programming problems:

$$(CMP)\quad \text{minimize}\{f_0(x) + f_1(x) \cdot f_2(x)\ |\ x \in X\},$$

where $f_0(\cdot), f_1(\cdot)$ and $f_2(\cdot)$ are convex functions defined on a polytope X.

We showed in [11] that the combination of parametrization and a branch and bound procedure gives an efficient method for solving (CM-P)'s when $f_0(\cdot), f_1(\cdot)$ and $f_2(\cdot)$ are positive for all $x \in X$.

The purpose of this paper is to propose algorithms for globally minimizing yet another important class of nonconvex function, i.e., the product of two linear fractional functions over a polytope:

$$(1.1) \quad \text{minimize} \left\{ \frac{d_1^t x + d_{10}}{c_1^t x + c_{10}} \cdot \frac{d_2^t x + d_{20}}{c_2^t x + c_{20}} \mid x \in X \right\}.$$

This problem has its origin in the bond portfolio optimization model [7]. Associated with a bond portfolio are several indices to measure its performance. Some of these indices such as average coupon rate, average maturity, average yield to maturity, etc, can be represented as a linear fractional function of the variables representing the rates of funds invested into each bond. The problem is usually multi-objective and a bond manager wants to minimize or maximize the geometric mean of several indices. We thus have a problem of the form (1.1) when the number of objectives is two.

2 Preliminaries

Let us consider a pair of nonlinear programming problems:

$$(2.1) \quad \begin{array}{l} \text{minimize} \quad f(x) = \dfrac{d_1^t x + d_{10}}{c_1^t x + c_{10}} \cdot \dfrac{d_2^t x + d_{20}}{c_2^t x + c_{20}} \\ \text{subject to} \quad Ax = b, \quad x \geq 0, \end{array}$$

and

$$(2.2) \quad \begin{array}{l} \text{maximize} \quad f(x) = \dfrac{d_1^t x + d_{10}}{c_1^t x + c_{10}} \cdot \dfrac{d_2^t x + d_{20}}{c_2^t x + c_{20}} \\ \text{subject to} \quad Ax = b, \quad x \geq 0. \end{array}$$

It is well known that a linear fractional function is both quasi-concave and quasi-convex [12]. Hence a linear fractional programming problems can be solved by a variant of the simplex algorithm [2]. Unfortunately, however, a product of linear fractional functions are neither quasi-convex nor quasi-concave as shown by the example below:

Example 2.1

Let

$$g(x_1, x_2, x_3, x_4) = (x_3/x_1) \cdot (x_4/x_2).$$

Also let $x^1 = (1, 1, 10, 1), x^2 = (1, 1, 1, 10), x^3 = (10, 1, 1, 1)$ and $x^4 = (1, 10, 1, 1)$. Then for $\lambda = 1/2$,

$$g(\lambda x^1 + (1-\lambda)x^2) = \frac{121}{4} > 10 = \max\{g(x^1), g(x^2)\},$$

$$g(\lambda x^3 + (1-\lambda)x^4) = \frac{4}{121} < 10 = \min\{g(x^3), g(x^4)\}. \quad \square$$

Thus a global optimum of the problems (2.1) and (2.2) may not be obtained by standard methods.

To solve (2.1) and (2.2), we first partition the feasible region by the signs of the quasi-linear functions $(d_j^t x + d_{j0})/(c_j^t x + c_{j0}), j = 1, 2$, and partition problems (2.1) and (2.2) into four subproblems each. Associated with these subproblems are additional constraints of the form:

(2.3) $\quad (c_1^t x + c_{10})(d_1^t x + d_{10}) \geq 0, \quad (c_2^t x + c_{20})(d_2^t x + d_{20}) \geq 0.$

(2.4) $\quad (c_1^t x + c_{10})(d_1^t x + d_{10}) \geq 0, \quad (c_2^t x + c_{20})(d_2^t x + d_{20}) \leq 0.$

(2.5) $\quad (c_1^t x + c_{10})(d_1^t x + d_{10}) \leq 0, \quad (c_2^t x + c_{20})(d_2^t x + d_{20}) \geq 0.$

(2.6) $\quad (c_1^t x + c_{10})(d_1^t x + d_{10}) \leq 0, \quad (c_2^t x + c_{20})(d_2^t x + d_{20}) \leq 0.$

Constraints of the form (2.6) can be converted to the type (2.3) by redefining $(d_1, d_{10}) := -(d_1, d_{10})$ and $(d_2, d_{20}) := -(d_2, d_{20})$. Similarly, constraints of the form (2.5) can be converted to the form (2.4). Thus we only need to consider the constraints of the form (2.3) and (2.4). However, problem (2.2) with form (2.4) constraints is equivalent to problem (2.1) with form (2.3) constraints and vice versa. This implies that we only need to consider the problems of the form:

(2.7) $\quad \text{minimize}\{f(x) \mid x \in \tilde{X}\},$

(2.8) $\quad \text{maximize}\{f(x) \mid x \in \tilde{X}\},$

where

$$\tilde{X} = X \cap \{x \in R^n \mid c_1^t x + c_{10} \geq 0,\ c_2^t x + c_{20} \geq 0,$$
$$d_1^t x + d_{10} \geq 0,\ d_2^t x + d_{20} \geq 0\}.$$

The latter problem (2.8) is equivalent to the minimization of $1/f(x)$ over $\tilde{X}$, which is solvable by the algorithm for solving problem (2.1) since $1/f(x)$ is again a product of two linear fractional functions.

We conclude from this that any algorithm that can solve (2.7), which is problem (2.1) under the assumption

(2.9) $\quad c_1^t x + c_{10} \geq 0,\ c_2^t x + c_{20} \geq 0,\ {}^\forall x \in X,$
$\quad\quad\ d_1^t x + d_{10} \geq 0,\ d_2^t x + d_{20} \geq 0,\ {}^\forall x \in X$

can solve the pair of problems (2.1) and (2.2) under general linear constraints.

In the following section, we will propose two algorithms for solving (2.1) under the assumption

(2.10) $\quad c_1^t x + c_{10} > 0,\ c_2^t x + c_{20} > x,\ {}^\forall x \in X.$

This assumptions is needed to avoid unnecessary technical difficulties.

Once we assume (2.10), we can assume further that

(2.11) $\quad d_1^t x + d_{10} > 0,\ d_2^t x + d_{20} > 0,\ {}^\forall x \in X.$

To see this, let us assume that there exists an $\tilde{x} \in X$ for which either $d_1 \tilde{x} + d_{10} = 0$ or $d_2 \tilde{x} + d_{20} = 0$. Then $\tilde{x}$ is obviously an optimal solution of (2.1) and can be found by solving two linear programs.

Remark: The problems (2.1) and (2.2) can be viewed as nonlinear fractional programming problems. It is well known that a nonlinear fractional function $f_1(x)/f_2(x)$ where $f_1(\cdot)$ is quasi-convex and $f_2(\cdot)$ is quasi-concave can be globally minimized [1, 6, 14]. Also, a nonlinear fractional function $f_1(x)/f_2(x)$ where $f_1(\cdot)$ is quasi-concave and $f_2(\cdot)$ is quasi-convex can be globally maximized. On the contrary, the objective functions of our problems (2.1) and (2.2) are the ratio of two quasi-concave functions under assumptions (2.10) and (2.11), so that these problems cannot be solved by existing methods.

Before concluding this section, we will cite a theorem which plays a crucial role in the following sections.

Let $f_1 : X \to R_+^1$, $f_2 : X \to R_+^1$, and let us consider the problem:

(2.12) $\quad \text{minimize}\{f_1(x) \cdot f_2(x) \mid x \in X\}$.

Associated with this problem is a master problem:

(2.13) $\quad \text{minimize}\{\xi f_1(x) + \dfrac{1}{\xi} f_2(x) \mid x \in X, \ \xi > 0\}$.

Theorem 2.1 *Let (x^*, ξ^*) be an optimal solution of (2.13). Then x^* is an optimal solution of (2.12).*

Proof. See [8]. □

3 Parametric Simplex Algorithm

In this section, we will propose a parametric simplex algorithm for solving the problem

(3.1) $\quad \begin{aligned} & \text{minimize} && \dfrac{d_1^t x + d_{10}}{c_1^t x + c_{10}} \cdot \dfrac{d_2^t x + d_{20}}{c_2^t x + c_{20}} \\ & \text{subject to} && Ax = b, \ x \geq 0 \end{aligned}$

under the following assumptions:

(3.2) $\quad X$ is non-empty and bounded,

(3.3) $\quad \begin{aligned} & c_1^t x + c_{10} > 0, && c_2^t x + c_{20} > 0, && \forall x \in X, \\ & d_1^t x + d_{10} > 0, && d_2^t x + d_{20} > 0, && \forall x \in X. \end{aligned}$

The first step is to apply the generalized Charnes-Cooper transformation [2] by introducing a new variable:

$$x = \dfrac{y}{y_0}, \quad (y_0 > 0).$$

Then (3.1) can be rewritten as follows:

(3.4) $\quad \begin{aligned} & \text{minimize} && \dfrac{d_1^t y + d_{10} y_0}{c_1^t y + c_{10} y_0} \cdot \dfrac{d_2^t y + d_{20} y_0}{c_2^t y + c_{20} y_0} \\ & \text{subject to} && Ay - b y_0 = 0, \ y \geq 0, \ y_0 \geq 0. \end{aligned}$

Remark: We may replace the condition $y_0 > 0$ by $y_0 \geq 0$ since we assumed that X is bounded. (Note that if there exists a vector $\tilde{y}$ for which $A\tilde{y} = 0$, then X is obviously unbounded.)

Let $w = (y, y_0) \in R^n$ and let $\tilde{A} = (A, -b)$, $\tilde{c}_1^t = (c_1^t, c_{10})$, $\tilde{c}_2^t = (c_2^t, c_{20})$, $\tilde{d}_1^t = (d_1^t, d_{10})$, $\tilde{d}_2^t = (d_2^t, c_{20})$. The problem (3.4) is now rewritten as follows:

$$(3.5) \quad \begin{vmatrix} \text{minimize} & \dfrac{\tilde{d}_1^t w}{\tilde{c}_1^t w} \cdot \dfrac{\tilde{d}_2^t w}{\tilde{c}_2^t w} \\ \text{subject to} & \tilde{A}w = 0, \quad w \geq 0. \end{vmatrix}$$

In this formulation, we may assume without loss of generality that

$$(3.6) \quad \tilde{c}_1^t w \cdot \tilde{c}_2^t w = 1.$$

To see this, let w^* be an optimal solution of (3.5). Let $\alpha = \tilde{c}_1^t w^* \cdot \tilde{c}_2^t w^*$. Then $w^{**} = w^*/\sqrt{\alpha}$ is also an optimal solution of (3.5) and $\tilde{c}_1^t w^{**} \cdot \tilde{c}_2^t w^{**} = 1$.

Now let us consider the problem:

$$(3.7) \quad \begin{vmatrix} \text{minimize} & \tilde{d}_1^t w \cdot \tilde{d}_2^t w \\ \text{subject to} & \tilde{A}w = 0, \quad w \geq 0, \\ & \tilde{c}_1^t w \cdot \tilde{c}_2^t w = 1, \end{vmatrix}$$

which, by virtue of Theorem 2.1 and the assumption (3.3), is equivalent to the following master problem:

$$(3.8) \quad \begin{vmatrix} \text{minimize} & \xi \tilde{d}_1^t w + \dfrac{1}{\xi} \tilde{d}_2^t w \\ \text{subject to} & \tilde{A}w = 0, \quad w \geq 0, \\ & \tilde{c}_1^t w = \eta, \quad \tilde{c}_2^t w = \dfrac{1}{\eta}, \\ & \xi > 0, \quad \eta > 0. \end{vmatrix}$$

Let us consider a subproblem:

$$(3.9) \quad P(\xi, \eta) \begin{vmatrix} \text{minimize} & \xi \tilde{d}_1^t w + \dfrac{1}{\xi} \tilde{d}_2^t w \\ \text{subject to} & \tilde{A}w = 0, \quad w \geq 0, \\ & \tilde{c}_1^t w = \eta, \quad \tilde{c}_2^t w = \dfrac{1}{\eta}. \end{vmatrix}$$

Let B be an optimal basis of the linear program $P(\xi,\eta)$ and let $\tilde{d}_{1B}$, $\tilde{d}_{2B}$ be the subvectors of $\tilde{d}_1$, $\tilde{d}_2$ corresponding to B, respectively. Also let N be the nonbasic matrix and let $\tilde{d}_{1N}$, $\tilde{d}_{2N}$ be the subvectors corresponding to N. Then we have a dictionary [3] corresponding to B:

(3.10)
$$\begin{aligned} \text{minimize} \quad & (\xi \tilde{d}_{1B} + \frac{1}{\xi}\tilde{d}_{2B})^t B^{-1} \begin{pmatrix} 0 \\ \eta \\ 1/\eta \end{pmatrix} \\ & + \left(\xi \bar{d}_{1N} + \frac{1}{\xi}\bar{d}_{2N}\right)^t x_N \\ \text{subject to} \quad & w_B = B^{-1}\begin{pmatrix} 0 \\ \eta \\ 1/\eta \end{pmatrix} - B^{-1} N w_N, \\ & w_B \geq 0, \quad w_N \geq 0. \end{aligned}$$

From this dictionary, we know that B is optimal for all (ξ,η) such that

(3.11) $\quad \xi \bar{d}_{1N} + \frac{1}{\xi}\bar{d}_{2N} \geq 0,$

(3.12) $\quad B^{-1}\begin{pmatrix} 0 \\ \eta \\ 1/\eta \end{pmatrix} \geq 0.$

Thus we obtain a rectangle

$$R(\underline{\xi},\bar{\xi};\underline{\eta},\bar{\eta}) = \{(\xi,\eta) \mid \underline{\xi} \leq \xi \leq \bar{\xi}, \quad \underline{\eta} \leq \eta \leq \bar{\eta}\},$$

in which B remains optimal. When ξ reaches $\bar{\xi}$, we will obtain an alternative optimal basis B' by a primal simplex pivoting. Also, when η reaches $\bar{\eta}$, we will obtain an alternative basis B'' by a dual simplex iteration.

We can thus partition the nonnegative orthant in (ξ,η) space by a finite number of rectangles by barring degeneracy appropriately (See [10] for details). Associated with each rectangle and a basis B, the solution can be presented as follows:

$$\begin{aligned} w_B &= a_1 \eta + a_2 \frac{1}{\eta}, \\ w_N &= 0. \end{aligned}$$

Hence the objective function in each rectangle can be written as follows:

$$p_1\xi + p_2\frac{1}{\xi} + q_1\eta + q_2\frac{1}{\eta} + r,$$

whose minimum over a rectangle can be calculated easily.

Algorithm Param

Stage1 (Initialization)

$\hat{\eta}_{\min} := \min\{\tilde{c}_1^t w \mid \tilde{A}w = 0, \; \tilde{c}_2^t w = 1, \; w \geq 0\}$;
Set $\eta := \sqrt{\hat{\eta}_{\min}}, \quad \xi := 0 \quad \mathcal{P} := \emptyset, \quad \mathcal{R} := \emptyset$;
Let B be an optimal basis of $P(\xi, \eta)$;

Stage2

Iterate:

  while $\xi < \infty$

    begin

      Calculate a pair of intervals $[\xi, \overline{\xi}], [\underline{\eta}, \overline{\eta}]$, and a new basis B such that B is an optimal basis of the problem $P(\xi, \eta)$;

      if $\eta = \overline{\eta}$ then goto Iterate ;

      $\mathcal{R} := \mathcal{R} \cup R(\xi, \overline{\xi}; \underline{\eta}, \overline{\eta})$;

      $\mathcal{P} := \mathcal{P} \cup \{(P(\xi, \overline{\eta}), B, \overline{\xi})\}$;

      (**Comment:** B is also an optimal basis of $P(\xi, \overline{\eta})$)

      $\xi := \overline{\xi}$;

    end

  if $\mathcal{P} \neq \emptyset$ then

    begin

      $(P(\xi, \eta), B, \hat{\xi}) := \min_{\eta} \mathcal{P}$;

      $\mathcal{P} := \mathcal{P} \setminus \{(P(\xi, \eta), B, \hat{\xi})\}$;

      goto Iterate;

    end

  else (**Comment:** $\mathcal{P} = \emptyset$)

    terminate   □

4 Branch and Bound Algorithm

Let us consider the problem:

(4.1)
$$\begin{aligned}\text{minimize} \quad & f(x) = \frac{d_1^t x + d_{10}}{c_1^t x + c_{10}} \cdot \frac{d_2^t x + d_{20}}{c_2^t x + c_{20}} \\ \text{subject to} \quad & Ax = b, \quad x \geq 0,\end{aligned}$$

under the same assumption as in Section 3. We will consider an alternative master problem:

(4.2)
$$\begin{aligned}\text{minimize} \quad & g(x;\xi) = \xi \frac{d_1^t x + d_{10}}{c_1^t x + c_{10}} + \frac{1}{\xi}\frac{d_2^t x + d_{20}}{c_2^t x + c_{20}} \\ \text{subject to} \quad & Ax = b, \quad x \geq 0, \quad \xi > 0.\end{aligned}$$

Let (x^*, ξ^*) be an optimal solution of this problem. Then x^* is an optimal solution of (4.1) by Theorem 2.1.

To solve (4.1), let us define a subproblem:

(4.3) $P(\xi)$
$$\begin{aligned}\text{minimize} \quad & g(x;\xi) = \xi \frac{d_1^t x + d_{10}}{c_1^t x + c_{10}} + \frac{1}{\xi}\frac{d_2^t x + d_{20}}{c_2^t x + c_{20}} \\ \text{subject to} \quad & Ax = b, \quad x \geq 0,\end{aligned}$$

and let $x(\xi)$ be an optimal solution of this problem. Also, let

$$h(\xi) = g(x(\xi), \xi).$$

We need to locate the global minimizer ξ^* of $h(\xi)$ over $\xi > 0$ and associated vector $x(\xi^*)$.

Let us note that $P(\xi)$ belongs to the class of problems (LFP) referred to in the Introduction, for which we have a very efficient algorithm [10]. Also, since

$$(d_j^t x + d_{j0})/(c_j^t x + c_{j0}) > 0, \quad j = 1, 2, \quad \forall x \in X,$$

we can apply the theorem proved in [15].

Theorem 4.1 *Let* $0 < \xi_1 < \xi_2$, *and let*

$$H(\xi) = \frac{h(\xi_2)\xi_2 - h(\xi_1)\xi_1}{\xi_2^2 - \xi_1^2}\xi + \frac{h(\xi_2)/\xi_2 - h(\xi_1)/\xi_1}{1/\xi_2^2 - 1/\xi_1^2}\frac{1}{\xi}.$$

Then

$$H(\xi) \leq h(\xi), \quad \forall \xi \in (\xi_1, \xi_2).$$

Proof. See [15]. □

We can thus construct a branch and bound algorithm using $H(\xi)$ as an underestimator of the function $h(\xi)$.

Algorithm Branch

stage 1 (Initialization)
  Set $\xi_{\min}$ and $\xi_{\max}$ sufficiently small and large, respectively;
  Compute $h(\xi_{\min})$ and $h(\xi_{\max})$ by solving subproblems $P(\xi_{\min})$ and
    $P(\xi_{\max})$,respectively;
  Set $z := \infty$;
stage 2
  Call Procedure $\mathbf{BBP}(1, z, \xi_{\min}, h(\xi_{\min}), \xi_{\max}, h(\xi_{\max}))$ □

Procedure $\mathbf{BBP}(j, z, \xi_s, h(\xi_s), \xi_t, h(\xi_t))$
  Compute $H(\xi)$ and set $H(\hat{\xi}) := \min\{H(\xi) \mid \xi_s \leq \xi \leq \xi_t\}$;
  if $H(\hat{\xi}) > z$ **then return;**
  else if $\hat{\xi} = \xi_s$ or $\hat{\xi} = \xi_t$ **then** $\xi^* := \hat{\xi};\ z := H(\hat{\xi});$ **return;**
  else
  begin
    Compute $h(\hat{\xi})$ by solving subproblem $P(\hat{\xi})$;
    if $h(\hat{\xi}) = H(\hat{\xi})$ **then** $\xi^* := \hat{\xi};\ z := H(\hat{\xi});$ **return;**
    else
    begin
      Set $\xi_j := \hat{\xi}$;
      Call procedure $\mathbf{BBP}(j+1, z, \xi_s, h(\xi_s), \xi_j, h(\xi_j))$;
      Call procedure $\mathbf{BBP}(j+1, z, \xi_j, h(\xi_j), \xi_t, h(\xi_t))$;
    end
  end □

5 Computational Experiments

Let us report our results of the computational experiments with the algorithms presented in Section 3 and Section 4. We solved the problem of the form:

(5.1) $\left|\begin{array}{ll} \text{minimize} & \dfrac{d_1^t x + d_{10}}{c_1^t x + c_{10}} \cdot \dfrac{d_2^t x + d_{20}}{c_2^t x + c_{20}} \\ \text{subject to} & Ax \geq b, \quad 1^t x \leq M, \quad x \geq 0, \end{array}\right.$

where $x \in R^n$, $A \in R^{(m-1) \times n}$, $c_1, c_2, d_1, d_2 \in R^n$, $b \in R^{m-1}$. All elements of A, b, c_1, c_2, d_1, d_2 were randomly generated from the range 0,100 and $M = 10^3$. This implies that every problem has a bounded feasible region and that they have an optimal solution. In the case of the Algorithm Branch, we set $\xi_{\min} = 10^{-5}$, $\xi_{\max} = 10^5$. Ten problems were solved for each size of m, n. The program was coded in C language and tested on a SUN4/280 workstation.

Table 5.1 shows the results of the Algorithm Param presented in Section 3. We see from this table that the total amount of computation time is about eight times as much as that for solving an associated linear programming problem.

Table 5.2 shows the results of the Algorithm Branch presented in Section 4. This table shows that the average number of branchings remains more or less the same, regardless of the size of the problem. Moreover the amount of total computation time increases slower than that of the Algorithm Param as the size of the problem increases.

The computation time of the Algorithm Param grows exponentially as the size of the problem becomes larger since the number of rectangle regions increases as the number of extreme points of the polyhedron increases. On the other hand, the number of branchings in the Algorithm Branch does not depend on the size of the problem. We see from these results that the Algorithm Branch would be more efficient than the Algorithm Param for large scale problems.

Acknowledgments

We are grateful to Professor T. Kuno for helpful discussions and to the financial support of the Energy Research Institute of Hitachi, Co.Ltd.

Table 5.1: Parametric Method

m	100	100	100	150	150	150	200	200	200
n	80	100	120	120	150	180	180	200	220
Average CPU time in Second (standard deviation)									
Stage 1	9.36	11.89	13.00	36.05	38.86	46.58	113.06	104.06	109.44
	(0.99)	(1.73)	(1.81)	(5.27)	(3.99)	(5.85)	(17.90)	(6.82)	(7.53)
Stage 2	46.06	63.42	93.44	175.55	258.75	288.63	501.03	652.11	688.38
	(11.48)	(15.77)	(29.41)	(51.48)	(64.14)	(51.59)	(72.94)	(166.39)	(155.01)
Total	55.42	75.31	106.44	211.61	297.61	335.20	614.09	756.17	797.82
	(11.66)	(16.07)	(28.79)	(53.49)	(62.82)	(49.79)	(71.10)	(164.53)	(154.55)
Average Number of Regions (standard deviation)									
	918.8	1078.5	1385.7	1718.7	2156.2	2050.5	2675.3	3212.7	3118.9
	(253.87)	(291.58)	(464.30)	(558.72)	(572.02)	(397.59)	(420.61)	(868.98)	(746.67)

Table 5.2: Branch and Bound Method

m	100	100	100	150	150	150	200	200	200
n	80	100	120	120	150	180	180	200	220
Average CPU time in Second (standard deviation)									
Stage 1	9.58	12.07	12.86	35.53	40.09	45.58	106.85	102.88	109.75
	(1.40)	(2.01)	(1.83)	(5.28)	(5.50)	(5.82)	(16.71)	(6.58)	(8.12)
Stage2	58.87	43.04	97.98	117.28	174.01	193.36	363.11	328.41	441.79
	(30.54)	(16.81)	(39.89)	(53.45)	(84.36)	(57.30)	(200.95)	(164.28)	(244.89)
Total	68.45	55.11	110.85	152.82	214.10	238.95	469.96	431.29	551.54
	(31.02)	(15.83)	(38.75)	(54.00)	(84.61)	(60.60)	(191.15)	(165.26)	(244.77)
Average Number of Branchings (standard deviation)									
	24.3	14.4	22.3	18.2	18.3	17.8	21.7	16.7	21.0
	(12.20)	(5.46)	(7.93)	(9.45)	(9.10)	(5.71)	(12.06)	(9.18)	(10.75)

References

[1] Cambini, A., Martein, L. and Schaible, S. (1989), "On Maximizing a Sum of Ratios," *J. Information & Optimization Sciences*, **10**, 65–79.

[2] Charnes, A. and Cooper, W. W. (1962), "Programming with Linear Fractional Functionals," *Naval Research Logistics Quarterly*, **9**, 181–186.

[3] Chvátal, V. (1983), *Linear Programming*, W. H. Freeman and Company.

[4] Falk, J. E. and Hoffman, K. R. (1976), "A Successive Underestimation Method for Concave Minimization Problems," *Math. Oper. Res.*, **1**, 251–259.

[5] Horst, R. and Tuy, H. (1990), *Global Optimization*, Springer-Verlag.

[6] Katoh, N. and Ibaraki, T. (1987), "A Parametric Characterization and an ε-approximation Scheme for the Minimization of a Quasiconcave Program," *Discrete Applied Math.*, **17**, 39–66.

[7] Konno, H. and Inori, M. (1988), "Bond Portfolio Optimization by Bilinear Fractional Programming," *J. Oper. Res. Soc. of Japan*, **32**, 143–158.

[8] Konno, H. and Kuno, T. (1989), "Linear Multiplicative Programming," IHSS Report 89-13, Institute of Human and Social Sciences, Tokyo Institute of Technology. (to appear in *Mathematical Programming, Series A*)

[9] Konno, H. and Yajima, Y. (1990), "Solving Rank Two Bilinear Programs by Parametric Simplex Algorithms," IHSS Report 90-17, Institute of Human and Social Sciences, Tokyo Institute of Technology.

[10] Konno, H., Yajima, Y. and Matsui, T. (1991), "Parametric Simplex Algorithm for Solving Special Class of Nonconvex Minimization Problems," *J. Global Optimization*, **1**, 65–81.

[11] Kuno, T. and Konno, H. (March, 1990, Revised and Extended, October 1990), "A Parametric Successive Underestimation Method for Convex Multiplicative Programming Problems," IHSS Report 90-19, Institute of Human and Social Sciences , Tokyo Institute of Technology. (to appear in the *J. of Global Optimization*)

[12] Mangasarian, O. L. (1969), *Nonlinear Programming*, McGraw-Hill Book Co,., New York.

[13] Rosen, J. B. and Pardalos, P. M. (1986), "Global Minimization of Large-Scale Constrained Concave Quadratic Problems by Separable Programming," *Mathematical Programming*, 34, 163–174.

[14] Schaible, S. (1977), "A Note on the Sum of a Linear and Linear-Fractional Function," *Naval Research Logistics Quarterly*, 24, 691–693.

[15] Tanaka, T., Thach, P. T. and Suzuki, S., "Methods for an Optimal Ordering Policy for Jointly Replenished Products by Nonconvex Minimization Techniques,"(to appear in the *J.Oper. Res. Soc. of Japan*).

[16] Yajima, Y. and Konno, H. (1990), "Efficient Algorithms for Solving Rank Two and Rank Three Bilinear Programming Problems," IHSS Report 90-21, Institute of Human and Social Sciences , Tokyo Institute of Technology. (to appear in the *J. of Global Optimization*)

Numerical Methods for Global Optimization

Yu. G. Evtushenko[1], M.A. Potapov[1], V.V. Korotkich[1]

Abstract

The purpose of this paper is to survey some of the numerical methods for global optimization available today. We mainly discuss the investigations that have been made in this field at the Computer Center of the USSR Academy of Sciences. Our sequential deterministic approach is based on the non-uniform space covering technique as a general framework, in which many apparently unrelated algorithm structures are embraced. The approach has proven to be appropriate for generalizing various methods: stripwise covering, branch-and-bound algorithms, bisection procedures and chain covering. The global optimization technique is used for nonlinear programming and multicriterion optimization.

Key words. Global optimization, multiextremal optimization, optimization algorithms, branch-and-bound algorithms, Lipschitzian optimization, nonlinear programming, multicriterion optimization.

§ 1. Introduction

We consider the global optimization problem

$$f_* = \min_{x \in X} f(x), \qquad (1.1)$$

where $f : R^n \to R^1$ is a continuous real valued objective function and $X \subset R^n$ is a compact feasible set.

As a special case, we consider the situation where X is a right parallelepiped P with sides parallel to the coordinate axes (a box in the sequel):

---
[1]USSR Academy of Sciences, Moscow USSR.

$$P = \{ x \in R^n : a \leq x \leq b, a \in R^n, b \in R^n \}. \quad (1.2)$$

Here and below, the vector inequality $a \leq b$, where $a, b \in R^n$, means that the componentwise inequalities $a^i \leq b^i$ hold for all $i = 1, \ldots, n$.

The set of global minimum points of the function f (solution set) and the set of ε-optimal solutions are defined as follows

$$X_* = \{x_* \in X : f_* = f(x_*) \leq f(x), \ \forall x \in X\}, \quad (1.3)$$

$$X_*^\varepsilon = \{x_\varepsilon \in X : f(x_\varepsilon) \leq f_* + \varepsilon\}. \quad (1.4)$$

These sets are nonempty because of the assumed compactness of X. The globally optimal value of f is denoted by $f_* = f(x_*)$, $x_* \in X_*$. Our goal is to find a point $x_\varepsilon \in X_*^\varepsilon$, then any value $f_\varepsilon = f(x_\varepsilon)$, where $x_\varepsilon \in X_*^\varepsilon$ is called an ε-optimal value of f on X.

Let $N_k = \{x_1, x_2, \ldots, x_k\}$ be a set of k points in X. After evaluating the objective function values at these points, we define the record value

$$R_k = \min_{1 \leq i \leq k} [f(x_1), f(x_2), \ldots, f(x_k)] = f(x_r), \quad (1.5)$$

where $r \in [1 : k]$, any such point x_r is called a record point.

We say that a numerical algorithm solves the problem (1.1) after k evaluations if a sequence N_k is such that $R_k \leq f_* + \varepsilon$, or equivalently $x_r \in X_*^\varepsilon$. The algorithm is defined by a rule for constructing such a sequence N_k.

We introduce Lebesque level sets in X

$$K(l) = \{x \in X : l - \varepsilon \leq f(x), \ l \in R^1\}. \quad (1.6)$$

Theorem 1.1 Let N_k be a sequence of feasible points such that

$$X \subseteq K(R_k), \quad (1.7)$$

then any current record point x_r must belong to X_*^ε.

Proof. The solution set $X_* \subseteq X \subseteq K(R_k)$. Let a point x_* belong to X_*, then according to (1.6) and (1.7) we have $x_* \in K(R_k)$ and $f(x_r) = R_k \leq f(x_*) + \varepsilon$. It means that $x_r \in X_*^\varepsilon$. Q.E.D.

It is very difficult to implement this result as a numerical algorithm because the level set $K(R_k)$ can be very complicated; in general it is not compact and it requires a special program to store it in computer memory. Therefore, we have to weaken the statement of this theorem and impose an additional requirement on the function f. We suppose that for any point $z \in X$ and any level value l, where $l \leq f(z)$, it is possible to define sets $B(z, l)$ and $Q(z)$ as follows

$$B(z, l) = \{x \in G(z): l - \varepsilon \leq f(x)\}, Q(z) = B(z, f(z)), \quad (1.8)$$

where $G(x) \subset R^n$ is a neighborhood of the point x such that $mes(G(x)) \geq 0$, and $x \in G(x)$.

Theorem 1.2 Let N_k be a set of feasible points such that

$$X \subseteq \bigcup_{x_j \in N_k} B(x_j, R_k), \quad (1.9)$$

then any record point x_r must belong to X_*^ε. If the set X is compact and there exists a number $c > 0$ such that $mes(Q(z)) \geq c$ for any $z \in X$, then a finite set N_k satisfying the condition of the theorem exists.

Proof. It is obvious that

$$B(z, f(z)) \subseteq K(f(z)), \bigcup_{x_j \in N_k} B(x_j, R_k) \subseteq K(R_k).$$

Therefore from condition (1.9) follows inclusion (1.7), that proves the theorem. Q.E.D.

This theorem suggests a constructive approach for solving the problem (1.1). The sequence N_k solves the problem (1.1), if inclusion (1.9) holds, i.e. the sequence of sets $B(x_j, R_k)$ covers the feasible set. Such an approach is called a space covering technique. Many optimization methods have been developed on the basis of this idea. The construction of numerical methods is split in two parts: 1) definition of the set $B(\ ,\ )$, 2) definition of covering rule.

Let us consider the first part. Assume that the function f satisfies a Lipschitz condition on R^n with constant L. It means that for any $x, z \in R^n$, we have

$$|f(x) - f(z)| \leq L \| x - z \|. \quad (1.10)$$

In this case, we can write that

$$B(z, l) = \{x : \| x - z \| \leq r = [\varepsilon + f(z) - l]/L\}, \quad (1.11)$$

i.e. $B(z, l)$ is a ball of radius r and a center z. If $x \in B(z, l)$ and $l \leq f(z)$, then from (1.10) we have that condition $l - \varepsilon \leq f(x)$ holds. If $f(z) = l$, then the balls $B(z, f(z)) = Q(z)$ have minimal radius $r = \varepsilon/L$.

Suppose the function f is differentiable and for any x and z of the convex compact set X we have

$$\| f_x(x) - f_x(z) \| \leq M \| x - z \|,$$

where M is a constant. In this case, $B(z, l)$ can be constructed as a ball centered at $\bar{z}$ with radius $\bar{r}$:

$$B(z, l) = \{x : \| x - z \| \leq \bar{r}\}, \bar{z} = z + f_x(z)/M,$$

$$\bar{r}^2 M^2 = \left[\| f_x(z) \|^2 + 2M [f(z) + \varepsilon - R_h] \right].$$

These formulas are given in [1]. Other more complicated cases were considered in [2, 3].

§ 2. Covering Algorithms

Numerical methods for seeking global solutions of multivariable problems, in spite of their practical importance, have been rather poorly developed. This is, no doubt, due to their exceedingly great complexity. We do not detail all the available approaches to this problem. Instead, we shall concentrate on one most promising direction, which is based on the idea of a non-uniform covering of a feasible set. This approach has turned out to be quite universal and, as we shall show, can be used not only for seeking global extrema of multivariable functions but also for nonlinear programming problems, for solving systems of equations and, most importantly, for multicriterion optimization. Problems that are solvable in reasonable computer time must be of limited dimension (or order 10 to 20); however, the use of multiprocessors, parallel computing, and distributed processing will substantially increase the possibilities of this approach.

During the last 20 years, a library of algorithms for global search was developed in the Computing Center of the USSR Academy of Sciences. All these codes are included in the DISO-dialogue system for optimization problem solving. We describe briefly four main directions that were used for global optimization.

1. **Stripwise covering.** This approach was proposed and implemented in ALGOL-60 language in [1]. This paper is well-known; therefore, we will not describe it here. The extension of this method can be found in chapter 7 of the book [4].

2. **Branch-and-bound algorithm (BBA).** The set $B(z, l)$ is a decreasing function of the level value l. This means that if $l_1 \leq l_2 \leq f(z)$ then $z \in B(z, l_2) \subseteq B(z, l_1)$. If $z \in N_k$, $l \geq R_k$ then the set $B(z, l)$ can not contain any such point x that $f(x) \leq R_k < \varepsilon$. The global minimum search on this set cannot improve the record value R_k by more than ε. Therefore, the set $B(z, l)$ can be excluded from the subsequent consideration; we say that this set is covered. Now we can describe the idea of BBA.

1. Split (partition) X into some subsets X_i, $i \in I, \cup X_i \subset X$.

2. For each subset X_i define a point $x_i \in X_i \cup X$ and evaluate $f(x_i)$, define a record R_i.

3. If $X_i \subset B(x_i, R_i)$, then we exclude X_i from consideration and the computation process continues on $Z = X \setminus X_i$. If Z_i is empty, then stop.

4. Otherwise, we split the set X_i further in $X_{i1}, X_{i2}, \ldots, X_{ik}$. For each set X_{ij}, we follow the same procedure as before; define $x_{ij} \in X_{ij} \cap X$ and new record R_s. If $X_{ij} \subset B(x_{ij}, R_s)$, then we exclude X_{ij}; if $X_i \setminus X_{ij}$ is empty then $i := i + 1$ and go to step 2.

The idea of BBA was proposed by E. Volkov [2] in 1974. A number of various codes that implemented this approach were developed by M. Potapov [5] and V. Ratkin [3].

3. **Bisection algorithm.** If in the definition (1.8), we take the level value l low enough and if we admit that a neighborhood $G(z)$ can contain set X, then we can say that $X \subset B(z, l)$ for some $z \in X$. If, in addition, $l \geq R_k$ then the problem is solved. Otherwise, if $l < R_k$ then we divide the set $B(z, l)$ into smaller subsets $A_1, A_2, \ldots, A_p$. In the simplest variant, $p = 2$. For each A_i, we define a point $x_i \in A_i$ and levels $l_1, \ldots, l_p$ such that $A_i \subset B(z_i, l_i)$. If it turns out that $l_i \geq R_{k+p}$, then the set A_i is covered and it is excluded. We divide the remaining sets and continue the process until we cover all feasible set X. We describe several steps for this algorithm:

1. Define a point $x_i \in X$, put $R_i = f(x_i)$ and suppose that we

can find a level l_i such that $X \subset B(x_i, l_i)$. If $l_i \geq R_i - \varepsilon$, then stop.

2. Otherwise, we split X in two subsets X_1, X_2 (maybe more) and define $x_1 \in X_1$, $x_2 \in X_2$, $f(x_1), f(x_2)$, $R_2 = \min\{f(x_i), f(x_1), f(x_2)\}$.

3. We define l_1 and l_2 such that $X_1 \subset B(x_1, l_1), X_2 \subset B(x_2, l_2)$. If $l_1 \geq R_2 - \varepsilon$ then we exclude the set X_1; otherwise split X_1 into two subsets. For each subset, we take some points from it and evaluate f, to define new record value. After excluding the set X_1, we consider the set X_2.

This algorithm was developed and implemented in C language in [6]. We consider the case where the feasible set X and covering sets are boxes: $X = \{x : a \leq x \leq b\}, A_i = \{x : a \leq a_i \leq x \leq b_i \leq b\}$. For any set A_i we take the midpoint $x_i = (b_i + a_i)/2$. The main diagonal of the box A_i is denoted by $d_i = b_i - a_i$. If the function f satisfies the Lipschitz condition (1.10), then the main diagonal of the covering box $B(x_i, l_i)$ satisfies the following inequality $L \parallel d_i \parallel \leq 2[\varepsilon + f(x_i) - l_i]$. For the set diminution, we used a simple bisection procedure. A similar version was proposed by H. Ratschek [7].

4. Chain covering algorithm. This algorithm is based on reduction of n-dimensional global minimization problems to one-dimensional non-uniform covering problems. This algorithm uses two notions: symmetric chain decomposition and global optimization problems equivalence. We consider these notions in turn.

For simplicity, we suppose that f satisfies a Lipschitz condition (1.8) and

$$f_* = \min_{x \in P} f(x). \tag{2.1}$$

For the following, we will need some standard definitions and results from the theory of partially ordered sets (posets). Let $S_n = \{x \in P : x = (\lambda^1, \lambda^2, \ldots, \lambda^n)\}$ be a ε-grid in the box P, where $\lambda^i \in \Lambda^i = \{a^i + \varepsilon, a^i + 2\varepsilon + \varepsilon, \ldots, a^i + 2\varepsilon k^i + \varepsilon\}$, $k^i = \lfloor (b^i - a^i)L/2\varepsilon \rfloor$, $1 \leq i \leq n$, P is defined by (1.2), and $\lfloor z \rfloor$ denotes the integer part of z.

We say that the point $x_2 = (\lambda_2^1, \lambda_2^2, \ldots, \lambda_2^n) \in S_n$ follows directly the point $x_1 = (\lambda_1^1, \lambda_1^2, \ldots, \lambda_1^n) \in S_n$ and write $x_1 \to x_2$, if there exists q, $1 \leq q \leq n$ such that $\lambda_2^q = \lambda_1^q + 2\varepsilon$ and for all others i, $1 \leq i \leq n$, $i \neq q$ we have $\lambda_1^i = \lambda_2^i$. A poset is ranked if each point $x \in S_n$ has been assigned a positive integer $rank(x)$ such that $x_1 \to x_2$ implies that $rank(x_2) = rank(x_1) + 1$. We define $rank(x) = \sum_{i=1}^n (\lambda^i - a^i)/2\varepsilon$, $rank(S_n) = \sum_{i=1}^n k^i$, $x = (\lambda^1, \lambda^2, \ldots, \lambda^n) \in S_n$. A chain $s \in S_n$ is a sequence of k points $x_i \in S_n$, $1 \leq i \leq k$ such that each $x_i \to x_{i+1}$, $1 \leq i \leq k-1$. A symmetric chain in a ranked poset is a sequence of elements of S_n, $x_1 \to x_2 \cdots \to x_k$ in which $rank(x_1) + rank(x_k) = rank(S_n)$. A symmetric chain decomposition (SCD) of a ranked poset S_n is a covering of S_n by pairwise disjoint symmetric chains.

We need the following Property 1.

Property 1. If Q_1 and Q_2 are ranked posets and if Q_1 and Q_2 have SCD's, then $Q_1 \times Q_2$ also has an SCD, and it can be constructed explicitly from the two given SCD's [8].

The best way to explain this property is first to picture the cartesian product of two chains

$$s_{p_1}^{q_1} = \{\eta_0, \eta_1, \ldots, \eta_{q_1}\} \in S_{p_1}, \; s_{p_2}^{q_2} = \{\gamma_0, \gamma_1, \ldots, \gamma_{q_2}\} \in S_{p_2},$$

$$1 \leq p_j \leq n-1, j=1,2, \; \eta_i = (\eta_i^1, \eta_i^2, \ldots, \eta_i^{p_1}) \in R^{p_1}, \; 0 \leq i \leq q_1,$$

$$\gamma_i = (\gamma_i^1, \gamma_i^2, \ldots, \gamma_i^{p_2}) \in R^{p_2}, \; 0 \leq i \leq q_2$$

in rectangular form

$$(\eta_0, \gamma_0) \; (\eta_0, \gamma_1) \; \cdots \; (\eta_0, \gamma_{q_2})$$

$$(\eta_1, \gamma_0) \; (\eta_1, \gamma_1) \; \cdots \; (\eta_1, \gamma_{q_2})$$

$$\cdots\cdots\cdots\cdots\cdots\cdots\cdots\cdots\cdots$$

$$(\eta_{q_1}, \gamma_0) \; (\eta_{q_1}, \gamma_1) \; \cdots \; (\eta_{q_1}, \gamma_{q_2}).$$

The chains in the SCD of $Q_1 \times Q_2$ that are contributed by just this one chain $s_{p_1}^{q_1}$ from the SCD of Q_1 and the one chain $s_{p_2}^{q_2}$ from the SCD of Q_2 are the ones that are obtained by successively 'peeling off' the chain that is obtained from right to left along the top row and continuing all the way down the leftmost column, as shown in Figure 1.

An SCD of a cartesian product of two chains $s_{p_1}^{q_1}$ and $s_{p_2}^{q_2}$ is defined as follows. Its elements are the chains in the set

$$s_{p_1}^{q_1} \odot s_{p_2}^{q_2} = \bigcup_m \left(s_{p_1}^{q_1} \odot s_{p_2}^{q_2}\right)_m, \; 0 \leq m \leq \min\{q_1, q_2\}, where$$

$$\left(s_{p_1}^{q_1} \odot s_{p_2}^{q_2}\right)_m = \{(\eta_m, \gamma_0), (\eta_m, \gamma_1), \ldots,$$

$$\left(\eta_m, \gamma_{q_2-m}\right), \left(\eta_{m+1}, \gamma_{q_2-m}\right), \ldots, \left(\eta_{q_1}, \gamma_{q_2-m}\right)\},$$

and $\odot$ stands for an SCD of the cartesian product of two chains. The complete description of the chain $\left(s_{P_1}^{q_1} \odot s_{P_2}^{q_2}\right)_m$ is encoded by the integer m.

The procedure of constructing S_n symmetric chain decompositions. For S_n, we have

$$S_n = S_{n-1} \times \Lambda^n, S_{n-1} = S_{n-2} \times \Lambda^{n-1}, \ldots, S_2 = \Lambda^1 \times \Lambda^2,$$

where $S_p = \{x : x \in R^p, x = (\lambda^1, \lambda^2, \ldots, \lambda^p)\}, \lambda^i \in \Lambda^i, 1 \leq i \leq p$. By applying property 1 recursively, we obtain an SCD for S_n, denoted in what follows by $\overline{S}_n$. Figure 2 depicts an SCD of S_3, $\Lambda^i = \{a^i + \varepsilon, a^i + 2\varepsilon + \varepsilon, a^i + 4\varepsilon + \varepsilon\}, i = 1, 2, 3$. A chain $s \in \overline{S}_n$ is composed as a result of the following operations $s_2 = (\Lambda^1 \odot \Lambda^2)_{m^1}, s_3 = (s_2 \odot \Lambda^3)_{m^2}, \cdots, s_n = s = (s_{n-1} \odot \Lambda^n)_{m^{n-1}}$, where $s_p \in \overline{S}_p, 2 \leq p \leq n$. A sequence of $(n-1)$ integer numbers $m^1, m^2, \ldots, m^{n-1}$ is called a code of a chain s and is denoted by $c(s)$.

Now we discuss the second notion mentioned above and give a definition of equivalence of global optimization problems. Let

$$s = \{x_0, x_1, \ldots, x_q\} \in \overline{S}_n.$$

We connect every point x_i with a point $x_{i+1}, 0 \leq i \leq q-1$ by a straight line and include the segment from x_i to x_{i+1} into a curved line Γ. This line may be parameterized by a metric preserving mapping

$$\omega : I \to \Gamma, \ x = \omega(t), x \in \Gamma, \ t \in I,$$

$$I = \{t : t \in R^1, 0 \leq t \leq d\},$$

d is a length of I and chain s. The length of a chain s is denoted by $v(s) = d$ and equals to the length of a curved line Γ. The mapping ω carries $f, x \in \Gamma$ into a function $g : I \to R^1$ such that for all $t \in I$, $s = \omega(t) \in \Gamma$ holds $f(x) = g(t)$.

Let's consider the two problems

$$\min_{x \in \Gamma} f(x) \text{ and } \min_{t \in I} g(t). \tag{2.2}$$

In order to explain the connection between these two problems, the notion of equivalence is introduced, which makes use of the space covering technique. Let

$$B^1(t, l) = \{t' : t' \in R^1, |t' - t| \leq (\varepsilon + g(t) - l)/L\}.$$

A set I is covered by the finite intervals $B^1(t_i, R_k)$, a set Γ is covered by an n-dimensional sets $B(x_i, R_k), x_i = \omega(t_i), i = 1, 2, \ldots, k$. We say that the problems (2.2) are equivalent, if and only if, $B^1(t_i, R_k)$ covers I and implies that $B(x_i, R_k)$, $1 \leq i \leq k$ covers Γ. Pictorially, we illustrate this notion as on Figures 3-6. Figure 3 depicts the points x_i, $1 \leq i \leq 6$, which are not a chain as we define it. The curved line Γ is formed by straight line segment points x_i, x_{i+1}, $1 \leq i \leq 5$ connections. Note that Γ is a segment of the Peano curve. For example, let us consider that $f(x_1) = 4\varepsilon$, $f(x_6) = -2\varepsilon$, thus for g we have $g(t_1) = f(x_1) = 4\varepsilon$, $g(t_6) = f(x_6) = -2\varepsilon$. In (1.10) we fix $\|x\| = \sum_{i=1}^{n} |x_i|$ and $L = 1$. Figure 3 shows the situation that occurs after evaluating function f and g at points x_1, x_6 and t_1, t_6.

The line Γ is covered completely by the sets $B(x_i, R_2)$, $i = 1, 6$, $R_2 = \min\{f(x_1), f(x_6)\}$, which are indicated in Figure 3 by the shaded area. The interval I is not completely covered by the sets $B^1(t_i, R_2)$, $i = 1, 6$, because, as one can check, the interval $[t_5 - \varepsilon, t_5 + \varepsilon]$ is not covered. Thus the problems (2.2) (according to our definition) are not equivalent. To cover I completely g may be evaluated at point t_5. This g evaluation at point t_5 induces f evaluation at point x_5, which is redundant because the problem $\min_{x \in \Gamma} f(x)$ is already solved. This situation is shown in Figure 4. One of the major reasons for introducing problems equivalence is to exclude redundant function evaluations.

Consider the situation shown in Figures 5-6. The points x_i, $1 \leq i \leq 6$ form a chain of SCD and $\Gamma = w(I)$. As in the preceding case, $f(x_1) = 4\varepsilon$, $f(x_6) = -2\varepsilon$. Figure 5 shows the situation that occurs after functions f and g have been evaluated at points x_1, x_6 and t_1, t_6. The situation in Figure 6 is analogous to the situation in Figure 4, but, unlike the situation in Figure 3, the set Γ is not completely covered by the sets $B(x_i, R_2)$, $i = 1, 6$. In order to cover I completely, we must evaluate g at point t_5. This induces evaluation of f at the corresponding point x_5. After this evaluation, the curve Γ will be covered completely. The sets I and Γ are covered simultaneously and this implies that there are no redundant evaluations. It is apparent that

$$\cup B(x_i, R_j) \cap \Gamma = w\left(\cup_i B^1(t_i, R_j)\right),$$

where $i = 1$ if $j = 1$; $i = 1, 6$ if $j = 2$; and $i = 1, 5, 6$ if $j = 3$. The problems (2.2) turn out to be equivalent.

The motivation for the chain decomposition approach is the fact that the global optimization problem on a chain is equivalent to the one-dimensional global optimization problem. The SCD chains number of S_n is minimal over all class of S_n chain decompositions. Because of this fact, SCD is useful in connection with global optimization [9].

The main idea of the approach discussed is to realize the space covering technique on the SCD set. SCD has nice properties with respect to both algorithmic and metric description. The important property of SCD is the following: for any $x \in s$, $s \in S_n$, $v(s) = d$, all points in its 2ε-neighborhood belong to the chains $s' \in \overline{S}_n$ such that $v(s')$ equals $d - 4\varepsilon$, d, $d + 4\varepsilon$.

The SCD structure may be described as follows. We define a partial order on the set $\overline{S}_n$: let $s_1, s_2 \in \overline{S}_n$,

$$c(s_1) = (m_1^1, m_1^2, \ldots, m_1^{n-1}), \quad c(s_2) = (m_2^1, m_2^2, \ldots, m_1^{n-1});$$

then we say that a chain s_2 follows directly a chain s_1 and write $s_1 \to s_2$, if there exists a number q, $1 \le q \le n-1$ such that $m_1^i = m_2^i$, $1 \le i \le q-1$, $m_1^q = m_2^q + 1$ and $m_1^i = m_2^i = 0$, $q+1 \le i \le n-1$.

Theorem 2.1 For every chain $s_1 \in \overline{S}_n$, apart from the chain with the code $(0, 0, \ldots, 0)$, there exists a chain s_2 that follows directly s_1, $v(s_2) = v(s_1) + 4\varepsilon$ and for any $x_1 \in s_1$ it holds

$$\max_{x_2 \in s_2} \sum_{i=1}^{n} |x_1^i - x_2^i| = 2\varepsilon,$$

and for any $x_2 \in s_2$

$$\max_{x_1 \in s_1} \sum_{i=1}^{n} |x_2^i - x_1^i| = 4\varepsilon.$$

This theorem produces many space covering schemes. As an example we consider the following method [10]. Let $\overline{S}_n = \cup H_d$, where $H_d = \{s \in \overline{S}_n : v(s) = d\}$ is a set consisting of chains of length d, and $\hat{f}(d) = \min_{s \in H_d} \min_{x \in s} f(x)$. We say that a chain $s_1 \in \overline{S}_n$ is initial if there exists no chain $s_2 \in \overline{S}_n$ such that $s_2 \to s_1$.

Theorem 2.2 If at all initial chains, the minimal function value is set up then for all admissible d_1, d_2 it holds that

$$|\hat{f}(d_2) - \hat{f}(d_1)| \le L |d_2 - d_1|. \tag{2.3}$$

The condition (2.3) is analogous to the condition (1.10). Therefore, the algorithms for global optimization of univariate Lipschitz functions may be used for $\hat{f}(d)$ minimization. The set $B(\ ,\ )$ on $\overline{S}_n$ is defined as follows

$$B(H_d, l) = \left\{ H_{d'} : |d' - d| \le (\varepsilon + \hat{f}(d) - l)/L \right\}.$$

As an illustration, we consider a function f in Figure 7, where S_n is the Boolean lattice B^4. Its SCD is composed of two chains of length 0, three chains of length 4ε and one chain of length 8ε. On the upper level, the method starts with H_0 (the chains of zero length). The function value $\hat{f}$ at point $d = 0$ is defined using the function evaluations of f at points of chain s, $v(s) = 0$. Then on the upper level the method covers the set of chains of length 8ε. The problem $\min_{x \in \Gamma} f(x)$ is equivalent to the univariate global optimization problem $\min_{t \in I} g(t)$, $f(x) = g(t)$, $x = w(t)$, $x \in \Gamma$, $t \in I$, $I = \{t : t \in R^1, 0 \leq t \leq 8\varepsilon\}$; in which Γ is a curved line corresponding to the chain s of length 8ε, $\Gamma = w(I)$.

The saw-tooth cover after three evaluations of g is

$$\varphi(t) = \min_{i=1,2,3}\{g(t_i) + |t - t_i|\}$$

and is represented in Figure 8, where $x_i = w(t_i)$, $i = 1, 2, 3$, $x_1 = (0,0,0,0)$, $x_2 = (1,1,1,0)$, $x_3 = (1,1,1,1)$. After these evaluations of g, the set I is covered. Due to equivalence of the problems $\min_{x \in \Gamma} f(x)$ and $\min_{t \in I} g(t)$, the sets I and Γ are covered simultaneously. In view of the Lipschitz property (2.3), the saw-tooth cover is $\hat{\varphi}(d) = \min_{i=1,2} \{\hat{f}(d_i) + |d - d_i|\}$ and is represented in Figure 9, where $d_1 = 0$, $d_2 = 8\varepsilon$. From Figure 9 we see that the set $\overline{S}_n = \cup_{d=0, 4\varepsilon, 8\varepsilon} H_d$ is covered by the sets $B(H_{d_1}, R_2)$, $B(H_{d_2}, R_2)$, $R_2 = \min_{i=1,2} \hat{f}(d_i)$. The set $H_{4\varepsilon}$ may be excluded from the subsequent consideration, because $\hat{f}(4\varepsilon) = \min_{s \in H_{4\varepsilon}} \min_{x \in s} f(x) \geq R_2 + \varepsilon$.

Now the whole global optimization problem is solved.

Concluding remark: the covering set $B(z, l)$ essentially depends on the value of the level l, and this set is greatest for $l = f_*$, the value f_* usually not being known and therefore, instead of f_*, we use current record value R_i. To extend the set $B(z, R_i)$, it is desirable that the record value R_i be as close as possible to f_*. The sequence N_k is chosen so as to guarantee the global solution, while to extend covering set $B(z, R_i)$, we use the auxiliary procedures of finding local minimum in the problem (2.1). If $x_i \in P$ and $f(x_i) \leq R_{i-1} - \varepsilon$, $\varepsilon > 0$, then one goes to the program of local search of the minimum; if one thereby obtains a point $\overline{x} \in P$, at which $f(\overline{x}) < f(x_i)$, then as R_i one takes quantity $f(\overline{x})$. This technique substantially expedites the computation time and accuracy.

§3. Solution of Nonlinear Programming Problems

The approach described in the preceding sections carries over to solving nonlinear programming problems. The feasible set X can be nonconvex and not easily connected. Very often, it is difficult to realize algorithmically a covering of it by balls or boxes. Therefore, it is easier to cover a box P that contains set X. The additional condition $x \in X$ will accelerate the covering procedure. Suppose the global minimum is sought:

$$f_* = \min_{x \in P \cap X} f(x), \quad X = \{ x \in R^n : \psi(x) \leq 0 \}, \qquad (3.1)$$

P being an n-dimensional parallelepiped defined by (1.2) and $\psi : R^n \to R^1$. the scalar function $\psi(x)$ is equal to zero everywhere on X, and more then zero outside X. We suppose that functions f and $\psi(x)$ satisfy the Lipschitz condition on P with the same constant:

$$|\psi(x) - \psi(y)| \leq L \| x - y \|, \text{ where } x, y \in P. \qquad (3.2)$$

Consider the case where the feasible set is defined by inequality type constraints

$$X = \{x \in R^n : h(x) \leq 0\}, \quad h : R^n \to R^c.$$

Suppose that each component of h satisfies the Lipschitz condition with constant L. In this case, we define

$$\psi(x) = \max_{i \in [1:c]} [h^i(x), 0].$$

Using Theorem 1.5.2. from [4], it is easy to show that $\psi(x)$ satisfies a Lipschitz condition (3.2). In the more general case, the feasible set may be defined by equality and inequality-type constraints

$$X = \{x : h(x) \leq 0, g(x) = 0\}, \quad g : R^n \to R^m. \qquad (3.3)$$

We can define the function $\psi(x)$ in various ways. For example

$$\psi(x) = \| h_+ \|_p + \| g(x) \|_p,$$

where $\| z \|_p$ is the Holder norm of z, $h_+ = [h_+^1, \ldots, h_+^c]$, $h_+^i = \max[0, h^i]$.

As before, we denote by X, the global solutions set of problem (3.1). We define the ε - feasible set X_ε and the set of approximate global solutions X_ε^* of the problem (3.1):

$$X_\varepsilon = \{x : \psi(x) \leq \varepsilon\}, \quad X_\varepsilon^* = \{x \in P \cap X_\varepsilon : f(x) - \varepsilon \leq f_*\}. \qquad (3.4)$$

Let $N_k = \{x_1, \ldots, x_k\}$ be a set of k points in P. The current record point x_r and record value R_k are defined similarly to (1.5):

$$R_k = \min_{x \in N \cap X_\varepsilon} f(x_i) = f(x_r),$$

with each point $x_s \in N_k$ we associate a ball B_{sk} centered in x_s and with radius P_{sk}:

$$B_{sk} = \{x : \| x - x_s \| \leq P_{sk}\}, \ P_{sk} = \max[\overline{P}_{sk}, \tilde{P}_{sk}],$$

$$\overline{P}_{sk} = (f(x_s) - R_k + \varepsilon)/L, \ \tilde{P}_{sk} = (\varphi(x_i) - \delta)/L, \ 0 < \delta < \varepsilon.$$

Theorem 3.1. Suppose that $P \cap X \neq \emptyset$, the functions f and φ satisfy the Lipschitz condition (1.10), (3.2) and the set N_k of the points from P is such that $P \subset \cup_{i=1}^{k} B_{ik}$, then any record point x_r satisfies $x_r \in X_\varepsilon^*$.

It is worthwhile to compare problem (2.1) of finding the minimum of $f(x)$ on P with the problem (3.1) under the additional constraints $x \in X$. At first glance, it seems paradoxical (although it is true) that finding the global solution in (3.1) is simpler than in (2.1). The constraint $x \in X$ provides an additional possibility to increase the radii of the covering balls on $P \backslash X$. Hence, the additional constraints merely simplify the problem of finding global solutions. The auxiliary procedures of local search in the problem (3.1) are not much more complex than in problem (2.1), if we are concerned with the problem of finding the global minimum. In that case, the employment of the penalty function method becomes absolutely irrational for finding the global solution of (3.1). We exclude a rather unusual case where the upper estimate of the Lagrange multipliers is known to be sufficiently precise at the global minimum point and the problem reduces to a one-step minimization of the exact penalty function. In the general case where the penalty function method is used, passing from (3.1) to the multiple solution of the problem (2.1) substantially complicates the computation.

Thus, the penalty function method, though a very effective tool for finding local solutions, is not advantageous for finding global solutions. The same can be said about other local methods that require multiple minimization in x of auxiliary functions (the cost-function parametrization method, and the method of modified Lagrangians, among others). We point out once more that all these methods play an important, but only auxiliary role, in finding global solutions.

The approach presented here is developed in [11,12,4].

§4. Numerical Solution of Multicriterion Problem

Suppose that $F(x) = [F^1(x), \ldots, F^m(x)]$, and $F : R^n \to R^m$. The problem of multicriterion minimization of the vector-valued function $F(x)$ on an admissible set X is denoted by

$$\min_{x \in X} F(x). \tag{4.1}$$

The set X_* of solutions of this problem is defined as follows:

$$X_* = \{x \in X : \text{ if } w \in X \text{ and } F(w) \leq F(x), \text{ then } F(w) = F(x)\}.$$

To solve problem (4.1) is to find the set X_*. In papers on multicriterion optimization, X_* is usually called the set of effective solutions, and its image $F(x_*)$, the Pareto set. The sets X and X_* are assumed to be nonempty. The structure of X_* turns out to be very complicated even for the simplest problems. It often happens that this set is not convex and not simply connected, and every attempt to describe it with the help of approximation formulas is extremely difficult to realize. Therefore, we define the concept of an ε-optimal solution and give a rule for finding it.

Definition. A set $A \subset X$ is called an ε-optimal solution of problem (4.1) if

1. for each point $x_* \in X_*$ there exists a point $z \in A$ such that $F(z) - \varepsilon \bullet e \leq F(x_*)$, e means the m-dimensional vector of ones,

2. the set A does not contain two distinct points x and z such that $F(x) \leq F(z)$.

Numerical methods for finding global extrema of functions of several variables can be used for construction ε-optimal solution of problem (4.1). Let $N_k = \{x_1, \ldots, x_k\}$ be a set of k points in X. We shall define a sequence of sets $A_k \subset N_k$ as k increases, while trying in the final analysis to find ε-optimal solution.

The rule for constructing A_k. The set A_1 consists of the single point $x_1 \in N_1$. Suppose that N_k, N_{k+1}, and A_k are known. We compute the vector $F(x_{k+1})$ at the point $x_{k+1} \in N_{k+1}$. Three cases are possible.

1. If it turns out that among the elements $x_i \in A_k$ there are some such that $F(x_{k+1}) \leq F(x_i)$, $F(x_{k+1}) \neq F(x_i)$, then they are all removed from A_k, the point x_{k+1} is included in A_k, and the set is denoted by A_{k+1}.
2. If it turns out that there exists at least one element $x_i \in A_k$ such that $F(x_i) \leq F(x_{k+1})$, then x_{k+1} is not included in A_k, and the set A_k is denoted by A_{k+1}.
3. If the conditions of the two preceding cases do not hold, then the point x_{k+1} is included in the set A_k, which is denoted by A_{k+1}.

The definition of the Lebesque set (1.6) is replaced now by

$$K(l) = \{ x : l - \varepsilon e \leq F(x), \, l \in R^m \}.$$

Theorem 4.1. Let the finite set A_k of admissible points be such that

$$X \subseteq \bigcup_{x_j \in A_k} K(F(x_j)).$$

Then the set A_k determined by the rule for constructing A_k forms an ε-optimal solution of the multicriterion problem (4.1).

In a manner similar to the first section we suppose that for any point $z \in X$ and level vector $l \in R^m$ where $l \leq F(z)$, it is possible to define the set

$$B(z, l) = \{x \in G(z) : l - \varepsilon e \leq F(x)\}.$$

From the inclusion $B(z, l) \subseteq K(F(z))$ it follows that the set A_k is ε-optimal solution if

$$X \subseteq \bigvee_{x_s \in A_k} B(x_s, F(x_s)).$$

We assume that each component of the vector-valued function $F(x)$ satisfies a Lipschitz condition on X with one and the same constant L. Therefore, for any x and z in X, we have the vector condition

$$F(z) - e \cdot L \cdot \| x - z \| \leq F(x).$$

In this case we can use the following covering balls

$$B(x_j, F(x_i)) = \{x \in R^n : \| x - x_j \| \leq r_{jk}\}, where$$

$$r_{jk} = [\varepsilon + h_{jk}]/L, \quad h_{jk} = \max_{x_i \in A_k} \min_{q \in [1:m]} [F^q(x_j) - F^q(x_i)]. \qquad (4.2)$$

As in the third section, we can take into account constraint restrictions. Suppose that problem (4.1) is replaced by the following

$$\min_{x \in Z} F(x), \quad Z = X \cap P. \qquad (4.3)$$

We extend the admissible set by introducing the set $Z_\varepsilon = X_\varepsilon \cap P$ (see 3.4).

The definition of an ε-optimal solution carries over to the case of problem (4.3) with the following changes: instead of the condition $A \subset X$, it is required that $A \subset Z_\varepsilon$, and the first ε-optimality condition is satisfied for every $x_* \in Z_*$. The rule for determining A_k is also changed. Suppose now that N_k is a set of k points belonging to P.

The rule for constructing A_k. Suppose that N_k, N_{k+1}, and A_k are known (A_k may be empty). At the point $x_{k+1} \in N_{k+1}$, it is checked to see whether $x_{k+1} \in Z_\varepsilon$. If not, then x_{k+1} is not included in A_k, and A_k is then denoted by A_{k+1}; otherwise, the same arguments as in the construction of A_k are carried out with a check of the three cases. Let

$$\bar{r}_{jk} = (1/L) \max [\varepsilon + h_{jk}, \varphi(x_j) - \delta],$$

$$B_{jk} = \{x \in R^n : \| x - x_j \| \leq \bar{r}_{jk}\},$$

where $0 < \delta < \varepsilon$ and h_{jk} is given by (4.2).

Theorem 4.2. Suppose that $Z \neq \emptyset$, and the vector-valued function F and the function φ satisfy Lipschitz conditions on P. Let the set N_k of points in P be such that

$$P \subset \bigcup_{j=1}^{k} B_{jk}.$$

The the set A_k constructed by the rule for constructing A_k forms an ε-optimal solution of the multicriterion problem (4.3).

Any radius $\bar{r}_{jk}$ cannot be less that the quantity $(\varepsilon - \delta)/L > 0$. Therefore, finite sets N_k and A_k satisfying the conditions of Theorem 4.2 exist in the case when the set Z is compact.

Now, for an approximate solution of the problem, it suffices to realize a covering of an admissible set P by a nonuniform net. This is an essential advantage of such an approach in comparison with the convolution method and the method of successive concessions, which require, for their realization, a multiple search for a global extremum of a function of several variables. The sets A_k obtained by computer calculations are transmitted to the user (the engineer solving the multicriterion problem). As the final solution, the user chooses a concrete point from the set A_k, starting from the specifics of the problem, or from some additional considerations not reflected in the statement of the problem (4.1).

The main result of this section is that, for the constructive solution of multicriterion optimization problems, it is possible to use the non-uniform covering technique created in searches for finding global extrema of the functions of several variables. The approach presented here is developed in [4,12-16].

Conclusion

In this paper, we briefly present some results in the field of global optimization that were published during the last twenty years, mainly in Russia. We did not consider the global approach to game problems, which is contained in [17,4,12] and to the solution of nonlinear systems, which are described in [2,3,4,12]. We did not give proofs of the theorems. We intend to publish a detailed description of our results in the future in the Journal of Global Optimization.

References

[1] Yu.G.Evtushenko. Numerical methods for finding global extrema (case of non-uniform mesh) Zh. Vychisl. Mat. i Mat. Fiz. 1971, vol 11, 1390-1403; English transl. in USSR Comput. Math. and Math. Phys.

[2] E.A.Volkov. Approximate and exact solutions of system of nonlinear equations. Proceeding of Mathematical Institute of USSR Academy of Sciences, 1974, vol 131, 64-80 (in Russian).

[3] V.A. Ratkin. Non-uniform covering technique for the solution of nonlinear equations. Moscow. Computer Center of the USSR Academy of Sciences, 1989, 36 (in Russian).

[4] Yu.G.Evtushenko. Numerical optimization technique. New York, Optimization Software Inc., Springer-Verlag. 1985.

[5] M.A. Potapov. Non-uniform covering methods and their use in solving global optimization problems in a dialogue mote. Author's Summary of Candidate's Dissertation, Moscow, 1984 (in Russian).

[6] Yu.G.Evtushenko, V.A. Ratkin. Bisection method for global optimization. Izvestija Akademii Nauk AN USSR, Tehnicheskaya Kibernetika, 1987, No. 1, 119-128 (in Russian).

[7] H. Ratschek. Inclusion functions and global optimization. - Mathematical Programming. Vol. 33, No. 3, 1985, 300-317.

[8] N.G. de Bruijn, C. Van E.Tenbergen, and D. Kruyswijk. On the set of divisors of a number, Nieuw. Arch. Wisk., 1952, vol. 23, No. 2, 191-193.

[9] V.V. Korotkich. On connection between the problems of optimal reconstruction of a functional class and decoding of monotone k-valued logic functions. Research Methods for Complex Systems, Moscow, VNIISI, 1983, 18-26 (in Russian).

[10] V.V. Korotkich. Multilevel dichotomy algorithm in global optimization. Proceedings of the 14th. IFIP Conference on System Modeling and Optimization, Leipzig, 1989, 34-36.

[11] Yu.G.Evtushenko. The methods for global search. Operations Research (models, systems, solutions), Moscow, Computer Center of the USSR Academy of Sciences, 1974, No. 4, 39-68 (in Russian).

[12] Yu.G.Evtushenko, M.A. Potapov. Global search. In book "Methods for Solution of Operational Control Problems". Moscow, VNIPOU, 1984, 128-152 (in Russian).

[13] Yu.G.Evtushenko, M.A.Potapov. Nondifferentiable optimization: Motivation and applications (Laxenburg, 1984), Lecture Notes in Economics and Math. Systems, Springer-Verlag, 1985, vol. 255, 97-102.

[14] Yu.G.Evtushenko, M.A. Potapov. Methods of numerical solution of multicriterion problems. Soviet Math. Dokl. Vol. 34 (1987), No. 3, 420-423.

[15] Yu.G.Evtushenko, M.A. Potapov. Numerical solution of multicriterion problem. Cybernetics and Computational Technique. Moscow 1987, No. 3, 209-218 (in Russian).

[16] Yu.G.Evtushenko, V.A.Ratkin, V.P.Mazourik. Multicriteria optimization in the DISO system. Lecture Notes in Control and Information Sciences, Springer-Verlag, 1988, 113, 231-240.

[17] Yu.G. Evtushenko. A numerical methods for finding best guaranteed estimates. Zh. Vychisl. Mat. i Mat. Fiz; 1971, vol. 12, 89-104. English transl. in USSR Comput. Math and Math. Phys.

Figure 1

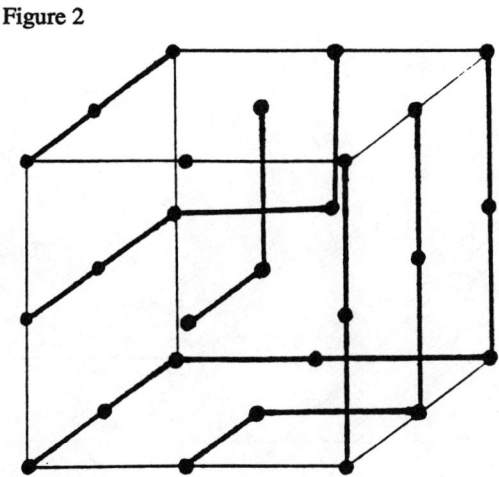

Figure 2

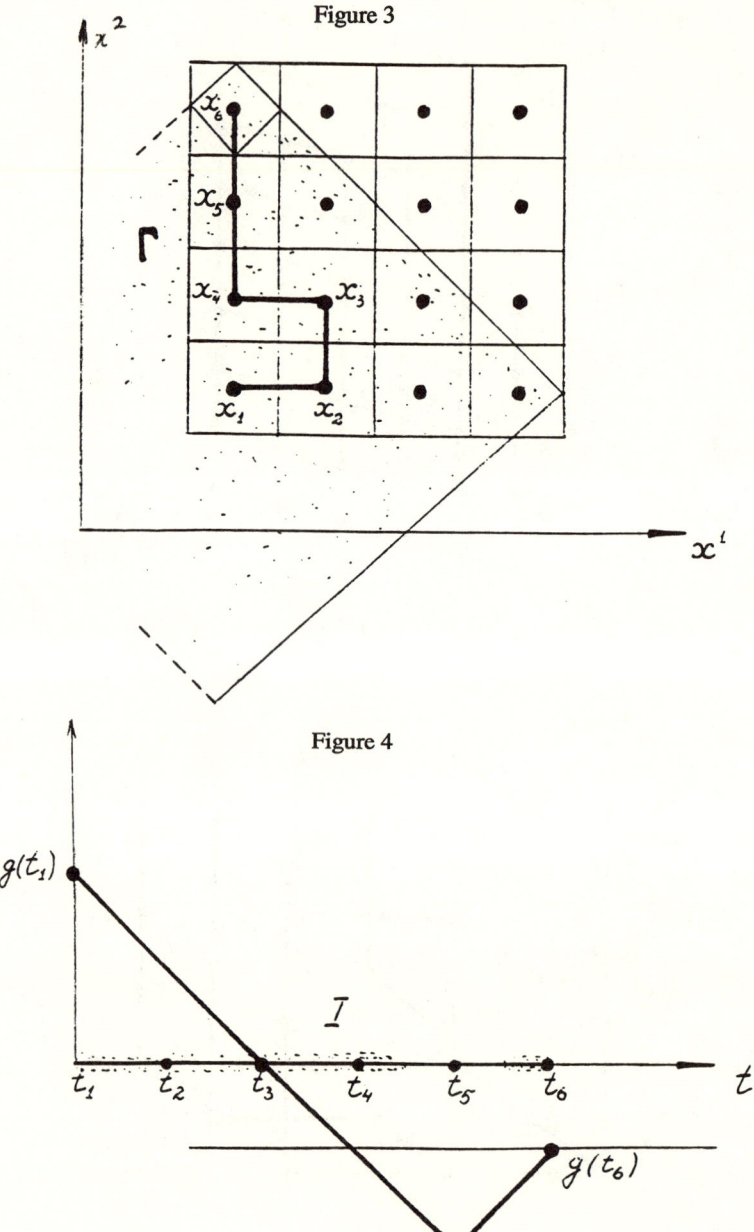

Figure 3

Figure 4

NUMERICAL METHODS

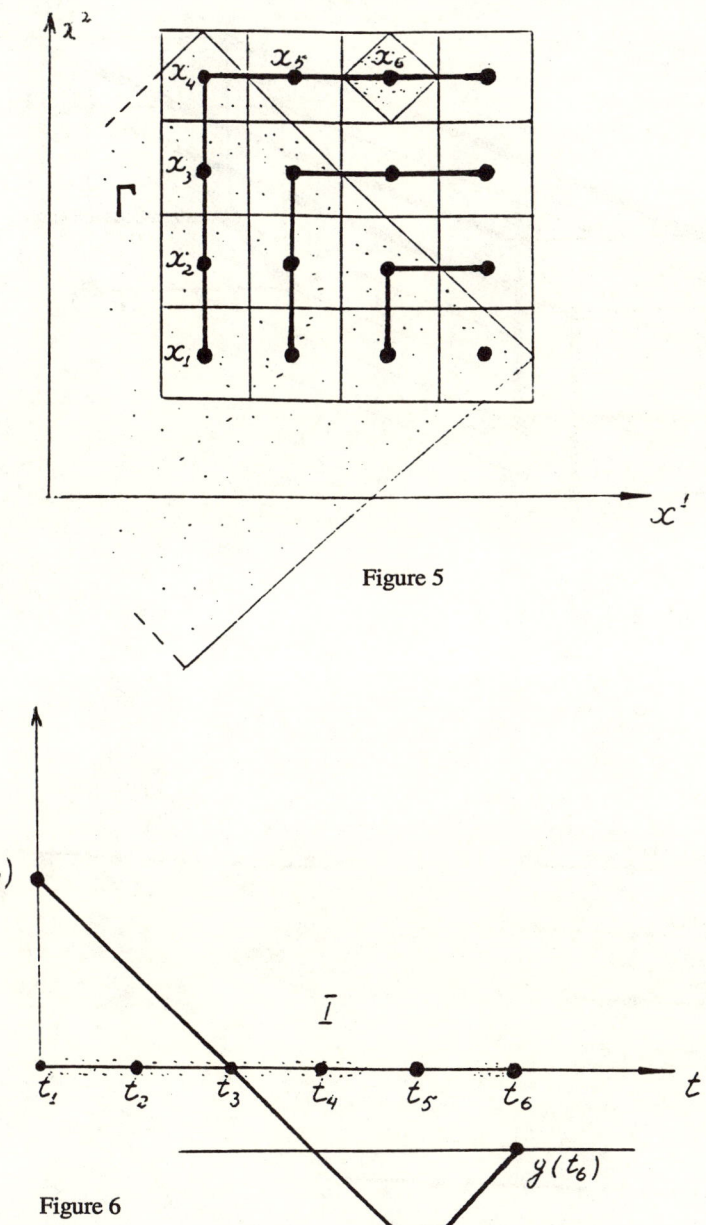

Figure 5

Figure 6

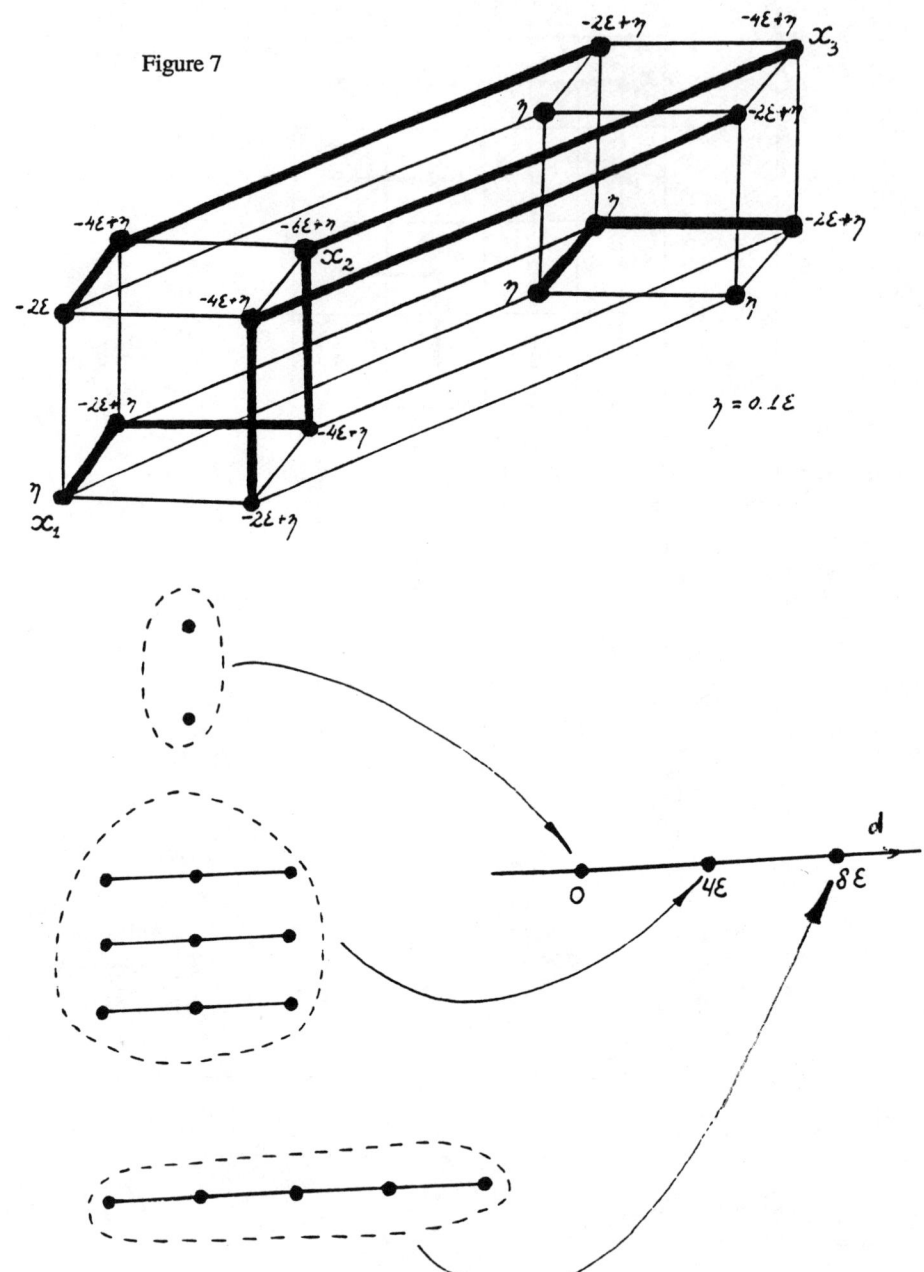

Figure 7

NUMERICAL METHODS

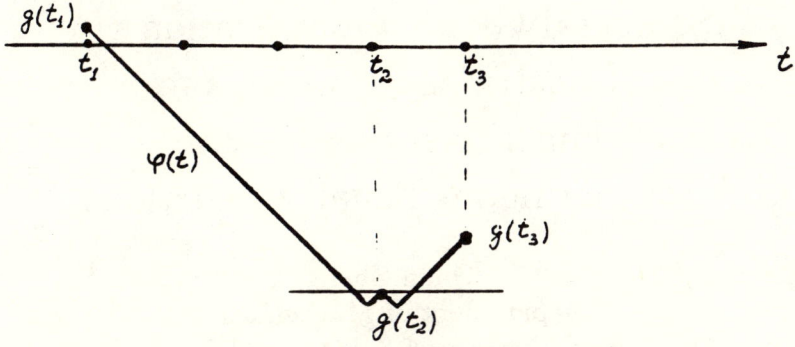

Figure 8

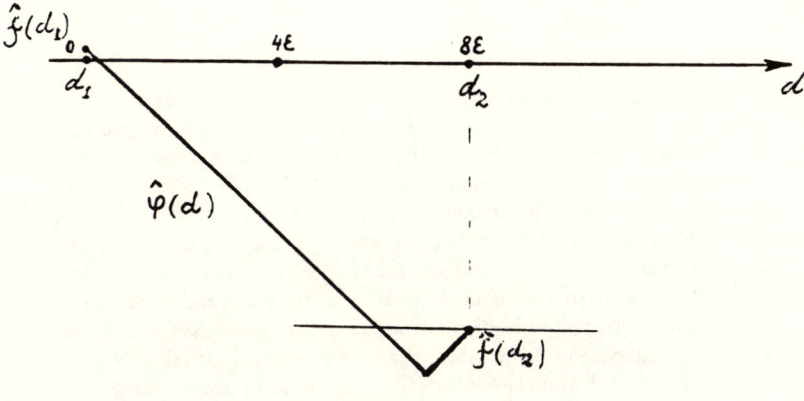

Figure 9

Integral Global Optimization of Constrained Problems in Functional Spaces with Discontinuous Penalty Functions

Quan Zheng
Department of Mathematics
Shanghai University of Science and Technology
Shanghai, China
and
Deming Zhuang [*]
Department of Mathematics and Computer Studies
Mount Saint Vincent University
Halifax, Nova Scotia, Canada B3M 2J6

Abstract

Problems from calculus of variations and optimal control require minimizing an objective function defined on a functional space over a constraint set. Numerically, we can only find an approximation to the solution in a finite dimensional space. Furthermore, it is difficult to characterize and to find such a solution when nonconvex state constraints are presented. In this paper the integral global optimization technique is applied to characterize the optimalities of such problems in a sequence of finite dimensional spaces with discontinuous penalty functions defined on these subspaces. A variable measure penalty algorithm is proposed to find solutions. A numerical example is presented which shows that the algorithm is efficient.

key words: Integral global optimization, discontinuous penalty function, optimal control with nonconvex state constraints.

1 Introduction

Let U be a topological space, S a subset of U and J a real-valued function on U. The problem considered here is to find the infimum

$$c^* = \inf_{u \in S} J(u) \tag{1}$$

---

[*]Research supported partially by NSERC grant and Mount St Vincent University internal research grant

and the set of global minima:

$$H^* = \{u \in S \mid J(u) = c^*\}. \tag{2}$$

Under the assumption

(A): J is lower semi-continuous, S is closed and there is a real number b such that the set $H_b = \{u \in S \mid J(u) \leq b\}$ is a nonempty compact set,

the set H^* is nonempty.

Problems from calculus of variations and optimal control, as well as differential games, require one to consider the case that the underlying space U is infinite-dimensional. But, in general, it is difficult to find the global minimum value c^* and the set of global minima H^* numerically when U is an infinite dimensional space. We usually can only find approximations to them:

$$c_n^* = \min_{u \in S^n} J(u) \text{ and } H_n^* = \{u \in S^n \mid J(u) = c_n^*\}, \tag{3}$$

where S^n is a subset of S. The set of such approximing solutions is usually a subset of a subspace U_n of U. In [16] we examine the problem of constructing the sequence $\{S^n\}$ such that the sequence $\{c_n^*\}$ converges to the global minimum value c^* and $\{H_n^*\}$ "converges" to the set of global minima H^* with the integral global optimization approach. On the other hand, a non-sequential penalty method is proposed in [14] which is proved to be efficient numerically for constrained minimization problems. However, this penalty algorithm is restricted to a fixed space U.

In this paper, we will consider the problem of approximating a constrained minimization problem in a functional space with the non-sequential penalty technique in a sequence of finite dimensional spaces $\{U_n\}$. In Section 2, several concepts and results in the integral global minimization are summarized, which will be useful for later investigation. The concept of a discontinuous penalty function is introduced in Section 3; the constrained global minimum in a functional space can be approximated by a sequence of global minima of associated penalized unconstrained minimization problems in finite dimensional spaces. Penalty optimality conditions are given in Section 4. With these results a variable measure penalty algorithm is proposed in Section 5; a convergence theorem is also proved in this section. An interesting example from calculus of variations and optimal control with a nonconvex state constraint is given in Section 6 to show that, numerically, the algorithm is efficient.

2 Integral Global Optimization in Infinite Dimensional Spaces

In this section we will summarize several concepts and properties of the integral global minimization developed in [15,10-13,2], which will be utilized in the following sections.

2.1 Robust Sets and Robust Functions

A set D in a topological space U is *robust* if

$$\text{cl } D = \text{cl int } D \tag{4}$$

An open set G is robust since $G = \text{int } G$. The empty set is a trivial robust set. A closed set may be robust or nonrobust. The concept of the robustness of a set is closely related to a topological structure of the space U. For instance, the set $D = \{1,2\}$ is nonrobust on R^1 but it is robust in $Z = \{$ set of all integers with the discrete topology $\}$.

The interior of a nonempty robust set is nonempty. A union of robust sets is robust. An intersection of two robust sets may be nonrobust; but the intersection of an open set and a robust set is robust. If A is robust in U and B is robust in V, then $A \times B$ is robust in $U \times V$ with the product topology. A convex set D in a topological vector space is robust if and only if the interior of D is nonempty.

A robust set consists of robust points of the set. A point $u \in \text{cl } D$ is said to be robust to D (or a robust point of D) if for each neighbourhood $N(u)$ of u, $N(u) \cap \text{int } D \neq \emptyset$. A set D is robust if and only if each point of D is robust to D.

A function J defined on a topological space U is *robust* if the set

$$F_c = \{u \mid J(u) < c\} \tag{5}$$

is robust for each real number c.

An upper semi continuous (u.s.c.) function is robust since (5) is open for each c; so is a probability function on R^n. The infimum of a family of robust functions is robust. A sum or a product of two robust functions may be nonrobust; but the sum of a robust function and an u.s.c. (for the product case nonnegativity is required) function is robust.

A function J is robust if and only if it is robust at each point or by each point; J is robust at a point u if $u \in F_c$ implies u is robust to F_c; J is robust by a point u if there is a neighbourhood $N(u)$ of u such that $N(u) \cap F_c$ is robust for each c. An example of a nonrobust function on R^1 is

$$J(u) = \begin{cases} 0, & u = 0, \\ 1, & u \neq 0. \end{cases}$$

J is nonrobust at $u = 0$.

We can investigate a robust function by its epigraph. A function J is robust if and only if its epigraph

$$\text{epi}(J) = \{(u,c) \mid J(u) \leq c\}$$

is robust in the product space $U \times R^1$.

When we consider a constrained minimization problem, the concept of relative robustness is needed. Let S be a given set in a topological space U and $u_o \in \text{cl } S$. A function J is said to be relatively robust to S at u if for each c, $u \in F_c$ implies u is robust to $F_c \cap S$. A function J is relatively robust at $u \in S$ if and only if J is robust at u with the relative topology on S. If J is relatively robust to S at each point u in S, then J is called a relatively robust function on S; or we simply say that J is robust on S.

For more detail about robustness, see [12,13].

In the following consideration we always suppose that there is a global minimum point $u^* \in S$ such that J is relatively robust to S at this point x^*. Or we simply make the following assumption:

(R): J is robust on S.

2.2 Q-Measure Spaces and Integrations

In order to investigate a minimization problem with an integral approach, a special class of measure spaces, which are called Q-measure spaces, should be examined.

Let U be a Hausdorff space, Ω a σ-field of subsets of U and μ a measure on Ω. A triple (U, Ω, μ) is called a Q-measure space iff

(i) Each open set in U is measurable;

(ii) The measure $\mu(G)$ of each nonempty open set G in U is positive: $\mu(G) > 0$;

(iii) The measure $\mu(K)$ of a compact set K in U is finite.

The n-dimensional Lebesgue measure space (R^n, Ω, μ) is a Q-measure space; a nondegenerate Gaussian measure μ on a separable Hilbert space H with Borel sets as measurable sets constitutes an infinite dimensional Q-measure space. A specific optimization problem is related to a specific Q-measure space which is suitable for consideration in this approach. In [5,6] we construct Q-measure spaces for problems from optimal control and differential games.

The construction of a Q-measure space in an infinite dimensional space is in general not trivial. For instance, it has been shown that for each $r > 0$ there exists on the space l_∞ a nondegenerate Gaussian measure μ such that the measure of an arbitrary ball with radius r is zero [9,pp 49-50].

Once a measure space is given we can define integration in a standard procedure.

Since the interior of a nonempty open set is nonempty, the Q-measure of a measurable set containing a nonempty robust set is always positive.

This is an essential property we need in the integral approach of minimization. Hence, the following assumption is usually required:

(M): (U, Ω, μ) is a Q-measure space.

The following lemma gives us a sufficient condition for the global minimum.

Lemma 2.1. Suppose that the conditions (A), (M) and (R) hold, and $H_c \cap S \neq \emptyset$, where $H_c = \{u \mid J(u) \leq c\}$ is the level set of J. If

$$\mu(H_c \cap S) = 0, \tag{6}$$

then c is the global minimum value of J over S and $H_c \cap S$ is the set of global minima.

In the following application, we need another form of the above lemma.

Lemma 2.2. Suppose that the conditions (A), (M) and (R) hold. If $c > c^* = \min_{u \in S} J(u)$, then

$$\mu(H_c \cap S) > 0.$$

2.3 Integral Optimality Conditions for Global Minimization

We now proceed to define the concepts of mean value, variance and higher moments of J over its level set. These concepts are closely related to optimality conditions and algorithms for global minimization.

Suppose that the assumptions (A), (M) and (R) hold, and $c > c^*$ and $c^* = \min_{u \in S} J(u)$. We define the mean value, variance, modified variance and m-th moment (centred at a), respectively, as follows:

$$M(f, c; S) = \frac{1}{\mu(H_c \cap S)} \int_{H_c \cap S} J(u) d\mu,$$

$$V(f, c; S) = \frac{1}{\mu(H_c \cap S)} \int_{H_c \cap S} (J(u) - M(J, c; S))^2 d\mu,$$

$$V_1(J, c; S) = \frac{1}{\mu(H_c \cap S)} \int_{H_c \cap S} (J(u) - c)^2 d\mu,$$

$$M_m(J, c; a; S) = \frac{1}{\mu(H_c \cap S)} \int_{H_c \cap S} (J(u) - c)^m d\mu,$$

$$m = 1, 2, \ldots.$$

By Lemma 2.2, they are well defined. These definitions can be extended to the case $c \geq c^*$ by a limit process. For instance,

$$M_m(J, c; a; S) = \lim_{c_k \downarrow c} \frac{1}{\mu(H_{c_k} \cap S)} \int_{H_{c_k} \cap S} (J(u) - a)^m d\mu,$$

$$m = 1, 2, \ldots.$$

The limits exist and are independent of the choice of $\{c_k\}$. The extended concepts are well defined and consistent with the above definitions.

Remark. Sometimes, we wish to emphasize these concepts for a particular Q-measure μ. The measure μ can then be included in the notation. For example,

$$M(J, c; S; \mu) = \frac{1}{\mu(H_c \cap S)} \int_{H_c \cap S} J(u) d\mu.$$

With these concepts we characterize the global optimality as follows:

Theorem 2.1. Under the assumptions (A), (M) and (R), the following statements are equivalent:

(i) $u^* \in S$ is a global minimizer of J over S and $c^* = J(u^*)$ is the global minimum value;

(ii) $M(J, c^*; S) = c^*$ (the mean value condition);

(iii) $V(J, c^*; S) = 0$ (the variance condition);

(iv) $V_1(J, c^*; S) = 0$ (the modified variance condition);

(v) $M_m(J, c^*; c^*; S) = 0$, for one of positive integers $m = 1, 2, \ldots$ (the higher moment conditions).

2.4 Q-Convergence of Measures

As we wish to investigate the approximation to an optimal solution in an infinite-dimensional space by a sequence of solutions in finite-dimensional spaces, it is natural to consider a sequence of measure spaces and examine its convergence. There are already several concepts of convergence of measures in the theory of probability and stochastic processes. But, these concepts of convergence are not suitable for working with Q-measure spaces. Thus, the concept of Q-convergence is introduced as follows:

Let $(\Omega_1, U_1, \mu_1), \ldots, (\Omega_n, U_n, \mu_n)$ and (Ω, U, μ) be Q-measure spaces, where U_n is a subspace of U and $\Omega_n = \{A \cap U_n \mid A \in \Omega \}$.

Definition 2.1. A sequence of Q-measures $\{\mu_n\}$ defined on measurable spaces $\{(U_n, \Omega_n)\}$ is said to be Q-convergent to a Q-measure μ defined on (U, Ω) if for each open set $G \subset U$

$$\mu_n(G \cap U_n) \to \mu(G) \quad as \quad n \to \infty. \tag{7}$$

and the Q-convergence is denoted by $\mu_n \xrightarrow{Q} \mu$.

Remark. In this work, we concentrate our attention to minimization problems in infinite-dimensional spaces. Thus, we always assume a finite measure μ defined on a measurable space (U, Ω), i.e., $\mu(U) < +\infty$.

The following theorem exhibits equivalent conditions for Q-convergence which are useful in the sequel. The proof is in [16].

Theorem 2.2. Suppose $(U_1, \Omega_1, \mu_1), \ldots, (U_n, \Omega_n, \mu_n), \ldots$, and (U, Ω, μ) are Q-measure spaces, where U_n is a subspace of U, $n = 1, 2, \ldots$. Then the following statement are equivalent:

(i) For each bounded lower-semidefined on U, $\int_{U_n} J d\mu_n \to \int_U J d\mu$;

(ii) For each bounded upper-semi continuous function J defined on U, $\int_{U_n} J d\mu_n \to \int_U J d\mu$;

(iii) For each open set G in U, $\mu_n(G \cap U_n) \to \mu(G)$;

(iv) For each closed set F in U, $\mu_n(F \cap U_n) \to \mu(F)$.

Remark. Recall that a sequence of measures $\{\mu_n\}$ on the Borel sets of a metric space U is said to be weakly convergent to a measure μ if for each bounded *continuous* function $J : U \to R$, $\int_U J d\mu_n \to \int_U J d\mu$. The requirement for the Q-convergence is more than that of the weak convergence.

The following proposition give us another equivalent condition for Q-convergence which is utilized in the sequel. (See the proof of Theorem 4.1.)

Proposition 2.3. Under the assumption of Theorem 2.1, $\mu_n \xrightarrow{Q} \mu$ if and only if for each closed set F in U and for each bounded l.s.c. function J one has

$$\int_{U_n \cap F} J d\mu_n \to \int_F J d\mu. \tag{8}$$

2.5 Finite-Dimensional Approximation

For a minimization problem (1), we usually can only find an approximation solution when U is a functional space:

$$c_n^* = \min_{u \in S^n} J(u) \text{ and } H_n^* = \{u \in S^n \mid J(u) = c_n^*\},$$

where S^n is a subset of S, and lies in a subspace U_n of U.

Let $\{U_n\}$ be a sequence of subspaces of U, let (U_n, Ω_n, μ_n), $n = 1, 2, \ldots$ be Q-measure spaces, $\{\mu_n\}$ is Q-convergent to μ and $S^n = S \cap U_n, n = 1, \ldots$ with the property that

$$\mu_n(D \cap U_n) \geq \mu(D), \forall D \in \Omega, n = 1, 2, \ldots. \tag{9}$$

Condition (9) is essential in our development. In [16], we demostrate, for a problem from control theory and a problem from differential game theory, the constructions of sequences of Q-measures which satisfy (9).

Since for $n = 1, 2, \ldots$, each U_n is a subspace of the space U,

$$\mathrm{cl} \bigcup_{n=1}^{\infty} U_n \subset U.$$

In order to approximate optimal solutions in the space U, it is natural to assume that no gap exists in the above set inclusion. Thus, we assume that

(F): $\mathrm{cl} \bigcup_{n=1}^{\infty} U_n = U$, and there is a sequence of Q-measure spaces $\{(U_n, \Omega_n, \mu_n)\}$ such that $\mu_n \xrightarrow{Q} \mu$ and μ is a Q-measure extended by $\{\mu_n\}$, and (9) is satisfied.

Remark. Since $\{U_n\}$ is an increasing set sequence, the statement cl $\bigcup_{n=1}^{\infty} U_n = U$ is equivalent to cl $\bigcup_{n=k}^{\infty} U_n = U$, for any $k = 1, 2, \ldots$.

Theorem 2.4. Suppose $c_n \downarrow c \leq c^* = \min_{u \in S} J(u)$, $\mu_n \xrightarrow{Q} \mu$ and (9) holds. Then,

(i) $\lim_{n \to \infty} \frac{1}{\mu_n(H_{c_n} \cap S^n)} \int_{H_{c_n} \cap S^n} J(u) d\mu_n = M(J, c; S; \mu)$;

(ii) $\lim_{n \to \infty} \frac{1}{\mu_n(H_{c_n} \cap S^n)} \int_{H_{c_n} \cap S^n} (J(u) - M(J, c_n; S^n; \mu_n))^2 d\mu_n = V(J, c; S; \mu)$;

(iii) $\lim_{n \to \infty} \frac{1}{\mu_n(H_{c_n} \cap S^n)} \int_{H_{c_n} \cap S^n} (J(u) - c_n)^2 d\mu_n = V_1(J, c; S, \mu)$;

(iv) $\lim_{n \to \infty} \frac{1}{\mu(H_{c_n} \cap S^n)} \int_{H_{c_n} \cap S^n} (J(u) - a)^m d\mu_n = M_m(J, c; S; a; \mu), m = 1, 2, \ldots$.

Under assumptions (A), (R) and (F), a variable measure algorithm is proposed as follows[16]:

Step 1: Take $c_o > \min_{u \in S^1} J(u); \epsilon > 0; n := 0$;

Step 2: Calculate the mean value

$$c_{n+1} = \frac{1}{\mu_n(H_{c_n} \cap S^n)} \int_{H_{c_n} \cap S^n} J(u) d\mu_n;$$

Step 3: Calculate the variance

$$v_{n+1} = \frac{1}{\mu_n(H_{c_n} \cap S^n)} \int_{H_{c_n} \cap S^n} (J(u) - c_n)^2 d\mu_n.$$

If $v_{n+1} \geq \epsilon$, then $n := n+1$, so go to Step 2; otherwise go to Step 4;

Step 4: $c^* \Leftarrow c_{n+1}; H^* \Leftarrow H_{c_{n+1}}$; Stop.

Here, $\epsilon > 0$ is the accuracy given in advance. Let $\epsilon = 0$ in the above algorithm, then either the algorithm may stop in a finite number of iterations, and we find global minima, or we obtain two monotone sequences:

$$c_o > c_1 > \cdots > c_n > c_{n+1} > \cdots \geq c^* \tag{10}$$

$$H_{c_1} \cap S^1 \supset \cdots \supset H_{c_n} \cap S^n \supset H_{c_{n+1}} \cap S^{n+1} \supset \cdots \supseteq H^*. \tag{11}$$

Theorem 2.5 Applying the above algorithm with $\epsilon = 0$ we have

$$\lim_{n \to \infty} c_n = c^* = \min_{u \in S} J(u) \tag{12}$$

and

$$\emptyset \neq \bigcap_{k=1}^{\infty} \text{cl} \bigcup_{n=k}^{\infty} (H_{c_n} \cap S^n) \subset H^*. \tag{13}$$

If J is continuous then,

$$\emptyset \neq \bigcap_{k=1}^{\infty} \text{cl} \bigcup_{n=k}^{\infty} (H_{c_n} \cap S^n) = H^*. \tag{14}$$

Corollary. Let $c_n^* = \min_{u \in S^n} J(u)$ and $H_n^* = \{u \in S^n \mid J(u) = c_n^*\}, n = 1,\ldots$, then

$$\lim_{n \to \infty} c_n^* = c^* \text{ and } \emptyset \neq \bigcap_{k=1}^{\infty} \text{cl}(\bigcup_{n=k}^{\infty} H_n^*) \subset H^*. \tag{15}$$

If J is continuous, then the last set inclusion becomes a equality.

3 Discontinuous Penalty Functions

In this section we will define a discontinuous penalty function for a constrained minimization problem and consider a sequence of associated penalized minimization problems in finite dimensional spaces. The global minima of these problems converge to the global minimum of the constrained one. Note that no Q-measure is involved in this section.

Suppose U is a metric space, S is a closed robust subset of U and J is a real-valued function on U. Under the assumption (A), the set of global minima of the constrained problem

$$c^* = \min_{u \in S} J(u) \tag{16}$$

is nonempty. Moreover, (A) also implies that J is bounded below on U, i.e., there is a constant L such that

$$J(u) \geq L, \text{ for all } u \in U.$$

The minimizers of the constrained problem (16) can be approximated by a sequence of solutions of associated penalized unconstrained problems in finite dimensional spaces.

A penalty function associated with the constraint set S is defined as follows.

Definition 3.1. A function $p(u)$ on U is a penalty function for the constraint set S if

(i) p is l.s.c.;

(ii) $p(u) = 0$ if $u \in S$;

(iii) $\inf_{u \in S_\beta} p(u) > 0$, where $S_\beta = \{u \mid \|u - v\| \leq \beta, \forall v \in S\}$ and $\beta > 0$.

Remarks. 1. In the above definition we relax the requirement of continuity from the traditional definition [3,4] as we wish to utilize discontinuous penalty functions. 2. It is expected that the penalty will be increasing when the distance of a point u to the constraint set S is getting larger. We replace the traditional property

$$p(u) > 0, \text{ if } u \notin S$$

by (iii).

Definition 3.2. A penalty function p for the constraint set S is *exact* for (16) if there is a real number $\alpha_o > 0$ such that for each $\alpha \geq \alpha_o$ we have
$$\min_{u \in U}\{J(u) + \alpha p(u)\} = \min_{u \in S} J(u) = c^* \tag{17}$$
and
$$\{u \mid J(u) + \alpha p(u) = c^*\} = \{u \in S \mid J(u) = c^*\} = H^*. \tag{18}$$

In [14] a class of discontinuous penalty functions is proposed as follws:
$$p(u) = \begin{cases} 0, & u \in S, \\ \delta + d(u), & u \notin S, \end{cases}$$
where δ is a positive number and $d(u)$ is a penalty-like function. For example, for the inequality-constraint set,
$$S = \{u \mid g_i(u) \leq 0, i = 1, \ldots, r\}.$$
we can take
$$d(u) = \sum_{i=1}^{r} \|\max(g_i(u), 0)\|^\rho \text{ or } d(u) = \max_{i} \|\max(g_i(u), 0)\|^\rho$$
where $\rho > 0$.

With a penalty function p, we examine a sequence of penalized unconstrained minimization problems in the finite dimensional spaces associated with (16)
$$\min_{u \in U_n}\{J(u) + \alpha_n p(u)\}, n = 1, 2, \ldots, \tag{19}$$
where $U_n \subset U_{n+1}, n = 1, 2, \ldots$ are finite dimensional subspaces of U, and $\alpha_n > 0, n = 1, 2, \ldots$ are penalty parameters. Under the assumption (A),
$$H_b^{\alpha_n} = \{u \mid J(u) + \alpha_n p(u) \leq b\}$$
is a nonempty closed subset of H_b. Thus $H_b^{\alpha_n}$ is compact in U, and then it is also compact in U_n. It follows that the minimum of (19) for each n also exists. Let c_n^* be the global minimum of $J(u)$ over $S \cap U_n$, that is $c_n^* = \min_{u \in S \cap U_n} J(u)$, then
$$\min_{u \in U_n}\{J(u) + \alpha_n p(u)\} \leq \min_{u \in S \cap U_n}\{J(u) + \alpha_n p(u)\}$$
$$= \min_{u \in S \cap U_n} J(u) = c_n^*$$

Remark. If we take an exact penalty function p in the penalized problem, then, for $\alpha_n \geq \alpha_0$, we also have that (by (17))
$$\min_{u \in U_n}\{J(u) + \alpha_n p(u)\} = \min_{u \in S \cap U_n} J(u).$$

We will construct two sequences $\{\alpha_n\}$ and $\{c_n\}$ so that $\alpha_n \uparrow \infty$ (for the exact penalty function case, $\alpha_n \geq \alpha_o$) and $c_n \downarrow c$ ($\geq c^*$, assuming $b > c$) as $n \to \infty$ with the property that

$$\min_{u \in H_n \cap U_n} \{J(u) + \alpha_n p(u)\} \to c^*, \text{ as } n \to \infty,$$

where, in order to simplify the notation, we denote

$$H_n = \{u \mid J(u) + \alpha_n p(u) \leq c_n\}. \tag{20}$$

Proposition 3.1. If $c_n \downarrow c \geq c^*$ then

$$\lim_{n \to \infty} H_n = \bigcap_{n=1}^{\infty} H_n = H_c \cap S. \tag{21}$$

Proof. We show first that $\{H_n\}$ is a monotone sequence. It follows that the limit in (21) exists and equals to the intersection. Suppose $u \in H_{n+1}$. Since $\alpha_{n+1} \geq \alpha_n$ and $c_{n+1} \leq c_n$,

$$J(u) + \alpha_n p(u) \leq J(u) + \alpha_{n+1} p(u) \leq c_{n+1} \leq c_n.$$

Therefore $u \in H_n$. This proves $H_{n+1} \subset H_n$. Now we show that

$$\bigcap_{n=1}^{\infty} H_n = H_c \cap S.$$

If $u \in H_c \cap S$, then $p(u) = 0$ and $J(u) + \alpha_n p(u) = J(u) \leq c \leq c_n$, $\forall n = 1, 2, \ldots$. Hence, $u \in H_n$, for $n = 1, 2, \ldots$. This proves

$$H_c \cap S \subset \bigcap_{n=1}^{\infty} H_n.$$

On the other hand, suppose $u \in \cap_{n=1}^{\infty} H_n$; then $J(u) \leq J(u) + \alpha_n p(u) \leq c_n$, for $n = 1, 2, \ldots$. Letting $n \to \infty$, we have $J(u) \leq c$, i.e., $u \in H_c$. If $u \notin S$ then $p(u) > 0$, and $J(u) + \alpha_n p(u) \to \infty$ as $n \to \infty$. (In the case of $p(u)$ being exact, we have $J(u) + \alpha_n p(u) \geq c_1 + L$, where L is the lower bound of J.) Since J is bounded below this contradicts that $J(u) + \alpha_n p(u) \leq c_n \leq c_1$, for $n = 1, 2, \ldots$. Hence, $u \in H_c \cap S$. This proves

$$\bigcap_{n=1}^{\infty} H_n \subset H_c \cap S.$$

The proof of Proposition 3.1 is completed.

Remark. We will use the concepts of mean value and variance to study a global minimization problem. If $c < c^* = \min_{u \in S} J(u)$, then $H_c \cap S = \emptyset$. From the above proposition, there is an integer N such that $H_n = \emptyset$ for $n \geq N$. In this case we cannot even define mean values and variances on U_n. Thus this situation will never be allowed to happen in the integral algorithm.

The following proposition shows that in the above framework the global minimum value of a constrained problem is the limit of the global minimum values of the penalized problems.

Proposition 3.2. Suppose that $\{\alpha_n\}$ is a positive increasing sequence which tends to infinity as $n \to \infty$, (or $\alpha_n \geq \alpha_o$, for the exact penalty function case). $\{c_n\}$ is a decreasing sequence which tends to $c \geq c^*$ as $n \to \infty$. Under assumption (A) we have

$$\min_{u \in U_n}\{J(u) + \alpha_n p(u)\} = \min_{u \in H_n \cap U_n}\{J(u) + \alpha_n p(u)\}$$
$$= a_n \to c^* = \min_{u \in S} J(u). \tag{22}$$

Proof. Since J and p are l.s.c., $H_n \cap U_n$ is closed and thus compact as it is bounded abd lies in a finite dimension space. Therefore,

$$\min_{u \in H_n \cap U_n}\{J(u) + \alpha_n p(u)\}$$

exists for each n. Since $H_c \cap S \subset H_n$ we have

$$\min_{u \in H_n \cap U_n}\{J(u) + \alpha_n p(u)\} \leq \min_{u \in H_n \cap S \cap U_n}\{J(u) + \alpha_n p(u)\}$$
$$= \min_{u \in H_n \cap S \cap U_n} J(u)$$
$$= \min_{u \in S \cap U_n} J(u) = c_n^*.$$

Hence, by the corollary of Theorem 2.4,

$$\limsup_{n \to \infty} \min_{u \in H_n \cap U_n}\{J(u) + \alpha_n p(u)\} \leq \lim_{n \to \infty} c_n^* = c^*. \tag{23}$$

We now prove

$$\liminf_{n \to \infty} \min_{u \in H_n \cap U_n}\{J(u) + \alpha_n p(u)\} = \hat{c} \geq c^*. \tag{24}$$

Suppose, on the contrary, $\hat{c} < c^*$. Let $c^* - \hat{c} = 2\eta > 0$; then there is a subsequence of $\{a_n\}$ (we denote it with the same notation) and an integer N such that $a_n \to \hat{c}$ and $a_n < c - \eta, \forall n \geq N$. Let $\hat{u}_n \in U_n \cap H_n$ be a global minimizer of $\min_{u \in H_n \cap U_n}\{J(u) + \alpha_n p(u)\}$, then

$$J(\hat{u}_n) \leq J(\hat{u}_n) + \alpha_n p(u) \leq c^* - \eta, n = 1, 2, \ldots.$$

We now have $\hat{u}_n \in H_{c^*-\eta} \cap H_n$, $n = N+1, N+2, \ldots$. Because of the monotonicity of $\{H_n\}$, $H_{c^*-\eta} \cap H_n \neq \emptyset$ implies that $H_{c^*-\eta} \cap H_k \neq \emptyset, k = 1, \ldots, n-1, n$. Hence the intersection of these decreasing closed (compact) sets is also nonempty:

$$\bigcap_{n=1}^{\infty}(H_{c^*-\eta} \cap H_n) = H_{c^*-\eta} \bigcap_{n=1}^{\infty} H_n = H_{c^*-\eta} \cap H_c \cap S \neq \emptyset. \tag{25}$$

Therefore, we have a point $\hat{u}$ which is in both S and $H_{c^*-\eta}$. This contradicts the fact that c^* is the global minimum value of J over S.

Combining (23) and (24), we obtain (22).

4 Penalty Optimality Conditions

In this section we will examine the concepts of penalized mean value, variance and higher moments with a variable measure model. Let S be a subset of U, J a real-valued function on U and p a penalty function for the constraint set S. Suppose that U_n is a subspace of U, and $U_n \subset U_{n+1}, n = 1, 2, \ldots$. The penalized mean value variance and higher moments are defined on a finite dimensional space with a Q-measure on it.

Definition 4.1. Let $c_n > \min_{u \in S \cap U_n} J(u)$. We define the penalty mean value, variance, modified variance and m-th moment (centred at a), respectively, of $J + \alpha_n p$ over the penalized level set

$$H_n = \{u \mid J(u) + \alpha_n p(u) \leq c_n\}$$

with a measure μ_n on U_n as follows

$$M(J, c_n; p; \mu_n)$$
$$= \frac{1}{\mu_n(H_n \cap U_n)} \int_{H_n \cap U_n} [J(u) + \alpha_n p(u)] d\mu_n,$$

$$V(J, c_n; p; \mu_n)$$
$$= \frac{1}{\mu_n(H_n \cap U_n)} \int_{H_{c_n} \cap U_n} [J(u) + \alpha_n p(u) - M(J, c_n; p; \mu_n)]^2 d\mu_n,$$

$$V_1(J, c_n; p; \mu_n)$$
$$= \frac{1}{\mu_n(H_n \cap U_n)} \int_{H_n \cap U_n} [J(u) + \alpha_n p(u) - c_n]^2 d\mu_n,$$

$$M_m(J, c; p; a; \mu_n)$$
$$= \frac{1}{\mu_n(H_n \cap U_n)} \int_{H_n \cap U_n} [J(u) + \alpha_n p(u) - a]^m d\mu_n,$$
$$m = 1, 2, \ldots.$$

Under the assumptions (A), (R) and (M), they are well defined.

Now we consider the convergence properties of the penalized mean value, variance and higher moments as $n \to \infty$. As usual, we assume that

$$c_n \downarrow c \geq c^* = \min_{u \in S} J(u). \tag{26}$$

We first prove a lemma.

Lemma 4.1. Suppose that $c \geq c^*$ and $c_n \downarrow c$. If $\mu_n \xrightarrow{Q} \mu$ then

$$\lim_{n \to \infty} \mu_n(H_n \cap U_n) = \mu(H_c \cap S). \tag{27}$$

Proof. Since $H_c \cap S \subset H_n, n = 1, 2, \ldots$ and $\mu_n \xrightarrow{Q} \mu$, we have

$$\mu_n(H_n \cap U_n) \geq \mu_n(H_c \cap S \cap U_n) \to \mu(H_c \cap S).$$

It follows that
$$\liminf_{n\to\infty} \mu_n(H_n \cap U_n) \geq \mu(H_c \cap S). \tag{28}$$

On the other hand, let $n \geq k$ for fixed k. Then $H_n \subset H_k$ and $\mu_n(H_n \cap U_n) \leq \mu_n(H_k \cap U_n)$. Thus
$$\limsup_{n\to\infty} \mu_n(H_n \cap U_n) \leq \lim_{n\to\infty} \mu_n(H_k \cap U_n) = \mu(H_k).$$

Letting $k \to \infty$ we then obtain, from the continuity of the measure μ and by Proposition 3.1,
$$\limsup_{n\to\infty} \mu_n(H_n \cap U_n) \leq \lim_{k\to\infty} \mu(H_k) = \mu(H_c \cap S). \tag{29}$$

Combining (28) and (29), we obtain (27).

To consider the limit process, we need assumption (F) instead of (M).

Theorem 4.1. Suppose that the assumptions (A), (R) and (F) are satisfied for $J + \alpha p$. Under the conditions of Lemma 4.1 we have, for $c \geq c^*$,

(i) $\lim_{n\to\infty} M(J, c_n; p; \mu_n) = M(J, c; S; \mu)$;

(ii) $\lim_{n\to\infty} V(J, c_n; p; \mu_n) = V(J, c; S; \mu)$;

(iii) $\lim_{n\to\infty} V_1(J, c_n; p; \mu_n) = V_1(J, c; S; \mu)$;

(iv) $\lim_{n\to\infty} M_m(J, c_n; p; c_n; \mu_n) = M_m(J, c; p; c; \mu)$, $m = 1, 2, \ldots$.

Proof. We first prove that when $c > c^*$, (i) holds. Since $\mu(H_c \cap S) > 0$ we have, by Lemma 4.1, that $\mu_n(H_n \cap U_n) > 0$ for n sufficiently large. Thus,

$$\begin{aligned}
& M(J, c_n; p; \mu_n) - M(J, c; S; \mu) \\
&= \frac{1}{\mu_n(H_n \cap U_n)} \int_{H_n \cap U_n} [J(u) + \alpha_n p(u)] d\mu_n \\
&\quad - \frac{1}{\mu(H_c \cap S)} \int_{H_c \cap S} J(u) d\mu \\
&= \left(\frac{1}{\mu_n(H_n \cap U_n)} - \frac{1}{\mu(H_c \cap S)}\right) \int_{H_n \cap U_n} [J(u) + \alpha_n p(u)] d\mu_n \\
&\quad + \frac{1}{\mu(H_c \cap S)} \left(\int_{H_n \cap U_n} [J(u) + \alpha_n p(u)] d\mu_n - \int_{H_c \cap S \cap U_n} J(u) d\mu_n\right) \\
&\quad + \frac{1}{\mu(H_c \cap S)} \left(\int_{H_c \cap S \cap U_n} J(u) d\mu_n - \int_{H_c \cap S} J(u) d\mu\right) \\
&= I_1 + I_2 + I_3. \tag{30}
\end{aligned}$$

Since
$$|I_1| \leq \left|\frac{1}{\mu_n(H_n \cap U_n)} - \frac{1}{\mu(H_c \cap S)}\right| \cdot A \cdot \mu_n(H_n \cap U_n),$$

where A is a bound of $|J(u)+\alpha_n p(u)|$ on $H_n \cap U_n$, $n=1,2,\ldots$ (e.g. $A=\max(c_1,|L|)$, and $\mu_n(H_n \cap U_n) \leq \mu_n(U_n) < \infty$ because $\mu_n(U_n) \to \mu(U) < \infty$. It follows from Lemma 4.1 that $I_1 \to 0$ as $n \to \infty$.

Next, we have

$$|I_2| \leq \frac{1}{\mu(H_c \cap S)} \cdot 2A \cdot |\mu_n(H_n \cap U_n) - \mu_n(H_c \cap S \cap U_n)|$$

$$\leq \frac{2A}{\mu(H_c \cap S)} (|\mu_n(H_n \cap U_n) - \mu(H_c \cap S)|$$
$$+ |\mu(H_c \cap S) - \mu_n(H_c \cap S \cap U_n)|),$$

which tends to zero as $n \to \infty$ because of Lemma 4.1 and because $\mu_n \xrightarrow{Q} \mu$. $I_3 \to 0$ follows from Proposition 2.3. We complete the proof of (i) for the case $c > c^*$.

When $c = c^*$, since $J(u) + \alpha_n p(u) \leq c_n$ on H_n, $\forall n$, we have,

$$M(J, c_n; p; \mu_n) = \frac{1}{\mu_n(H_n \cap U_n)} \int_{H_n \cap U_n} [J(u) + \alpha_n p(u)] d\mu_n$$
$$\leq c_n, \quad n = 1, 2, \ldots.$$

It follows that

$$\limsup_{n \to \infty} M(J, c_n; p; \mu_n) \leq \lim_{n \to \infty} c_n = c = c^*. \tag{31}$$

We now prove

$$\liminf_{n \to \infty} \frac{1}{\mu_n(H_n \cap U_n)} \int_{H_n \cap U_n} [J(u) + \alpha_n p(u)] d\mu_n \geq c^*. \tag{32}$$

Suppose, on the contrary, that (32) does not hold. Then there is a subsequence of

$$\frac{1}{\mu_n(H_n \cap U_n)} \int_{H_n \cap U_n} [J(u) + \alpha_n p(u)] d\mu_n,$$

(which we denote with the same notation, such that

$$\lim_{n \to \infty} \frac{1}{\mu_n(H_n \cap U_n)} \int_{H_n \cap U_n} [J(u) + \alpha_n p(u)] d\mu_n = \hat{c} < c^*.$$

Let $2\eta = c^* - \hat{c} > 0$. Thus there is a positive integer N such that for $n \geq N$

$$\frac{1}{\mu_n(H_n \cap U_n)} \int_{H_n \cap U_n} J(u) d\mu_n$$
$$\leq \frac{1}{\mu_n(H_n \cap U_n)} \int_{H_n \cap U_n} [J(u) + \alpha_n p(u)] d\mu_n$$
$$\leq c^* - \eta.$$

This implies that
$$H_{c^*-\eta} \cap H_n \neq \emptyset, \text{ for } n \geq N$$
and hence
$$H_{c^*-\eta} \cap H_{c^*} \cap S \neq \emptyset.$$
That is to say, we have points both $H_{c^*-\eta}$ and S. This contradicts the assumption that c^* is the global minimum value of J over S.

We now prove that (iv) holds. When $c > c^*$ or when $c = c^*$ and m is odd, the proof is similar to that of the mean value case. Suppose $m = 2r$ is even and $c = c^*$. We consider the case $r = 1$ first. Since
$$V_1(J, c_n; p; \mu_n) \geq 0, n = 1, 2, \ldots,$$
it follows that
$$\liminf_{n \to \infty} V_1(J, c_n; p; \mu_n) \geq 0. \tag{33}$$
We prove that $\limsup_{n \to \infty} V_1(J, c_n; p; \mu_n) = 0$. Suppose, on the contrary, it does not hold. Then there is a subsequence (for which we keep the same notation) such that
$$V_1(J, c_n; p; \mu_n) \to 2\eta > 0$$
. Thus there is an integer N such that
$$V_1(J, c_n; p; \mu_n) > \eta, \text{ when } n \geq N.$$

Since J is bounded below on $H_1 \cap S$, there exists a real number $g \geq 0$ such that
$$J(u) + \alpha_n p(u) + g \geq 0, \forall u \in H_1. \tag{34}$$
Therefore
$$\begin{aligned}
&V_1(J, c_n; p; \mu_n) \\
&= \frac{1}{\mu_n(H_n \cap U_n)} \int_{H_n \cap U_n} [J(u) + \alpha_n p(u) - c_n]^2 d\mu_n \\
&= \frac{1}{\mu_n(H_n \cap U_n)} \{ \int_{H_n \cap U_n} [J(u) + \alpha_n p(u) + g]^2 d\mu_n \\
&\quad + \int_{H_n \cap U_n} (g + c_n)^2 d\mu_n \\
&\quad - 2(g + c_n) \int_{H_n \cap U_n} [J(u) + \alpha_n p(u) + g] d\mu_n \} \\
&> \eta.
\end{aligned}$$
It follows that
$$(c_n + g)^2 + (g + c_n)^2 > 2(g + c_n) \cdot (g + c_n^*) + \eta > 2(g + c_n) \cdot (g + c^*) + \eta.$$
(Recall, c_n^* is the global minimum value of $J(u)$ over $S \cap U_n$). Letting $n \to \infty$ in the above inequality, we obtain
$$(c^* + g)^2 + (g + c^*)^2 \geq 2(g + c^*) \cdot (g + c^*) + \eta,$$

and have a contradiction: $0 \geq \eta > 0$. Therefore, when $c = c^*$ the limit of (iv) exists for $m = 2$ and the limit is equal to 0. But, according to the variance condition, c^* is the global minimum value if and only if $V_1(J, c^*; S; \mu) = 0$. Hence, when $c = c^*$, we have

$$\lim_{n \to \infty} V_1(J, c_n; p; \mu_n) = 0 = V_1(J, c^*; S : \mu).$$

When $m = 2r$ and $r > 1$, $M_m(J, c_n; p; c_n; \mu_n) \geq 0$. Thus,

$$\liminf_{n \to \infty} M_m(J, c_n; p; c_n; \mu_n) \geq 0. \tag{35}$$

On the other hand,

$$M_m(J, c_n; p; c_n; \mu_n) \tag{36}$$
$$= \frac{1}{\mu_n(H_n \cap U_n)} \int_{H_n \cap U_n} [J(u) + \alpha_n p(u) - c_n]^m d\mu_n$$
$$\leq A^{2(r-1)} \frac{1}{\mu_n(H_n \cap U_n)} \int_{H_n \cap U_n} [J(u) + \alpha_n p(u) - c_n]^2 d\mu_n$$
$$= A^{2(r-1)} V_1(J, c_n; p; \mu_n) \to 0, \text{ as } n \to \infty, \tag{37}$$

where $\mid J(u) + \alpha_n p(u) - c_n \mid \leq A, \forall u \in H_1 \cap S$.

Therefore, we have proved that

$$\lim_{n \to \infty} M_m(J, c_n; p; c_n; \mu_n) = 0 = M_m(J, c^*; ; S; c^*; \mu). \tag{38}$$

The last equality holds because of the higher moment conditions for
global minimization. The equalities (ii) and (iii) follow from (iv) immediately. The above theorem, in fact, also gives us necessary and sufficient conditions for global minimization with variable measures and penalty functions.

Theorem 4.2. Under the assumptions of Theorem 4.1, $c^*(c_n \downarrow c = c^*)$ is the global minimum value of J over S if and only if one of the following conditions holds:

(i) $\lim_{n \to \infty} M(J, c_n; p; \mu_n) = c^*$;

(ii) $\lim_{n \to \infty} V(J, c_n; p; \mu_n) = 0$;

(iii) $\lim_{n \to \infty} V_1(J, c_n; p; \mu_n) = 0$;

(iv) $\lim_{n \to \infty} M_m(J, c_n; p; c_n; \mu_n) = 0, m = 1, 2, \ldots$.

5 A Variable Measure Penalty Algorithm

In the previous section the concepts of penalized mean value and variance with variable measures were introduced. The optimality conditions were then presented. In this section we propose a variable measure penalty algorithm in terms of these concepts. We then prove that the algorithm produces a sequence which converges to the global minimum.

We now proceed to describe the proposed algorithm. Take a real number
$$c_1 > \min_{u \in S \cap U_1} J(u),$$
an exact penalty function $p(u)$ and a penalty parameter α_1. Let
$$c_2 = M(J, c_1; p; \mu_1) = \frac{1}{\mu_1(H_1 \cap U_1)} \int_{H_1 \cap U_1} [J(u) + \alpha_1 p(u)] d\mu_1.$$

Replace c_1 by c_2, α_1 by $\alpha_2 = \alpha_1 \cdot \beta$ (where $\beta \geq 1.0$ is a pre-specified constant), U_1 by U_2 and μ_1 by μ_2 to get the next iteration.

Lemma 5.1. If $\mu_1(H_1 \cap U_1) > 0$ then $\mu_2(H_2 \cap U_2) > 0$.

Proof. By the definition of H_1, we see that $c_2 \leq c_1$. If $c_2 = c_1$ then $\mu_2(H_2 \cap U_2) > 0$. Indeed, suppose, on the contrary, that $\mu_2(H_2 \cap U_2) = 0$; then c_2 is the global minimum of $J + \alpha_2 p$ in U_2. But
$$c_1 > \min_{u \in S \cap U_1} J(u) \geq \min_{u \in S \cap U_2} J(u) = \min_{u \in U_2}[J(u) + \alpha_2 p(u)] = c_2.$$

The last equality holds because we have an exact penalty function. This contradicts that $c_1 = c_2$.

Now suppose $c_2 < c_1$ and suppose, on the contrary, that $\mu_2(H_2 \cap U_2) = 0$; then c_2 is the global minimum of $J + \alpha_2 p$ in the subspace U_2, i.e.,
$$J(u) + \alpha_2 \geq c_2, \forall u \in U_2.$$
Since $\mu_1(H_1 \cap U_1) > 0$, there exists $\epsilon > 0$ such that $0 < \epsilon < c_1 - c_2$, and
$$\mu_1(G_\epsilon \cap U_1 \cap S) > 0,$$
where
$$G_\epsilon = \{u \mid c_2 + \epsilon < J(u) \leq c_1\};$$
otherwise c_1 would be the global minimum of J over $S \cap U_1$. Hence,
$$\begin{aligned}
c_2 &= M(J, c_1; p; \mu_1) \\
&= \frac{1}{\mu_1(H_1 \cap U_1)} \int_{H_1 \cap U_1} [J(u) + \alpha_1 p(u)] d\mu_1 \\
&= \frac{1}{\mu_1(H_1 \cap U_1)} \{ \int_{H_1 \cap U_1 \setminus G_\epsilon \cap U_1 \cap S} [J(u) + \alpha_1 p(u)] d\mu_1 \\
&\quad + \int_{G_\epsilon \cap U_1 \cap S} J(u) d\mu_1 \} \\
&\geq \frac{c_2}{\mu_1(H_1 \cap U_1)} (\mu_1(H_1 \cap U_1) - \mu_1(G_\epsilon \cap U_1 \cap S)) \\
&\quad + (c_2 + \epsilon) \frac{\mu_1(G_\epsilon \cap U_1 \cap S)}{\mu_1(H_1 \cap U_1)} \\
&= c_2 + \epsilon \cdot \frac{\mu_1(G_\epsilon \cap U_1 \cap S)}{\mu_1(H_1 \cap U_1)} > c_2.
\end{aligned}$$

This is a contradiction. The proof is now complete.

Continuing the process described above, we obtain a sequence of reals c_n which converges to the global minimum of $J(u)$ on $S \cap U$.

It may happen that $\mu_n(H_n \cap U_n) = 0$, for some integer n. **Lemma 5.2.** If $\mu_n(H_n \cap U_n) = 0$, then c_n is the global minimum value of J over S and $H_{c_n} \cap S$ is the set of global minima.

Proof. If $\mu_n(H_n \cap U_n) = 0$, then c_n is the global minimum value of $J + \alpha_n p$. Because of the exactness of the penalty function, we have

$$c_n = c_n^* = \min_{s \in S \cap U_n} J(u),$$

$$H_n \cap U_n = H_n^* = \{u \in S \cap U_n \mid J(u) = c_n^*\}.$$

Furthermore, by the construction of the measure μ, (9) holds. Thus

$$0 = \mu_n(H_n \cap U_n) = \mu_n(H_{c_n} \cap U_n \cap S) \geq \mu(H_{c_n} \cap S)$$

and so

$$\mu(H_{c_n} \cap S) = 0.$$

It follows from Lemma 2.1 that c_n is the global minimum value of J over S and $H_n^* = H_{c_n} \cap S$ is the set of global minima.

Remark. Usually, the set $H_n \cap U_n$ is a subset of the set of global minima because it may happen that $H_n \cap U_n \subset H_n^*$ and $H_n \cap U_n \neq H_n^*$.

A variable measure algorithm is proposed as follows:

Step 1: Take $c_o > \min_{u \in S \cap U_1} J(u)$; $\epsilon > 0; n := 0; \beta \geq 1.0$;

Step 2: Calculate mean value

$$c_{n+1} = \frac{1}{\mu_n(H_n \cap U_n)} \int_{H_n \cap U_n} [J(u) + \alpha_n p(u)] d\mu_n; \qquad (39)$$

Step 3: Calculate the variance

$$v_{n+1} = \frac{1}{\mu_n(H_n \cap U_n)} \int_{H_n \cap U_n} (J(u) + \alpha_n - c_n)^2 d\mu_n.$$

If $v_{n+1} \geq \epsilon$, then $n := n+1$ and $\alpha_{n+1} = \alpha_n \cdot \beta$ so go to Step 2; otherwise go to Step 4;

Step 4: $c^* \Leftarrow c_{n+1}; H^* \Leftarrow H_{c_{n+1}}$; Stop.

Remark. The algorithm may stop in a finite numbers of iterations, in which case we let $c_{n+k} = c_n$ and $H_{n+k} = H_n, k = 1, 2, \ldots$. Applying this algorithm with $\epsilon = 0$, we obtain a decreasing sequence

$$c_1 \geq c_2 \geq \cdots \geq c_n \geq c_{n+1} \geq \cdots \qquad (40)$$

and a sequence of sets

$$H_n \cap U_n = \{u \in U_n \mid J(u) + \alpha_n p(u) \leq c_n\}, n = 1, 2, \ldots \qquad (41)$$

Theorem 5.2 With this algorithm, we have

$$\lim_{n \to \infty} c_n = c^* = \min_{u \in S} J(u) \qquad (42)$$

and
$$\emptyset \neq \bigcap_{k=1}^{\infty} \text{cl} \bigcup_{n=k}^{\infty} (H_n \cap U_n) \subset H^*. \tag{43}$$

Proof. We need only prove the case that $\mu_n(H_n \cap U_n) > 0$, $n = 1, 2, \ldots$. According to the algorithm, we know that $c_n \geq c^*$ for $n = 1, 2, \ldots$, and the sequence $\{c_n\}$ is decreasing. Thus the limit
$$\lim_{n \to \infty} c_n = \hat{c} \geq c^* \tag{44}$$
exists. Letting $n \to \infty$ in (38), we obtain
$$\hat{c} = \lim_{n \to \infty} c_{n+1} = \lim_{n \to \infty} M(J, c_n; p; \mu_n) = M(J, \hat{c}; S; \mu). \tag{45}$$

It follows from Theorem 4.2 that $\hat{c}$ is the global minimum value of J over S, i.e. $\hat{c} = c^*$.

Furthermore, since $\cup_{n=k}^{\infty} H_n \cap U_n \subset H_k, k = 1, 2, \ldots$, we have
$$\text{cl} \bigcup_{n=k}^{\infty} (H_n \cap U_n) \subset H_k, k = 1, 2, \ldots.$$

It follows that
$$\bigcap_{k=1}^{\infty} \text{cl} \bigcup_{n=k}^{\infty} (H_n \cap U_n) \subset \bigcap_{k=1}^{\infty} H_k = H_{c^*} \cap S = H^*. \tag{46}$$

The sets $\text{cl} \cup_{n=k}^{\infty}(H_n \cap U_n), k = 1, 2, \ldots$ are closed sets contained in a compact set H_1. Thus they are deceasing compact sets, and the intersection of these sets is nonempty. This completes the proof of the theorem.

Corollary. Let $c_n^* = \min_{u \in S \cap U_n}[J(u) + \alpha_n p(u)]$ and $H_n^* = \{u \in S \cap U_n \mid J(u) + \alpha_n p(u) = c_n^*\}, n = 1, 2, \ldots$. Then
$$\lim_{n \to \infty} c_n^* = c^* \tag{47}$$

$$\emptyset \neq \bigcap_{k=1}^{\infty} \text{cl}(\bigcup_{n=k}^{\infty} H_n^*) \subset H^*. \tag{48}$$

Remark. In this corollary the value c_n^* and the set H_n^* are independent of the choice of the measure μ_n. Only the existence of this kind of Q-measure is required. In practice, we usually take a fixed integer n and find c_n^* and H_n^* as the approximations of c^* and H^* respectively.

6 Numerical Example

Consider the *brachistochrone* problem with an obstacle in the state space which is formulated to a constrained optimal control problem as follows:

$$\min J = \min[-x_1(t_T)]$$

subject to

$$\dot{x}_1(t) = x_3(t)\cos[u(t)], \; x_1(0) = 0,$$
$$\dot{x}_2(t) = x_3(t)\sin[u(t)], \; x_2(0) = 0, \qquad (49)$$
$$\dot{x}_3(t) = g\sin[u(t)], \; x_3(0) = 0.2,$$

and

$$(x_1(t) - a)^2 + (x_2(t) - b)^2 \geq (0.2)^2, \qquad (50)$$

where $g = 1.0, t_T = 2.0$, and the point (a,b) is the centre of the circle constraint (49). Once a control $u(t)$ is given, the dynamic system (48) can be solved. The states x_1, x_2 and x_3 are, in fact, functionals of the control $u(t)$:

$$x_i(t) = x_i[u(.), t], i = 1, 2, 3,$$

and thus the objective J is also a functional of $u(.)$.

The appearance of an obstacle in the state space for an optimal control problem is quite popular in applications, but the nonconvex state constraint (49) couses serious difficulties in optimal control theory, as well as numerical computation. Conventional optimality conditions, such as the maximal principle and the Hamilton-Jacobi-Bellnam equation, cannot characterize the global minimum. Gradient-based numerical optimization algorithms cannot be applied to find the optimal solution.

For people who wish to find an optimal control for a space program with an obstacle, such as the moon, this kind of examples would be of significance.

We use the integral global minimization method with a discontinuous penalty function to transform this problem to be a unconstrained one as we described in section 5. Here we use the penalty function

$$p(u) = \begin{cases} 0, & u \in S, \\ \delta + d(u), & u \notin S, \end{cases}$$

with

$$d(u) = \int_0^{t_T} |(0.2)^2 - (x_1(t) - a)^2 - (x_2(t) - b)^2| dt.$$

We have written a computer program to solve this problem with different centres for the circle. The optimal trajectory may make a detour above or below the obstacle depending on the location of the centre of the circle. We select two of the optimal solutions which are shown in the Table. In case I, the global optimal control $u(.)$ makes more effort to push the trajectory down in order to pass through below the obstacle circle. On the contrary, in case II, the global optimal control $u(.)$ does not keep decreasing to make a detour above the circle. In some cases, the control $u(.)$ even changes its sign to "pull" the trajectory up in order to pass through above the obstacle.

t	x_1	x_2	x_3	u
\multicolumn{5}{c}{case I: $a = 0.3, b = 0.43$}				
0.0	.000000	.000000	.200000	1.49941
0.1	.001426	.019949	.299745	1.47650
0.2	.004249	.049790	.399301	1.44159
0.3	.009394	.089388	.498468	1.40210
0.4	.017763	.138527	.597048	1.38050
0.5	.029056	.197154	.695243	1.34635
0.6	.044536	.264933	.792732	1.31915
0.7	.064275	.341709	.889583	1.27715
0.8	.090024	.426860	.985302	1.25679
0.9	.121486	.520232	1.08007	.829381
1.0	.194426	.599888	1.15382	.630321
1.1	.287636	.667895	1.21759	.553057
1.2	.390832	.731600	1.26529	.491333
1.3	.502393	.791296	1.31247	.415840
1.4	.622455	.844315	1.35286	.351809
1.5	.749455	.890934	1.38732	.283104
1.6	.882665	.929687	1.41526	.207265
1.7	1.02116	.958811	1.43584	.140669
1.8	1.16333	.978942	1.44986	.070126
1.9	1.30796	.989101	1.45686	.005579
2.0	1.45364	.989768	1.45732	.000000
\multicolumn{5}{c}{Case II: $a = 0.9, b = 0.85$}				
0.0	.000000	.000000	.200000	1.39611
0.1	.003476	.019696	.298478	1.31747
0.2	.010956	.048591	.395287	1.22616
0.3	.024312	.085795	.489406	1.13724
0.4	.044872	.130276	.590154	1.07283
0.5	.072582	.181177	.668010	.974923
0.6	.110073	.236466	.750776	.869268
0.7	.158527	.293814	.827161	.795840
0.8	.216402	.352911	.898606	.706513
0.9	.284753	.411247	.963525	.609347
1.0	.363764	.466393	1.02076	.525300
1.1	.452077	.517581	1.07091	.430519
1.2	.549396	.562274	1.11264	.352472
1.3	.653819	.600685	1.14716	.275865
1.4	.764198	.631931	1.17440	.162782
1.5	.880086	.650964	1.19061	.202513
1.6	.996713	.674911	1.21072	.231528
1.7	1.11455	.702693	1.23367	.165590
1.8	1.23623	.723028	1.25015	.075861
1.9	1.36089	.732502	1.25773	.004701
2.0	1.48666	.733094	1.25820	.000000

Table: Solutions of the brachistochrone problem.

References

[1] R. B. Ash, *Real Analysis and Probability*, Academic Press, New York, 1972.

[2] S. H. Chew and Q. Zheng, *Integral Global Optimization*, Lecture Notes in Economics and Mathematical Systems, No. 298, Springer-Verlag, 1988.

[3] G. Di Pillo and L. Grippo, Exact penalty functions in constrained optimization, *SIAM Journal, Control and Optimization*, 27(1989),1333-1360.

[4] A. V. Fiacco and G. P. McCormick, *Nonlinear Programming: Sequential Unconstrained Minimization Techniques*, John Wiley and Sons, New York, 1968.

[5] E. Galperin and Q. Zheng, Integral global optimization method for differential games with application to pursuit-evasion games, *International Journal, Computer Mathematics with Applications*, 18(1989), 209-243.

[6] E. Galperin and Q. Zheng, Variation-free iterative method for global optimal control, *International Journal of Control*, ,50(1989), 1731-1743.

[7] G. Leitman, *The Calculus of Variations and Optimal Control*, Plenum Press, New York, 1983.

[8] D. L. Russell, Penalty functions and bounded phase coordinate control, *SIAM Journal, Control, Ser. A*, 2(1965),409-422.

[9] N. N. Vakhania, *Probability Distributions on Linear Spaces*, North Holland, New York, 1981.

[10] Q. Zheng, Optimality conditions for global optimization (I), *Acta Mathematicae Applicatae Sinica*, 1985,2 pp 63-78.

[11] Q. Zheng, Optimality conditions for global optimization (II), *Acta Mathematicae Applicatae Sinica*, 1985,3, pp 118-134.

[12] Q. Zheng, Robust analysis and global minimization (I) *Acta Mathematicae Applicatae Sinica*, 1990.

[13] Q. Zheng, Robust Analysis and global minimization (II), *Acta Mathematicae Applicatae Sinica*, 1990.

[14] Q. Zheng, Integral minimization of constrained problems with discontinuous penalty functions, to be submitted.

[15] Q. Zheng, B. C. Jiang and S. L. Zhuang, A method for finding global extreme, *Acta Mathematicae Applicatae Sinica*, 2(1978), 146-153 (in chinese).

[16] Q. Zheng and D. Zhuang, Finite dimensional approximation to solutions of minimization problems in functional spaces, submitted.

Rigorous Methods for Global Optimization

Ramon Moore*, Eldon Hansen[†], and Anthony Leclerc*

Abstract

Rigorous methods are presented for enclosing the set of all global minimizers in a list of boxes and the minimum value in an interval. The most basic algorithm does not even use differentiability. Improvements in efficiency are shown for differentiable problems, using interval Newton methods and other techniques. Finally, the algorithms are parallelized for further improvements in efficiency. The techniques are illustrated by numerical examples.

1 Introduction

A very active area of research concerns methods for global nonlinear optimization. A number of new international journals on the subject are planned or have just begun [7, 13]. Existing journals are publishing special issues on global optimization [8, 9].

It has long been held that rigorous global optimization is impossible in the nonlinear case, and this is probably true using only evaluations of functions at points. At the very least we need to be able to compute upper and lower bounds on the ranges of functions over sets. Interval arithmetic, for example, provides upper and lower bounds on ranges of values of functions over continua [10, 11, 14, 15, 16, 17, 18, 21].

Interval computation is an active area of research. It is coming into play in a variety of applications such as: computer-aided proofs in analysis [5, 17]; global optimization [10, 11, 16, 18, 21]; computer simulations of the onset of turbulent flow [6]; etc.

Recent developments of software systems for variable-precision interval computation, such as PBASIC [1] and VPI (written in C++) [6], provide interval results of arbitrarily narrow widths containing exact solutions to mathematical problems. The efficiency of these systems can be improved.

Lengthy computations are required to solve large scale problems no matter whether floating-point or interval arithmetic is used. For such problems,

---

*Department of Computer and Information Science, The Ohio State University
[†]654 Paco Drive, Los Altos, California 94024

extensive use has been made of parallel computation with floating-point arithmetic. A number of ideas are under investigation for parallel computing with interval arithmetic. These include implementations for a vector-supercomputer (CRAY Y-MP 8/64) and for a distributed network of computers (100 or more SUN workstations).

We do not generally need or want great accuracy in final results. However, for large problems or for small ill-conditioned problems, in order to obtain even modest accuracy in final results, it can be necessary to carry many digits in intermediate computations. This is the case whether we are using floating point arithmetic (with unknown accuracy in the solution), or interval arithmetic (with guaranteed bounds on the solution).

In this paper, we consider the application of interval methods to some difficult problems. In section 2, we discuss some simple-looking global optimization problems which are relatively difficult to solve because the functions involved are not differentiable. Therefore, only simple (and relatively inefficient) procedures are applied. In section 3 we discuss other procedures which are applicable when the functions are differentiable. In section 4, we consider parallel methods. Section 5 contains the results of numerical testing.

2 A Basic Algorithm for Global Optimization

A *single* evaluation of a function using interval arithmetic provides upper and lower bounds on the range of values of the function over a *set* of values (a continuum) of the arguments. This fact alone provides a basic algorithm for global optimization without using derivatives. See [10, 11, 14, 15, 17, 18], and especially [20].

Using interval arithmetic, we can test—on a computer—the truth of a relation such as:

$$f(x) = f(x_1, x_2, \ldots, x_n) \leq 0 \text{ for } all \text{ points } x \text{ in } B.$$

Here B is any "box", i.e. an n-dimensional interval:

$$\begin{aligned} B &= \{x : a_i \leq x_i \leq b_i \text{ for } all \text{ } i = 1, 2, \ldots, n\}. \\ &= ([a_1, b_1], \ldots, [a_n, b_n]). \end{aligned}$$

We can do this for any programmable function using interval software, because a computer can find, simply by evaluating f with argument B, numbers L and U such that

$$L \leq f(x) \leq U \text{ for } all \text{ } x \text{ in } B.$$

If $U \leq 0$, then it follows that $f(x) \leq 0$ for all x in B.

Even if infinite precision interval arithmetic were used, the bounds L and U would not generally be sharp (unless B is degenerate). With fixed precision, these bounds will be slightly less sharp because of rounding errors. In practice, interval arithmetic is done with "outward" rounding (left endpoint to the left, right end-point to the right) so that the bounds are always correct even when they are not sharp.

Note that a given region can be covered by a set of boxes which can be subdivided so that each sub-box is as small as we like. Because of this, there are two facts which make it possible to compute arbitrarily sharp bounds on the ranges of values of functions:

1. as a sub-box Y gets small, the over-estimation of $f(Y)$ shrinks and we converge toward the actual range of values (limited only by the number of digits carried) [16, 20];

2. if, as Y gets small, we increase the number of digits carried, we can come arbitrarily close to the exact range of values $f(Y)$. See [16].

And, of course, the range of values of f over a box B is contained in the union of the ranges of values of f over a covering of B by sub-boxes.

In this section, we present a simple algorithm for optimization problems of the form:
minimize (globally) $f(x)$
subject to x in B and

$$p_i(x) \leq 0, i = 1, \ldots, k. \tag{1}$$

That is, we wish to "find" (bound arbitrarily tightly), within a box B, the set X^* of all global minimizers x^* lying in the feasible region defined by the inequality constraints (1). We also wish to "find" the global minimum value f^* of the given objective function f. That is, we wish to bound x^* and f^* such that

$$f(x^*) = f^* \text{ and } f^* \leq f(x) \text{ for all } x \text{ in } B \text{ satisfying (1).}$$

In this section, the functions f, p_1, ..., p_k are assumed to be programmable, but not necessarily differentiable. For extensions of the algorithm to unbounded feasible regions and to problems involving equality constraints, see [21].

Now to the basic global optimization algorithm.

The *width* of a box is defined as the maximum edge length over all the coordinate directions.

The basic algorithm will find a list of small boxes whose union *contains* the set X^* of all feasible global minimizers x^*. We terminate when the maximum box width in the list is less than a prescribed tolerance $EPSB$.

The algorithm also finds lower and upper bounds on the minimum value $f^* = f(x^*)$.

The algorithm proceeds by *deleting* parts of the initial box B which *cannot* contain a global minimizer, leaving a list of sub-boxes (of B) whose union still contains the set of all global minimizers of $f(x)$.

We now describe how parts of a box are deleted, and how we find lower and upper bounds on the minimum value of $f(x)$ for feasible points x.

1. feasibility/infeasibility test: Let X be a sub-box of B. A point in B can be represented as a degenerate box. We can evaluate (in interval arithmetic) the constraint functions $p_1(X), \ldots, p_k(X)$ and make the following tests:

    (a) if $p_i(X) \leq 0$ for all $i = 1, \ldots, k$, we say that X is *certainly feasible*.

    If X is certainly feasible, then every point x in X is feasible. This is guaranteed despite the presence of rounding errors because of the outward rounding used in the computer implementation of interval arithmetic.

    (b) if, for some $i = 1, \ldots, k$, we find that $p_i(X) > 0$, we say that X is *certainly infeasible* (it contains *no* feasible points), and X may be deleted from further consideration;

2. midpoint test: Let mX be the midpoint (or any other point) of a sub-box X of B. If mX is certainly feasible (see (1a)), then we evaluate the objective function at mX to obtain an interval $f(mX) = [LFMX, UFMX]$. It is certainly the case that $UFMX \geq f^*$, that is, $UFMX$ is an upper bound on the minimum value of $f(x)$ over the feasible region. If we evaluate f over another sub-box Y of B to find $f(Y) = [LFY, UFY]$, and we can make the following test. Let UF^* be the smallest $UFMX$ yet found.

    (a) if $LFY > UF^*$ then Y cannot contain a feasible global minimizer in B, so we can delete Y from further consideration.

Using these tests, we can formulate a very simple algorithm for global optimization with inequality constraints. The algorithm is valid whether or not the functions involved are differentiable.

The "list" of boxes in the algorithm to follow is technically a "queue". We LIST things at the end and UNLIST from the beginning. Every time we UNLIST a box, we bisect it in a coordinate direction of maximum width. We test each half. If it cannot be deleted, then we put it at the end of the list.

As a result, the widest box remaining is always the first one. If it is narrower than *EPSB* then so are all the rest. At any stage of the process, the remaining boxes on the list contain all the feasible global minimum points.

ALGORITHM 1

INPUT INITIAL BOX: B, BOX-WIDTH-TOLERANCE: (EPSILON B) EPSB
LIST B, LFB ; UPDATE UF*
MAIN PROGRAM:
    UNLIST FIRST BOX X ON THE "LIST"
    BISECT IN COORDINATE DIRECTION OF MAX WIDTH: X = X1 U X2
    DELETE X1, OR LIST X1 AND LFX1 AT END OF "LIST";
        IF X1 IS LISTED, THEN UPDATE UF* IF MX1 IS FEASIBLE
    DELETE X2, OR LIST X2 AND LFX2 AT END OF "LIST";
        IF X2 IS LISTED, THEN UPDATE UF* IF MX2 IS FEASIBLE
    IF FIRST BOX ON "LIST" HAS WIDTH ¡ EPSB, THEN GO TO OUTPUT
    OTHERWISE GO TO UNLIST AT BEGINNING OF MAIN PROGRAM
END MAIN PROGRAM
    OUTPUT:
    GRAPH (OPTIONALLY PRINT) BOX "LIST"
    PRINT LF* = MIN(LFX) OVER BOXES X ON THE "LIST"
    PRINT CURRENT UF*

Some remarks are in order. We can, of course, display lots of other things on the screen as the computation proceeds. For instance, we can graphically display the bisection and deletion steps as they take place in any two coordinate directions we may wish. We can keep count of the number of function evaluations and the total elapsed real time during the computation, etc.

We will have $LF^* \leq f^* \leq UF^*$ when we stop. The union of boxes in the list will certainly contain all the feasible global minimizers of f in B. We "update" UF^* by replacing it with any smaller $UFMX$ found (see "midpoint test" (2)).

Suppose that a sub-box Y of B contains a local (but non-global) minimizer. We delete Y if $LFY > UF^*$ (see "midpoint test" (2)). The smaller Y, the closer the lower bound LFY is to being exact. Therefore, the smaller $EPSB$, the more likely we are to delete a local minimum. Compare Figures 1 and 2.

As an illustration of what the algorithm can do, consider the following non-differentiable optimization problem [1]

We want to minimize

$$f(x) = (|x_1^2 + x_2^2 - 1| + .001)|x_1^2 + x_2^2 - 0.25|$$

---
[1] This problem was suggested by Devin Moore.

subject to

$$p(x) = \max\{\ (1 - \max\{|x_1|/0.6, |x_2|/0.25\}),$$
$$(1 - \max\{|x_1|/0.25, |x_2 - 0.4|/0.3\})\} \leq 0$$

and x in the initial box $B = ([-1.2, 1.2], [-1.2, 1.2])$, (that is: $-1.2 \leq x_1 \leq 1.2$ and $-1.2 \leq x_2 \leq 1.2$).

The results we obtain depend on a number of things, for example on the number of digits carried in the arithmetic and on the final box-width tolerance, $EPSB$.

Using interval arithmetic with outward rounding at the 11th decimal digit, we obtained the following results (carried out on an Apple MacIntosh with all the programming written in MS-BASIC).

For $EPSB = 0.2$, there were 74 boxes in the final list (see Figure 1). For $EPSB = 0.1$, there were 28 boxes in the final list (see Figure 2).

For this problem there are three disconnected continuua of global optimizers consisting of arcs, on the circle $x_1^2 + x_2^2 = 0.25$, which lie outside the union of two rectangles cutting through the circle. The entire unit circle, $x_1^2 + x_2^2 = 1$, consists of local minimizers. The local (but not the global) minimizers are eliminated merely by reducing the final box width tolerance, $EPSB$. Compare Figures 1 and 2.

It is clear that standard (non-interval) optimization methods using only function evaluations at sample points will not be able to solve problems of this type in the same sense that we are doing it here. The interval "solution" consists of a list of boxes whose union contains the set of all global minimizers. Other points are also contained, of course; but we can come as close as we please to the actual set of minimizers by doing enough computing and by carrying enough digits in the interval arithmetic. See Figure 3.

As alternative stopping criteria we might use, for instance:

1. STOP IF (UF*-LF*) < EPSILON, or

2. STOP IF (UF*-LF*) < EPSILON *AND* MAX BOX WIDTH < EPSB.

However, it seems necessary to scan the existing list for the minimum lower bound LFX (see MAIN PROGRAM) each time such a test is applied, so that this would require some additional time.

For discussion of the "complexity" of the type of algorithm presented in this section, see [21]. An accurate a-priori estimate of computing time seems difficult, even for the two-dimensional example given above, as the following table of results shows. We list the initial box B, the stopping tolerance $EPSB$ on the maximum width of boxes in the final list, a count of the number of objective function evaluations $FCNT$ and the actual computing time in

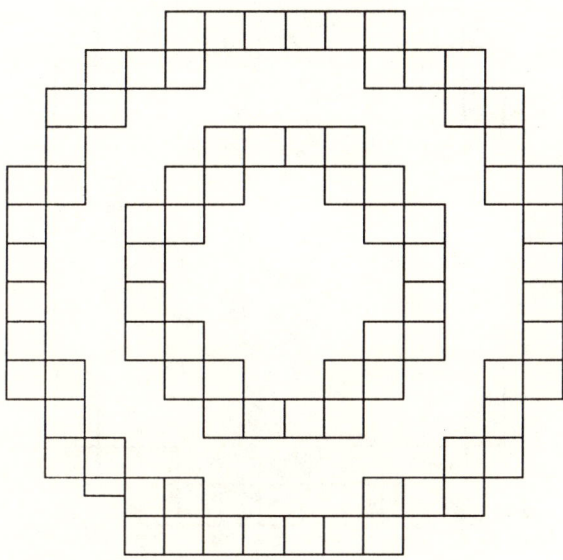

Figure 1: **Smiley face with** $EPSB = 0.2$

$SECONDS$ (in the slow, interpretive language MS-BASIC on the MacIntosh). In each case, the final list of boxes contained the global minimizers in B, but not the local minimizers on the unit circle.

B	$EPSB$	$FCNT$	$SECONDS$
$[-1.2, 1.2], [-1.2, 1.2]$	0.1	898	337
$[-1.2, 1.2], [-1.2, 1.2]$	0.05	1143	445
$[0, 1.1], [0, 1.1]$	0.025	444	183
$[0, 1], [0, 1]$	0.025	943	346
$[0, 1], [0, 1]$	0.0125	1137	417
$[0, 2], [0, 2]$	0.025	1020	392
$[0, 10], [0, 10]$	0.025	629	308
$[0, 100], [0, 100]$	0.025	546	201
$[0, 10^6], [0, 10^6]$	0.025	798	298

An explanation of these timing results is difficult. They depend on the details of the deletion process. In particular, note that the cases involving the initial box $([0, 1], [0, 1])$, in the example at hand, take longer than some of the much larger initial boxes.

It is likely that this happened, for this example, because the unit circle of local minimizers is tangent to the edges of the initial box $([0, 1], [0, 1])$. A similar thing happens for the initial box $([0, 2], [0, 2])$ after the first couple of

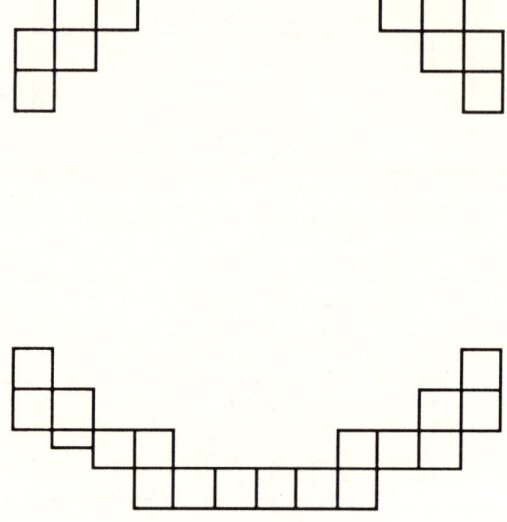

Figure 2: **Smiley face with** $EPSB = 0.1$

bisections. This type of curious behavior has been observed by the authors for other examples using quite different interval algorithms.

It is remarkable that the much larger initial boxes ($[0, 100], [0, 100]$) and even ($[0, 10^6], [0, 10^6]$) run faster than ($[0, 1], [0, 1]$). This is because, upon bisection, they never get cut down to any sub-box with edges tangent to the circle of local minimizers. Thus, an accurate a-priori complexity analysis for this algorithm must take into account the geometry of the solution set relative to the boundary of the feasible region. If this is not impossible, it is perhaps a research problem for the future.

More to the point, many improvements in efficiency are possible, particularly for differentiable problems, as we show in the next section.

3 Differentiable Optimization Problems

When the objective function and the constraint functions are differentiable, we can use more efficient methods such as those discussed in this section.

When the objective function is differentiable, we can make use of local monotonicity. We can also make more efficient use of the upper bound UF^* on the globally minimum value f^* of $f(x)$. When the objective function is twice differentiable, we can make use of local convexity, and we can use interval Newton methods. See subsection 3.4.

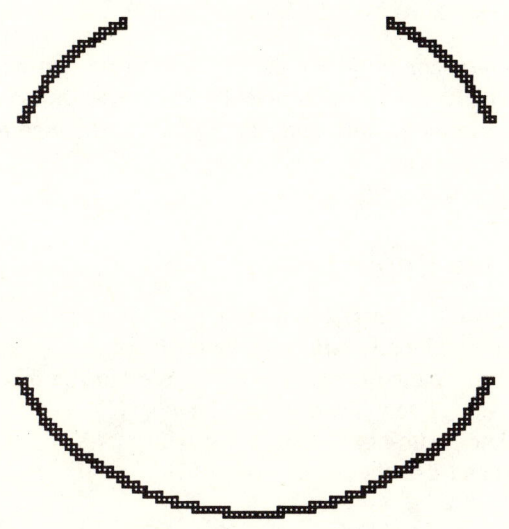

Figure 3: **Smiley face with** $EPSB = 0.0125$

If the constraints are also continuously differentiable, then we can apply an interval Newton method to solve the John or the Kuhn-Tucker conditions which must be satisfied at a solution point.

For simplicity in discussing the procedures of this section, we now assume that the constraint functions $p_i(x)$ are continuously differentiable and the objective function $f(x)$ is twice continuously differentiable.

3.1 Monotonicity

Consider the case in which a box B is certainly strictly feasible. Suppose we evaluate the gradient g over a sub-box X of B. If 0 is not in $g_i(X)$ for some $i = 1, \ldots, n$, then the gradient is not zero in X. Therefore the global minimum cannot occur in X, and we can delete X.

3.2 Nonconvexity

Again consider a certainly strictly feasible box B and consider a sub-box X of B. If a global solution point, x^* occurs in B, then the Hessian of $f(x)$ must be non-negative definite at x^*. If we can show that the Hessian is not non-negative definite anywhere in X, then we can delete X.

The diagonal elements of the Hessian are given by

$$h_{ii}(x) = \partial^2 f(x)/\partial x^2, i = 1, \ldots, n.$$

One necessary condition for the Hessian to be non-negative definite is that its diagonal elements be non-negative. Suppose we evaluate h_{ii} over X, using interval arithmetic, and find that $h_{ii}(X) < 0$ for some i, then we can delete X. We could use other necessary conditions for non-convexity, but here we use only this simplest one.

3.3 Using the Upper Bound

In the basic algorithm described in section 2, we considered use of an upper bound UF^* on the globally minimum value f^* of $f(x)$. We pointed out that a box X could be deleted if $f(X) > UF^*$. We can use this bound in a more sophisticated way. Even if we do not have $f(X) > UF^*$, it may still be possible to delete a sub-box X' of X for which $f(X') > UF^*$.

We can expand f in the form

$$f(y) \in f(x) + (y - x)^T g(X).$$

See, for example [14]. If

$$f(x) + (y - x)^T g(X) > UF^* \qquad (2)$$

for all y in X', then $f(y) > UF^*$ for all y in X'; therefore, we can delete X'.

We can solve the inequality (2) for y to determine such a sub-box X'. See [11] for details.

We can also use a second order expansion of f in the same way. Again, see [11] for details.

3.4 An Interval Newton Method

The most effective procedure for use in solving differentiable global optimization problems is the interval Newton method.

Consider a vector function $g(x)$ which might be the gradient of f.

Suppose we seek the zero(s) of g in a box X. Let J denote the Jacobian of g and let x be a point in X. For any point y in X, the expansion

$$g(y) \in g(x) + J(X)(y - x)$$

holds [14]. If y is a zero of g, then y is in the solution set of

$$g(x) + J(X)(y - x) = 0. \qquad (3)$$

We can find a box Y containing the set of solution points y of (3). We then replace X by $X \cap Y$.

Note that this intersection could be empty. If so, we simply delete X. See [12] for details.

If X is certainly strictly feasible, then any minimum of $f(x)$ in X is at a stationary point. Therefore, $g(x) = 0$ at such a point, where $g(x)$ is the gradient of $f(x)$. In such a case, we can apply the interval Newton method to solve $g(x) = 0$ in X.

If X is not certainly strictly feasible, then a global minimum may not occur where $g(x) = 0$. However, it will occur where the John or Kuhn-Tucker conditions are satisfied. In this case, we can apply the interval Newton method to solve the set of equations expressing the John or Kuhn-Tucker conditions.

For the problem given by (1), these conditions are

$$g(x) + \sum_i u_i \nabla p_i(x) = 0$$

$$u_i p_i(x) = 0, i = 1, \ldots, k$$

where u_i is a Lagrange multiplier. See, for example [21].

4 Parallelization

We gave a basic interval global optimization algorithm (ALGORITHM 1) for possibly non-differentiable functions in section 2. In section 3 we gave a list of more "powerful" box and sub-box deletion tests which could be incorporated into ALGORITHM 1 when the objective function and the constraint functions are differentiable.

We can further improve the efficiency of ALGORITHM 1, or augmented variations of ALGORITHM 1 which use the tests given in section 3, by parallelizing. The parallelization of a sequential algorithm is not an obvious or even unique task for most problems. In particular this is true for the global optimization algorithms mentioned in this paper. For what follows, we will not differentiate among the possible variations of ALGORITHM 1. We will simply refer collectively to these algorithms which we intend to parallelize as "the algorithm".

In order to better qualify our parallelization intent, we will use the terms *fine grain parallelism* and *coarse grain parallelism*. By fine grain parallelism we mean the parallelism in an algorithm at the instruction level. For example, consider the addition of two intervals, $A = [al, ar]$ and $B = [bl, br]$ to obtain $C = [cl, cr] = [al + bl, ar + br]$. We could implement this simultaneously on two processors, one of which calculates $cl = al + bl$ while the other performs $cr = ar + br$.

The *speedup* of this parallel interval addition over the sequential interval addition is at most 2. Furthermore, this maximum speedup can only be realized on a *tightly-coupled* parallel machine such as the Butterfly, Hypercube, or Transputer network, in which communication/shared-memory overhead is minimal. In a similar vein, the other basic operations $(-,*,/)$ could also be fine grain parallelized.

If we are to consider arbitrary precision intervals such as those used in VPI [6], we can gain much greater speedups over the sequentially coded version by implementing the basic arithmetic operations as vector operations on a vector processor such as the CRAY Y-MP 8/64. For this paper, however, we are using fixed precision intervals and wish to achieve speedups on the order of N where N is the number of processors used.

In addition, we have available more than 250 SPARC station SLCs on a distributed ethernet network which we desire to use. We do not consider fine grain parallelism in this paper, but instead investigate *coarse grain parallelism* in which a different *partition* of the initial input data is *mapped* onto various processors.

One possible parallel implementation would designate one processor as the queue processor, QP. QP would be responsible for maintaining the *global queue*. All the other processors would queue/dequeue boxes by consulting QP and then perform the rest of the algorithm in parallel.

Whenever an improved upper bound UF^* on the global minimum (see *midpoint test* in section 2) is discovered by any processor, this *new* UF^* would be broadcast to all other processors so that all could make sharper midpoint deletion tests. Note that at any given instant in time, some of the processors might be performing a midpoint test with an *older* UF^*. This does not affect the correctness of the algorithm, but rather in the worst case, the box for which the test is being applied will not be deleted, but instead bisected. When each bisected half later comes to the foreground of the queue, the *newer* UF^* will be available.

This parallel version possesses at least two short-comings. First, as the total number of processors involved in the computation increases, so does the demand for services from QP. Quickly, QP will become a *bottleneck* with all other processors waiting unreasonable amounts of time in order to access the global queue. The second handicap is that the size of the global queue is limited to the amount of memory available on a single processor. From considerable experience with the sequential algorithm, we believe this to be too constraining for real world problems. Therefore, we consider a coarse grain distributed parallel global optimization algorithm in which each processor maintains its own *local queue*.

Our distributed parallel algorithm has three steps:

1. Initialize/Startup all processors

2. Perform "the algorithm" in parallel
   - Dynamic partitioning and load balancing
   - Broadcasting *new UF**

3. Terminate all processors
   - Detect global termination
   - Compute final solution list

The steps are defined and discussed below. Before doing so, we define the pair of terms *partitioning* and *mapping* in the context of our parallel program. Partitioning refers to the manner in which the input data is divided-up among each of the processors. Mapping is concerned with the particular feasible assignment (with respect to the processor interconnection topology) of processor to process which minimizes communication costs.

Since we are operating in a distributed network environment, we have a fully connected system in which the interprocessor communication time is virtually homogeneous. Therefore, in the succeeding algorithm description, the reader can assume that any process can be mapped to any processor, and we will not address the mapping issue further. We now describe our parallel algorithm.

4.1 Initialize/Startup All Processors

Observing the initial step for ALGORITHM 1 (see section 2), we note 4 steps:

1. Input initial box, B

2. Input initial box width tolerance, *EPSB*

3. Queue the tuple, (B, LFB), on the *box queue*

4. Update UF^*

These first 4 steps are performed only on the main processor, namely P_0. Next, P_0 will attempt to spawn $N-1$ process copies of itself on $N-1$ remote processors, $P_i, 0 < i < N$. P_0 will then wait until it has received from each P_i a *local state message*, *LSM*, indicating the status of the attempted spawn (many errors can occur when attempting to spawn a remote process on a distributed network). Each of the *LSM*s are compiled, along with the sending processor's unique identification number, domain name, and Ethernet address, into a *global state message*, *GSM*. After all $N-1$ *LSM*s have been received, P_0 sends a copy of the *GSM* to all P_is. All processors now have the necessary information to communicate with any other *living* processor involved in the parallel computation.

4.2 Perform "The Algorithm" In Parallel

Once step 4.1 above has been completed, only P_0 has a box on the box queue. How do the other $N-1$ processors proceed? This brings us to the issue of *dynamic partitioning* and *load balancing*.

4.2.1 Dynamic Partitioning and Load Balancing

Whenever any processor, P_j, has an empty box queue, it begins sending *box request messages*, *BRM*s, to a random $P_i, i \neq j$. If there are boxes available on P_i's box queue, then P_i sends P_j a *box message*, *BM*, containing half of its queued boxes, but no more than *NUMBOXES* (we don't want to send arbitrarily large messages).

Otherwise, P_i sends back a short *no boxes available message*, *NBM*, indicating that it has no boxes available. If P_j receives a *NBM*, it then sends requests to processors, $P_{(i+1) mod N}, P_{(i+2) mod N}, P_{(i+3) mod N}, \ldots, P_{(k) mod N}$ until it receives a *BM* or until $k = N - 1$ (see section 4.3.1).

This partitioning scheme is dynamic and demand driven. Our hope is that by sending half of the workload to each box requesting processor, we can balance our work load (number of boxes) among all processors.

4.2.2 Broadcasting *New UF**

As each processor executes "the algorithm" in parallel, eventually (assuming there exists a point, $x \in B$ (the initial input box) such that $f(x) < LFB$) an improved upper bound UF^* on the global minimum will be discovered by a given processor, P_j. At this point, P_j will send this new UF^*, NUF^*, to all other processors $P_i, i \neq j$. When a given P_i receives this NUF^* it compares it with its local UF^*. If $NUF^* < UF^*$, P_i updates UF^*. Otherwise, P_i must have received a lower NUF^* from some other processor or calculated a lower UF^* itself during the time it took to receive P_j's NUF^*. In this case, P_i's UF^* is not updated.

4.3 Termination

With the sequential version of "the algorithm", we were guaranteed that if the first box on the box queue had width less than *EPSB*, then so did all the other remaining queued boxes. However, in the parallel case, if the first queued box on the box queue of given processor, P_i, has width less than *EPSB*, then this does *not* necessarily imply that all the remaining boxes on the $N-1$ other processors' box queues will have width less than *EPSB*.

Indeed, P_i may very well only have found a local minimum. What is P_i to do in this case? If P_i simply prints its output as described in ALGORITHM 1 and then terminates, then we very likely have outputted an uninteresting local solution, and moreover, have lost a valuable worker processor.

Our solution, for the moment, is to maintain a second queue, called the *possible solution queue*, *PSQ*, on every processor. Now, if the width of the first queued box on P_i's box queue is less than *EPSB*, then all of the boxes on P_i's box queue are placed on *PSQ*. P_i then behaves as in 4.2.1 for a processor with no queued boxes. Furthermore, whenever P_i determines a *new UF^**, it checks all boxes on *PSQ* and discards those boxes which fail the midpoint test (see section 2) using the *new UF^**. Global termination now becomes a question of detecting when *every* processors' box queue is empty. We have chosen to detect such a state with a simple centralized algorithm. A distributed algorithm using either a ring [3] or a tree [23] would be more efficient and fault tolerant.

4.3.1 Detect Global Termination

If a given P_i does not receive a *BM* after sending $N - 1$ *BRM*s, P_i then sends P_0 a *possible global termination message*, *PGTM*. P_i then waits for either a *BM* or a *terminate message*, *TM*, from P_0. If P_0 receives a *PGTM* and has boxes on its box queue, then P_0 simply sends P_i a *BM*. If P_0 receives a *PGTM* while it has no boxes on its box queue, then P_0 logs P_i's *PGTM*. When P_0 receives $N - 1$ *PGTM*s, P_0 sends a *TM* to all other processors. Additionally, if P_0 is sending *BRM*s and receives a *BM*, P_0 must send *BM*s to all processors for which a *PGTM* was logged.

4.3.2 Compute Final Solution List

When a processor, P_i, receives a *TM*, it prints out all boxes on its *PSQ* and terminates. P_0 does the same as soon as it detects global termination and has sent $N - 1$ *TM*s. The final solution list is obtained by combining the output from each terminated processor. As in the sequential version, the union of all the boxes on the final solution list will contain the set of all global minimizers.

5 Numerical Results On Test Problems

We will now exemplify the methods described in this paper. All examples were run on one or more SPARC station SLCs connected on an Ethernet network. In addition, all examples utilized the midpoint, monotonicity, and Newton tests described in sections 2 and 3. Termination was effected when the width of each remaining box containing a solution was less than 10^{-6}.

5.1 Kowalik Problem

For our first example, we solved the Kowalik problem mentioned by Walster and Hansen [24]. For this unconstrained global optimization problem we desired to minimize $f(x_1, x_2, x_3, x_4)$ defined as follows:

$$f(x_1, x_2, x_3, x_4) = \sum_{i=1}^{11} \left(a_i - x_1 \frac{b_i^2 + b_i x_2}{b_i^2 + b_i x_3 + x_4} \right)^2$$

i	a_i	$1/b_i$
1	0.1957	0.25
2	0.1947	0.5
3	0.1735	1
4	0.1600	2
5	0.0844	4
6	0.0627	6
7	0.0456	8
8	0.0342	10
9	0.0323	12
10	0.0235	14
11	0.0246	16

The correct solution as reported by Walster and Hansen is as follows:

$x_1^* = [0.1928334529823, 0.1928334529827]$
$x_2^* = [0.190836238780, 0.190836238785]$
$x_3^* = [0.123117296277, 0.123117296279]$
$x_4^* = [0.135765989980, 0.135765989983]$

$f^* = [0.00030748598779, 0.00030748598781]$

Given an initial input box of $[-0.2892, 0.2893]$ in all dimensions and running on 1 processor, our program reported the following results:

```
convex hull of boxes on the queue:

x1 = [0.19283345267383470, 0.19283345332189572]
x2 = [0.19083623185103013, 0.19083624496089555]
x3 = [0.12311729477860661, 0.12311729761465269]
x4 = [0.13576598850717606, 0.13576599113717366]

f(hull): [0.00030748584424075611, 0.00030748613137046295]
max width of 1 box on the queue = 1.31099e-08
```

```
performed 545490 function evaluations.
performed 492923 Jacobian evaluations.
performed 200627 Hessian evaluations.

performed  51263 midpoint deletions.
performed  88999 monotonicity deletions.
performed  49741 Newton deletions.
performed  34635 Newton reductions.

computation time = 4057067 milleseconds
```

5.2 Photoelectron Spectroscopy Problem

Our second example concerns a *real world* problem which arises in the field of chemistry. More specifically, chemists performing *photoelectron spectroscopy* collide photons with atoms or molecules. These collisions result in the ejection of photoelectrons [4, 2, 22]. The chemist is left with a *photoelectron spectrum* which is a plot of the number of photoelectrons ejected as a function of the kinetic energy of the photoelectron. A typical spectrum consists of a number of overlapping peaks of various shapes and intensities. The chemist desires to resolve the individual peaks.

One method for isolating each peak attempts to "fit" the spectrum as the sum of *peak functions*. Peak functions are functions of variables which convey information regarding the peak's *position, intensity, width, function type,* and *tail characteristics*. Various types of functions have been used for this purpose, but the most common are Gaussian and/or Lorentzian.

For a specific test problem, we constructed our own spectral curve as the sum of two Gaussian functions (see Figure 4). The function definition is as follows:

$$x_i = 4.0 + 0.1(i+1), \ i = 1, 2, \ldots, n$$
$$y_i = a_1 e^{-[\frac{x_i - u_1}{s_1}]^2} + a_2 e^{-[\frac{x_i - u_2}{s_2}]^2}$$

$a_1 = 130.89$ $a_2 = 52.6$
$u_1 = 6.73$ $u_2 = 9.342$
$s_1 = 1.2$ $s_2 = 0.97$

We then attempted to "fit" this curve by recovering $a_1, a_2, u_1, u_2, s_1,$ and s_2. Given $n = 81, (x_i, y_i)$, and the initial input box, B, defined below, our task was to minimize f defined as follows:

$$f(a_1, a_2, u_1, u_2, s_1, s_2) = \sum_{i=1}^{n} \left(a_1 e^{-[\frac{x_i - u_1}{s_1}]^2} + a_2 e^{-[\frac{x_i - u_2}{s_2}]^2} - y_i \right)^2$$

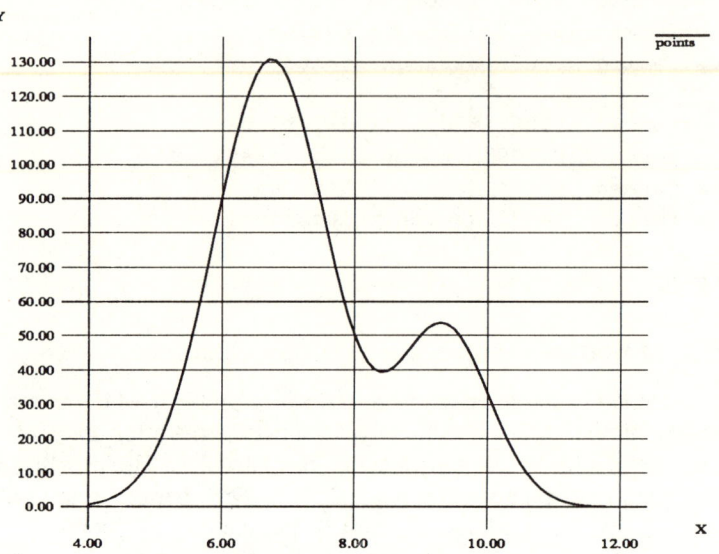

Figure 4: **Graph of points**

$$B : \begin{array}{l} a_1 = [130, 135] \\ a_2 = [50, 55] \\ u_1 = [6, 8] \\ u_2 = [8, 10] \\ s_1 = [1, 2] \\ s_2 = [0.5, 1] \end{array}$$

Our results were as follows:

```
convex hull of boxes on the queue:

a1 = [130.889999624668920, 130.890000237423440]
a2 = [52.5999994426222910, 52.6000003353821410]
u1 = [6.72999999580056230, 6.73000000523584680]
u2 = [9.34199999170696670, 9.34200000792551850]
s1 = [1.19999999502502950, 1.20000000672384770]
s2 = [0.96999998507893725, 0.97000001469388031]

f(hull): [6.3015390640982946e-13, 9.9696829305332294e-11]
max width of 31 boxes on the queue = 5.57378e-07
computation time on 1 processor = 109240 seconds
```

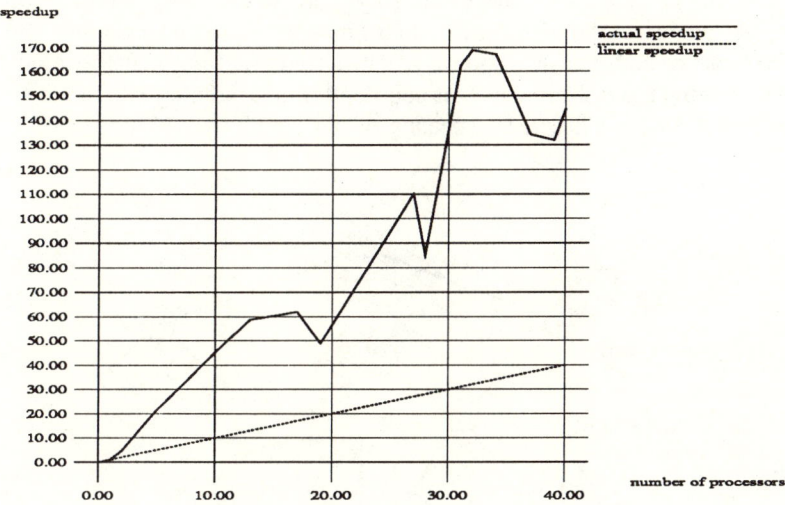

Figure 5: **Speedup graph**

We ran our parallel algorithm on upto 40 processors and achieved superlinear speedup as indicated by Figure 5. In order to explain this superlinear speedup, we will consider the progress of our parallel algorithm in the form of a binary tree.

At the beginning of our algorithm, we are usually given a single initial input box (denoted as the root of the tree). We either eliminate this box or divide it in half giving us two new boxes (which we depict as child nodes). Likewise these two new boxes can be eliminated or split. Continuing in this manner, we gradually create what we call a *binary progress tree*.

A portion of one possible binary progress tree is given in Figure 6. The rectangularized regions represent sets of boxes which would be deleted using the current upper bound UF^* on the global minimum. An improved upper bound NUF^* on the global minimum exists within box $B30$.

In the single processor case, boxes are tested in the order $B1, B2, \ldots, B31$. The reason for this is the fact that boxes are queued based upon the time in which they were generated. This progress amounts to a breadth first search of the entire tree for a "small" enough box containing a solution. Because of this searching strategy, the NUF^* within box $B30$ would require 22 tests before being discovered.

In the two processor case (P_1 initially getting $B2$ and P_2 initially getting $B3$), each processor would employ a breadth first search on its respective half of the tree. Therefore P_2 would discover the NUF^* in 6 tests (nearly 1/4 the Number of tests it took in the single processor case). Furthermore, P_2 would broadcast the NUF^* to P_1 thus allowing P_1 to make sharper midpoint tests earlier and possibly "pruning" other subtrees from consideration. It is this combination of breadth first and depth first searching which we believe accounts for the superlinear speedup of our parallel algorithm.

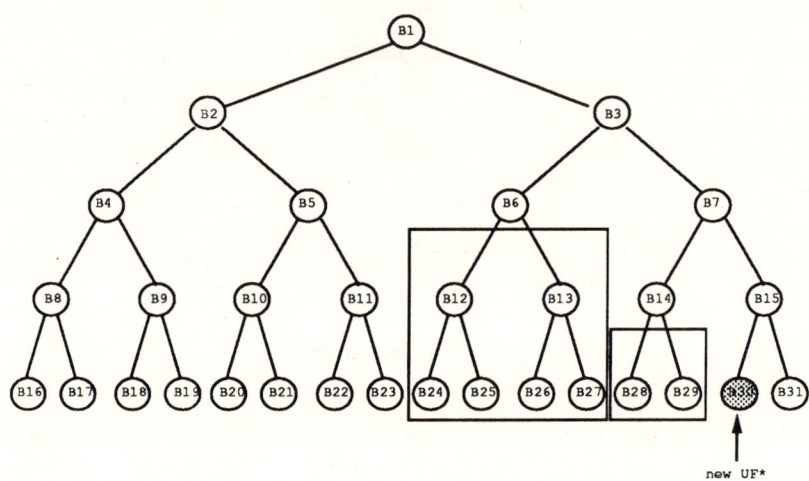

Figure 6: **Portion of one possible binary progress tree**

References

[1] O. Aberth. *Precise Numerical Analysis.* Wm. C. Brown, Dubuque, 1988. with variable precision software on a floppy disk.

[2] D. Briggs and M. P. Seah. *Practical Surface Analysis: by Auger and X-ray Photoelectron Spectroscopy.* Wiley, Chichester, New York, 1983.

[3] E.W. Dijkstra, W.H.J. Feijen, and A.J.M. van Gasteren. Derivation of a termination detection algorithm for distributed computations. *Information Processing Letters*, 16(5):217–219, June 1983.

[4] Russell S. Drago. *Physical Methods in Chemistry.* Saunders Golden Sunburst Series. W.B. Saunders Company, Philadelphia, PA 19105, 1977.

[5] J. P. Eckmann, H. Kock, and P. Wittner. A computer-assisted proof of universality for area-preserving maps. *Memoirs of the American Math-*

ematical Society, page 67, 1984. Using interval analysis, the authors observe "rigorous functional analytic estimates are possible on a computer".

[6] J. S. Ely. *Prospects for Using Variable Precision Interval Software in C++ for Solving Some Contemporary Scientific Problems*. PhD thesis, The Ohio State University, 1990.

[7] E. A. Galperin. Control and games. *International Journal of Global Optimization*. To appear, personal communication.

[8] E. A. Galperin. Global optimization, control and games. *Special Issue of the International Journal of Computers and Mathematics*, 1990.

[9] E. A. Galperin. Global optimization, control and games. *Second Special Issue of the International Journal of Computers and Mathematics*, 1991. To appear.

[10] E. R. Hansen. *Global Optimization Using Interval Analysis*. To be published.

[11] E. R. Hansen. Global optimization using interval analysis — the multidimensional case. *Numer. Math.*, 34:247–270, 1980.

[12] E. R. Hansen and R. I. Greenberg. An interval newton method. *Appl. Math. Comp.*, 12:89–98, 1983.

[13] R. Horst. Global optimization. To appear, personal communication.

[14] R. E. Moore. *Methods and Applications of Interval Analysis*. SIAM Studies in Applied Mathematics. SIAM, Philadelphia, 1979.

[15] R. E. Moore. *Reliability in Computing*. Academic Press, 1988. See especially the papers by E. Hansen 289-308, G. W. Walster 309-324, H. Ratschek 325-340, and W. A. Lodwick 341-354.

[16] R. E. Moore. Global optimization to prescribed accuracy. *Computers Math. Applic.*, 21:25–39, 1991.

[17] R. E. Moore. Interval tools for computer aided proofs in analysis. *The IMA Volumes in Mathematics and Its Applications*, 28:211–216, 1991.

[18] R. E. Moore and H. Ratschek. Inclusion functions and global optimization II. *Mathematical Programming*, 41:341–356, 1988.

[19] P. M. Pardalos. Computer Science Department, The Pennsylvania State University, personal communication.

[20] H. Ratschek and J. Rokne. *Computer Methods for the Range of Functions*. Ellis Horwood and John Wiley, 1984.

[21] H. Ratschek and J. Rokne. *New Computer Methods for Global Optimization*. Ellis Horwood and John Wiley, 1988.

[22] Peter E. Sobol. A comparison of techniques for compositional and chemical analysis of surfaces. *Perkin-Elmer*, 11(2):2–5, Winter 1989.

[23] Rodney W. Topor. Termination detection for distributed computations. *Information Processing Letters*, 18(1):33–36, January 1984.

[24] G. W. Walster, E. R. Hansen, and S. Sengupta. *Test Results for a Global Optimization Algorithm*, pages 272–287. Numerical Optimization. SIAM, 1985. Boggs, Byrd, Schnabel (eds.).

Global Optimization of Composite Laminates Using Improving Hit and Run [*]

Zelda B. Zabinsky [†]   Douglas L. Graesser [‡]
Mark E. Tuttle [‡]   Gun-In Kim [‡]

Abstract

We formulate a composite laminate design problem as a global optimization problem, and solve it using a random search algorithm called Improving Hit and Run. Experiments are run to describe the performance of the algorithm, and a test function is created to simulate the original formulation. The test function allows experiments in a controlled domain, and we make some inferences to the original problem.

1 Introduction

In recent years random search algorithms have been used during optimization studies associated with a number of disciplines, including image processing, biology, physics and chemistry. A relatively new application area is the optimal design of

---

[*] This research was funded in part by the Boeing Commercial Airplane Company and the NASA-Langley Research Center.
[†] Industrial Engineering Program, FU-20, University of Washington, Seattle, WA 98195.
[‡] Mechanical Engineering Department, FU-10, University of Washington, Seattle, WA 98195.

laminated composite materials [6,7,9,10,11]. Laminated composites are composed of several thin layers called plies, which are bonded together to form a composite laminate. A single ply consists of long reinforcing fibers (e.g., graphite fibers), embedded within a relatively weak matrix material (e.g., epoxy). All fibers within an individual ply are oriented in one direction. Composite laminates are usually fabricated such that fiber angles vary from ply to ply.

Formulation of an optimization problem for a composite laminate with fiber angles as decision variables leads to a complex objective function. An objective function will be developed in a following section. The objective function is typically global in nature due to the trigonometric functions involved in the stress/strain relationships [11,3], and may involve a large number of dimensions. Since recent research[8,13,12] indicates random search algorithms may be very efficient for global optimization problems in large dimensions we have applied a random search algorithm called "Improving Hit and Run" (IHR) [12] to the problem of optimal composite design.

IHR can be considered to be a special case of simulated annealing [2,4] where only improving points are accepted (i.e. the cooling parameter is zero). Theoretical results for the performance of IHR[12] show that, for a spherically symmetric class of global optimization problems, it is an efficient algorithm whose complexity is linear in dimension. However, the composite design problem does not fall within the class of problems for which this theoretical result holds. Our experience[3] nevertheless indicates that IHR can be very effectively applied to a composite design problem, and leads us to hypothesize that IHR is nearly linear in dimension for a broader class of problems than that considered in [12].

The objective of the present paper is to numerically investigate this hypothesis. It would be ideal to investigate the performance of IHR using the objective function for a composite design problem. A difficulty is that, in general, the optimal solutions to composite design problems are not known. Therefore, we construct a test function for experimentation, whose global solution is known and whose mathematical form is similar to the composite design problem.

The remainder of the paper is organized as follows. In the following section we briefly describe the objective function used in a composite design problem, to demonstrate the mathematical form and global nature of the composite objective function. Two constructed global objective functions selected for study will then be described, and numerical results presented. Results of the study are then summarized and discussed in a concluding section.

2 The Composite Design Problem

One of the advantages of composite laminated materials to a design engineer is that the material itself can be tailored to the directional stiffness and strength required for the overall structure. Thus composite materials offer a great deal of flexibility, but new design procedures are needed to exploit the advantages offered by composite materials. This motivated our formulation of composite design as an optimization problem.

The variables associated with composite laminate design include (but are not limited to) the number of plies, ply fiber orientations, ply thickness, and ply stacking sequence. As shown schematically in Figure 1a, the fiber angles may vary from ply to ply. We define the decision variables x_k to be the fiber orientation angle in the kth ply for $k = 1, \ldots, n$, where n equals the number of plies in the laminate, i.e. let $x_k = \theta_k$, as shown in Figure 1. We have not included ply thickness as a design variable in our studies because in most practical instances the ply thickness is defined by the type of composite material available for use, and hence cannot be directly controlled by the designer. This approach is in contrast to that of Vanderplaats and Weisshaar[11] and Schmit and Farshi[9], who considered ply thickness as the design variable, and pre-assigned ply angles.

Figure 1 also illustrates the convention for defining the reference coordinate systems. Each individual ply has a 1,2 coordinate system where the 1 coordinate direction is aligned with the fiber, and the 2 coordinate direction is perpendicular to the fiber. The x,y coordinate system is defined for the entire laminate. The loads applied to the laminate, as illustrated in Figure 1b, are defined relative to the x,y coordinate system.

The general problem is to find the fiber angles that maximize the strength of the material. This is equivalent to minimizing the strains associated with a given set of loading conditions. There are three types of normal and shear strains that can be calculated for each ply k, $k = 1, \ldots, n$ and each loading condition j, $j = 1, \ldots, m$:

$\epsilon_{1(j)}^k$ is the calculated normal strain in the 1 direction for the kth ply induced by loading condition j (function of x),

$\epsilon_{2(j)}^k$ is the calculated normal strain in the 2 direction for the kth ply induced by loading condition j (function of x), and

$\gamma_{12(j)}^k$ is the calculated shear strain for the kth ply induced by loading condition j (function of x),

The calculations for the normal and shear strains are based on classical lamination theory [5]. The calculated normal and shear strains are a function of fiber angles, x, as well as a user-defined loading condition. The calculations are summarized in Appendix 1.

The theory includes a maximum strain failure criterion which predicts failure will occur if any calculated ply strain

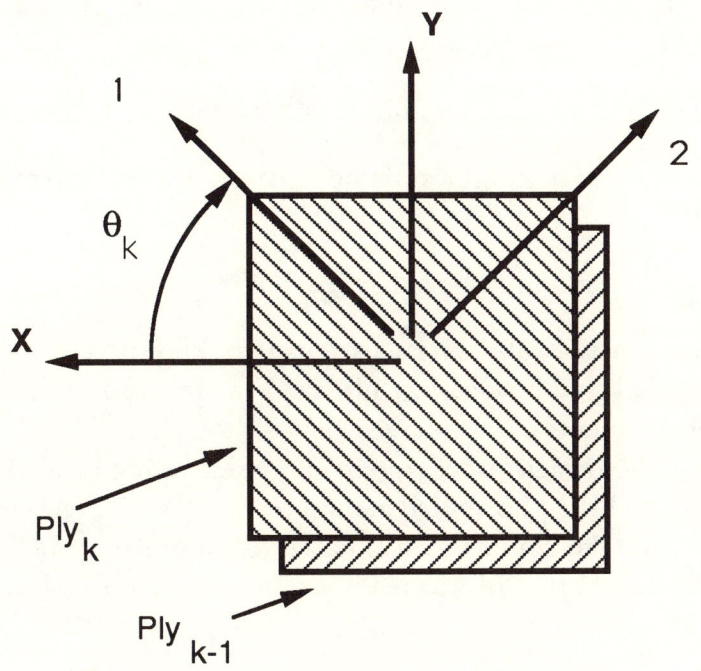

Figure 1a. Composite Laminate Ply Orientation Coordinate Systems

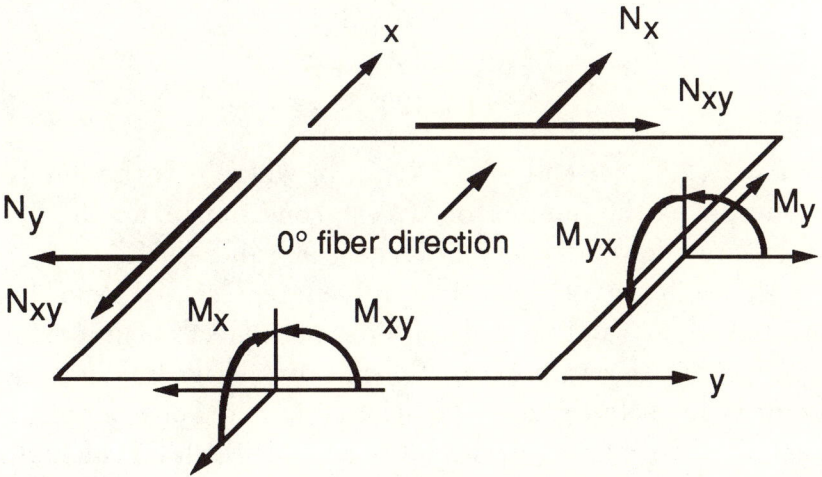

Figure 1b. Loading Condition Coordinate System

exceeds the maximum allowable (failure) strain. There are three types of maximum allowable strains:

ϵ_1^{allow} is the maximum allowable fiber strain in the 1 direction,

ϵ_2^{allow} is the maximum allowable fiber strain in the 2 direction, and

γ_{12}^{allow} is the maximum allowable shear fiber strain.

The maximum allowable strains are usually known values that depend on material properties, and also may incorporate safety factor requirements.

During the design process, one or more loading conditions are defined which the entire laminate must safely support. Thus the problem could be formulated as minimizing the sum of the absolute value of strains with the constraints that no ply strains exceed the allowables:

(P.1)
$$\begin{aligned}
\text{minimize} \quad & \sum_{j=1}^{m} \sum_{k=1}^{n} \left(|\epsilon_{1(j)}^k| + |\epsilon_{2(j)}^k| + |\gamma_{12(j)}^k| \right) \\
\text{subject to} \quad & |\epsilon_{1(j)}^k| \leq \epsilon_1^{allow} \\
& |\epsilon_{2(j)}^k| \leq \epsilon_2^{allow} \\
& |\gamma_{12(j)}^k| \leq \gamma_{12}^{allow} \\
& -90° \leq x_k \leq 90°
\end{aligned}$$

for $k = 1, 2, \ldots, n$ and $j = 1, 2, \ldots, m$ where n is the number of plies, m is the number of loading conditions, and the fiber angles are allowed to vary between $+90°$ and $-90°$.

Although the minimization problem represented by (P.1) is intuitive and straightforward, it does not reflect one important aspect of the design process. Specifically, design engineers attempt to identify the most "efficient" design. For example, an ideal composite laminate would be one in which all calculated ply strains were exactly equal to the allowable strains. This is considered very efficient because the laminate is tailored to support the user-defined loading condition, but no more.

We have therefore modified our objective function to reflect this aspect of design. First, we consider the sum of normalized differences between actual strains and maximum allowable strains instead of the original sum of strains. The three normalized terms for each ply k and loading condition j are:

$$\frac{|\epsilon_{1(j)}^k| - \epsilon_1^{allow}}{\epsilon_1^{allow}}, \quad \frac{|\epsilon_{2(j)}^k| - \epsilon_2^{allow}}{\epsilon_2^{allow}}, \quad \frac{|\gamma_{12(j)}^k| - \gamma_{12}^{allow}}{\gamma_{12}^{allow}}.$$

Second, we combine the normalized terms with an exponential transformation to produce a function that is very steep in the neighborhood of critical values. Also, squaring the terms makes the objective function steeper in the region close to the maximum allowable strains, and shallower farther away. The revised objective function is:

$$F_{P.2}(x) = \sum_{j=1}^{m} \left[\sum_{k=1}^{n} \left[\left(\exp \frac{|\epsilon_{1(j)}^k| - \epsilon_1^{allow}}{\epsilon_1^{allow}} \right)^2 \right.\right.$$
$$+ \left(\exp \frac{|\epsilon_{2(j)}^k| - \epsilon_2^{allow}}{\epsilon_2^{allow}} \right)^2$$
$$\left.\left.+ \left(\exp \frac{|\gamma_{12(j)}^k| - \gamma_{12}^{allow}}{\gamma_{12}^{allow}} \right)^2 \right] \right].$$

The revised formulation is:

$(P.2)$
$$\begin{aligned}
\text{minimize} \quad & F_{P.2}(x) \\
\text{subject to} \quad & |\epsilon_{1(j)}^k| \leq \epsilon_1^{allow} \\
& |\epsilon_{2(j)}^k| \leq \epsilon_2^{allow} \\
& |\gamma_{12(j)}^k| \leq \gamma_{12}^{allow} \\
& -90° \leq x_k \leq 90°
\end{aligned}$$

for $k = 1, 2, \ldots, n$ and $j = 1, 2, \ldots, m$.

The last step is to define a magnification factor, δ, which is used to identify ply strains that exceed maximum allowable strains. The magnification factor is applied only when the normalized term indicates failure:

$$\delta = \begin{cases} 3*m*n & \text{if numerator is } > 0 \\ 1 & \text{if numerator is } \leq 0. \end{cases}$$

The final objective function is defined as:

$$F_{P.3}(x) = \left(\frac{1}{3*m*n}\right) \sum_{j=1}^{m} \left[\sum_{k=1}^{n} \left[\left(\delta * \exp \frac{|\epsilon_{1(j)}^{k}| - \epsilon_{1}^{allow}}{\epsilon_{1}^{allow}}\right)^2 \right.\right.$$
$$+ \left(\delta * \exp \frac{|\epsilon_{2(j)}^{k}| - \epsilon_{2}^{allow}}{\epsilon_{2}^{allow}}\right)^2$$
$$\left.\left.+ \left(\delta * \exp \frac{|\gamma_{12(j)}^{k}| - \gamma_{12}^{allow}}{\gamma_{12}^{allow}}\right)^2 \right]\right].$$

Our final formulation can be stated as:

$$(P.3) \quad \begin{array}{l} \text{minimize} \quad F_{P.3}(x) \\ \text{subject to} \quad -90° \leq x_k \leq 90° \end{array}$$

for $k = 1, 2, \ldots, n$ and $j = 1, 2, \ldots, m$.

The entire summation is divided by the number of terms, $3*m*n$, so that any value of $F_{P.3}(x)$ greater than 1.0 indicates laminate failure. It is straightforward to show formulations $(P.2)$ and $(P.3)$ are mathematically equivalent in the sense that any global optimum of $(P.2)$ is a global optimum of $(P.3)$, and any global optimum of $(P.3)$ with $F_{P.3}(x) \leq 1.0$ is a global optimum of $(P.2)$. It is interesting to note that the level set of $(P.3)$ at 1.0, i.e. $\{x : F_{P.3}(x) \leq 1.0, -90° \leq x_k \leq 90°\}$ corresponds exactly to the feasible region of $(P.2)$. The final formulation is easier to solve directly because of the simple box constraints. Another advantage to the final formulation occurs if an optimum is found with $F_{P.3}(x) > 1.0$, while $(P.2)$ would be found to be infeasible. The optimum to $(P,3)$ may provide useful insight to a design engineer.

As stated in previous studies [11,14], the design space associated with laminated composite materials may contain many local minima. As a specific example, consider the following simple design problem of finding the optimal ply angles that minimize strain ($F(x)$) in an 8-ply laminate that is subjected to a uniaxial load, N_x. In this case the optimum solution is known; the optimum solution has all plies in the 0° direction, so that the fibers are in line with the loading direction. The design space associated with this problem has many local minima. A sample of these minimum points can be visualized using a plot of the objective function. We have found that objective function plots are much easier to interpret if the negative of the objective function (i.e., $-1 * F(x)$) is plotted, such that minimum points become "peaks" in the plotted objective function. Figure 2(a) shows an objective function plot in which only one design variable (i.e., one ply fiber angle) is allowed to change, all others are fixed at 0°. The optimal point is clearly at $x_k = 0°$, as expected. In Figure 2(b) two variables are allowed to change. The optimal point is at $x_1 = x_2 = 0°$, again as expected. The plots display the global nature of the design problem, since many local minima are apparent. Of course, if additional design variables were allowed to change then many more local minima would be present in the n-dimensional design space.

3 Performance of IHR

Improving Hit and Run [12] can be summarized as a sequential random search algorithm that, given an initial feasible point, samples a direction vector according to a uniform distribution on a hypersphere, and then randomly samples points along the feasible portion of the line segment in that direction. If an improving point is found, the current point is updated with the new point and an iteration is complete. If no improving point

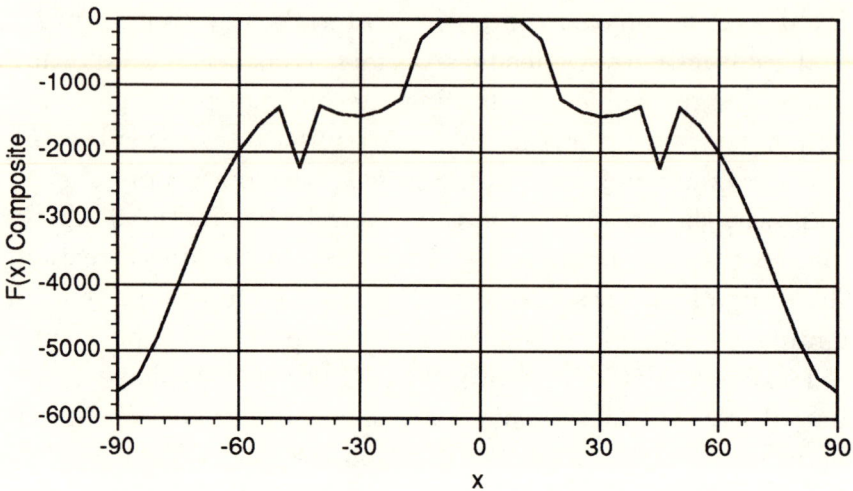

Figure 2a. Composite Objective Function in 1 Dimension

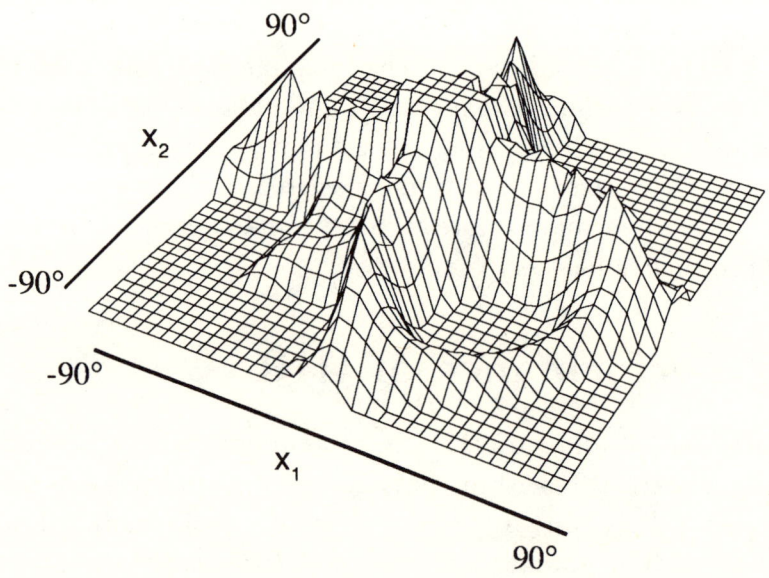

Figure 2b. Composite Objective Function in 2 Dimensions

is found after a user-defined number of tries, a new direction is selected. Several stopping criterion have been implemented, including reaching a maximum number of directions without improvement, a maximum number of iterations, or convergence to a point reflected by a very small change in objective function value. In the experiments reported later in the paper, we used a stopping criterion of 100-fold improvement from the initial point for objective functions with known solutions, or a maximum number of iterations. The reason we use IHR in the composite design problem is because we expect large dimensional problems of a global nature, and a recent theoretical analysis of IHR shows it is an efficient algorithm in terms of dimension for a class of optimization problems [12]. In theory, the number of improving points needed to achieve m-fold improvement with $1 - \alpha$ certainty ($m > 0, 0 < \alpha < 1$) is shown to be bounded by a linear function of dimension for a class of spherically symmetric problems[12]. The class of spherically symmetric problems include unimodal objective functions, not necessarily convex, with a property of having hyperspherical level sets. The composite design problem is not included in this class of problems, yet we have experienced similar performance. We will present experimental evidence that linearity in the number of improving points holds for more general objective functions, however, IHR is still limited in its ability to find the global optimum with 100% confidence.

Although application of IHR to the composite design problem has been quite successful, there are still occasions when IHR gets trapped in a local minimum that is not the global minimum [3]. In an attempt to understand the conditions under which IHR fails to identify the global minimum, we have experimented with an objective function with a known solution, but which is similar in nature to the composite objective function.

The general form of the test objective function is made up

of terms involving trigonometric functions, as follows:

$$F(x) = -\left[A \prod_{k=1}^{n} \sin(x_k) + \prod_{k=1}^{n} \sin(5x_k)\right]$$

where A is a constant. The test problem is defined as:

Minimize $F(x)$
subject to : $0° \leq x_k \leq 180°$ $\quad k = 1, \ldots, n.$

This test problem was chosen because it has a known global minimum at the point $(90°, 90°, \ldots, 90°)$, has multiple local minima, is easily generalized to n dimensions, and is similar in shape to the composite objective function. Also, the degree of difficulty in identifying the global minimum is in part associated with the relative depth of local minima. The relative depth can be easily adjusted by selecting different values of A. As the constant A gets large the test problem approaches a spherically symmetric program. Thus it is anticipated that IHR will perform better for large values of A.

We have conducted our analyses using two values of A. In the first case (objective function 1) A was taken to be 2.5, while in the second case (objective function 2) A was taken to be 5. One and two dimensional plots of objective functions 1 and 2 are shown in Figures 3 and 4, respectively (as in Figure 2, the negative of the objective functions have been plotted in Figures 3 and 4). Notice the rough similarity to the composite objective function shown in Figure 2. The constructed objective functions have many local minima that grow rapidly in dimension. The number of local minima is given by,

$$\sum_{i=0}^{\lfloor n/2 \rfloor} \left(\frac{n!}{(n-2i)!(2i)!} * 3^{n-2i} * 2^{2i}\right)$$

where $\lfloor n/2 \rfloor$ in the summation is rounded down to the nearest integer, and the constants 3 and 2 reflect the periodicity due to the coefficient 5 in the sine term of $F(x)$. Since the relative depth of the local minimum is much greater in objective function 1, it is farther from a spherically symmetric problem, and we anticipate the solution to function 1 will be more difficult to identify using IHR than for objective function 2.

4 Experiments On Objective Functions

Experimental results for IHR on the constructed objective functions are summarized in Figures 5-8. In the experiments, we ran IHR using two different methods of selecting the random direction. The "HD" version uses a uniform distribution on a n-dimensional hypersphere, while the "CD" version uses a uniform distribution on the n coordinate directions. These techniques are similar to those presented by Berbee et.al.[1] for Hit-and-Run methods. We conducted analyses using dimensions ranging from 5 to 25. Ten separate runs using random starting points were used for each dimension.

The experiments were run until either 100-fold improvement from the initial point was reached, or when a maximum number of iterations ($10n$) was reached. In the test problem, a 100-fold improvement will distinguish the global minimum from the suboptimal local minima. Unfortunately, sometimes 100-fold improvement was not reached after $10n$ iterations. Although theoretically IHR would eventually reach 100-fold improvement with HD, there are examples where CD will never reach 100-fold improvement. As an estimate of the rate of success at "finding the global minimum", the percentage of time IHR achieved 100-fold improvement within the maximum number of iterations was recorded.

Figure 5 shows the percentage of time IHR achieved 100-fold improvement within the stopping criterion for objective

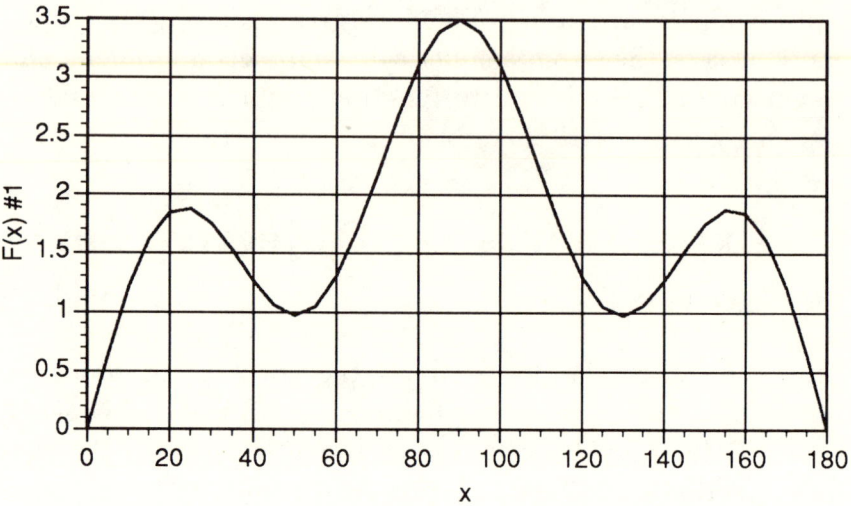

Figure 3a. Constructed Objective Function #1 in 1 dimension, A = 2.5

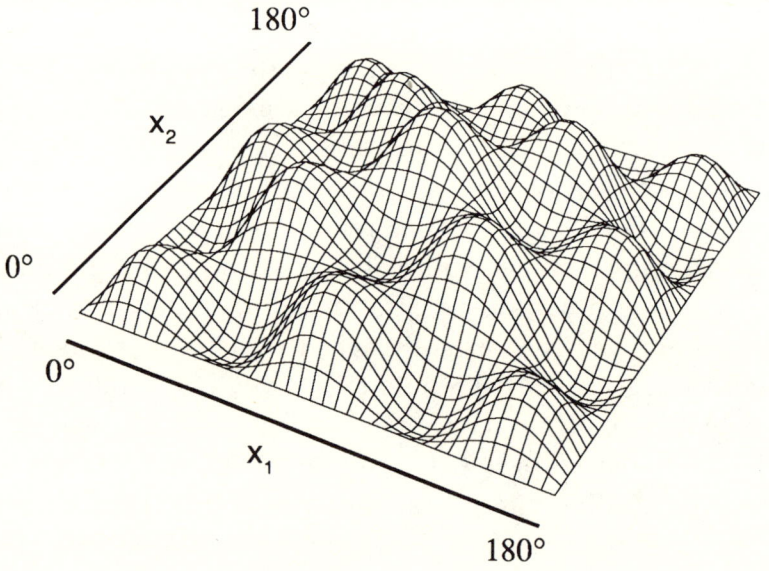

Figure 3b. Constructed Objective Function #1 in 2 dimensions, A = 2.5

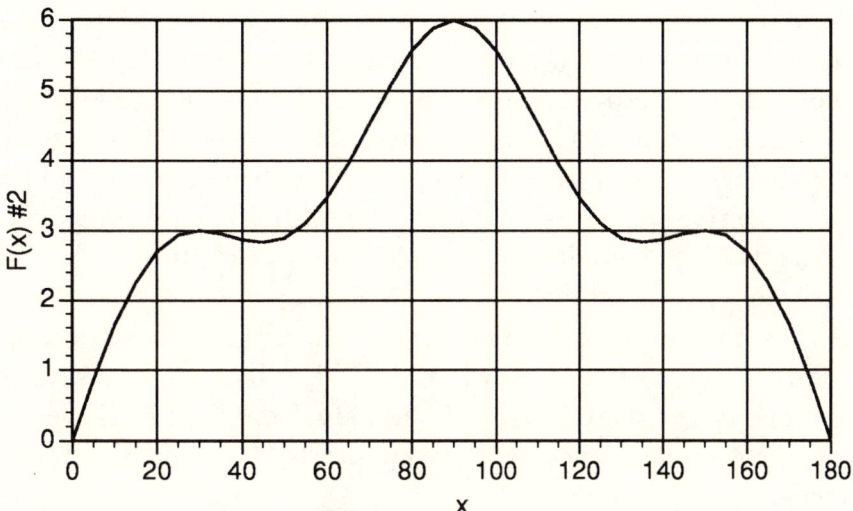

Figure 4a. Constructed Objective Function #2 in 1 dimension, A = 5

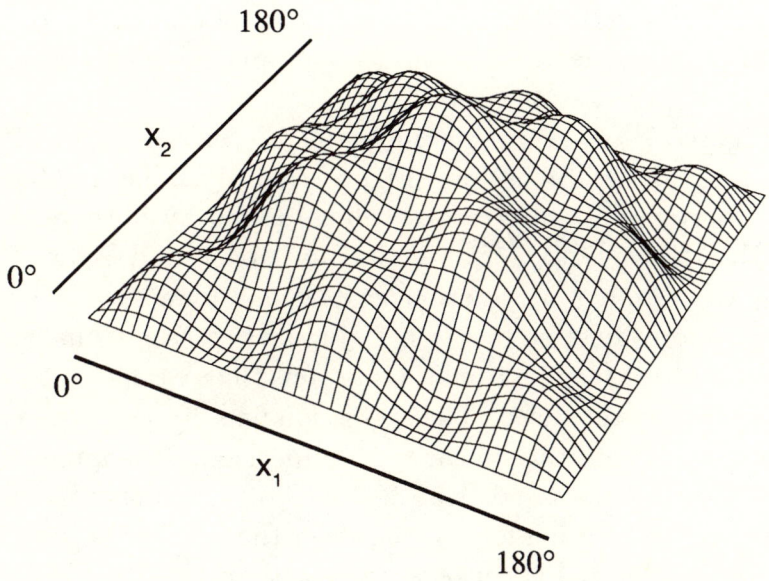

Figure 4b. Constructed Objective Function #2 in 2 dimensions, A = 5

function 1 with both HD and CD techniques. Figure 6 shows similar information for objective function 2. As expected, objective function 1 is a harder problem to solve. Using the HD method, the estimated probability of achieving 100-fold improvement to objective functions 1 and 2 was 0.50 and 0.66, respectively. Using the CD method, the estimated probability for objective functions 1 and 2 was 0.42 and 0.58, respectively. Thus, the HD technique is a modest improvement over the CD technique, at least for the problems considered. One might expect the problem to increase in difficulty as dimension increases, due to the larger number of local minima as well as the decreased relative size of the global minimum. However, the experimental success rates appear to be insensitive to dimension.

As the relative depth of the local minima go to zero, the problem approaches a spherically symmetric problem, for which the theoretical result of linearity in dimension holds. We now turn to investigate how complexity varies in dimension for the constructed objective functions that are in a more general class of global optimization problems.

Figures 7 and 8 present measures of performance of IHR as a function of dimension. Figure 7(a) plots the number of new improving points (record values) found by both versions of IHR (HD and CD) for objective function 1, and Figure 7(b) plots the number of function evaluations found by HD and CD also for objective function 1. Figure 8 presents analogous plots for objective function 2. Only runs that attained 100-fold improvement were used. The regression lines for the number of points clearly indicate linearity in dimension. It is interesting to notice that the number of new improving points for CD is consistently less than for HD, however the number of function evaluations for HD is consistently less than for CD.

Similar experiments testing the behavior of IHR were run on the composite design problem defined earlier. Only 4, 5 and

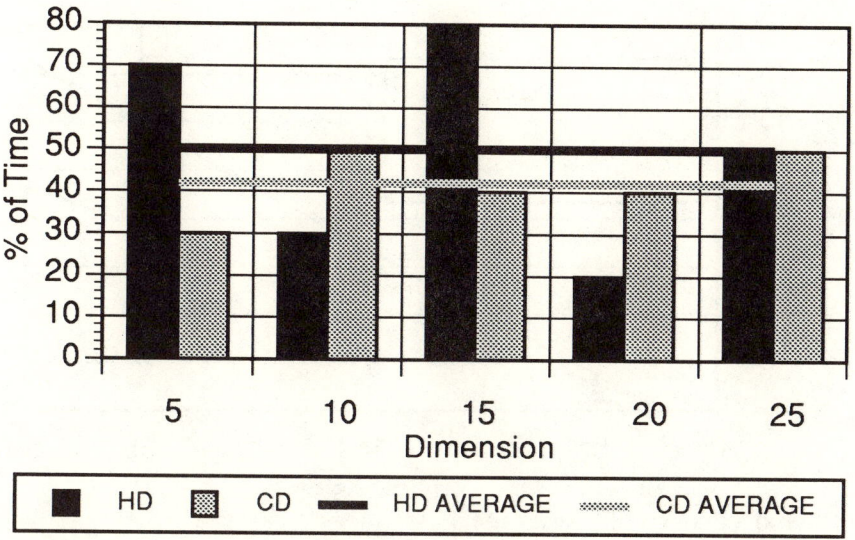

Figure 5. Percentage of Time Global Optimum was Found on Objective Function #1

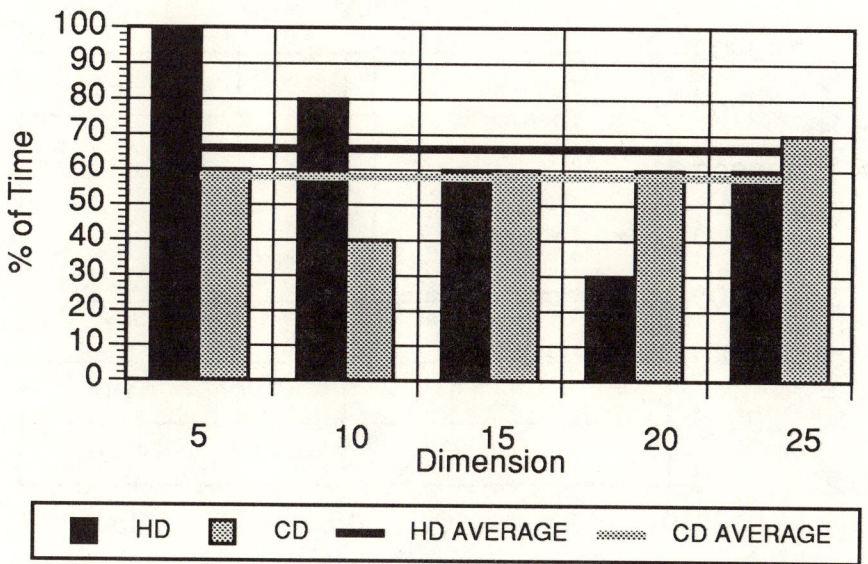

Figure 6. Percentage of Time Global Optimum was Found on Objective Function #2

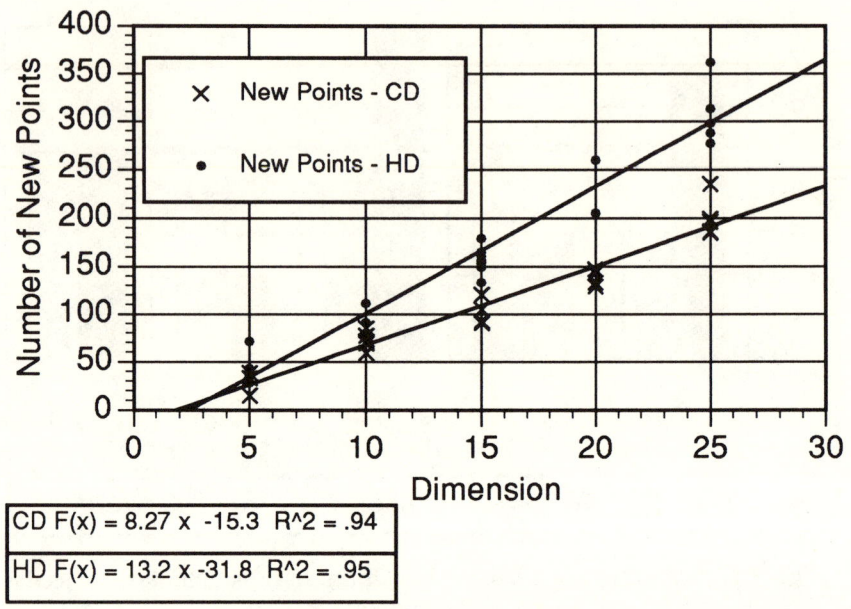

Figure 7a. Number of New Points for Objective Function #1

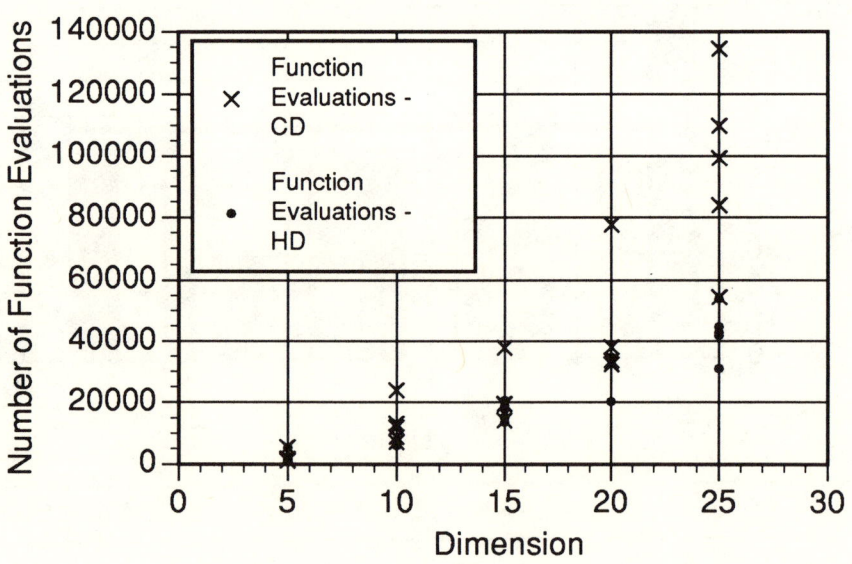

Figure 7b. Number of Function Evaluations for Objective Function #1

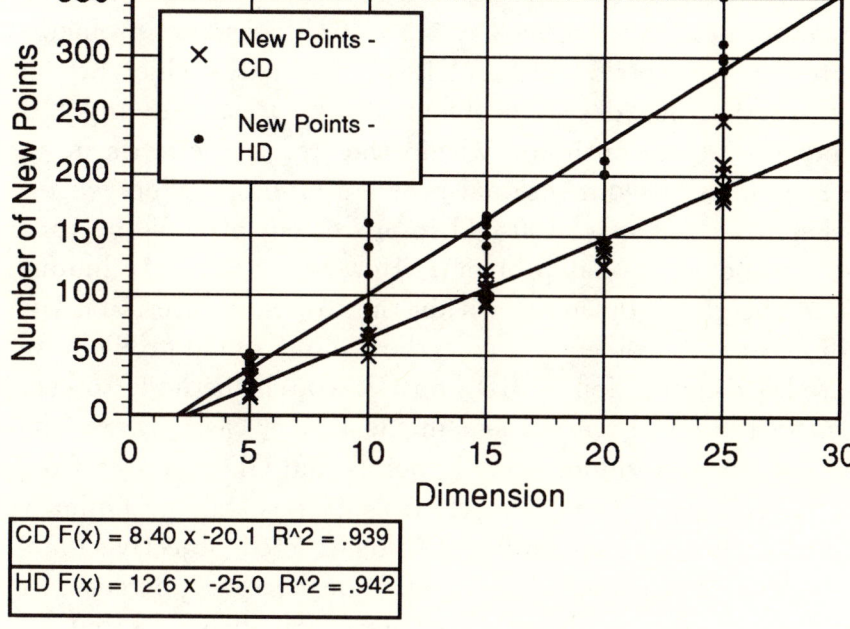

Figure 8a. Number of New Points for Objective Function #2

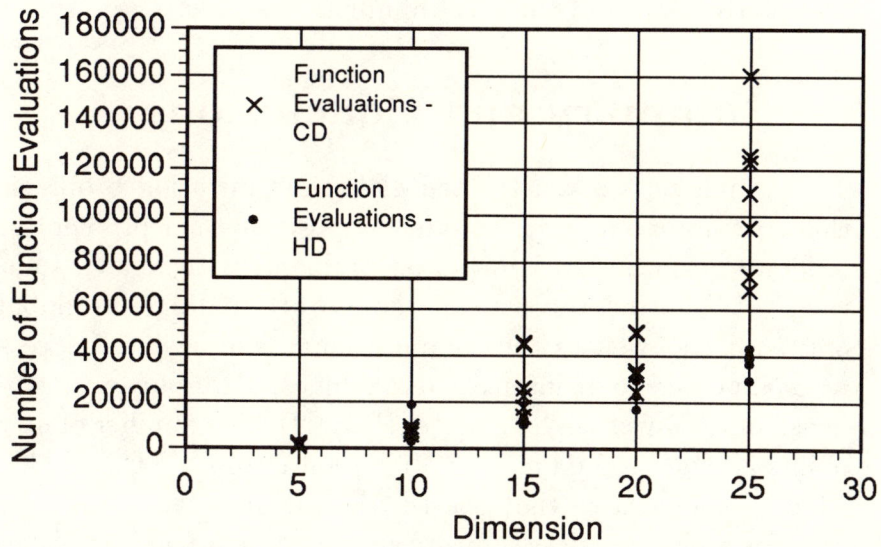

Figure 8b. Number of Function Evaluations for Objective Function #2

6-dimensional problems were conducted because CPU times were much greater (roughly 10 times) than for the constructed objective functions. These 4, 5 and 6-dimensional problems represent symmetric 8, 10 and 12-ply laminates which are on the order of our design problem [3,14]. Both HD and CD found the global optimum within the stopping criteria in all 10 runs for the 4-dimensional problem (100%). Averaged together, they achieved 100-fold improvement 60% of the time for the 5-dimensional problem. However, neither technique found the global optimum within the stopping criteria for the 6-dimensional problem. In fact, the 6-dimensional problem required a combination of HD and CD search methods to ever achieve 100-fold improvement in the 10 test cases, where HD was used from random starting points and CD was used from the stopping points of the HD runs. Thus, our preliminary results indicate that, unlike the constructed objective functions, the composite objective function becomes increasingly difficult to solve in higher dimensions. We speculate that increases in dimension for the composite objective function not only increases the number of local minima but also increases the relative depth of the local minima.

5 Summary and Conclusions

To summarize, we constructed global optimization problems that are similar in form to our composite design problem in order to perform experiments on problems with known solutions. We tested the rate of success of "finding the global optimum", as well as the computational complexity. Success rates appeared to be insensitive to dimension for our test problems, and HD was slightly better than CD. The number of new points appears to be linear in dimension for both HD and CD, which demonstrates that the theoretical linearity result may hold for a broader class of problems. Although CD was better than HD in terms of number of new points, HD was better than CD in terms of number of function evaluations. In experimenting with a simple composite objective function with a known solution, we have discovered that success rate decreases with dimension. Thus, we conclude that a modification to the algorithm, and possibly to the form of the objective function is necessary to achieve a better optimization technique for composite design.

References

[1] Berbee, H.C.P., Boender, C.G.E., Rinnooy Kan, A.H.G., Scheffer, C.L., Smith, R.L., and Telgen, J. "Hit-and-Run Algorithms for the Identification of Nonredundant Linear Inequalities," *Mathematical Programming 37* (1987) 184-207.

[2] Bélisle, C.J.P., Romeijn, H.E., and Smith, R.L., "Hide-and-Seek: A Simulated Annealing Algorithm for Global Optimization," Technical Report No. 90-25, Department of Industrial and Operations Engineering, University of Michigan, Ann Arbor, MI, Sept. 1990.

[3] Graesser, D.L., Zabinsky, Z.B., Tuttle, M.E., and Kim, G.I., "Designing Laminated Composites Using Random Search Techniques" *Journal of Composite Structures*, forthcoming.

[4] Johnson, M.E., "Simulated Annealing (SA) & Optimization", *American Journal of Mathematics and Management Sciences 18*, 1988.

[5] Jones, R.M., *Mechanics of Composite Materials*, Scripta Book Company, Washington D.C., 1975.

[6] Massard, T.N., "Computer Sizing of Composite Laminates for Strength," Journal of Reinforced Plastics and Composites, 3, pp. 300-345, 1984.

[7] Park, W.J., "An Optimal Design of Simple Symmetric Laminates Under the First Ply Failure Criteria," *Journal of Composite Materials 16*, pp. 341-355, 1982.

[8] Patel, N.R., Smith, R.L., and Zabinsky, Z.B., "Pure Adaptive Search in Monte Carlo Optimization," *Mathematical Programming 43*, pp. 317-328, 1988.

[9] Schmit, L.A., Farshi, B., "Optimum Laminate Design for Strength and Stiffness," *International Journal for Numerical Methods in Engineering 7*, pp. 519-536, 1973.

[10] Tauchert, T.R., Adibhalta, S., "Design of Laminated Plates for Maximum Stiffness," *Journal of Composite Materials 18*, pp. 58-69, 1984.

[11] Vanderplaats, G.N., Weisshaar, T.A., "Optimum Design of Composite Structures," *International Journal for Numerical Methods in Engineering 27*, pp. 437-448, 1989.

[12] Zabinsky, Z.B., Smith, R.L. and McDonald, J.F., "Improving Hit and Run for Global Optimization," working paper. Department of Industrial and Operations Engineering, University of Michigan, Ann Arbor, MI, 1990.

[13] Zabinsky, Z.B., and Smith, R.L., "Pure Adaptive Search in Global Optimization," *Mathematical Programming*, forthcoming.

[14] Zabinsky, Z.B., Tuttle, M.E., Graesser, D.L., Kim. G.I., and Hatcher, D., "Multi-Parameter Optimization Tool for Low-Cost Commercial Fuselage Crown Designs," *First NASA Advanced Composites Technology (ACT) Conference Proceedings*, October 1990.

Appendix 1
Summarize Calculation of Ply Strains

This appendix summarizes the calculations needed to determine ply strains using classical lamination theory [5] as needed in the objective function $F(x)$. The appendix provides the equations for a single set of loads, as may be specified by loading condition j. The subscript j is dropped in the appendix for ease in reading. Notice that a single loading condition may include values for all of the following: N_x, N_y, N_{xy}, M_x, M_y, and M_{xy}.

User specified parameters:

n	number of plies
t	ply thickness
N_x, N_y, N_{xy}	In plane loads for loading condition j (see Figure 1)
M_x, M_y, M_{xy}	Out of plane bending moments for loading condition j (see Figure 1)
$Q_{11}, Q_{12}, Q_{22}, Q_{66}$	Material specific stiffness parameters as functions of Young's Modulus and Poisson's Ratio

Decision variable:

$x_k = \theta_k$ fiber angle for ply k, where $k = 1, 2, \ldots, n$

Calculate the individual k^{th} ply lamina stiffness matrix, $\left[\overline{Q_{ij}^k}\right]$, for $i, j = 1, 2, 6$ and $k = 1, 2, \ldots, n$:

$$\overline{Q_{11}^k} = Q_{11}\cos^4\theta_k + 2(Q_{12} + 2Q_{66})\cos^2\theta_k \sin^2\theta_k + Q_{22}\sin^4\theta_k$$
$$\overline{Q_{12}^k} = (Q_{11} + Q_{22} - 4Q_{66})\cos^2\theta_k \sin^2\theta_k + Q_{12}(\cos^4\theta_k + \sin^4\theta_k)$$

COMPOSITE LAMINATES

$$\overline{Q_{22}^k} = \overline{Q_{21}^k}$$
$$\overline{Q_{22}^k} = Q_{11}\sin^4\theta_k + 2(Q_{12} + 2Q_{66})\cos^2\theta_k\sin^2\theta_k + Q_{22}\cos^4\theta_k$$
$$\overline{Q_{16}^k} = (Q_{11} - Q_{12} - 2Q_{66})\cos^3\theta_k\sin\theta_k$$
$$\quad + (Q_{12} - Q_{22} + 2Q_{66})\cos\theta_k\sin^3\theta_k$$
$$= \overline{Q_{61}^k}$$
$$\overline{Q_{26}^k} = (Q_{11} - Q_{12} - 2Q_{66})\cos\theta_k\sin^3\theta_k$$
$$\quad + (Q_{12} - Q_{22} + 2Q_{66})\cos^3\theta_k\sin\theta_k$$
$$= \overline{Q_{62}^k}$$
$$\overline{Q_{66}^k} = (Q_{11} + Q_{22} - 2Q_{12} - 2Q_{66})\cos^2\theta_k\sin^2\theta_k$$
$$\quad + Q_{66}(\cos^4\theta_k + \sin^4\theta_k)$$

Calculate the combined n ply laminate stiffness matrix, $\begin{bmatrix} A_{ij} & B_{ij} \\ B_{ij} & D_{ij} \end{bmatrix}$, for $i, j = 1, 2, 6$:

$$A_{ij} = \sum_{k=1}^{n} \overline{Q_{ij}^k}(z_k - z_{k-1})$$
$$B_{ij} = \frac{1}{2}\sum_{k=1}^{n} \overline{Q_{ij}^k}(z_k^2 - z_{k-1}^2)$$
$$D_{ij} = \frac{1}{3}\sum_{k=1}^{n} \overline{Q_{ij}^k}(z_k^3 - z_{k-1}^3)$$

where z_k is the interface location of plies k and $k-1$ measured from the geometric midplane of the laminate, $z_0 = nt/2$, and $z_k = z_{k-1} + t$.

Calculate the midplane laminate strains ϵ_x^0, ϵ_y^0, γ_{xy}^0, and curvatures κ_x, κ_y, κ_{xy}, created by loading condition j:

$$\begin{bmatrix} \epsilon_x^0 \\ \epsilon_y^0 \\ \gamma_{xy}^0 \\ \kappa_x \\ \kappa_y \\ \kappa_{xy} \end{bmatrix} = \begin{bmatrix} A_{ij} & B_{ij} \\ B_{ij} & D_{ij} \end{bmatrix}^{-1} \begin{bmatrix} N_x \\ N_y \\ N_{xy} \\ M_x \\ M_y \\ M_{xy} \end{bmatrix}$$

Calculate the individual k^{th} ply strains ϵ_x^k, ϵ_y^k, γ_{xy}^k, for $k = 1, 2, \ldots, n$:

$$\begin{bmatrix} \epsilon_x^k \\ \epsilon_y^k \\ \gamma_{xy}^k \end{bmatrix} = \begin{bmatrix} \epsilon_x^0 \\ \epsilon_y^0 \\ \gamma_{xy}^0 \end{bmatrix} + z_k \begin{bmatrix} \kappa_x \\ \kappa_y \\ \kappa_{xy} \end{bmatrix}$$

Calculate the individual k^{th} ply strains ϵ_1^k, ϵ_2^k, γ_{12}^k, in the **1,2 coordinate system** for $k = 1, 2, \ldots, n$:

$$\begin{bmatrix} \epsilon_1^k \\ \epsilon_2^k \\ \gamma_{12}^k \end{bmatrix} = \begin{bmatrix} \cos^2 \theta_k & \sin^2 \theta_k & 2\cos \theta_k \sin \theta_k \\ \sin^2 \theta_k & \cos^2 \theta_k & -2\cos \theta_k \sin \theta_k \\ -\cos \theta_k \sin \theta_k & \cos \theta_k \sin \theta_k & \cos^2 \theta_k - \sin^2 \theta_k \end{bmatrix} \begin{bmatrix} \epsilon_x^k \\ \epsilon_y^k \\ \gamma_{xy}^k \end{bmatrix}$$

Finally, the **1,2 coordinate system** strains ϵ_1^k, ϵ_2^k, γ_{12}^k, for plies k, $k = 1, 2, \ldots, n$ **are used to calculate the objective function** $F(x)$ **as defined in the text.**

Stochastic Minimization of Lipschitz Functions

Regina Hunter Mladineo[*]

Abstract: A random search method for minimizing Lipschitz continuous functions is presented. Convergence to the global minimum with probability one is proved and computer test results for several functions are presented.

I. Introduction.

This paper presents a stochastic modification of a deterministic global optimization algorithm. The introduction of stochastic elements into the algorithm, which in original form [3] converges with relatively few function evaluations, is for the purpose of decreasing overhead (computational effort in each iteration excluding function evaluations).

The problem we consider is:

$$\text{Min } f \text{ subject to } x \in I^N$$

where $I^N = \{ x=(x^1,\ldots x^N) \mid 0 \leq x^i \leq 1 \}$

We assume furthermore that f satisfies the Lipschitz condition that, for all x, y in I^N,

$$|f(x) - f(y)| \leq L \|x - y\|,$$

L a real positive constant, $\|\ \|$ Euclidean distance.

The deterministic algorithm previously developed by the author has been tested and results have

---

[*]Management Sciences Department, Rider College, Lawrenceville, NJ 08648.

been published [3, 4]. Briefly, it builds an approximating function, a lower envelope of downward right circular cones at each iterate, x_j. The cone at x_j:

$$\{(x,z) \mid z = f(x_j) - L \|x - x_j\|\}.$$

These cones bound the graph of the objective function from below. Other algorithms which also assume a Lipschitz condition and build a lower envelope from cones at sample points are by Wood [8] who uses a simplex-shaped cone and by Breiman and Cutler [1] who use a cone with quadratic vertical sections.

The complexity of random searches for minima of certain convex functions was obesrved by Solis and Wets [7] to increase linearly in dimension when complexity is measured as the number of iterations which improve the function value. Zabinsky and Smith [9,10] proved this result and extended it to arbitrary Lipschitz continuous functions in the context of global optimization. They use in [9] the term "record" for an improving function evaluation and "record value" for the function value at a record. We will use these terms in this paper, since iteration here means function evaluation and not all function evaluations are records. Computational results in [10] exhibit the theoretical linearity for two convex functions.

One purpose of the research presented here was to devise an algorithm which exhibits linearity in records (not necessarily in function evaluations) and which eliminates some overhead associated with the deterministic method. The algorithm decreases function evaluations by restricting the region from which sample points are drawn. It is thus an attempt to approximate the level set of f at each iteration. It uses randomly generated points for evaluation thus reducing computational overhead found in the original algorithm. Another purpose was to test the algorithm on a set of more difficult Lipschitz functions which are of the same level of difficulty in different dimensions so that comparisons in dimension could be made. In [6] the functions used were difficult (had several local optima) in lower dimensions, but "smoothed" out in higher, so the seeming linearity is misleading. Improvements in the algorithm of [6] in this paper include fewer parameters and a better (more uniform) random point generator. A consistently difficult test function GR7 was devised to test the algorithm.

Section II presents the stochastic algorithm and section III restates it for the purpose of proving convergence to the global minimum with probability one. Computational results are presented in section IV and future research and conclusions are discussed in section V.

II. The Algorithm.

We use the following to define the algorithm.
f = the objective function.
$F_j(x) = \max_{k \leq j} \{ f(x_k) - L \|x - x_k\| \}$, the lower envelope
at x at iteration j.
$M_j = (x_j, z_j)$ is the jth iterate,
where $x_j \in I^N$ and $z_j \in \mathbb{R}^1$ are randomly generated
UB_j = the upper bound on the global minimum of f at iteration j
LB_j = the estimated lower bound on the global minimum of f at iteration j

Parameters set by input to the program:
$\varepsilon > 0$, real-valued, small
$\Delta > 0$, real-valued, small
$R \geq 1$, integer-valued
LIM = integer limit on the number of function evaluations
D = integer limit on the number of random points generated whether evaluated or not

Initialization:
  Evaluate f at $(0,\ldots,0)$.
  $UB_0 = f(0,\ldots,0)$.
  $LB_0 = F_0(1,\ldots,1)$.
  $j = 0$.
  $x_1 = (1,1,\ldots,1)$
  $z_1 = LB_0$

Loop:
  Step 1:
    $j = j + 1$
    Evaluate $f(x_j)$

$$UB_j = \begin{cases} f(x_j), & \text{if } f(x_j) < UB_{j-1} \\ UB_{j-1}, & \text{otherwise} \end{cases}$$

$LB_j = LB_{j-1}$. (LB_j is changed in step 6 after a new set of random points is generated and accepted or in step 2, when no more points can be generated.)

Step 2:
If out of time or
j = LIM or
$UB_j - LB_j < \varepsilon$, then stop.
If D = 0 then
    $LB_j = z_{j+1}$
    go to step 1
    (The points M_J, $J > j$ will be evaluated without any new points being generated.)

Step 3:
If $z_j - LB_j < \Delta$ and if $j > 1$, then go to step 1.

Step 4:
Generate R random points, $\{r_k = (u_k, \xi_k)\}_{k=1}^{R}$, uniformly distributed in $I^N \times [LB_j, UB_j]$.
D = D - R

Step 5:
Loop: for k = 1 to R
    If $\xi_k < F_j(u_k)$, then eliminate r_k.
End loop.
If all points eliminated
    $LB_j = z_{j+1}$
    go to step 1.

Step 6:
Sort points remaining from step 5 and merge into array M in order of increasing ξ_k. Thus there is a mapping from these points to the array M:

$$u_k \rightarrow x_J$$
$$\xi_k \rightarrow z_J, \text{ some } k, \text{ some } J > j.$$

$LB_j = z_{j+1}$.
Go to step 1.

This algorithm is a "simulation" of the increasing lower envelopes. Figure 1 shows the lower envelope after five iterations (J = 0 to J = 4). Note that LB_1 is updated, after new points (the x's) are generated and accepted, to equal z_2. Thus LB_2 also equals z_2, the N+1st coordinate of M_2. A similar updating occurs for LB_4 which increases as shown.

Without the sort, the parameter R could be lowered to 1, in imitation of the usual random searches of [7,9,10], but doing so in practice did not improve results. In fact, the implementation was designed to use large R, in order to be able to update LB_j and save running time. As will be seen in the next section, for R = 1, in order to assure convergence, LB is not updated. Updating LB_j, when using large R is essentially an heuristic.

III. Convergence for Fixed LB.

We will show that the above algorithm, modified to fix LB, the lower bound on the global minimum, and with R = 1, converges to the global minimum with probability equal to one.

Let α be the global minimum of f in I^N. We define the ε-optimal region

$$\Re_\varepsilon = \{x \in I^N \mid f(x) < \alpha + \varepsilon \}.$$

Let U be the probability measure corresponding to the uniform distribution on I^N. Let ν be Lebesgue measure on $\mathbb{R}^{N+1}$.

Step 0:
  $x_0 = (0,0,\ldots,0)$
  $x_1 = (1,1,\ldots,1)$
  Evaluate f at x_0 and x_1
  $UB_1 = \min \{ f(x_0), f(x_1) \}$
  $LB = F_0(1,1,\ldots,1)$

Step 1:
  Generate u_j from the uniform distribution on I^N.
  Generate ξ_j from the uniform distribution on $[LB,UB_j]$.

Step 2:
  Accept u_j if (u_j,ξ_j) lies above the lower envelope else repeat step 1.

Step 3:
  Evaluate $f(x_j)$.
  $UB_j = \min\{UB_{j-1}, f(x_j)\}$.
  (Note that LB is not updated.)
  Go to Step 1.

This defines a probability measure μ_j on I^N as
  follows: Let A be a Borel subset of I^N.
  Define

$$C_j(A) = \{(a,z) \mid a \in A,\ z \in [F_j(a), UB_j]\ \text{and}\ UB_j > F_j(a)\}$$

$$\mu_j(A) = \nu(C_j(A))/\nu(C_j(I^N))$$

We call $C_j(I^N)$ the bracket at iteration j.

Theorem:

$$\lim_{j \to \infty} \mu_j(I^N - \mathfrak{R}_\varepsilon) = 0$$

Proof: We assume f is not constant, since in this case $\mathfrak{R}_\varepsilon = I^N$.

Since $\mu_j(\mathfrak{R}_\varepsilon) > 0$ and U (the uniform measure) has the property that for any set A of non-0 Lebesgue measure

$$\lim_{j \to \infty} (1-U(A))^j = 0 \qquad (1)$$

then the probability is 1 that there is J such that $x_J \in \mathfrak{R}_\varepsilon$, so we have

$$UB_j < \alpha + \varepsilon,\ \forall\ j \geq J.$$

Furthermore, there are at most J-1 iterates, thus a finite number of disjoint regions in $C_J(I^N)$. Thus $C_J(I^N - \mathfrak{R}_\varepsilon)$ is compact.

Let $(x,z) \in C_J(I^N - \mathfrak{R}_\varepsilon)$. Define, for $j \geq$,

$$A_j = \{(a,b) \mid b \leq f(x_j) - L\|a - x_j\|\}.$$

Int(A_j) is the "removal" cone - the region removed from the bracket as a result of a function evaluation at x_j and the Lipschitz condition. We will show that (x,z) is eventually covered by a removal cone and thus eliminated from the bracket.

If x is in the interior of a connected subset, say B, of $C_J(I^N - \mathfrak{R}_\varepsilon)$ projected to I^N, then by (1), in a finite number of iterations it will be removed.

So, at worst, (x,z) is an isolated limit point of the boundaries of the removal cones, thus on the

boundary of B, so that $z \geq UB_j$, $j \geq J$.
   Now, either in a finite number of iterations, say J', $UB_{J'} < z$ and (x,z) is removed or $UB_j \geq z \; \forall \; j > J$. Suppose the latter.
   Let $\{A_i\}$, $i=1,\ldots$, be the sequence of removal cones (and their boundaries) wwhich have (x,z) as a limit point. Let (v_i, z_i) be the corresponding vertices. Then the v_i's become arbitrarily close to x, thus by continuity, $f(v_i)$ gets close to $f(x) > \varepsilon + \alpha > z$ (by assumption). Thus, for some i, $f(v_i) > z + L \|x-v_i\|$. But then that removal cone covers x.

<u>Corollary</u>:

$$\lim_{j \to \infty} P(x_j \in \Re_\varepsilon) = 1.$$

Proof: $P(x_j \in \Re_\varepsilon) = 1 - P(x_j \in (I^N - \Re_\varepsilon)) = 1 - \mu(I^N - \Re_\varepsilon)$. The result follows immediately from the theorem.

IV. Computational Results.

Each test function was run with R=50, NUMIT=1000, where R is the number of random points generated in a block and NUMIT is the limit of iterations (function evaluations). If the run converged too far from the minimum, then NUMIT was raised to 20,000, a convenient size for the large memory queue on the VAX4000, then to 80,000, essentially as large as possible on this machine. For larger N and for some functions as necessary R was raised to 5000 with NUMIT = 20,000 or 80,000. For all test runs, the stopping criteria used ε = .00001.

The test function GR7 (Table 1) was designed to maintain several local optima (Figure 2, which shows -f(x)) as N increases, unlike previous related test functions as GR1 - GR5 [6], which tended in higher dimensions to flatten out and have only one local minimum. It also has the global minimum located at randomly scattered coordinates so there is less likelihood of convergence due to a non-uniform bias in the random generation of sample points.
   The function CONVEX1 (Table 2) is from Solis and Wets [7], also from Zabinsky and Smith [10]. The other convex conical function tested in both of those papers yielded poor results here. The program consistently ended early, before significant progress

to the minimum was attained.

GP (Table 3) is the Goldstein, Price function [2] and illustrates how varying R and NUMIT affect the behavior of the algorithm. The "spike" function, SQRIN5 also from [2], on the other hand, yielded poor results, getting no closer than within .4 of the minimum in all variations of R and NUMIT and taking up to an hour of CPU time for 10,000 iterations and 7 records.

Zabinsky and Smith [9] obtain a theoretical upper bound for the number $N(\varepsilon)$ of records to reach a function value within ε of the global minimum. They show the expected number of records,

$$E(N^*(\varepsilon)) \leq 1 + \ln(LD/\varepsilon) \cdot N,$$

where L is the Lipschitz constant and D is the diameter of the domain. It is clear from tables 1-3 that the algorithm of this paper stays far below that upper bound for NREC, the number of records. Note that we are not attempting to show linearity of NREC as a function of N for GR7, since L and D are both increasing functions of N for that test function.

The computational overhead per iteration which is in the original algorithm but not in the algorithm presented here consists primarily of the calculation of the cone intersections (See [3].)
At iteration j this involves the following three routines: first, j intersections of the cone at the current iterate with the previous cones at x_0 through x_{j-1} are computed. Each intersection results in an N-plane equation, so the number of operations is proportional to jN. The second routine, with approximately j^2 repetitions, eliminates some intersections which are clearly below the lower envelope, leaving say j' planes. Then $\binom{j'}{N}$ solutions of N plane equations and one quadratic cone equation are found. This gives an approximate total number of operations (ops) :

$$ops = O(jN + j^2 + \binom{j'}{N}N^2)$$

For obvious reasons this was unsatisfactory for N >> 4 so a later modification [4] uses a heuristic which eliminates pairwise intersections outside a neighborhood of the current iterate (and involves a sort - the jlogj below). This yields :

$$ops = O(jN + j\log j + N^2)$$

All of this overhead is eliminated by using a random point generator to find "corners" or points that are within the lower envelope.

Table 4 compares the deterministic and stochastic algorithms with respect to CPU times and accuracy for four standard test functions found in Dixon and Szegö [2]. Though it is not possible to equate exactly iterations for both programs, all runs were limited to 1000 iterations - equal to function evaluations in the stochastic case, usually many more than the number of function evaluations in the deterministic case (and thus number of records is not a meaningful measure for comparison). However, this equated as closely as possible the overhead the two algorithms have in common, that is, other than the overhead discussed above. In addition, as can be seen in Table 4, the errors for both were nearly the same, so that the difference in CPU times is a fairly accurate indication of the savings in overhead using the stochastic method.

V. Future Directions and Conclusions.

In order to apply the approach presented here to functions with larger numbers of variables, more work is necessary to decrease the overhead - the number of function evaluations and the CPU time - used in estimating the sampling region. The most promising next step appears to be to decrease the computations involved in determining the lower envelope, i. e. in deciding whether or not a random point is above the lower envelope. Another area that needs attention is the distribution on $[LB_j, UB_j]$ - whether or not changing it from the uniform diistribution to some other will yield better results.

The results presented here indicate that using Lipschitz cones with random search may be a fruitful technique. The combination results in low numbers of records because the use of the cones to give a lower envelope decreases the region from which random samples are drawn. Furthermore a substantial amount of overhead associated with the deterministic algorithm is eliminated with the stochastic approach.

Acknowledgement: The author wishes to thank G.R. Wood for alerting her to the Zabinsky and Smith results.

Bibliography

[1] L.Brieman and A.Cutler, "A deterministic algorithm for global optimization", Technical Report #224, Department of Statistics, University of California, Berkeley, CA (1989).

[2] Dixon, L.C.W. and G.P. Szegö, <u>Towards Global Optimisation 2</u>, North-Holland, Amsterdam (1978).

[3] R.H.Mladineo, "An algorithm for finding the global maximum of a multimodal, multivariate function", <u>Math. Prog.</u> 34 (1986) pp.188-200.

[4] R.H. Mladineo, "Supercomputers and global optimization", <u>Impacts of Recent Computer Advances on Operations Research</u>, R. Sharda et al. editors, Elsevier Science Publishing, Inc. (1989).

[5] R.H. Mladineo, "Vectorized random search vs. deterministic global optimization", Working Paper , Rider College, Lawrenceville, NJ (1990).

[6] R.H.Mladineo, "Randomized Cone Algorithm", preprint.

[7] F.J.Solis and R.J.-B.Wets, "Minimization by random search techniques", <u>Math. of Operations Research</u> 6 (1981) pp.19-30.

[8] G.R,Wood, "Multidimensinal bisection applied to global optimisation", <u>Math. Prog.</u> (1991).

[9] Z.B.Zabinsky and R.L.Smith, "Pure adaptive search in global optimization", <u>Math. Prog.</u> to appear (1991).

[10] Z.B.Zabinsky and R.L.Smith, "An adaptive random search with linear complexity in dimension", Technical Report 90-15, University of Washington (1990).

TABLE 1

GR7

$$f(x) = \frac{1}{N} \sum_{i=1}^{N} x_i^2 - \prod_{i=1}^{N} \cos(10\log(i+1)x_i) + 1$$

$-r_i \leq x_i \leq 1 - r_i$,
r_i, a random number,
x_i scaled to $[0,1]$.

N	L	ε	NREC	E	time	fe
2	5	.0003	9	21	2	310
3	7	.0064	8	24	11	745
4	9	.0064	6	33	395	5132
5	10	.0640	13	30	178	71358
6	12	.1040	15	36	1050	79645

N = number of variables
L = Lipschitz constant
ε = distance from minimum value
NREC = number of records attained by the algorithm
E = theoretical upper bound for NREC, with diameter
   $D = \sqrt{N}$
time = CPU time in seconds
fe = number of function evaluations

TABLE 2

CONVEX1

$$f(x) = \sum_{i=1}^{N} x_i^2 ,$$

$-10 \leq x_i \leq 10$, scaled to $[0,1]$

N	ε	NREC	E	time	fe
2	.00095	9	19	1	280
3	.0050	9	23	8	721
4	.0086	14	28	25	1283
5	.0310	17	29	181	3431
6	.0470	16	32	352	4413

Note: The Lipschitz constant was $L = 8/\sqrt{N}$.

TABLE 3

GP

R	NUMIT	ε	fe	time
50	1000	.00001	1000	7
50	20000	.000001	7234	434
5000	20000	.0000001	19976	75
5000	80000	.0000001	756	1207

R = number of random numbers generated at step 4
NUMIT = limit on number of iterations

TABLE 4

Comparison of Deterministic and Stochastic Algorithms

function	Deterministic		Stochastic	
	ε	time	ε	time
GP	.000008	94	.000008	7
RCOS	.0002	100	.0002	13
HART	.1	284	.09	7
SQRIN	1.8	555	1.65	3

Note: All runs were limited to 1000 iterations on the VAX4000.

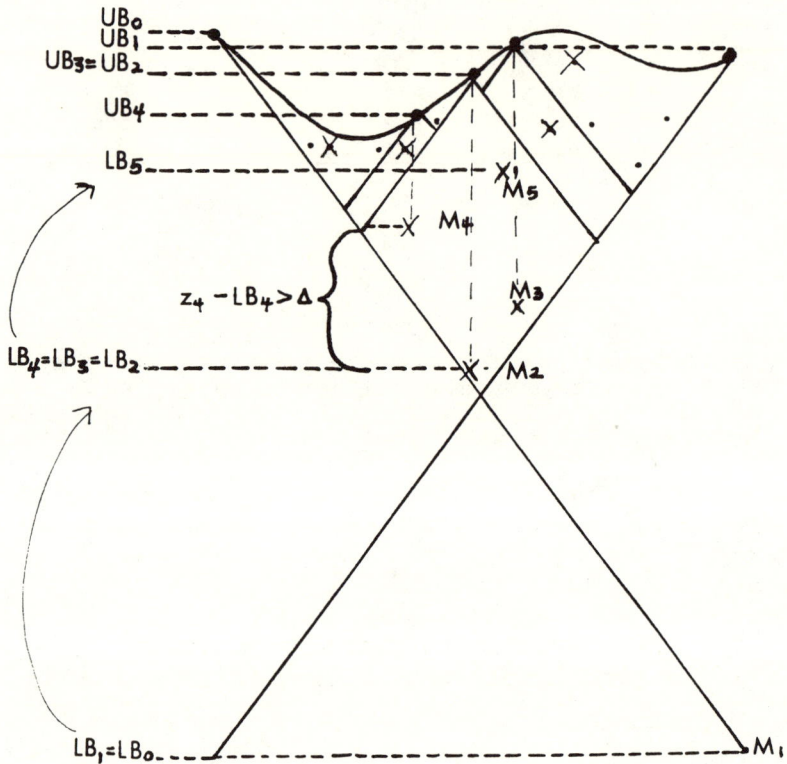

Figure 1

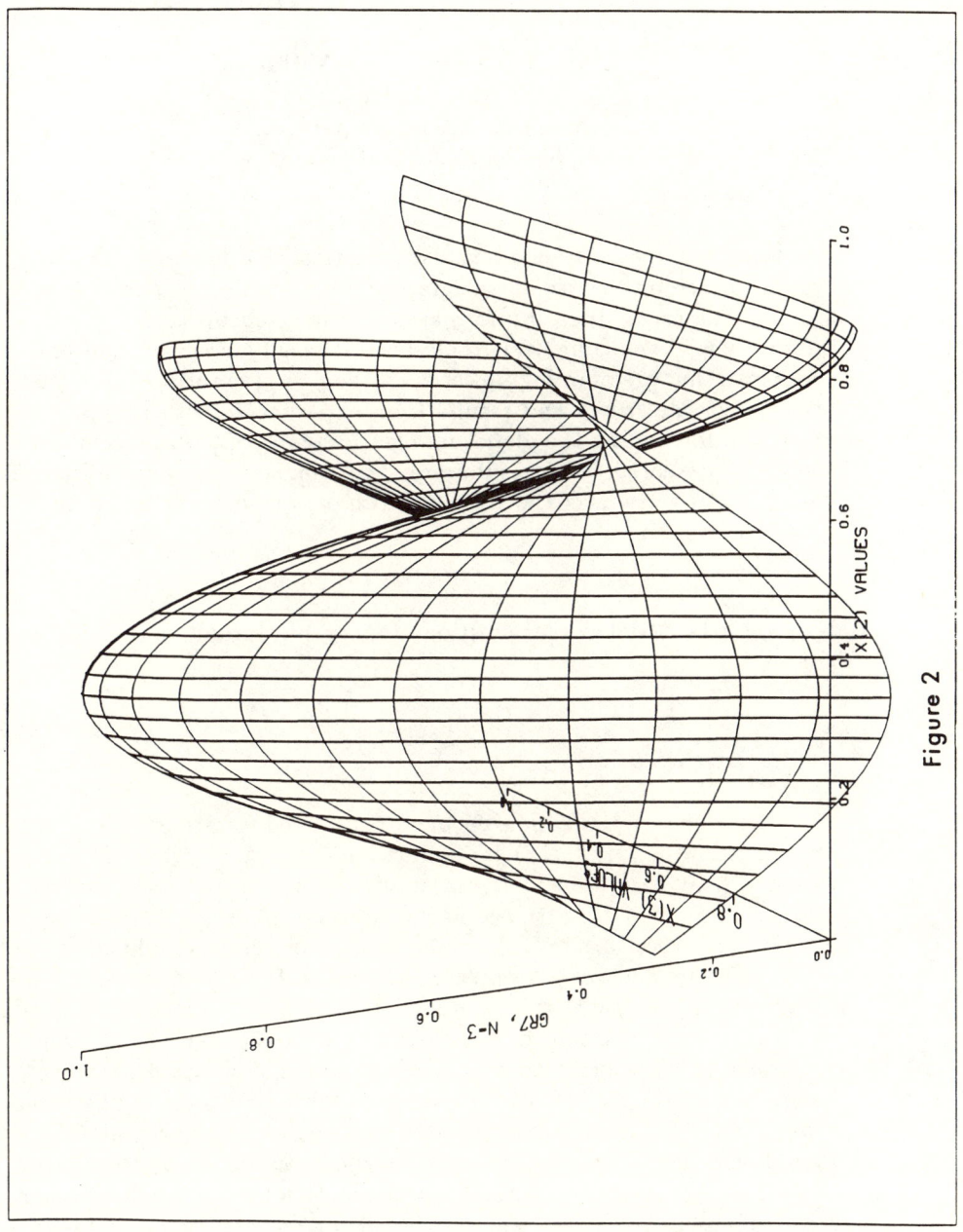

Figure 2

Topographical Global Optimization

Aimo Törn and Sami Viitanen
Åbo Akademi
Computer Science Department
SF-20520 ÅBO, Finland

Abstract: A method for global optimization of a function $f(x)$, $x \in A \subset R^n$ based on the use of topographical information of f in A is presented. The topographical information is extracted by sampling points uniformly in A and computing the k-nearest-neighbour graph containing those arcs that are pointing upwards on the $y = f(x)$ surface. The basic algorithm is described and results from experiments with a sequential algorithm for finding the global minima for some often used test functions are presented. The results are then compared to those obtained with parallel versions of the algorithm run on a transputer based multi-processor system.

Key words: Global optimization, topography graph, parallel algorithms.

1 Introduction

The global optimization problem considered here is characterized by the following. Let $f(x)$ be a function from $A \subset R^n$ to R. It is assumed that the problem is essentially unconstrained, i.e., that the global minimum of f is attained in the interior of A and has a basin with positive measure. Normally A would be a hyperrectangle indicating the region of interest but any A would do as long as A can be subscribed by a hyperrectangle H and that it is reasonably probable that a point sampled at random in H will fall in A, i.e., that the average number of points to sample in H in order to obtain one point in A is not prohibitively large. Based on computational experiences with an algorithm of similar type [Törn 1974], it can be expected that the algorithm presented here should work also in many cases with the global minimum on the closure of A. The success is then normally dependent on the local minimization algorithm used. The constraints A may be implemented as a user written boolean function, which returns the value **true** if $x \in A$. This function is then consulted both in the sampling and the local minimization step.

An inefficient but obvious way to tackle this problem is to start a local minimization from each sampled point. One strategy to solve the problem in a more efficient way is to try to find points in the basins of local minima and then start a local minimization from just one point in each identified basin. The key problem is how to transform the set of sampled points in such a way that identifiable subsets of points concentrated to the basins are obtained.

One class of methods of this type is the so called clustering methods. In these sampled points are concentrated to possible basins, these concentrations are identified (the points are clustered) and their corresponding local minima are determined by using some local optimization algorithm. The techniques for concentrating are: either leaving out unpromising points [Becker and Lago 1970; Törn 1974] or using a few steps of local optimization to bring the points nearer to the local minima 'attracting' them [Törn 1974]. A review of clustering methods can be found in [Törn and Žilinskas 1989].

The way to identify possible basins used here is based on representing topographical information about the objective function by a directed graph. The graph is connecting neighbouring points to each other by directed arcs pointing towards points of higher (alternatively lower) function values. If such a graph is covering enough it would be sufficient to start a local optimizer from each local minimum in the graph, i.e., from a node with no incoming arcs (alternatively a node with no outgoing arcs).

In this paper we investigate such an approach and present the basic algorithm, a parallelization of it and results obtained for some of the standard test functions [Törn and Žilinskas 1989].

The idea of exploring the topography of the objective function by connecting sampled points to aid in global optimization was first suggested as a topic worth further consideration in a review of clustering methods presented at the Sopron Symposium on Global Optimization in 1985 [Törn 1986]. A preliminary version of the method was presented at the Sopron Symposium in 1990 based on [Törn 1990].

2 The Basic Topographical Algorithm

The key problem to be solved in topographical global optimization is how to represent the topographical information of the objective function by a graph. The solution means answers to the questions of how to construct the graph and how to decide that the graph is good enough.

2.1 Constructing the Topography Graph. The construction of the topography graph is made in the following way. N points are sampled uniformly (either at random or even more uniformly) in A. The points are given the id:s $1, 2, ..., N$. For each point a reference list (a list of id:s) with references to the other points is constructed by ordering the points into nearest neighbour order. The list is further complemented by indicating if the reference is to a point with larger or smaller function value by giving the refrence a plus or minus sign respectively. The N reference lists constitute an $N \times (N-1)$ matrix, the *t-matrix* of the objective function given the points $1, 2, ..., N$. The submatrix $(N \times k)$ obtained by considering only the k nearest neighbours is called the *k-t-matrix*. The corresponding graph where arcs are drawn to the reference points with plus sign is called the k^+-*topograph*. The minima in the graph are all nodes with no incoming arcs.

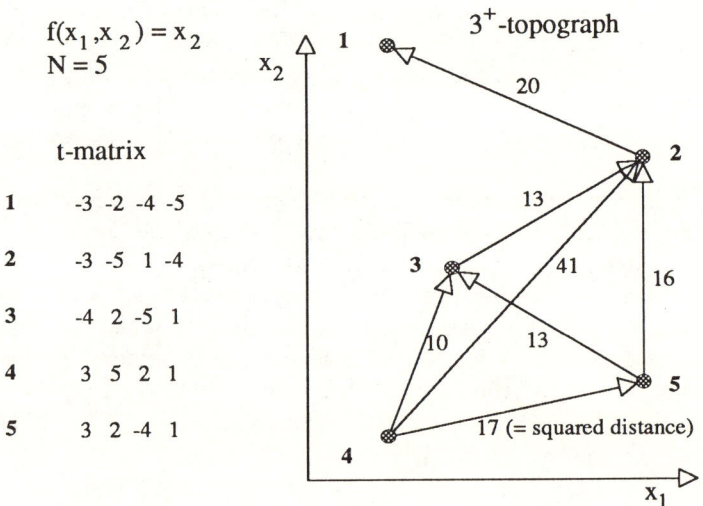

Figure 1. A t-matrix and its 3^+-topograph

In Figure 1 a t-matrix and its corresponding 3^+-topograph of the function $f(x_1, x_2) = x_2$ is shown. There are $N = 5$ points numbered 1–5. There is only one minimum, namely point 4. Looking at the 3-t-matrix we see that point 4 has only positive references, i.e., its k nearest neighbours have larger function values, and that no positive reference to point 4 exists in the 3-t-matrix, i.e. no other point has point 4 as a point with larger function value among its k nearest neighbours. Choosing $k = 2$ there are two positive reference vectors, namely for points 4 and 5 but in point 4 there is a positive reference to point 5 in the 2-t-matrix, which means that the only graph minimum also in the 2^+-topograph is point 4. In the corresponding 1^+-topograph there are two graph minima, namely point 4 and point 5.

Table 1
Number of graph minima, L and pos ref minima, L^* as a function of k

k	L	L^*	Max dist
1	43	53	2.3
2	21	24	2.5
3	10	15	2.9
4	6	(10)	3.1
5	6	8	4.1
6	6	7	4.1
7	6	6	4.2
8	5	6	4.5
9	4	5	5.4
10	4	4	5.5
20	2	3	7.4
1	35	48	2.1
2	9	16	2.5
3	7	(7)	3.3
4	4	4	3.3
20	3	3	7.4

The number of minima in the graph is thus dependent on k. Checking only the first condition, i.e., picking out points with positive references would possibly adds false graph minima to the true ones. However, the computations needed to find the graph minima is smaller when just considering positive references. A possible way to reduce the number of false minima would be to use a larger k. In order to investigate the dependence of the number of minima on k and the consequence of making the choice of graph minima based on just positive references a number of experiments for the test functions used in the optimization experiments were conducted. The results of the experiments were very similar. In Table 1 typical results are presented.

It can be seen from the table that the number of graph minima decreases rapidly for the first few values of k, and the simpler technique (positive references) really can be compensated by using a larger k. One can also see that the number of false minima decreases rapidly with growing k. The fourth column shows the maximum distance to a referenced point in the k-t-matrix. As expected this distance grows with k. The first 11 rows refers to an experiment were 100 points were sampled at random, and the last 5 rows to an experiment with a more uniform covering using 100 points. It can be seen that the more uniform distribution of points in the second experiment results in a more rapid decrease in the number of minima when k increases.

In Figure 2 and 3 the result from two experiments in constructing a topography graph for the Branin function

$$f_B(x) = \left(x_2 - \frac{5.1}{4\pi^2}x_1^2 + \frac{5}{\pi}x_1 - 6\right)^2 + 10\left(1 - \frac{1}{8\pi}\right)\cos x_1 + 10$$

in the region $A_B = (-5 \leq x_1 \leq 10, 0 \leq x_2 \leq 15)$ is shown.

The three local minima are marked M, and the positive reference graph minima L. Figure 2 shows the 4^+-topograph for 100 points sampled at random and Figure 3 the 3^+-topograph for 100 points resulting from sampling points at random but discarding any point with nearest neighbour within the distance 1.2. The distribution of points in Figure 3 is more uniform than in Figure 2 but the procedure required 1223 points to be sampled. A possible way to efficiently utilize the later technique is outlined in Section 4. The number of positive reference graph minima are 10 and 7 corresponding to the circled numbers in Table 1. In both cases there are graph minima in the basins of the three local minima of $f_B(x)$ in A_B.

2.2 Stopping Condition. In topographical global optimization the sampling phase aims at covering the region of interest as uniformly as possible.

How should the sampling be terminated? When only few points are sampled the resulting graph is rather crude and the graph minima will only represent some of the local minima. For an increasing number of sampled points the graph will approximate the objective function better and better and for resonably smooth functions it is expected that the graph minima will at some stage represent all essential local minima.

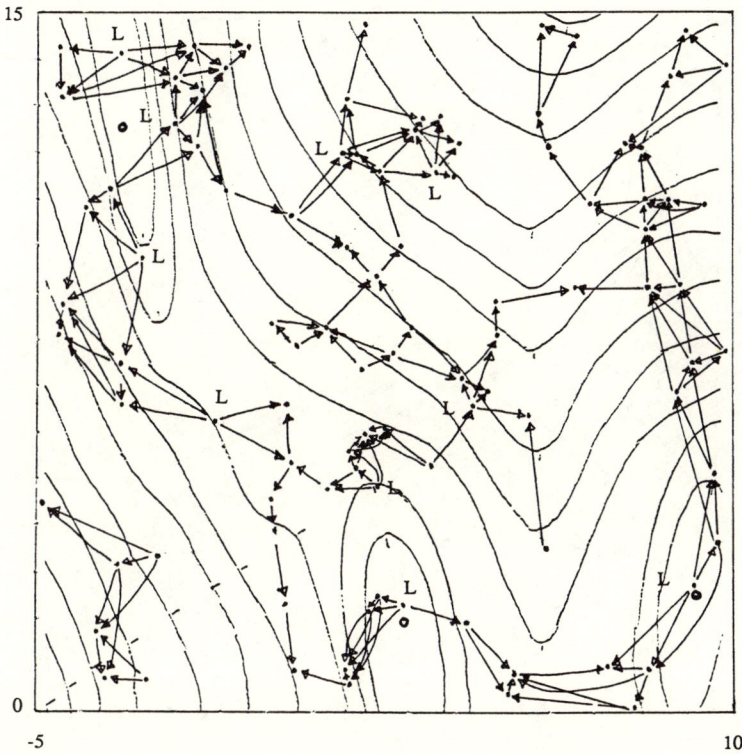

Figure 2. A 4⁺-topograph for Branin with 100 random points

The stopping condition should thus be based on the information that all parts of the region of interest have been explored, i.e., that the sampled points well cover the region. The number of points that need to be sampled in order for the probability, that at least one point will be in L to be no less than p is dependent on the relative size of this set. The probability is also dependent on how uniformly the sampled points cover the region of interest – random sampling could require several times more points than some more uniform technique for the probability to be the same [Törn 1974]. However, the size of L is normally not known and therefore this technique to determine the number of points to sample requires that an assumption about the size of L is made. Knowing the size of L the number of points to sample could be determined for instance as in [Törn 1978].

In order to cover the region well stratified sampling could be used [Törn 1978]. Another simple technique would be to specify a thresh-

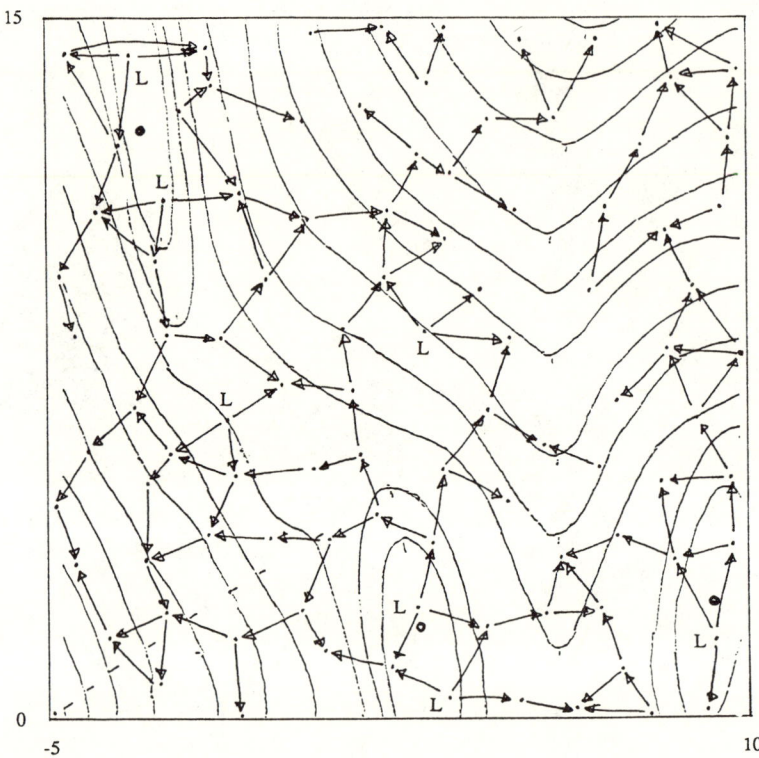

Figure 3. A 3^+-topograph for Branin, uniform covering, 100 points

old distance and discard any sampled point within this distance to an existing point. As the points cover the region better and better it will be increasingly difficult to find a point that will be accepted. Therefore a maximum number of discarded points could be specified. When this maximum is achieved the sampling could be terminated. Further, the user could be given the option to change the threshold value and start the sampling of additional points. By this the user would be given the possibility of successively obtaining a finer and finer uniform covering of the region of interest. This technique requires that a large number of distance computations are made. We will discuss later how this may be avoided.

All these stopping conditions determine the effort that will be applied in searching for a solution. For functions expensive to evaluate the effort will be measured in number of function evaluations. Normally the information about the optimiztion problem is insufficient to establish the relation of this number to the probability that a point in the basin of the global minimum is obtained. In a real application the stopping condition therefore in many cases would be given as the maximum number of affordable function evaluations (available resources). The resources available would be dependent on the savings expected in obtaining a good solution.

In the preliminary experiments rapported here the discarding technique was applied. The parameters were adjusted in such a way that the number of points in the graph were in the range 80-100. Rather few points were discarded (ten) so the points are just just slightly more uniformly distributed than uniformly random.

2.3 Local Optimization. On terminating the sampling the local minima of the graph are presented and and the local optimization stage is entered by starting local minimizations from a number of graph minima.a. Minimization is started from the three best minima in the graph and the corresponding minima found by the local minimization algorithm are presented. The user has then the option to start additional local minimizations from other local minima.

In the exploration of the topographical method the choice of local minimization algorithm is not a central question. Any local minimization algorithm could have been chosen. The minimization algorithm used in the experiments was IMSL's Uminf, which is a finit-difference gradient method.

2.4 Experimental Results. The resulting algorithm was implemented in Fortran 77 and experiments for the test functions Branin, Shekel 5 and Hartman 6 [Törn and Žilinskas 1989] were conducted [Juselius 1989]. In Table 2 a summary of the results are given. The success % is the number of runs giving the global minimum. The total number of function evaluations is the sum of those needed to construct the graph (= number of points in the graph) and those needed for local minimizations started from the three best local minima in the graph. T_1 is the 'standard time', i.e., the execution time divided by the time of 1000 Shekel 5 function evaluations. This standard time was introduced in [Dixon and Szegö 1968]in order to make the results obtained on different computers comparable.

No systematic experimentation in order to 'optimize' the method or to automate the choice of parameter values to be fixed by the user like the choice of threshold distance was performed. Compared to 23 other methods used to solve these problems (we exclude Solis 81 which stops when the known global minimum is reached) the number of function

Table 2
Average results with the basic topographical method based on 100 runs for three test functions

	Test function f		
	Branin	Shekel 5	Hartman 6
Dimensionality of f	2	4	6
Treshold distance	0.45	1.5	0.3
Points in graph	80	95	90
Local minima in graph	11	14	16
Success (%)	100	87	98
Function evaluations	104	170	168
T_1 (unit: 10^3 Shekel 5)	6.9	10.8	11.6

evaluations needed for topographical global optimization is the least (in most cases many times much smaller) with exception of one method for Shekel 5 [Törn and Žilinskas 1989].

When comparing methods one must remember that the stopping conditions used in different methods cannot be related to the probability that the global solution is found. Therefore the finding does not necessarily mean that our method is superior to the other methods. However we may make the conclusion that the method works and that it would be worth further testing and development.

One of the draw backs of the method is obviously the increasing work in adding new points to the graph. Some hierarchical approach [Betrò and Rotondi 1984] where the nearest neighbours to a new point would be searched in the subgraph of the closest local minimum in the graph could reduce the work. In the next section we explore another approach, namely to use a parallel computer to speed up the work.

3 Parallel Topographical Algorithms

In the basic algorithm the work to find the k nearest neighbours of sampled points grows for each sampled point. On a parallel computer the distance computations could be pipelined by distributing the sampled points evenly over the available processors. Also the local optimizations could be carried out in parallel.

3.1 Parallel Machines.
With the spreading supply of massively parallel machines super computing power becomes available at an order of magnitude cheaper than with the classical super computers. The reason for this is the use of a large number of identical relatively cheap processors and other components. A further important cost reducing factor is that these machines are operated more like a micro computer than like a large mainframe and therefore do not need operating staff.

The machine available at our department for the experiments was a configurable 100 transputer (T800) parallel computer HATHI-2. The HATHI-2 is available on a LAN connecting the computers of three local universities. The connection is through a SUN computer with a T800, the root transputer. The operating system on a T800 is a timesharing system which means that several tasks can be performed interleaved on a T800.

3.2 Parallel Algorithm Design.
The efficiency of the topographical method will mainly be determined by the time needed to compute the function values, the time needed to compute the t-matrix and the time needed for local minimizations. The function evaluations are independent of each other and can therefore be easily parallelized. The same is true for the local minimizations. The computation of the t-matix must be performed on the processor storing the points in the graph. One way to achieve parallelism in the t-matrix computation is to distribute the points evenly over a number of processors. By this arrangement the time needed to compute the t-matrix could be brought down by a factor equalling the number of processors used.

The programming language used to implement our parallel algorithms was Occam-2.

3.3 Basic Parallel Algorithm.
The configuration used for the parallel algorithm was a ring of 8, 16 or 32 transputers (nodes) starting from and ending up in the root, meaning that in total 1+8, 1+16 and 1+32 processors were utilized. The crude design of the algorithm is presented in the form of pseudo code for the root transputer and the node transputers, see below. SEQ means that the processes shown in the code with the same indentation are performed in sequence, whereas the processes indented after PAR are performed interleaved.

The pseudo code for the root transputer is [Viitanen 1991]:

```
SEQ
    PAR
        GENERATOR : Generates points in the region of interest.
```

SENDER : Has two functions. 1) Sends a generated point on the ring for k nearest neighbours (k-nn) determination. 2) Sends an accepted point on the ring for storing on successive nodes (node 1 follows after node n).
RECEIVER : Accepts k-nn results and sends an accepted point x to the process SENDER, and the point x and the k-nn information to the process POINTER.SETUP.
POINTER.SETUP : Calculates the function value $f(x)$ and updates the k-t-matrix.
RETRIEVE.MINIMA : Scans the distance table to find the local minima.
SORT: Sorts the minima, best first.
SEND.TO.NODES : Sends the local minima of the graph to the nodes in the ring to be used as starting points for local minimizations.
GET.LOCAL.MINIMA : Accepts results from the local minimizations on the nodes.
DISPLAY : Prints the result.

The pseudo code for the transputers in the ring is:

SEQ
  PAR
    CALCULATE.STORE : Calculates distances from a point received to all points stored on the node. Updates the k-nn information. Stores a point adressed to the node.
    RECEIVER : Receives data from its predecessor.
    SENDER : Sends data to its successor in the ring.
  GET.START.POINT : Accepts a starting point for local minimization.
  LOCAL.MIN : Performs a local minimization starting from the starting point received.
  REPORT.MIN : Sends the result of the local minimization to the root.

This description should be detailed enough for understanding how the basic parallel algorithm is realized.

3.4 Results. The test functions are the same as in Section 2. The results are reported for a configuration utilizing 8 nodes. No further speedup was obtained for configurations with 16 and 32 nodes. The explanation is that the distance calculations are rather cheap and therefore the time to distribute the tasks to be performed by the nodes becomes restricting in that not all processors can be kept working all the time.

This is nothing special for our problem it is a well known fact in parallel computation.

The results for 8 nodes are shown in Table 3. It can be seen that a rather good speedup has been obtained. Calculating the speedup as the ratio between the single processor solution time (Table 2), and the 1+8 processor solution time (Table 3), we obtain the figures shown in the last row of Table 3. The figures should be compared to the theoretical maximum, which should be in the range 8-9+.

Table 3
Average results with the basic parallel algorithm on 1+8 nodes based on 100 runs for three test functions

	Test function f		
	Branin	Shekel 5	Hartman 6
Dimensionality of f	2	4	6
Treshold distance	0.45	1.5	0.3
Points in graph	84	97	94
Local minima in graph	12	14	16
Success (%)	100	96	100
Function evaluations	142	177	208
T_{1+8} (unit: 10^3 Shekel 5)	1.2	1.6	1.7
Speedup (T_1/T_{1+8})	5.8	6.7	6.8

3.5 Modified Parallel Algorithm. In the basic parallel algorithm described above all the function evaluations were computed on the root. For functions expensive to evaluate the function evaluations should be performed in parallel. The basic algorithm was therefore modified in the following way [Viitanen 1991]:

- Points are sampled and accepted as in the basic algorithm. For each accepted point the corresponding pointer information is stored in the knn-matrix on the root.
- Then function evaluations are performed in parallel for the points stored on the nodes.
- The function values and their point-ids are sent to the root when ready.

- The k-t-matrix is computed using the knn-matrix and the function values. The graph minima are determined.
- The rest of the algorithm is unaltered.

The time with the altered algorithm for the test function Hartman 6 was further reduced from 1.7 to 1.5, which means a speedup compared to the sequential algorithm of 7.7.

4 Summary and Some Ideas

A new method for global optimization, topographical global optimization is presented. It consists of three conceptual steps.

Uniform random sampling of points in the region of interest A is used to explore the function in each subregion of A as well as possible. Ideally we want the sampling to be such that at least one of the sampled points lies in the basin of the global minimum. In order to force the sampled points to cover the region better than what would be the case for uniform random sampling a threshold value is defined. Any sampled point falling within his threshold distance to an existing point is discarded. When some predefined number of points have been discarded for a given threshold value this threshold value may be reduced and the sampling continued.

The second conceptual step is to determine a subset S of the accepted sampled points such that the points in this set represent the basins of all local minima 'detected' in the sampling step. This is done by constructing the topography graph which contains topographical information of the objective function represented by the accepted points. The topography graph is a directed graph with directed arcs connecting the accepted sampled points on a k-nearest-neighbour bases. The minima in the graph, i.e., those points that have no better neighbours are part of this topographical information. We let these minima be the set S. This is motivated by the observation that for an increasing number of accepted points and decreasing threshold these points would better and better approximate all local minima of any reasonably smooth objective function.

The third step is to determine the local minima represented by the set S. This is done by using the points in S (or a subset thereof) as starting points for local minimizations. The best point so determined is then given as the result of the global optimization effort.

Successful experiments with a sequential and a parallel realization of topographical global optimization were undertaken. In the parallel version the task of determining of nearest neighbours, the objective function evaluations task and the local minimizations task were all parallelized.

The following idea may be worth further consideration. The t-matrix used to find graph minima is mainly dependent on the sampled points. The function values of the points will only affect the sign of the references. This means that given a number of points covering the region of interest the matrix without signs could be computed once and for all and stored for later use. The points could either be stored or be regenerated for the function evaluations needed to obtain the complete t-matrix including also the signs. Such a procedure requires that the region of interest is in some standard form e.g. the unit hypercube. Also different matrices are needed for different selections of dimensions n and number of points N. For applications where similar global optimization problems have to be solved repeatedly such an approach seems to be practical. When applied in this way the overhead of the method is neglible even for large N.

Acknowledgements

The work of Christina Juselius in developing and implementing the sequential algorithm is gratefully acknowledged.

References

[Becker and Lago 1970] R.W. Becker and G.V. Lago, *A global optimization algorithm*, In: Proceedings of the 8th Allerton Conference on Circuits and Systems Theory (Monticello, Illinois), 3-12.

[Betrò and Rotondi 1984] B. Betrò and R. Rotondi, *A Bayesian algorithm for global optimization*, Annals of Op. Res. 1, 111-128.

Dixon, and Szegö 1978, L.C.W. Dixon and G.P. Szegö (Eds.), *Towards global optimization 2*, (North-Holland, Amsterdam), 363 p.

[Juselius 1989] C. Juselius, *A topographical method for global optimization*, M.Sc. Thesis, Computer Science Department, Åbo Akademi University, 113 pp (In Swedish).

[Törn 1974] A.A. Törn, *Global optimization as a combination of global and local search*, Ph.D. Thesis, Åbo Akademi, HHÅA A:13, 65 p.

[Törn 1978] A.A. Törn, *A search-clustering approach to global optimization*, In: [Dixon and Szegö 1978], 49-62.

[Törn 1986] A.A. Törn, *Clustering methods in global optimization*, Preprints of the Second IFAC Symposium on Stochastic Control, Vilnius, 19-23 May, Part 2, 138-143.

[Törn 1990] A. Törn, *Topographical global optimization*, Åbo Akademi, Reports on Computer Science & Mathematics Ser. A, No 119, 8 p.

[Törn and Žilinskas 1989] A. Törn and A. Žilinskas, *Global Optimization*, Lecture Notes in Computer Science 350 (Springer-Verlag, Berlin), 255 p.

[Viitanen 1990] S. Viitanen, *Parallelization of the topographical method for global optimization*, M.Sc. Thesis Project, Computer Science Department, Åbo Akademi University, (In Swedish).

Lipschitzian Global Optimization: Some Prospective Applications[*]

János Pintér[+]

Abstract

A great number of practically important issues in sciences, economics and engineering leads to constrained global optimization problems (CGOP's). In this paper a particular attention is paid to such GOP's, in which the available structural information may be rather limited (so that only the *Lipschitz-continuity* of the defining functions can be postulated); furthermore, in which the objective function and/or some constraints are to be evaluated by a computationally intensive procedure. The discussion, however, is not restricted to these problem-types. Following the brief presentation of a general problem statement and solution approach, a number of existing and prospective applications of Lipschitz optimization are highlighted.

---

[*] A significant portion of the research surveyed in the present paper was supported by and carried out at the following institutions:

Water Resources Research Centre (VITUKI), Budapest, Hungary;
Delft University of Technology (TU Delft), The Netherlands;
National Institute for Inland Water Management and Waste Water Treatment (Rijkswaterstaat RIZA), Lelystad, The Netherlands;
Wageningen Agricultural University (LU Wageningen), The Netherlands.

The cooperation with many colleagues from the above institutions - that has led also to a number of joint works referred to - is gratefully acknowledged. The author also wishes to thank two anonymous referees for their valuable clarifying comments.

[+] Present (temporary) address: National Institute for Inland Water Management and Waste Water Treatment (Rijkswaterstaat RIZA) P.O.Box 17, 8200 AA Lelystad, The Netherlands.

1 Introduction

In recent years, a broad spectrum of strategies has been proposed to solve the *global optimization problem (GOP)*, stated in the general form

(1.1) $\min f(x)$
        $x \in D \subset R^n$

under its diverse (further) specifications. The monographs and surveys provided by Dixon and Szegö,eds. (1975, 1978), Strongin (1978), Fedorov, ed. (1985), Pardalos and Rosen (1987), Ratschek and Rokne (1988), Törn and Žilinskas (1989), Horst and Tuy (1990) and the numerous references therein describe a great variety of GO approaches. These differ considerably with respect to their *prior models* of the GOP, the *search/optimality criteria* applied and the corresponding *solution algorithms* derived.

In this paper we shall consider the following specification of (1.1):

(1.2) $\min f(x)$
        $x \in D = \{ a \leq x \leq b : g_j(x) \leq 0 \quad j = 1,...,J \} \subset R^n$

under the conditions postulated below.

C1. The set D is *robust (body-like)*, i.e. it is the closure of a bounded non-empty open set of the Euclidean n-space R^n.
C2. The objective function $f \equiv g_0$ and all constraints g_j are *Lipschitz-continuous* on the n-dimensional interval $D_0 = [a,b]$; that is, for an ar-

bitrary pair of points $x_1, x_2 \in D_0$ there hold the inequalities

(1.3) $\quad |g_j(x_1) - g_j(x_2)| \leq L_j \|x_1 - x_2\| \qquad j = 0, 1, \ldots, J$

here $L_j = L_j(g_j, D_0)$ is the Lipschitz-constant of g_j on D_0 and $\|.\|$ is the Euclidean norm. (Note that other ℓ_p-norms ($1 \leq p \leq \infty$) are also used in certain cases, when postulating the relation (1.3).)

C3. The objective function f is (potentially) *multiextremal*; it has at most a *countable* number of isolated global optimizers on D. The set of globally optimal solutions will be denoted by X^*.

That is, one seeks the absolute (*global*) minimum of the possibly multi-extremal function f on the feasible region D: here D is defined by the given (componentwise) explicit bounds $a, b \in R^n$ which determine a finite interval region in R^n, and by a finite number of *Lipschitzian* inequalities. In the frames of the present paper we shall be particularly interested in GOP's (1.2), in which the objective function (f) and/or constraint (g_j) evaluations may be "expensive" (as being produced by some "black box" operations e.g. via running an imbedded algorithm, a program package, or by accomplishing numerical differentiation/integration, Monte Carlo simulation etc.). Under such circumstances, the available structural information with respect to (1.2) may also be fairly limited. Consequently, most of the specifically "tailored" GO approaches may be difficult or even impossible to apply, as their structural requirements can not be readily verified, or simply are not fulfilled. At the same time, the assumptions C1-C3 (including the *Lipschitz-continuity* supposition concerning the defining functions f, g_j) frequently hold. In such cases, the problem at hand can be treated in the frames of Lipschitzian global optimization.

In a sequence of papers (Pintér, 1983, 1986 a,b,c, 1987, 1988, 1990c) a general class of *adaptive partition strategies* was suggested for solving diverse special cases of the Lipschitzian GOP statement presented above. Strong theoretical convergence properties (the *identity* of the accumulation point set of the search sequence generated and the set X^*) were established; moreover, implementation aspects were investigated in detail. One can list a large number of related theoretical studies, special cases (algorithm instances), modifications and extensions to solve GOP's with *Lipschitzian* (Basso (1985), Boender (1984), Danilin-Piyavskii-Shubert (1967,1972), Fujii-Ichida-Ozasa (1986), Galperin (1988), Hansen (1979,1980), Meewella-Mayne (1988,1989), Mladineo (1986), Ratschek (1985), Schoen (1982), Shen-Zhu (1987), Strongin (1989), Wood (1985)) or *continuous* (Kushner-Archetti-Betró (1964,1979), Mockus-Tiesis-Žilinskas (1978), Žilinskas (1981)) objective function. Moreover, comprehensive convergence investigations of *deterministic* GO algorithm-classes were recently accomplished also by Hansen, Jaumard and Lu (1989), Horst (1986, 1988), Horst and Tuy (1987, 1990). Therefore all theoretical details will be omitted here, and the primary emphasis is placed on highlighting several existing and prospective applications of Lipschitzian GO.

2 Some Possible Applications of Lipschitzian Optimization

2.1 Nonlinear Approximation

The "closest" approximation of a given function by means of a parameterized family of functions is an essential problem-type in numerical analysis, cf. e.g. Braess (1986). Below we shall consider this problem in

a fairly general setting and show the relevance of Lipschitzian global optimization for its solution.

Let $S \subset R^m$ denote the set on which the Lipschitz-continuous function $h(y)$ is to be approximated. We shall assume that S is non-empty, closed and bounded, being defined by a finite number of inequalities:

(2.1) $S = \{ y \in R^m : g_j(y) \leq 0 \quad j = 1,\ldots,J \}$

where all functions g_j are *Lipschitzian* in some neighbourhood of S. The approximation $H(x,y) \simeq h(y)$ $H : R^{n+m} \to R^1$ is sought in a Lipschitzian n-parameter family of functions: $x \in D \subset R^n$ is the parameter-vector to be chosen from the Lipschitzian robust set D. Assuming that the discrepancy between H and h is expressed by a given L_p-norm (for some fixed $1 \leq p \leq \infty$), the *nonlinear approximation problem* - specified by h, H, S, D and p - can be stated as

(2.2)
$$\min_{x \in D} L_{p,S}\{H(x,y), h(y)\}, \quad \text{where}$$

$$L_{p,S}\{H(x,y), h(y)\} = \left(\int_S |H(x,y) - h(y)|^p \, dy\right)^{1/p}$$

Note that well-known special cases of the general problem (2.2) include *Chebyshev-approximation* (corresponding to $p=\infty$) or the *least squares estimation problem* ($p=2$).

As it is known, the formulation (2.2) does not imply the *uniqueness* of the optimal solution (see e.g. Braess (1986) or Diener (1987a)). More precisely, in general, an unknown number of *locally optimal* approximations may exist: thus most "traditional" numerical methods are, at best, capable of finding one of these. Turning therefore to the selection of appropriate GO strategies, without further detailed analysis it may be difficult (if at all possible) to find out or check special structural features of the objective function

$$L_{p,S}\{H(x,y), h(y)\}$$

(such as e.g. convexity, concavity, differential convexity etc.). Therefore it is often impossible to apply the corresponding, more specific optimization approaches. At the same time, it is not difficult to verify that the objective function in (2.2) is *Lipschitz-continuous* with respect to x. This way, Lipschitzian global optimization methodology can be directly applied to solve (in principle) *arbitrary* instances of the general problem-class described by (2.1)-(2.2), *without stating additional structural requirements*. Analogous reasoning applies with respect to a number of special nonlinear approximation problems and other (similarly complicated) GO tasks that will be shortly discussed below.

2.2 Solution of Systems of Nonlinear Equations and Inequalities

Consider the system of Lipschitzian nonlinear equations

(2.3) $h_k(x) = 0$ $k = 1,\ldots,K$ $x \in D \subset R^n$

Evidently, (2.3) is a special case of the general approximation problem, defining formally $h \equiv 0$ (hence, the argument y is absent) and replacing the objective (discrepancy) function in (2.2) by

(2.4) $H(x) = \left(\sum_{k=1}^{K} | h_k(x) |^p \right)^{1/p}$ ($1 \leq p \leq \infty$ arbitrary)

Note that the *existence* or *uniqueness* of the solution to the system (2.3) is *not* postulated. Due to the paramount importance of this problem (cf. e.g. Forster, ed. (1980), Robinson, ed. (1980), Garcia and Zangwill (1981), Allgower and Georg (1983) or Dennis and Schnabel (1983)), many "classical" approaches - standard univariate methods, fixed point algorithms, continuation procedures, quasi-Newton methods or other *convex optimization* techniques - have been widely applied to its solution. It is well known, however, that the solution strategies listed (because of their local - as opposed to *global* - search scope) may not converge to the "best" solution(s) of (2.3).

Global optimization for solving nonlinear equations was proposed already by Branin (1972); numerical test results are also presented e.g. by Strongin (1978), Vysotskaya and Strongin (1983), Diener (1987b), Horst and Thoai (1988), Pardalos and Rosen (1988). In Pintér (1990a) several hundred *randomly generated* equation systems (up to five variables) were solved, via globally minimizing the corresponding *Lipschitzian* objective function (2.4) (setting p equal to 1 or 2). The first problem-class studied consists of trigonometric systems of equations with an additional quadratic term:

$$h_i(x) = \sum_{j=1}^{n} A_{ij}\sin^2[B_{ij}(x_j-x_j^*)] + \sum_{j=1}^{n} C_{ij}\cos^2[D_{ij}(x_j-x_j^*)+\pi/2] + (x_i-x_i^*)^2 = 0$$

(2.5) $\qquad i=1,\ldots,n$

$$x \in D = \{ x=(x_1,\ldots,x_n)^T \quad -\pi \leq x_i \leq \pi \quad i=1,\ldots,n \} \qquad n=1,2,3$$

(The parameters A_{ij}, B_{ij}, C_{ij}, D_{ij} and x^* are generated according to fixed pseudo-random rules; this way - besides the global optimum x^* - a realization-dependent number of locally best "solutions" with non-zero residuals exists.) The second problem-class analysed has the form:

$$h_i(x) = (x_i-x_i^*)^2 - \sum_{j=3(i-1)+1}^{3i} \frac{1}{\|x-d_j\|^2 + c_j} + k_i = 0 \qquad i=1,\ldots,n$$

(2.6)

$$x \in D = \{ x=(x_1,\ldots,x_n)^T \quad 0 \leq x_i \leq 10 \quad i=1,\ldots,n \} \qquad n=1,\ldots,5$$

(Again, the problems (2.6) are generated sequentially, applying a pseudo-random generator to obtain x^*, c_j, d_j and k_i : note that besides the unique global optimizer x^*, these problems have 3n locally best "solutions".) The numerical results presented in Pintér (1990a) indicate *robustness* of the global optimization methodology suggested; moreover, although on average an exponential increase of the computational demand could be observed, the solution *efficiency* seems to be acceptable. (A detailed description and evaluation of the numerical experiments can be found in the paper mentioned.)

Closing the above brief discussion, let us remark that (Lipschitzian) global optimization can be applied also to find the solution(s) of systems of *nonlinear equations and inequalities*, cf. e.g. Horst and Thoai (1988), Hendrix and Pintér (1991).

2.3 Calibration of Nonlinear Descriptive Models

One of the basic problems in scientific modelling is the *numerical calibration* of (preidentified) descriptive model structures, i.e. the selection of a set of model parameters in "best agreement" with the available measurements (see e.g. the related discussions and some recent applications in Beck (1987), Csendes (1988), Hendrickson, Sorooshian and Brazil (1988), Loehle (1988), with additional references therein). In Pintér (1990b) a detailed formulation and analytical investigation of the calibration problem is presented: the essence of that study and some applications are summarized below.

In mathematical terms, the *parameter estimation problem* is the following. Given a quantitative descriptive model of the (environmental, ecologic, economic etc.) system in question, find that parameterization x^* which assures that the model output values $\hat{y}_t(x^*)$ are in "best possible match" with the corresponding set of observations y_t t=1,...,T. Recalling our brief exposition on nonlinear approximation, this requirement can be formulated as a GOP:

$$(2.7) \qquad \min_{x \in D} f(x) \quad, \quad \text{where} \quad f(x) = \left(\sum_{t=1}^{T} |\hat{y}_t(x) - y_t|^p \right)^{1/p}$$

In (2.7) D is the region of *a priori* admissible parameter settings and f expresses the overall discrepancy between $\{\hat{y}_t(x)\}$ and $\{y_t\}$ (again, p determines an appropriate ℓ_p-norm). We assume that

1) the feasible model parameterizations belong to a finite interval region D of R^n (this assumption is typical in many practical calibration studies); and

2) the model output and the observations are *discretized*, *deterministic* and *scalar* quantities (in other words, the aspects of *stochasticity* and *multiobjectivity* are properly incorporated, or can be neglected).

In spite of the strong simplifying assumptions made, (2.7) is typically a hard optimization problem. The reasons of this are not only the frequent *multiextremality* of f(x), but also its *analytically unknown* form. In many cases, the values $\hat{y}_t(x)$ are to be produced by carrying out a large amount of "black box" operations (such as e.g. producing the approximate numerical solution of a system of differential equations, or running an "imbedded" computer program). Consequently, such global optimization techniques are to be applied which practically do *not* need structural information concerning the analytical properties of D. Recall now in this context, that if the discrepancy measure f depends smoothly on x and D is a compact (robust) subset of R^n, then the Lipschitzian solution approach can be directly used.

To mention some applications of Lipschitzian GO to model-fitting, reference is made first to the calibration of a sediment-water interaction model (cf. Pintér, Szabó and Somlyódy (1986) for details). This model serves for describing the temporal evolution of suspended solids concentration in the water of Lake Balaton (Hungary), as being influenced by varying wind conditions. The evaluation of the "goodness" of an arbitrary model parameterization (expressed by $f^2(x)$ for p=2 in (2.7)) involves in this case the approximate numerical solution of a differential equation. The model in question is highly nonlinear and has three important parameters to be selected. Applying Lipschitzian optimization, some 20 percent accuracy improvement was attained, when comparing the results to the "best" one obtained by repeated applications of the (frequently used, local scope) Levenberg-Marquardt algorithm.

As for a second example, let us also mention the calibration of the model system DELWAQ-IMPAQT for Lake Ketelmeer (The Netherlands). This model serves to describe the fate of micropollutants in aquatic systems. The program code of DELWAQ-IMPAQT consists of several thousand FORTRAN source lines and its *single* run - i.e. the evaluation of a parameterization chosen - takes minutes on a PC (depending on the accuracy of the numerical solution of the underlying descriptive model). DELWAQ-IMPAQT has several tens of parameters that, in principle, can all be subject to calibration. Given the computational demand indicated, "exact" model-fitting is prohibitively expensive. Therefore – on the basis of a preliminary sensitivity analysis – only the *four* most important parameters were optimized, in several consecutive runs. In this calibration study, the *average relative error* in ℓ_1-norm, i.e. the objective function

$$(2.8) \qquad f(x) = \frac{1}{T} \sum_{t=1}^{T} |\hat{y}_t(x) - y_t| * y_t^{-1} \qquad (y_t \neq 0 \quad t=1,\ldots,T)$$

was minimized. As a result of applying global optimization, the value of the objective function (2.8) improved by some 34 per cent, when compared to the result obtained using (previously applied) "expert parameter values". Note that further improvements could be expected, calibrating also the other – probably less significant – parameters. (For additional details concerning this study, see ten Hulscher, Bak and Pintér (1991).)

2.4 Parameterization of Statistical Models

A common problem in applied probability and statistics is the evaluation (fitting) of the parameters of a preselected probability distribution function (pdf) form, on the basis of a given sample (see e.g. Gourdin, Jaumard and MacGibbon (1990) for a recent discussion, with many references). Let $\xi = \xi(x)$ be a random variable (rv), its pdf $F_x(z) = P(\xi(x) < z)$ being parameterized by the n-vector x. Let z_s s=1,...,S be a finite sample drawn from the (unknown) pdf of ξ that serves as a basis to estimate x. (For simplicity, only the case of *scalar* ξ is discussed here.) As it is well-known, there exist several *non-equivalent* ways to formulate the pdf parameterization problem: three such approaches will be highlighted below. (With respect to the relevance of *Lipschitzian* GO, cf. the discussion of Section 2.1, *mutatis mutandis.*)

1) The *moment estimation technique*, in general, leads to a system of nonlinear equations:

$$(2.9) \qquad \frac{1}{S} \sum_{s=1}^{S} (z_s^k) = E\{\xi^k(x)\} \qquad x \in D$$

(here E is the expected value operator and D is the set of feasible parameterizations). Consequently, this formulation will, in general, induce a multiextremal optimization problem, although there exist more easily solvable important special cases.

2) The *maximum likelihood estimation method* leads to parameterizations that have statistically well established properties. This method is based on the maximization of the function

$$L_x(z_1,...,z_S) = \prod_{s=1}^{S} f_x(z_s)$$

where $f_x(z)$ is the *parameterized* density function of ξ. Again, although in a number of special cases the optimization problem

$$(2.10) \qquad \max_{x \in D} L_x(z_1,...,z_S)$$

has a single locally (thus also globally) optimal solution, this is not the case in general. To illustrate this point, note that e.g. the fitting of a three-parameter *lognormal* or *Weibull* pdf to empirical data leads to respective global optimization problems, cf. e.g. Wingo (1984) or Gourdin, Jaumard and Hansen (1990).

3) *Optimized pdf model parameterizations* can be based also on the use of an appropriate difference measure ρ that expresses the "discrepancy" between the *parameterized* theoretical pdf and the empirical (sample-based) pdf $F^{(e)}$. This approach leads to the general problem-type

(2.11) $$\min_{x \in D} \rho\{F_x(z), F^{(e)}(z)\}$$

The function ρ can be defined in a number of widely different ways: examples are the Kolmogorov-Smirnov, Sherman or von Mises statistics, L_p- -norm induced distances, relative entropy etc. (a number of appropriate functions ρ are collected in Pintér and Cooke (1987) and in Klafszky, Mayer and Terlaky (1989)). Note that the model form (2.11) can be applied e.g. in connection with environmental management problems, in which the *decision-dependent* pdf F_x of the resulting environmental state

has to approximate a target "ideal" pdf: see Pintér and Somlyódy (1986) for an application in lake water quality management.

2.5 Product (Mixture) Design

Consider a finite number of base materials (components), and denote (a fixed volumetric unit of) them by $v_0, \ldots, v_n$. Assuming that all mixtures of the form

(2.12) $$x = \sum_{j=0}^{n} a_j v_j \qquad a_j \geq 0 \qquad (\sum_{j=0}^{n} a_j > 0)$$

can, in principle, be produced, one may be interested in finding the "best" product mixed by the rule (2.12). We shall suppose that the quality and overall performance of each mixture product can be expressed by a given (analytically, algorithmically or experimentally computable) function f: that is, the "value" of mixture x equals f(x).

As it is seen from the above brief problem statement, there are no specific restrictions, concerning the definition of the components (they can be e.g. base materials used by some food processing industry or by a chemical company etc.). In mathematical terms, the only essential requirement is the admissibility of all possible mixtures: that is, the components $v_0,\ldots,v_n$ are modelled as elements of some (n+1)-dimensional linear space X. Suppose now that the quality of an arbitrary mixture x is identical to that of its "standardized" x_{st}, defined by

(2.13) $$x_{st} = \sum_{j=0}^{n} \alpha_j v_j \qquad \alpha_j = \frac{a_j}{\sum_{j=0}^{n} a_j}$$

This way, we can model the possible mixtures as points of a (non-degenerate) simplex S in an n-dimensional subspace of X:

(2.14) $$S = \{x \in X : x = \sum_{j=0}^{n} \alpha_j v_j , \quad \alpha_j \geq 0 , \quad \sum_{j=0}^{n} \alpha_j = 1\}$$

and the problem of finding the optimal product can be deduced to that of (say) minimizing f(x) on the simplex S.

It is important to remark that in a great number of practical cases, the exact analytical dependence of f on the components of the product, i.e. on the set of values $\{\alpha_j\}$, is not known, and the quality of the mixtures obtained is to be evaluated *experimentally*. Notwithstanding – on the basis of scientific or practical knowledge – the *smooth* behavior of the performance indicator f, with respect to the values $\{\alpha_j\}$, may frequently be known (while the use of further structural information concerning f may be tedious or impossible). In such cases, it seems to be rational to apply Lipschitzian GO for solving the mixture design problem.

As a special instance from this general problem-class, one can mention the work of Klafszky, Mayer and Terlaky (1989), in which the performance indicator function f(x) is determined by the "discrepancy" between mixture x and a prespecified "ideal (target) product" x_{ideal}. They investigate the case, in which the base materials and the target can be modelled by discrete probability distributions and the discrepancy measure is defined by one of several so-called statistical divergence functions (cf. Csiszár (1975)).

Let us remark additionally that the proper consideration of diverse restrictions on the feasible mixtures can also be important (for example, the rate of certain base materials has to be between given bounds, their unit rate applied has to be bounded etc.). In many cases, these constraints can be expressed by a finite number of Lipschitzian relations.

This way, one faces a constrained Lipschitzian optimization problem on an imbedding simplex region and the general partition algorithm framework, tailored to this problem, can be applied (Pintér, 1990c). A closely related application - finding a feasible mixture that has "satisfactory" properties (expressed by a system of indefinite quadratic inequalities) - is solved by Hendrix and Pintér (1991) along these lines.

2.6 Combining Negotiated Expert Opinions

There are different reasons to combine expert opinions, related e.g. to the location of some facilities, the potential hazard of a waste disposal site, the uncertain results (yield) of a natural resources exploration project, the risk of a new ("non-standard") environmental management option etc. First, as experience shows, the combination of different opinions generally produces improved results: more stable, accurate and reliable estimates, reduced random forecasting errors, more realistic ranges for the estimated values and so on. Second, a decision-maker often has to ask and consider the opinion of more than one individual (e.g. a committee or board etc.) and then, rather than deciding which opinion is the "best", combine them to arrive at some "generally acceptable" compromise. Third, lack of data often creates a situation when one has to rely, to a significant extent, on expert estimates. For a more complete exposition concerning this important subject, see e.g. Pintér and Cooke (1987) and the works cited therein. In that report a fairly general optimization approach is presented for aggregating individual "opinions" (modelled as point values, vectors or probability distributions) in some quantitatively "best" sense. The framework proposed subsumes a variety of context-dependent realizations. The short discussion below will be restricted to the combination of expert opinions, in order to arrive at a compromise solution: this will be called the "group decision problem" (as opposed to the "forecasting problem").

In mathematical terms, one can state a prototype group decision making problem as follows. Given the opinions $a_1,...,a_n$, the expert weights $w_1,...,w_n$, the pooling operator (formula) F and the difference measure ρ, find the pooling parameter vector $x=(x_1,...,x_n)$ in the set D that solves the optimization problem

(2.15) $\qquad \min_{x \in D} \| w_0 \rho(a_0, a_c),...,w_n \rho(a_n, a_c) \| \qquad a_c = F(a,x) \in A$

where $\|.\|$ is a fixed norm functional on R^n. In other words, for the optimal parameterization x^*, $a_c = F(a, x^*)$ minimizes in norm the weighted difference between the experts' opinions and the combined opinion. As it is seen from the general problem statement (2.15), its concretizations depend primarily on the selection of the norm $\|.\|$, the set of feasible opinions A, the corresponding discrepancy measure ρ and the pooling operator F: these specifications can be investigated separately (see our report mentioned above). Let us only remark that in a number of practically relevant cases - for example, when the parameters x can be interpreted as "weight factors" ($x_j \geq 0$ $\sum_0^n x_j = 1$) - one can deduce the formulation (2.15) to a (potential) global optimization problem on a simplex. In Pintér and Cooke (1987) illustrative numerical results are presented for solving group decision-making problems via Lipschitzian optimization, when the "opinions" are represented by real vectors from a (two-dimensional) interval region.

2.7 Data Classification (Cluster Analysis)

The "best" partition of a finite set of entities into groups of "similar" objects (called clusters) is often required in operations research and statistics. To mention a few examples, one may refer to facility location problems (cf. e.g. Florian, ed. (1984)), data classification (Gordon (1981), Kaufman and Rousseeuw (1990)) or to certain optimization algorithms (Rinnooy Kan and Timmer (1987a,b)).

Following the notation of Hansen and Jaumard (1987), let us define

$O = \{ O_k \quad k=1,\ldots,N \}$ the set of entities to be partitioned

$D = \{ d_{kl} \quad k,l=1,\ldots,N \}$ the set of dissimilarities $d_{kl} = d(O_k, O_l)$ between all pairs of entities

$P_M = \{ C_1, \ldots, C_M \}$ a partition of O into M clusters (assuming $C_j \neq \emptyset$, $C_i \cap C_j = \emptyset$, $\cup_j C_j = O$)

Π_M the set of all possible partitions P_M

With the notation above, a prototype problem of "optimal set partition" (OSP) can be verbally stated as follows:

Find a partition $P_M \in \Pi_M$ which provides a "most homogenous" (or, alternatively, "most discriminative") classification of the set of entities O, as expressed in terms of the dissimilarities D.

Clearly, the problem statement (OSP) can be quantitatively expressed in a number of *non-equivalent* ways, see e.g. Hansen and Jaumard (1987) or

Pintér and Pesti (1991) for several examples.

The *"exact"* solution approaches to the clustering problem (based on combinatorial optimization techniques) encounter serious numerical difficulties, for most realistic problem-sizes (cf. e.g. Brucker (1978) or Welch (1983)). In order to avoid the complications referred to, a large number of *heuristic* solution approaches have also been suggested: being typically based on problem decomposition and "myopic" classification rules, these often lead to strongly *suboptimal* decisions.

Assuming that the entities can be represented by real vectors, a new algorithmic solution strategy was proposed by Pintér and Pesti (1991). Following this approach – and supposing that the number of clusters is known (cf. e.g. Rousseeuw (1987) on the latter aspect) – the choice of "cluster seed points" is globally optimized. The evaluation of the seed point sets is accomplished by an imbedded algorithm (applying a simple rule such as e.g. "entities are classified to that cluster whose seed point is closest"). Again, if the quality of classification depends in a *smooth* manner on the seed point configuration F, and the sets F belong to a *compact* feasible region, then *Lipschitzian* GO methodology can be directly applied, in essence, independently of the concrete form (and further structural analysis) of the imbedded clustering rule.

The numerical results reported in Pintér and Pesti (1991) indicate the practical viability of the Lipschitzian GO approach. For illustration, let us mention that e.g. 200 pseudorandomly generated points (which could come from any of five different, truncated normal probability distributions defined in the unit square) were properly classified into five clusters in less, than 7 minutes (using an IBM PC/AT compatible machine).

2.8 Black Box Design of Engineering Systems

As already indicated by several examples, in many complex decision problems a number of analytically non-tractable, uncertain, unforeseen or not completely known factors may play a significant role. Therefore the *performance* of a *decision alternative* depends also on these factors and the decision variants are often to be evaluated by a computationally intensive procedure (running an imbedded algorithm or program code, accomplishing numerical approximation methods or stochastic simulation etc.). This way, the system in question often can be conceived as some sort of "oracle" or "black box". Many such problems arise, *inter alia*, in the fields of

- mechanical, chemical, electrical, aerospace etc. engineering (cf. e.g. Strongin (1978), Wilde (1978), Johnson (1979), Henley and Kumamoto (1981), Dempster and Coupé (1982), Muroga (1982), Hu and Kuh, eds. (1988), Hansen, Jaumard and Lu (1988), Cooke and Pintér (1989), Törn and Žilinskas (1989));
- manufacturing processes (Boothroyd, Poli and Murch (1982), Kusiak, ed. (1986),Nitti and Speranza (1987), Jaumard,Lu and Sriskandarajah (1988));
- facility location, transportation analysis and planning (Jacobsen and Madsen (1980), Dempster, Fisher, Jansen, Lageweg, Lenstra and Rinnooy Kan (1981), Florian, ed. (1984));
- inventory planning (Chikán, ed. (1990));
- environmental analysis and management (Loucks, Stedinger and Haith

(1981), Haith (1982), Beck (1985), Somlyódy and Van Straten, eds. (1986), Kleindorfer and Kunreuther, eds. (1987), Richardson (1988)).

Needless to say that - depending on the actual model structure associated with the problem-types referred to - diverse (linear, integer/combinatorial, convex, nonconvex, stochastic etc.) optimization techniques can be applied to the solution of specific problems in many cases. Again, the point we attempt to make is the general relevance of Lipschitzian GO methodology to solving "vaguely structured, complicated, black box" problem instances. In this context, a few *environmental modelling and management* problems will be highlighted below.

The question of establishing efficient regional water quality (WQ) control strategies that meet preassigned environmental criteria, has been studied since several decades. Beck (1987) provides an overview of descriptive WQ modelling, referring also to several management issues. Among the numerous recent works devoted to *stochastic* river WQ modelling, one may cite e.g. Burn and McBean (1985), Chen and Papadopoulos (1988), Ellis (1987), Fujiwara, Gnanendran and Ohgaki (1987) or Tung and Hatthorn (1988).

Descriptive WQ models are based on the physical and biochemical mass balance of the water system considered. The classical Streeter-Phelps model (cf. e.g. Orlob, ed. (1983)) approximates the temporal coevolution of dissolved oxygen (DO) and biological oxygen demand (BOD) concentrations in a river section as a function of their initial values, and of the reaeration, oxidation and decomposition processes. Assuming steady--state conditions, the Streeter-Phelps description directly makes possible the determination of the *river section specific* minimal DO concentration (being an important threshold value for a number of WQ management purposes).

Because of the obvious *stochasticity* of hydrometeorological and biochemical conditions, the deterministic Streeter-Phelps description can be naturally extended: this leads to a stochastic BOD-DO model in which the occurring minimal DO values can be considered as random variables. In Boon, Pintér and Somlyódy (1989) the *chance constrained* optimization model below was formulated and solved:

$$\min \sum_{i=1}^{I} C_i(r_i) + w \{ \alpha - P(.) \}_+$$

(2.16) $\quad P\{ D_i^{min} \geq D_i^0 \quad i = 1,\ldots, I \} \geq \alpha$

$$r_{i,min} \leq r_i \leq r_{i,max} \quad i = 1,\ldots, I$$

In (2.16) r_i i=1,...,I denote the treatment plant efficiencies to be chosen on the subsequent river sections; $r_{i,min}$ and $r_{i,max}$ are explicit bounds for r_i; D_i^{min} is the - *treatment efficiency dependent, but also stochastically varying* - minimal DO concentration on river section i; D_i^0 is its prescribed bound (the WQ standard); P{.} is a shorthand notation for the left-hand side of the chance constraint; 0<α<1 is the reliability level required; $C_i(r_i)$ are the treatment plant (investment and discounted operations) cost functions; w>0 is a penalty multiplier (for penalizing unfavourable deviations from the target reliability, thus constraining the search towards reliable designs). Verbally stated, one is seeking for a regional WQ control strategy that is technologically feasible and assures acceptable WQ simultaneously on all river sections of interest, with a sufficiently high probability.

Although the treatment efficiency bounds in the model (2.16) define a simple rectangular region and the cost function terms in its objective function are typically convex, the chance constraint (and thus also the penalty term) are of *analytically unknown* nonlinear structure. (Let us note that the values D_i^{min} are determined numerically, solving a pair of differential equations with *randomly realized* coefficients and then solving a *randomly parameterized* algebraic equation.) Consequently, the model (2.16) may well be nonconvex and the use of higher level structural information is again out of question. Notwithstanding, *smooth* dependence on the decision variables r_i may be still assumed: this way, Lipschitzian optimization methodology can be applied.

In a sequence of numerical examples (based on real-world data), a four waste water treatment plants configuration was investigated. In this case the best design could be well approximated in a few hundred GO iterations, i.e. the sequential selection of options and their detailed performance analysis via Monte Carlo simulation. (The numerical results are described in the paper referred to above.)

Without going into more details, let us mention - as another application perspective of "black box" environmental engineering design - the risk assessment and management of permanent and accidental water pollution (see e.g. Henley and Kumamoto (1981), Richardson, ed. (1988), Gorelick (1990), Pintér, Benedek and Darázs (1990), De Lange and Pintér (1991), with many further references therein).

3 Concluding Remarks

In this paper several application possibilities of Lipschitzian global optimization are summarized, illustrating the practical relevance of this problem-class and methodology. As our numerical experience indicates, (so far) low-dimensional, but truly complicated Lipschitz GO problem instances can be successfully solved, even on PC's. Let us mention in this context that the PC computational limitations are primarily of *storage capacity* (as opposed to *algorithmic*) character: viz., since only a limited number of search steps can be accomplished in an often relatively high-dimensional feasible region, the numerically attained approximation of the globally optimal solution sometimes may be rather poor. Decompositional approaches (such as e.g. the two-phase search strategy applied by Pintér and Pesti (1991)), new algorithmic developments (e.g. parallel computational schemes, cf. Zenios, 1989)) and, certainly, the use of more powerful hardware facilities can help to mitigate this inherent difficulty. Moreover, although in the present paper the main emphasis was placed on *Lipschitzian* GO models and methodology, the full "repertoire" of global optimization is to be kept in mind. Thus - whenever the actual problem structure makes possible - specifically "tailored" GO methods can be preferred to or combined with the almost "universal" Lipschitzian approach.

References

Allgower, E.L. and Georg, K. (1983) Predictor-corrector and simplicial methods for approximating fixed points and zero points of nonlinear mappings, pp. 15-56. in: Bachem, A., Grötschel, M. and Korte,B., eds. Mathematical Programming. The State of the Art. Springer-Verlag, Berlin Heidelberg New York.

Archetti, F. and Betró, B.(1979) A probabilistic algorithm for global optimization. Calcolo 16, 335-343.

Basso, P. (1985) Optimal search for the global maximum of a function with bounded seminorm. SIAM Journal of Numerical Analysis 22,888-903.

Beck, M.B. (1985) Water Quality Management: A Review of the Development and Application of Mathematical Models. Springer-Verlag, Berlin Heidelberg New York.

Beck, M.B. (1987) Water quality modelling: A review of the analysis of uncertainty. Water Resources Research 23, 1393-1442.

Boender,C.G.E. (1984) The Generalized Multinomial Distribution: A Bayesian Analysis and Applications.Ph.D. Dissertation, Erasmus University, Rotterdam.

Boon, J.G., Pintér, J. and Somlyódy, L. (1989) A new stochastic approach for controlling point source river pollution, pp. 241-249. in: Publications of the International Association of Hydrologic Sciences No. 180. (Proceedings of the Third IAHS Assembly, Baltimore, May 1989).

Boothroyd, G., Poli,C. and Murch, L.E. (1982) Automatic Assembly. Marcel Dekker, New York.

Braess,D. (1986) Nonlinear Approximation Theory. Springer-Verlag, Berlin Heidelberg New York.

Branin, F.H. (1972) Widely convergent method for finding multiple solutions of simultaneous nonlinear equations. IBM Journal of Research and Development, 504-522.

Brucker, P. (1978) On the complexity of clustering problems, pp. 45-54. in: Beckmann, M. and Kunzi, H.P., eds. Optimization and Operations Research. Springer-Verlag, Berlin Heidelberg New York.

Burn,D.H. and McBean, E.A. (1985) Optimization modeling of water quality in an uncertain environment. Water Resources Research 21, 934-940.

Chen, K.-W. and Papadopoulos, A.S. (1988) A nonparametric method for estimating the joint probability density of BOD and DO. Ecological Modelling 41, 183-191.

Chikán, A. ed. (1990) Inventory Models. Akadémiai Kiadó, Budapest and Kluwer Academic Publishers, Dordrecht.

Cooke, R. and Pintér, J. (1989) Optimization in risk management. Civil Engineering Systems 6, 122-128.

Csendes, T. (1988) Nonlinear parameter estimation by global optimization - efficiency and reliability. Acta Cybernetica 8, 361-370.

Csiszár, I. (1975) I-divergence geometry of probability distributions and minimization problems. Annals of Probability 3, 146-158.

Danilin, Yu.M. and Piyavskii, S.A. (1967) An algorithm for finding the absolute minimum, pp.25-37.in: Theory of Optimal Decisions, Vol.2., Institute of Cybernetics of the Ukrainian Academy of Sciences, Kiev. (In Russian.)

Dempster, M.A.H., Fisher, M.L., Jansen, L., Lageweg, B.J., Lenstra, J.K. and Rinnooy Kan, A.H.G. (1981) Analytical evaluation of hierarchical planning systems. Operations Research 29, 707-716.

Dempster, M.A.H. and Coupé, G.M. (1982) Investigation of the stability of satellite large angle attitude manoeuvres using nonlinear optimization methods. Proceedings of the IFAC/ESA Symposium on Automatic Control in Space, ESTEC, Nordwijk, The Netherlands.

Dennis, J.E. and Schnabel, R.B. (1983) Numerical Methods for Nonlinear Equations and Unconstrained Optimization. Prentice-Hall, Englewood Cliffs, New Jersey.

Diener, I. (1987a) On non-uniqueness of nonlinear L_2-approximation. Journal of Approximation Theory 51, 54-67.

Diener, I. (1987b) On the global convergence of path-following methods to determine all solutions to a system of nonlinear equations. Mathematical Programming 39, 181-188.

Dixon, L.C.W. and Szegö, G.P., eds. (1975,1978) Towards Global Optimisation. Vols. 1-2, North-Holland, Amsterdam.

Ellis, J.H. (1987) Stochastic water quality optimization using imbedded chance constraints. Water Resources Research 23, 2227-2238.

Fedorov, V.V., ed. (1985) Problems of Cybernetics. Models and Methods in

Global Optimization. USSR Academy of Sciences, Moscow. (In Russian.)

Fields, D.E. and Miller,C.W. (1988) A methodology for deriving model input parameters from a set of environmental data. Ecological Modelling 40, 155-159.

Florian, M., ed. (1984) Transportation Planning Models. North-Holland, Amsterdam.

Forster, W., ed. (1980) Numerical Solution of Highly Nonlinear Problems. North-Holland, Amsterdam.

Fujii, V., Ichida, K. and Ozasa, M. (1986) Maximization of multivariable functions using interval analysis. In: Nickel, K.L.E., ed. Interval Mathematics 1985. Springer-Verlag, Berlin Heidelberg New York.

Galperin, E.A. (1988) Precision, complexity and computational schemes of the cubic algorithm. Journal of Optimization Theory and Applications 57, 223-238.

Garcia, C.B. and Zangwill, W.I. (1981) Pathways to Solutions, Fixed Points and Equilibria. Prentice-Hall, Englewood Cliffs, New Jersey.

Gordon, A.D. (1981) Classification: Methods for the Exploratory Analysis of Multivariate Data. Chapman and Hall, New York.

Gorelick, S.M. (1990) Large-scale nonlinear deterministic and stochastic optimization: Formulations involving simulation of subsurface contamination. Mathematical Programming 48, 19-39.

Gourdin, E., Jaumard, B. and Hansen, P. (1990) Global optimization for maximum likelihood estimation for the three parameter Weibull distribution. (Submitted for publication.)

Gourdin, E., Jaumard, B. and MacGibbon, B. (1990) Global optimization decomposition methods for bounded parameter minimax estimation. Research Report G-90-48, HEC-GERAD, University of Montréal.

Haith, D.A. (1982) Environmental Systems Optimization. Wiley, New York.

Hansen, E.R. (1979) Global optimization using interval analysis - the one-dimensional case. Journal of Optimization Theory and Applications 29, 331-344.

Hansen, E.R. (1980) Global optimization using interval analysis - the multidimensional case. Numerische Mathematik 34, 247-270.

Hansen, P. and Jaumard, B. (1987) Minimum sum of diameters clustering.

Journal of Classification 4, 215-226.

Hansen, P., Jaumard, B. and Lu, S.-H. (1988) A framework for algorithms in globally optimal design. Research Report G-88-17, HEC-GERAD, University of Montréal.

Hansen, P., Jaumard, B. and Lu, S.-H. (1989) Global optimization of univariate Lipschitz functions: I. Survey and properties. Research Report G-89-10, HEC-GERAD, University of Montréal.

Hendrickson, J.D., Sorooshian, S. and Brazil, L.E. (1988) Comparison of Newton-type and direct search algorithms for calibration of conceptual rainfall-runoff models. Water Resources Research 24, 691-700.

Hendrix, E.M.T. and Pintér, J. (1991) An application of Lipschitzian global optimization to product design. Technical Note 91-01, Wageningen Agricultural University. Revised version to appear in: Journal of Global Optimization.

Henley, E. and Kumamoto, H. (1981) Reliability Engineering and Risk Assesment. Prentice-Hall, Englewood Cliffs, New Jersey.

Horst, R. (1986) A general class of branch-and-bound methods in global optimization, with some new approaches for concave minimization. Journal of Optimization Theory and Applications 51, 271-291.

Horst, R. and Tuy, H. (1987) On the convergence of global methods in multiextremal optimization. Journal of Optimization Theory and Applications 54, 253-271.

Horst, R. (1988) Deterministic global optimization with partition sets whose feasibility is not known. Application to concave minimization, DC-programming, reverse convex constraints and Lipschitzian optimization. Journal of Optimization Theory and Applications 58, 11-37.

Horst, R. and Thoai, Ng.V. (1988) Branch-and-bound methods for solving systems of Lipschitzian equations and inequalities. Journal of Optimization Theory and Applications 58, 139-146.

Horst, R. and Tuy, H. (1990) Global Optimization - Deterministic Approaches. Springer-Verlag, Berlin Heidelberg New York.

Hu, T.C. and Kuh, E.S., eds. (1987) VLSI Circuit Layout: Theory and Design. IEEE Press, New York.

ten Hulscher, D., Bak, C. and Pintér, J. (1991) Calibration of the model system DELWAQ-IMPAQT for the Lake Ketelmeer. Research Report 91.001,

Rijkswaterstaat RIZA, Lelystad.

Jacobsen, S.K. and Madsen, O.B.G. (1980) A comparative study of heuristics for a two level routing-location problem. European Journal of Operational Research 5, 378-387.

Jaumard, B., Lu, S.-H. and Sriskandarajah, C. (1988) Design parameters selection and ordering of part-orienting devices. Research Report G-88-39, HEC-GERAD, University of Montréal.

Johnson, R.C. (1979) Optimum Design of Mechanical Elements. Wiley Interscience, New York.

Kaufman, L. and Rousseeuw, P. (1990) Finding Groups in Data: An Introduction to Cluster Analysis. Wiley, New York.

Klafszky, E., Mayer, J. and Terlaky, T. (1989) Linearly constrained estimation by mathematical programming. European Journal of Operational Research 34, 254-267.

Kleindorfer, P.R. and Kunreuther, H.C., eds. (1987) Insuring and Managing Hazardous Risks: From Seveso to Bhopal and Beyond. Springer-Verlag Berlin Heidelberg New York.

Kushner, H.J. (1964) A new method of locating the maximum point of an arbitrary multipeak curve in the presence of noise, Transactions of ASME, Series D, Journal of Basic Engineering 86, 97-105.

Kusiak, A., ed. (1986) Flexible Manufacturing Systems: Methods and Studies. North-Holland, Amsterdam.

De Lange, W.J. and Pintér, J. (1991) Groundwater quality assessment and management: A stochastic modelling approach. Research Report 91.009, Rijkswaterstaat RIZA, Lelystad.

Loehle, C. (1988) Robust parameter estimation for nonlinear models. Ecological Modelling 41, 41-54.

Loucks, D.P., Stedinger, J.R. and Haith, D.A. (1981) Water Resources Systems Planning and Analysis. Prentice-Hall, Englewood Cliffs, New Jersey.

Meewella, C.C. and Mayne, D.Q. (1988) An algorithm for global optimization of Lipschitz-functions. Journal of Optimization Theory and Applications 57, 307-323.

Meewella, C.C. and Mayne, D.Q. (1989) Efficient domain partitioning algorithms for global optimization of rational and Lipschitzian functions. Journal of Optimization Theory and Applications 61, 247-270.

Mladineo, R.H. (1986) An algorithm for finding the global maximum of a a multimodal, multivariate function. Mathematical Programming 34, 188-200.

Mockus, J., Tiesis, V. and Žilinskas, A. (1978) The application of Bayesian methods for seeking the extremum, pp. 117-129. in: Dixon, L.C.W. and Szegö, G.P., eds., Towards Global Optimisation, Vol. 2. North-Holland, Amsterdam.

Muroga, S. (1982) VLSI System Design. Wiley Interscience, New York.

Nitti, M.L. and Speranza, M.G. (1987) Heuristic control strategies in flexible flow lines, pp. 97-105. in: Kusiak, A.,ed. Modern Production Management Systems. North-Holland, Amsterdam.

Orlob, G.T., ed. (1983) Mathematical Modeling of Water Quality: Streams, Lakes and Reservoirs. Wiley, New York.

Pardalos, P.M. and Rosen, J.B. (1987) Constrained Global Optimization Algorithms and Applications. Springer-Verlag, Berlin Heidelberg.

Pardalos, P.M. and Rosen, J.B. (1988) Global optimization approach to the linear complementarity problem. SIAM Journal of Scientific and Statistical Computing 9, 341-353.

Pintér, J. (1983) A unified approach to globally convergent one-dimensional optimization algorithms. Research Report IAMI 83-5, Institute of Applied Mathematics and Informatics, CNR, Milano.

Pintér, J. (1986a) Globally convergent methods for n-dimensional multiextremal optimization. Optimization 17, 187-202.

Pintér, J. (1986b) Extended univariate algorithms for n-dimensional global optimization. Computing 36, 91-103.

Pintér, J. (1986c) Global optimization on convex sets. Operations Research Spektrum 8, 197-202.

Pintér, J. and Somlyódy, L. (1986) A stochastic lake eutrophication management model, pp. 501-512. in: Arkin, V.I., Shiriaev, A.N. and Wets, R., eds. Stochastic Optimization. Springer-Verlag, Berlin Heidelberg.

Pintér, J., Szabó, J. and Somlyódy, L. (1986) Multiextremal optimization for calibrating water resources models. Environmental Software 1, 98-105.

Pintér, J. (1987) Convergence qualification of partition algorithms in global optimization. Research Report 87-61, Department of Mathematics and Informatics, Delft University of Technology. Revised version to appear in: Mathematical Programming.

Pintér, J. and Cooke, R. (1987) Combining expert opinions: An optimization framework. Research Report 87-84, Department of Mathematics and Informatics, Delft University of Technology.

Pintér, J. (1988) Branch-and-bound methods for solving global optimization problems with Lipschitzian structure. Optimization 19, 101-110.

Pintér, J. (1990a) Solving nonlinear equation systems via global partition and search: Some experimental results. Computing 43, 309-323.

Pintér, J. (1990b) Model calibration: Problem statement, solution method and implementation. Research Report 90.024, Rijkswaterstaat RIZA, Lelystad.

Pintér, J. (1990c) Simplicial partition strategies for Lipschitzian global optimization. Working paper, Rijkswaterstaat RIZA, Lelystad.

Pintér, J. Benedek, P. and Darázs, A. (1990) Risk management of accidental water pollution: An illustrative application. Water Science and Technology 22, 265-274.

Pintér, J. and Pesti, G. (1991) Set partition by globally optimized cluster seed points. European Journal of Operational Research 51, 127-135.

Ratschek, H. (1985) Inclusion functions and global optimization. Mathematical Programming 33, 300-317.

Ratschek, H. and Rokne, J. (1988) New Computer Methods for Global Optimization. Ellis Horwood, Chichester.

Richardson, M.L.,ed. (1988) Risk Assessment of Chemicals in the Environment. The Royal Society of Chemistry, London.

Rinnooy Kan, A.H.G. and Timmer, G.T. (1987a) Stochastic global optimization methods. Part I. Clustering methods. Mathematical Programming 39, 27-56.

Rinnooy Kan, A.H.G. and Timmer, G.T. (1987b) Stochastic global optimization methods. Part II. Multi-level methods. Mathematical Programming 39, 57-78.

Robinson, S.M., ed. (1980) Analysis and Computation of Fixed Points.

Academic Press, New York.

Rousseeuw, P. (1987) Silhouettes: a graphical aid to the interpretation and validation of cluster analysis. Journal of Computational and Applied Mathematics 20, 53-65.

Schoen, F. (1982) On a sequential search strategy in global optimization problems. Calcolo 19, 321-334.

Shen, Z. and Zhu, Y. (1987) An interval version of Shubert's iterative method for the localization of the global maximum. Computing 38, 275-280.

Shubert, B.O. (1972) A sequential method seeking the global maximum of a function. SIAM Journal of Numerical Analysis 9, 379-388.

Somlyódy, L. and van Straten, G., eds. (1986) Modeling and Managing Shallow Lake Eutrophication. Springer-Verlag, Berlin Heidelberg New York.

Strongin, R. G. (1978) Numerical Methods for Multiextremal Problems. Nauka, Moscow. (In Russian.)

Strongin, R.G. (1989) The information approach to multiextremal optimization problems. Stochastics and Stochastics Reports 27, 65-82.

Törn, A. and Žilinskas, A. (1989) Global Optimization. Springer-Verlag, Berlin Heidelberg.

Tung, Y-K. and Hathhorn, W.E. (1988) Probability distribution for critical DO location in streams. Ecological Modelling 42, 61-74.

Vysotskaya, I.N. and Strongin, R.G. (1983) A method for solving nonlinear equations using prior probabilistic root estimates. Journal of Computational Mathematics and Mathematical Physics 23, 3-12. (In Russian.)

Welch, J.W. (1983) Algorithmic complexity: Three NP-hard problems in computational statistics. Journal of Statistical Computation and Simulation 15, 17-25.

Wilde, D.J. (1978) Globally Optimal Design. Wiley Interscience, New York.

Wingo, D.R. (1984) Fitting three-parameter lognormal models by numerical global optimization - an improved algorithm. Computational Statisics and Data Analysis 2, 13-25.

Wood, G.R. (1985) Multidimensional bisection and global minimization.

Research Report, University of Canterbury, New Zealand.

Zenios, S.A. (1989) Parallel numerical optimization: Current status and an annotated bibliography. ORSA Journal on Computing 1, 20-43.

Žilinskas, A. (1981) Two algorithms for one-dimensional multimodal minimization. Optimization 12, 53-63.

Packet Annealing: A Deterministic Method for Global Minimization. Application to Molecular Conformation

David Shalloway[*]

Abstract

We outline a new global minimization method in which the Gibbs distribution of the objective function is deterministically annealed by tracing the evolution of a multiple-Gaussian-packet approximation. Solutions are reached by iterative approximations with decreasing coarse-graining of both objective-function and spatial scales. Results from application of a partial implementation to the atomic-microcluster conformation problem are presented.

1 Introduction

The unnormalized Gibbs distribution at temperature T for a system governed by objective (energy) function H(R) over a continuous domain parametrized by R

$$p_T(R) = e^{-H(R)/k_B T} \qquad (1.1)$$

($k_B \equiv$ Boltzmann's constant)

converges at sufficiently low temperature T_{lo}, to the simple form[1]

$$\tilde{p}_{T_{lo}}(R) \approx e^{-|(R-R_g^*)/\Lambda_{lo}|^2} \qquad (1.2)$$

---

[*]Section of Biochemistry, Molecular and Cell Biology, Cornell University, Ithaca, NY 14853.

This research was supported in part by a Public Health Service Research Career Development Award (CA01139).

[1]Matrix notation is implicitly assumed throughout. That is

$$|R/\Lambda|^2 \equiv \sum_{i,j,k=1}^{N} r_i \Lambda_{ik}^{-1} r_j \Lambda_{jk}^{-1}$$

where N is the dimensionality of the space. In subsequent equations, R^2 denotes the outer product $r_i r_j$. The squared vector magnitude is denoted $|R^2|$; $|R^2| = \text{Tr}(R^2)$.

where the density is concentrated near the global minimum R^*_g. The simulated annealing algorithm (Kirkpatrick, 1983), when applied to continuous systems (e.g., Welle, 1986), can be viewed as a method for stochastically tracing this convergence by Monte Carlo simulation. Its success depends on the existence of a (non-linear) correlation between T and the size scale $\Lambda(T)$ of the Monte Carlo steps that produce the most rapid equilibration at each temperature. In practice, regions of decreasing extent are sampled as $T \rightarrow T_{lo}$.

This behavior is demonstrated in Fig. 1 which displays $p_T(R)$ for a hypothetical 2-dimensional objective function at different temperatures. At high temperature, $p_T(R)$ can be roughly modeled by a large Gaussian packet (Fig. 1A). The packet radius decreases with decreasing temperature (Fig. 1B) until a point is reached (Fig. 1C) where it branches into multiple packets. At yet lower temperatures [when Eq. (1.2) is valid], $p_T(R)$ is dominated by a single packet centered at R^*_g (Fig. lE). Simulated annealing in effect traces the approximate locations of the packets as they converge towards R^*_g by tracing the path of a sampling point through multiple stochastic runs. But its application to continuous problems of high dimensionality is limited by its requirement for large numbers of objective function evaluations. Our goal is to construct a deterministic annealing algorithm that converges to R^*_g with fewer objective function evaluations.

It is natural to begin by constructing $\tilde{\tilde{p}}_T(R)$, an approximation to $p_T(R)$, as a finite sum of Gaussian packets:

$$\tilde{\tilde{p}}_T(R) = \sum_{\{\alpha\}_T} p_\alpha(T) \, C[\Lambda_\alpha(T)] \, e^{-|[R-R^o_\alpha(T)]/\Lambda_\alpha(T)|^2} \qquad (1.3)$$

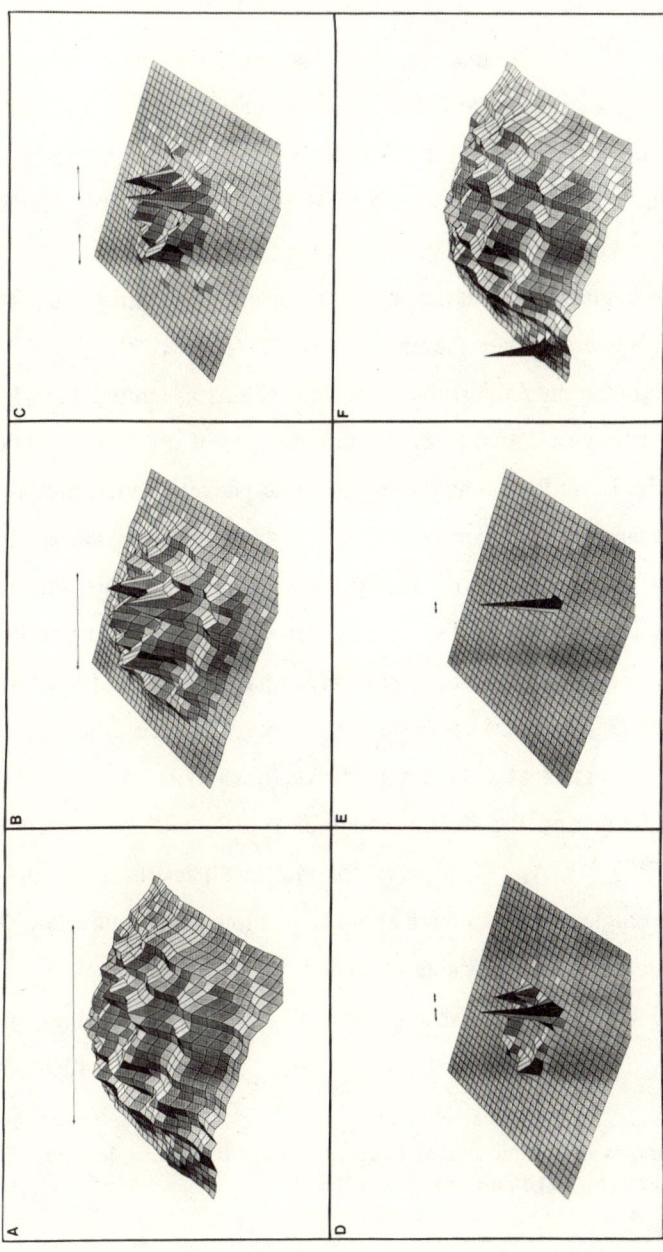

Fig. 1. Temperature and Size Scales in $p_T(R)$. The Gibbs distribution $p_T(R)$ of a hypothetical 2-dimensional system at decreasing temperatures are displayed in panels A-E (A = highest temperature, E = lowest temperature). The bars approximate the sizes of the appropriate packets at each temperature. (F) An anomalous objective-function that frustrates annealing methods.

where C is a normalization constant defined by [2]

$$C(\Lambda) \int e^{-|R/\Lambda|^2} dR = 1.$$

Each packet is assigned a unique index α. The sum spans the complete set of indices $\{\alpha\}_T$ for the packets that represent the dominant structure of $p_T(R)$. As illustrated in Fig. 1, this set grows and diminishes as T decreases. Each packet α is described by three parameters: $R^o{}_\alpha$ (packet center), Λ_α (packet width), and p_α (packet occupation probability). For simplicity, we begin the discussion by considering only isotropic packets even though this overestimates the spread of $\tilde{p}_T(R)$ (see Sec. 2).

As in simulated annealing, the procedure begins at temperature T_{hi}, where $k_B T_{hi}$ is large relative to the important scale of objective function values. It is initialized by constructing a single-packet approximation to $p_{T_{hi}}(R)$ that spans the dominant region of R (e.g., of size corresponding to the arrowhead bar in Fig.1A). Packet size and position are initially determined by iterative solution of a set of self-consistent constraints (see Sec. 4). The same constraints are used to adjust packet size and position as the temperature is lowered in small steps. The Λ_α will generally decrease with decreasing T. Packets will often subdivide (branch) as T is reduced (cf Figs. 1B and 1C), and the development of $\tilde{p}_T(R)$ will be traced by calculating $\{R^o{}_\alpha(T), \Lambda_\alpha(T), p_\alpha(T)\}$ for multiple packets until $\tilde{p}_T(R)$ converges to a single packet centered about $R^*{}_g$ at low temperature T_{lo}. To control computational effort, packets with small $p_\alpha(T)$ are discarded.

Branching and merging of packets is naturally accomodated and does not interfere with the tracing procedure (see Sec. 5). However, for some

---

[2] $\int dR$ denotes multidimensional integration over the entire domain. For simplicity we assume that there are no constraints.

objective functions, the global minimum may be located in a region where the Gibbs distribution is, on-average, small (e.g., Fig. 1F). The packet containing this minimum will not emerge from subdivision or shrinking of a larger packet and may not be detected. This difficulty is not particular to the packet approximation method; simulated annealing is also likely to fail (or to require inordinate time to detect the minimum) in such cases. None-the-less, there are many problems, particularly those where the objective function is the sum of a large number of partially-independent terms, where it is unlikely that such anomalous behavior will occur or where anomalous minima are not of primary interest (see Sec. 9).

In Sec. 2 we show that $\tilde{\tilde{p}}_T(R)$ is governed by an effective potential that is calculated by spatial averaging. A physical interpretation is presented in Sec. 3 and the packet annealing algorithm is outlined in Secs. 4 and 5. The method is applicable only to problems where the spatial averaging integrals can be approximated. The molecular conformation problem, described in Sec. 6, is of this type. We show how packet-annealing can be applied to this problem in Sec. 7 and present preliminary numerical results in Sec. 8.

2 Packet Expansion of $p_T(R)$

$\tilde{\tilde{p}}_T(R)$ is an effective distribution that does not approximate $p_T(R)$ in a uniformly convergent sense: $|\tilde{\tilde{p}}_T(R) - p_T(R)| \not< \varepsilon$. Instead, it provides a method for dissecting the spatial variations of $p_T(R)$ by size-scale. The expansion is constructed so that packets do not significantly overlap; that is,[3]

---

[3]The procedures that ensure that Eq. (2.1) is satisfied for all α and β are discussed in Sec. 5.

$$|R^o_\alpha - R^o_\beta| > \Lambda_\alpha + \Lambda_\beta . \qquad (2.1)$$

Within each region ($|R-R^o{}_\alpha| \le \Lambda_\alpha$), variations on spatial scales $< \Lambda_\alpha$ are absorbed into a single Gaussian approximation; variations on scales $> \Lambda_\alpha$ are represented by the positions and amplitudes of the Gaussians. Only the regions where $p_T(R)$ is significant are modeled since packets having small p_α are deleted. $\tilde{\tilde{p}}_T(R)$ is self-consistently defined by the requirement that, when sampled in region α by a function $S_\Lambda(R)$ having length scale $\Lambda \ge \Lambda_\alpha$, it yields results close to those obtained by sampling $p_T(R)$ directly:

$$\int p_T(R') S_\Lambda (R - R') \, dR' \approx \int \tilde{\tilde{p}}_T (R') S_\Lambda(R - R') \, dR' \qquad (2.2)$$

$|R - R'| < \Lambda_\alpha, \Lambda \ge \Lambda_\alpha,$

where the sampling function $S_\Lambda(R)$ has the properties

$$S_\Lambda (R) \to 0 \text{ for } |R|/\Lambda \to \infty$$
$$\int S_\Lambda (R) \, dR = 1$$

We use
$$S_\Lambda (R) = C(\Lambda) \, e^{-|R/\Lambda|^2} \qquad (2.3)$$

The integral of $p_T(R)$ over all conformation space gives the partition function $Z(T)$. We define the spatially-localized integral in the left-hand-side of Eq. (2.2) as the "local partition function" $\tilde{Z}_{\Lambda,T}(R)$ and, following Wilson (1975), define the "effective energy function" $\tilde{H}_{\Lambda,T}(R)$ by

$$e^{-\tilde{H}_{\Lambda,T}(R)/k_BT} \equiv \tilde{Z}_{\Lambda,T}(R) \equiv \int p_T(R') S_\Lambda (R - R') \, dR' =$$
$$C(\Lambda) \int e^{-H(R')/k_BT} \, e^{-|(R-R')/\Lambda|^2} \, dR' \qquad (2.4)$$

[The right-hand-side of Eq. (2.4) is obtained from Eqs. (1.1) and (2.3).]

$\tilde{Z}_{\Lambda,T}(R)$ satisfies the boundary condition

$$\lim_{\Lambda \to \infty} \tilde{Z}_{\Lambda,T}(R)/C(\Lambda) = Z(T) \qquad (2.5)$$

$\tilde{H}_{\Lambda,T}(R)$ is a non-rescaled spatial-domain renormalization group transform[4] of H(R) (Wilson, 1975) and satisfies the boundary condition

$$\lim_{\Lambda \to 0} \tilde{H}_{\Lambda,T}(R) = H(R) \qquad (2.6)$$

Eq. (2.2), with Eq. (1.3), can be expanded in a Taylor's series in $(R-R^o{}_\alpha)$ for each $R^o{}_\alpha$. When restricted to contributions from a single packet α, the first three terms yield constraints on p_α, $R^o{}_\alpha$, and Λ_α in terms of $\tilde{H}_{\Lambda,T}(R)$:

$$\tilde{Z}_{\Lambda,T}(R^o_\alpha) \equiv e^{-\tilde{H}_{\Lambda,T}(R^o_\alpha)/k_B T} \approx p_\alpha\, C(\Lambda)\, C(\Lambda_\alpha)/C[(\Lambda^{-2}+\Lambda_\alpha^{-2})^{-1/2}] \qquad (2.7)$$

$$-k_B T\, \frac{\partial \tilde{Z}_{\Lambda,T}(R)/\partial R}{\tilde{Z}_{\Lambda,T}(R)}\bigg|_{R=R^o_\alpha} = \partial\, \tilde{H}_{\Lambda,T}(R)/\partial R\,\bigg|_{R=R^o_\alpha} \approx 0 \qquad (2.8)$$

and

$$\mathrm{MINIMUM}\left\{ -k_B T\, \frac{\partial^2 \tilde{Z}_{\Lambda,T}(R)/\partial R^2}{\tilde{Z}_{\Lambda,T}(R)}\bigg|_{R=R^o_\alpha} \right\} =$$

$$\mathrm{MINIMUM}\,\{\partial^2 \tilde{H}_{\Lambda,T}(R)/\partial R^2\,|_{R=R^o_\alpha}\} \approx \frac{2 k_B T}{\Lambda^2 + \Lambda_\alpha^2} \qquad (2.9)$$

($\Lambda \geq \Lambda_\alpha$; single-packet approximation. The dependence of $R^o{}_\alpha$, Λ_α and p_α on T is implicit.)

---

[4] A complete renormalization group transformation would also rescale R and $\tilde{H}$. Rescalings are incorporated in the computational implementation but, for simplicity, are omitted in the discussion. Renormalization group transformations are usually defined in momentum space, but transformations in configuration space are more appropriate here. Eq. (2.4) corresponds to multiplying the Fourier transform of $p_T(R)$ by $\exp(-4k^2\Lambda^2)$, where k is the transform momentum variable. That is, it corresponds to a suppression of spatial-frequencies above cutoff $\sim O(2/\Lambda)$.

Eq. (2.9) is a matrix equation in the components of R. If non-isotropic Gaussian packets were used, Λ_α^{-2} would be the symmetric Gaussian coefficient matrix and Eq. (2.9) (without the MINIMUM[5] operator) would represent a matrix of constraints. However, if we restrict to isotropic packets, only one condition can be fixed. We fix Λ_α by the minimum eigenvalue of the Hessian of $\tilde{H}_{\Lambda,T}(R)$ so that the size of the Gaussian packet matches $p_T(R)$ in the least-localized direction. This conservative choice assures that an anisotropic concentration of $p_T(R)$ in region α will be contained within isotropic Gaussian packet α.

If $p_T(R)$ were well-approximated by a single packet, Eqs. (2.7)-(2.9) would be satisfied for all $\Lambda \geq \Lambda_\alpha$. However, in general, as Λ gets large, S_Λ will overlap multiple packets and (2.7)-(2.9) will not be valid. But, because of the non-overlap condition Eq. (2.1), Eqs. (2.7)-(2.9) will be approximately valid even in the presence of multiple packets when $\Lambda = \Lambda_\alpha$. Restricting Eqs. (2.7)-(2.9) by this condition generates a set of self-consistent conditions that fix p_α, R^o_α and Λ_α:

$$\tilde{Z}_{\Lambda_\alpha,T}(R^o_\alpha) \equiv e^{-\tilde{H}_{\Lambda_\alpha,T}(R^o_\alpha)/k_B T} = p_\alpha\, 2^{-N/2}\, C(\Lambda_\alpha) \qquad (2.10)$$

(where N is the dimensionality of the domain)

$$-k_B T\, \frac{\partial \tilde{Z}_{\Lambda_\alpha,T}(R)/\partial R}{\tilde{Z}_{\Lambda_\alpha,T}(R)}\bigg|_{R=R^o_\alpha} = \partial \tilde{H}_{\Lambda_\alpha,T}(R)/\partial R \bigg|_{R=R^o_\alpha} = 0 \qquad (2.11)$$

---

[5]MINIMUM and MAXIMUM operators refer to the eigenvalues of the matrix.

$$\text{MINIMUM}\left\{-k_B T \frac{\partial^2 \tilde{Z}_{\Lambda_\alpha,T}(R)/\partial R^2}{\tilde{Z}_{\Lambda_\alpha,T}(R)}\bigg|_{R=R_\alpha^o}\right\} =$$

$$\text{MINIMUM}\ \{\partial^2 \tilde{H}_{\Lambda_\alpha,T}(R)/\partial R^2\,|_{R=R_\alpha^o}\} = \frac{k_B T}{\Lambda_\alpha^2} \qquad (2.12)$$

In practice, Eqs. (2.11) and (2.12) will be solved by alternating iteration. Once an initial solution has been found, it can be iteratively propagated through small downward steps in T (see Sec. 4). Except for isolated discontinuities caused by branching and merging of packets (see Sec. 5), the changes in $R^o_\alpha(T)$ and $\Lambda_\alpha(T)$ at each step in T will be small and only one or a few iterations will be required.

3 Physical Interpretation

The significance of Eq. (2.10) is clear: p_α, the coefficient of packet α, is proportional to the integrated Gibbs density in the vicinity of R^o_α. The significance of Eqs. (2.11) and (2.12) is clarified by considering a physical model.

If the objective function H(R) is associated with an electric field, simulated annealing corresponds to tracking the thermal migration of a point charge. We can improve the convergence rate by sampling with an extended charged test-object of isotropic size Λ_α. Since it will sample H(R) over a finite region, it will migrate under the influence of a smoothed (to size-scale Λ_α) effective potential in which the high spatial frequency variation has been suppressed.

We construct the extended test-object by constraining a light point charge (at position R') to a heavy uncharged mass (located at R) by a linear spring with spring constant $2 k_B T/\Lambda_\alpha^2$. The energy function for this object in the field is

$$H(R') + \frac{k_B T}{\Lambda_\alpha^2} |R - R'|^2 \qquad (3.1)$$

The position of the light charged object equilibrates more rapidly than motions of R. Thus (ala Born-Oppenheimer) we integrate over R' to get the effective free energy function

$$\tilde{Z}_{\Lambda_\alpha,T}(R) \equiv e^{-\tilde{H}_{\Lambda_\alpha,T}(R)} \propto \int e^{-H(R')/k_B T - |R-R'|^2/\Lambda_\alpha^2} \, dR' \qquad (3.2)$$

Eq. (3.2) is identical to Eq. (2.4) except for an irrelevant multiplicative constant which corresponds to a constant added to $\tilde{H}_{\Lambda_\alpha,T}(R)$.

$\tilde{H}_{\Lambda_\alpha,T}(R)$ governs the stochastic motion of the extended test-object in the same way that H(R) governs the stochastic motion of a point charge. The central difference is that the stochastic fluctuations of R for the extended test-object will be small when compared to Λ_α, the test-object size, if Λ_α is sufficiently large. Then the test-object will move according to

$$\frac{dR}{dt} \propto \frac{-\partial \tilde{H}_{\Lambda_\alpha,T}(R)}{\partial R} \qquad (3.3)$$

(i.e., overdamped motion along a steepest-descent trajectory) until equilibrium is reached at a local minimum R^o_α

$$\frac{\partial \tilde{H}_{\Lambda_\alpha,T}(R)}{\partial R} \bigg|_{R=R^o_\alpha} = 0 \qquad (3.4)$$

This is identical to Eq. (2.11).

The condition that the stochastic fluctuations are relatively small is
$$\text{MAXIMUM} < (R - R^o_\alpha)^2 >_{\Lambda_\alpha,T} \leq \Lambda_\alpha^2 \qquad (3.5)$$

where

$$\langle (R - R_\alpha^o)^2 \rangle_{\Lambda_\alpha, T} \equiv \frac{\int (R - R_\alpha^o)^2 \, e^{-\tilde{H}_{\Lambda_\alpha, T}(R)/k_B T} \, dR}{\int e^{-\tilde{H}_{\Lambda_\alpha, T}(R)/k_B T} \, dR}$$

Approximating

$$\tilde{H}_{\Lambda_\alpha, T}(R) \approx \tilde{H}_{\Lambda_\alpha, T}(R_\alpha^o) + \frac{1}{2} \frac{\partial^2 \tilde{H}_{\Lambda_\alpha, T}(R)}{\partial R^2}\bigg|_{R=R_\alpha^o} (R - R_\alpha^o)^2 \qquad (3.6)$$

we have

$$\langle (R - R_\alpha^o)^2 \rangle_{\Lambda_\alpha, T} \approx \frac{k_B T}{\dfrac{\partial^2 \tilde{H}_{\Lambda_\alpha, T}(R)}{\partial R^2}\bigg|_{R=R_\alpha^o}} \qquad (3.7)$$

Eqs. (3.5) and (3.7) yield a relationship between Λ_α and the Hessian of $\tilde{H}_{\Lambda_\alpha, T}$ at the stability position $R^o{}_\alpha$:

$$\text{MAXIMUM} \left\{ \frac{1}{\dfrac{\partial^2 \tilde{H}_{\Lambda_\alpha, T}(R)}{\partial R^2}\bigg|_{R=R_\alpha^o}} \right\} \leq \Lambda_\alpha^2/k_B T \qquad (3.8)$$

Eq. (3.8) provides a lower bound on Λ_α.[6] Since maximal information is provided when the test-object is as small as possible without significant fluctuation in R, the equality in Eq. (3.8) fixes Λ_α. This is equivalent to Eq. (2.12).

Because (3.6) is only approximate, the fluctuations in R about $R^o{}_\alpha$ will only be small for a characteristic time period Δt that is small compared to the period Δt_{global} required for equilibration with nearby local minima. That is, Eqs. (3.3) and (3.4) are valid for

---

[6]In addition, using Eq. (3.2) we can show that

$$\Lambda_\alpha^2/2 \, k_B T \leq \text{MINIMUM} \left\{ [\partial^2 \tilde{H}_{\Lambda_\alpha, T}(R)/\partial R^2 \big|_{R=R_\alpha^o}]^{-1} \right\}$$

is always satisfied.

$$\Delta t_{point} \ll \Delta t \ll \Delta t_{global} \tag{3.9}$$

where Δt_{point} is the characteristic period required for local equilibration of the point charge. Thus, the $R^0{}_\alpha$ defined by Eq. (3.4) define metastable "states" of the system.

4 Packet Annealing Algorithm

4.1 Energy-Level Trajectories

The overall development of the annealing process can be followed by studying the trajectories described by the set of $F_\alpha(T) \equiv -k_B T \log[p_\alpha(T)]$.[7] In the physical interpretation of Sec. 3, these are the "free-energy levels" of the metastable "states" (self-consistent packet solutions) at different T. Fig. 2 displays a hypothetical energy level diagram where the trajectories of all states are displayed.[8] The salient points are: (1) there tend to be fewer states at higher T because the smoother effective potential surface supports fewer local minima,[9] (2) states may merge or branch, and (3) states may become unstable and disappear both as T increases and decreases. In this example, the position $R^*(T_{lo})$ of the global minimum at T_{lo} can be found by tracing a low-lying trajectory from the global minimum at T_{hi}. In this example, a low-lying trajectory can be traced from the global minimum at T_{hi},

---

[7] In principle, the trajectories of the $R_\alpha(T)$ could be traced, but this is impractical because of the high dimensionality of the space.

[8] In practice, trajectories with $F_\alpha(T) - F^*(T) > -k_B T \log(p_{min})$ [where $F^*(T)$ is the lowest energy state at T and $p_{min} \ll 1$] will not be calculated.

[9] Here we are referring to the total number of local minima of $\tilde{H}_{\Lambda,T}$ which will generally (although not strictly) tend to increase as T decreases. For practical reasons, we will only track low energy minima; their number may increase or decrease as T is lowered.

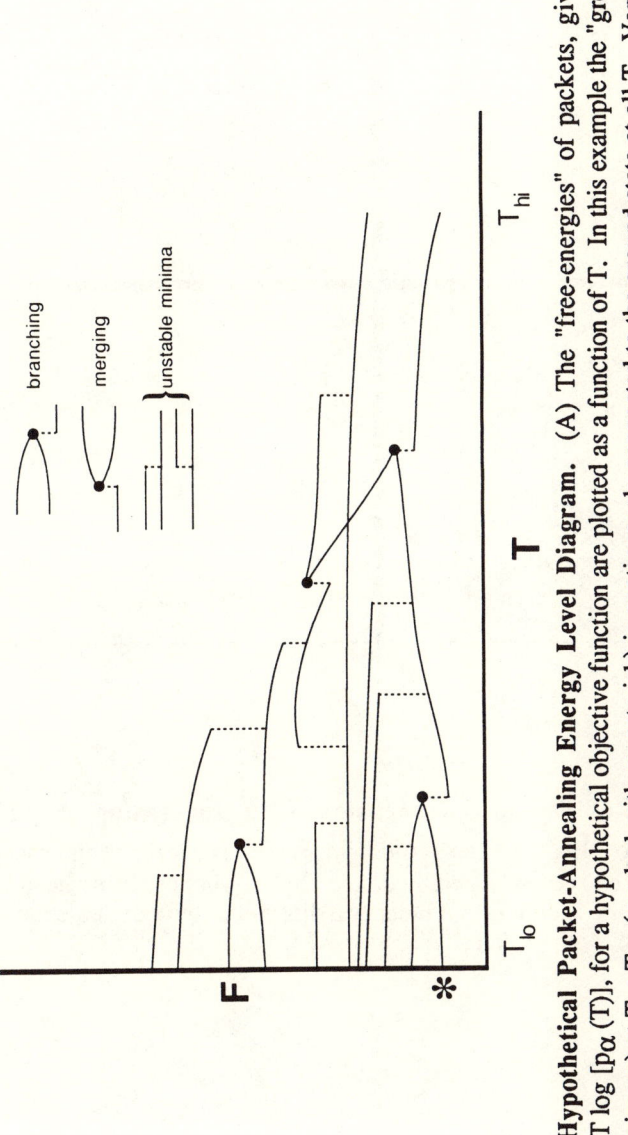

Fig. 2. Hypothetical Packet-Annealing Energy Level Diagram. (A) The "free-energies" of packets, given by $F\alpha(T) = -k_B T \log[p\alpha(T)]$, for a hypothetical objective function are plotted as a function of T. In this example the "ground state" (global minimum) at $T = T_{lo}$ (marked with an asterisk) is continuously connected to the ground state at all T. Vertical dotted lines show the discontinuous trajectory connections that occur when states become unstable. (B) The primary types of energy level trajectory singularities.[11]

---

[11]Branches and mergers between three or more packets are also possible; they are accomodated by the same procedures that handle the two-packet branches and mergers.

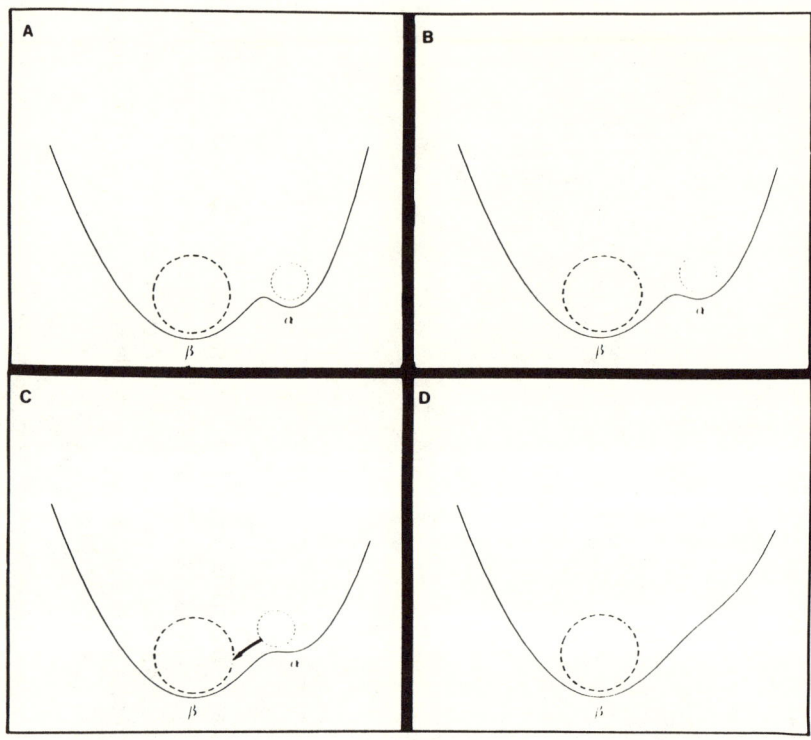

Fig. 3. Instability in 1-Dimension with Increasing Λ_α (T).
Panels A-D show $\tilde{H}_{\Lambda,T}(R)$ as T and Λ increase. The sizes of the packets (finite-test objects) are indicated by dotted circles. The instability in state α at the temperature shown in (C) induces a discontinuity in the trajectory as packet α merges with packet β.

through two branch points, to the global minimum at T_{lo}. Such "traceability" (which is not guaranteed) is required for success. It is not affected by mergers and branches (see Sec. 5) but may be disrupted by discontinuities associated with state instabilities. States can become unstable as T increases when a narrow local minimum in $\tilde{H}_{\Lambda_\alpha(T),T}(R)$ is averaged out as $\Lambda_\alpha(T)$ increases. As shown in Fig. 3, the α trajectory discontinuously merges into the trajectory of the catchment region that contains packet α at the instability point. Conversely these singularities correspond to points where new states spontaneously appear as T decreases. These states may not be detected, so success requires that a continuous path to $R^*(T_{lo})$ can be found without them. This requirement is not unreasonable since the new states will have relatively high energies when they first appear (cf packets α and β in Fig. 3).[10]

States can also become unstable as T (and Λ_α) decreases. This destabilization does not interfere with the trajectory tracing process since the trajectory can be (discontinuously) traced through the singularity (e.g., following the dotted lines in Fig. 2).

4.2 Algorithm

An important characteristic for computation is that the scales of significant variation of both spatial and objective-function values are regulated at all steps by Λ_α and T. These parameters (multiplied by small factors ε_R, ε_Λ, ε_H) provide natural measures for the accuracies needed in

---

[10] The relative energy of a new state, while initially high, may decrease as T is lowered so that it ultimately becomes the global minimum. This is precisely what occurs in an anomalous case like that displayed in Fig. 1F.

(iterative) solutions of Eqs. (2.11) and (2.12) so computational effort is controlled. The procedure terminates when a putative global minimum has been identified to accuracy $\leq \Lambda_{lo}$.

For convenience we redefine p_α, $R^0{}_\alpha$ and Λ_α to be functions of the iteration parameter τ ($\tau = 1, 2,...$) rather than as direct functions of T. The central features of the algorithm are:

(* INITIALIZE *)
$T = T_{hi}$ (* T_{hi} is sufficiently high to "melt" all structure in $p_{T_{hi}}(R)$ *)

$^{(0)}R = 0$ (* an arbitrary point in the domain is selected *)
$^{(0)}\Lambda = \Lambda_{hi}$ (* Λ_{hi} > the largest size scale of $p_{T_{hi}}(R)$; see Sec. 7*)

(*solve Eqn. (2.11) and (2.12) for the initial solution by iteration*)

i = 0

<u>repeat</u>

  i = i + 1

$$^{(i)}\Lambda^2 = \text{MAX}\left\{\left[\frac{\partial^2 \tilde{H}_{(i)_{\Lambda,T}}(R)}{\partial R^2}\bigg|_{R=^{(i-1)}R}\right]^{-1}\right\} \quad (4.1)$$

local minimization of $^{(i)}R$ starting from $^{(i-1)}R$ until

$$\frac{\partial \tilde{H}_{(i)_{\Lambda,T}}(R)}{\partial R} \leq \frac{\varepsilon_H T}{\Lambda^{(i)}} \quad (4.2)$$

<u>until</u> $|^{(i)}R - ^{(i-1)}R| \leq \varepsilon_R {}^{(i)}\Lambda$ AND $|\log {}^{(i)}\Lambda/^{(i-1)}\Lambda| < \varepsilon_\Lambda {}^{(i)}\Lambda$

$R^0{}_1(0) = {}^{(i)}R$
$\Lambda_1(0) = {}^{(i)}\Lambda$

(*ITERATE: decrement T, adjust Λ, locally minimize R^0, branch/merge *)

<u>repeat</u>

$\tau = \tau + 1$

$T(\tau) = [1 - \varepsilon_T(\tau)] T(\tau-1)$

<u>for</u> $\alpha \in \{\alpha\}_T$ <u>do</u>

   <u>begin</u>

     (* update $\Lambda_\alpha(\tau)$ using Eq. (2.12) *)

$$\Lambda_\alpha^2(\tau) = \text{MAX} \left\{ \left[\frac{\partial^2 \tilde{H}_{\Lambda_\alpha(\tau-1), T(\tau)}(R)}{\partial R^2} \bigg|_{R=R_\alpha^o(\tau-1)} \right]^{-1} \right\} \quad (4.3)$$

     local minimization starting from $R^o{}_\alpha(\tau-1)$ until

$$\frac{\partial \tilde{H}_{\Lambda_\alpha(\tau), T(\tau)}}{\partial R} \bigg|_{R=R_\alpha^o(\tau)} \leq \frac{\varepsilon_H T}{\Lambda(\tau)} \quad (4.4)$$

     Repeat iterations of (4.1) and (4.2) if necessary (see below)

     Test and branch/merge packets if necessary (* see Sec. 5 *)

     (* **update** $p_\alpha(\tau)$ **using Eq. (2.10)** *)

$$p_\alpha(\tau) = \frac{2^{-N/2}}{C[\Lambda_\alpha(\tau)]} e^{-\tilde{H}_{\Lambda_\alpha(\tau), T(\tau)}[R_\alpha^o(\tau)]}$$

   <u>end</u>

     (* **discard low probability packets** *)

$P_{tot} = \sum_{\{\alpha\}_T} p_\alpha(\tau)$ (* calculate probability normalization factor *)

   <u>for</u> $\alpha \in \{\alpha\}_T$ <u>do</u>

     <u>if</u> $p_\alpha(\tau) < p_{min} P_{tot}$ <u>then</u> $\{\alpha\}_T = \{\alpha\}_T - \alpha$

(* **test for termination** *)

<u>until</u> $\{\alpha\}_T$ = unique AND $|R^o{}_\alpha(\tau) - R^o{}_\alpha(\tau-1)| \leq \Lambda_{lo}$.

The maximum decrease in T at each step [determined by $\varepsilon_T(\tau)$] is limited by the requirement that $R^o{}_\alpha(\tau-1)$ lies within the catchment region of $R^o{}_\alpha(\tau)$. This ensures that $R^o{}_\alpha(\tau)$ can be found by local minimization. The maximum number of packets in the expansion at any time will be

$\leq 1/p_{min}$.

Multiple iterations of (4.1) and (4.2) during initialization are required to get the initial one-packet approximation. In some cases, this packet will be unstable (as described in Sec. 5) and will divide into subpackets when first tested for branching. During the main iterative loop, Eqs. (4.3) and (4.4) update Λ_α and $R^o{}_\alpha$ towards solutions of Eqs. (2.12) and (2.11). If the steps in T are sufficiently small, one update per cycle will be sufficiently accurate. This can be checked empirically and additional alternating iterations of (4.3) and (4.4) can be applied if necessary. Multiple iterations will definitely be required when a state becomes unstable and the packet moves discontinuously. This can be detected by testing for $|R^o{}_\alpha(\tau) - R^o{}_\alpha(\tau-1)| > \Lambda_\alpha(\tau-1)$. In this case, the packet will move to a lower energy state where it may merge with a pre-existing packet. This process is naturally handled by the algorithm and poses no difficulty.

5 Branching and Merging of Packets

The possibility that packets α and β may merge can be tested by checking for

$$|R^o_\alpha(T) - R^o_\beta(T)| \sim O[\Lambda_\alpha(T) + \Lambda_\beta(T)] \tag{5.1}$$

Packet merging is naturally accomodated by Eqs. (2.11) and (2.12). When the packets come sufficiently close, their separate solutions $[R^o{}_\alpha(T), \Lambda_\alpha(T)]$ and $[R^o{}_\beta(T), \Lambda_\beta(T)]$ will become unstable; only a single solution centered between them [at $R^o{}_\gamma \approx (R^o{}_\alpha + R^o{}_\beta)/2$ with width $\Lambda_\gamma \approx (\Lambda_\alpha + \Lambda_\beta)$] will exist. Conversely, packet branching can be detected by testing for the appearance of new solutions of Eqs. (2.11) and (2.12).

The merged (γ) packet remains a solution of Eqs. (2.11) and (2.12) throughout branches and mergers. However, when separate (α and β) solutions exist, they are included in expansion (1.3) and the parental solution is deleted. Thus, the size scale for approximation changes discontinuously. While Eqs. (2.11) and (2.12) can continuously propagate solutions between branch/merger points, special procedures are needed to handle these discontinuities.

The packet merging procedure is straightforward: alternating iteration of Eqs. (2.11) and (2.12) will propagate the separated solutions to the merged solution when the former become unstable. The branching procedure is more complex. In this case, we first test if a branched solution can exist at an appropriately reduced $\Lambda = \Lambda_b$. If so, iterative alternation of Eqs. (2.11) and (2.12), using Λ_b and $R^o\pm = R^o_\gamma \pm \Delta R^o_b$, is used to seek the branched packets. A prescription for specifying Λ_b and ΔR^o_b is needed.

These parameters can be explicitly calculated in the 1-dimensional case shown in Fig. 4 where a probability distribution that is well-represented by a single packet at large T changes to a distribution that must be represented by two packets at lower T.[12] For analytic simplicity, we model this function as[13]

$$p_T(x) = [e^{-[(x-a)/w(T)]^2} + e^{-[(x+a)/w(T)]^2}]/[2\sqrt{\pi}\, w(T)] \qquad (5.2)$$

[dw/dT > 0]

The effective energy can be analytically calculated from Eqs. (2.4) and (5.2):

$$e^{-\tilde{H}_{\Lambda,T}(x)} \equiv \tilde{Z}_{\Lambda,T}(x) = \frac{e^{-(x-a)^2/[\Lambda^2 + w^2(T)]} + e^{-(x+a)^2/[\Lambda^2 + w^2(T)]}}{2\sqrt{\pi}\sqrt{\Lambda^2 + w^2(T)}} \qquad (5.3)$$

The derivatives of $\tilde{H}_{\Lambda,T}$ are

---

[12] For this discussion is convenient to consider the variables as functions of T rather than τ.

[13] For the potential shown in Fig. 4, $w(T) \sim 2.25a\sqrt{T/T_o}$, where T_o is the temperature in panel A. The exact form of $w(T)$ is irrelevant.

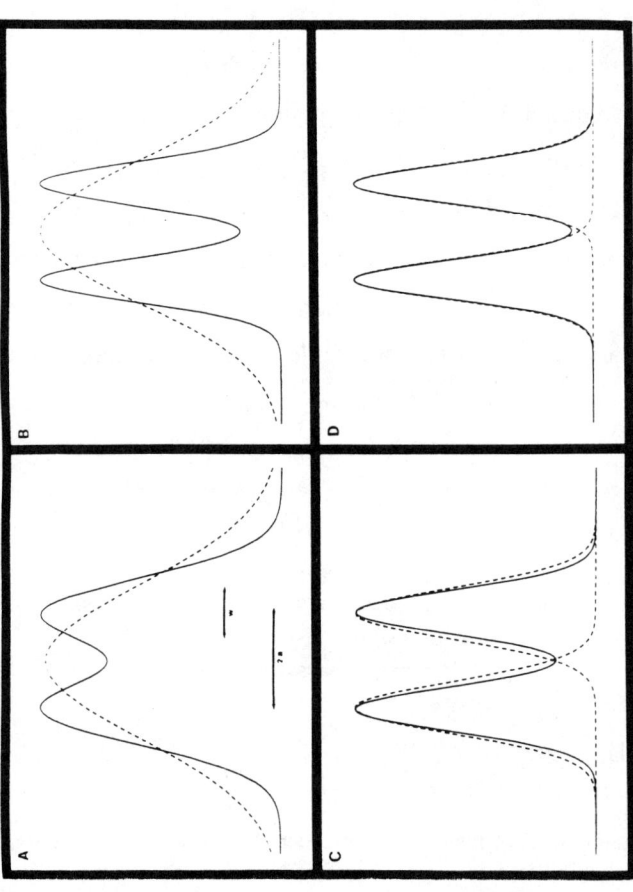

Fig. 4. Packet Branching. The bifurcated Gibbs distribution $p_T(x)$ of Eq. (5.2) (solid lines) and its packet approximations $\tilde{p}_T(x)$ determined by Eqs. (5.8) and (5.9) (dashed lines) are displayed at different temperatures. For $w(T) = a$, the single packet ($x^0{}_0 = 0$, $\Lambda_0 = 2.06\ a$) is the only solution to (5.8) and (5.9) (panel A). When $w(T)$ decreases to 0.63 a, the single packet approximation ($x^0{}_0 = 0$, $\Lambda_0 = 2.01\ a$) (panel B) bifurcates into two subpackets ($x^0{}_b = \pm 0.97\ a$, $\Lambda_b = 0.74$) (panel C). These packets closely approximate $p_T(x)$ as T is further reduced [$w(T) = 0.57\ a$; panel D].

$$\frac{\partial \tilde{H}_{\Lambda,T}(x)}{\partial x} = \frac{2 k_B T}{\Lambda'^{\,2}} \left[x - a \tanh\left(\frac{2ax}{\Lambda'^{\,2}}\right) \right] \quad (5.4)$$

$$\frac{\partial^2 \tilde{H}_{\Lambda,T}(x)}{\partial x^2} = \frac{4 k_B T}{\Lambda'^{\,4}} \left\{ \frac{\Lambda'^{\,2}}{2} - x^2 - a^2 + 2ax \tanh\left(\frac{2ax}{\Lambda'^{\,2}}\right) \right\} \quad (5.5)$$

where $\Lambda'^{\,2} \equiv \Lambda^2 + w^2$

Thus, Eqs. (2.11) and (2.12), replacing $R \to x$ for this one-dimensional case, become

$$\frac{2 k_B T}{\Lambda'^{\,2}} \left[x_\alpha^o - a \tanh\left(\frac{2 a x_\alpha^o}{\Lambda'^{\,2}}\right) \right] = 0 \quad (5.6)$$

$$\frac{4 k_B T}{\Lambda_\alpha'^{\,4}} \left[\frac{\Lambda_\alpha'^{\,2}}{2} - {x_\alpha^o}^2 - a^2 + 2 x_\alpha^o \tanh\left(\frac{2 a x_\alpha^o}{\Lambda_\alpha'^{\,2}}\right) \right] = \frac{k_B T}{\Lambda_\alpha^2} \quad (5.7)$$

where $\Lambda_\alpha'^{\,2} \equiv \Lambda_\alpha^2 + w^2$

Eqs. (5.6) and (5.7) imply

$$x_\alpha^o = \tanh\left(\frac{2 a x_\alpha^o}{\Lambda'^{\,2}}\right) \quad (5.8)$$

$$\Lambda_\alpha^2 = 2 a'^{\,2} + \sqrt{4 a'^{\,4} + w^4} \quad (5.9)$$

$$a'^{\,2} \equiv a^2 - (x_\alpha^o)^2$$

For $2a^2 \leq w(T)^2$, $p_T(x)$ has a single maximum at $x=0$ and Eqs. (5.8) and (5.9) have the unique solution

$$x_o^o = 0$$

$$\Lambda_o^2 = 2a^2 + \sqrt{4a^4 + w^4} \quad (2a^2 \leq w^2) \quad (5.10)$$

For $2a^2 > w(T)^2$, the maximum of $p_T(x)$ at 0 bifurcates. However,

$\tilde{p}_T(x)$ will not bifurcate until $w(T)$ becomes small enough so that solutions of Eqs. (5.8) and (5.9) with $x^o{}_\alpha = \pm \Delta x_b \neq 0$; $\Lambda_\alpha = \Lambda_b$ exist. The condition specifying $w(T_b)$, where T_b is the highest temperature where bifurcated solutions can exist, is obtained by substituting (5.9) into (5.8) to yield

$$\xi = \tanh\left\{\xi / \left[v^2 + (1 - \xi^2) + \sqrt{(1 - \xi^2)^2 + v^4}\right]\right\} \quad (5.11)$$

$$\xi \equiv \frac{\Delta x_b^o}{a}$$

$$v \equiv \frac{w^2}{2a^2}$$

Numerical analysis shows that Eq. (5.11) has real non-zero solutions for ξ only when

$$v^2 \leq 0.20 \quad (5.12)$$

That is, as T and $w(T)$ decrease, bifurcation is first possible when

$$w(T_b) = 0.63\,a \quad (5.13)$$

Eq. (5.11) implies that the bifurcated packets will be first be located at

$$\pm \Delta x_b^o = \pm\, 0.96\,a \quad (5.14)$$

with [using Eq. (5.9)]

$$\Lambda_b = 0.77\,a \quad (5.15)$$

Using (5.10), (5.13) and (5.15), we find that

$$\Lambda_o(T_b) = 2.0\,a \quad (5.16)$$

implying that

$$\Lambda_b = 0.38\,\Lambda_o(T_b) \equiv \gamma\,\Lambda_o(T_b) \quad (5.17)$$

$$\Delta x_b^o = 0.48\,\Lambda_o(T_b) \equiv \chi\,\Lambda_o(T_B) \quad (5.18)$$

This agrees with the intuitive idea that branching will occur when the width of each subpacket is about half the width of the parental packet. Using Eqs. (5.3) and (5.16) we get

$$\frac{\partial^2 \tilde{H}_{\gamma \Lambda_o, T_b}(x)}{\partial x^2}\bigg|_{x=x_o^o=0} = \frac{-8.3 k_B T}{\Lambda_o^2 (T_b)} \equiv \frac{-\rho k_B T}{\Lambda_o (T_b)} \tag{5.19}$$

In practice, a and w(T) will be unknown, but Eqs. (5.16)-(5.18) depend only on the known values $\Lambda_o (T_b)$ and $x^o{}_o (T_b)$ and provide the basis for a test-and-branch procedure in which we check to see if branched solutions to Eq. (2.11) and (2.12) can exist with $\Lambda = \gamma \Lambda_o$. In more general form (replacing $x \rightarrow R$ and $o \rightarrow \alpha$) the algorithm is:

(* Given $R^o{}_\alpha (\tau)$ and $\Lambda_\alpha (\tau)$ *)

$$\text{if MIN} \left\{ \frac{\partial^2 \tilde{H}_{\gamma \Lambda_\alpha (\tau), T(\tau)}(R)}{\partial R^2}\bigg|_{R=R_\alpha^o(\tau)} \right\} \leq \frac{-\rho k_B T}{\Lambda_\alpha^2 (\tau)} \tag{5.20}$$

then begin
  initialize $R^o = R_\alpha^o (\tau) \pm \chi \Lambda_\alpha (\tau)$ (5.21)

  $$\Lambda = \gamma \Lambda_\alpha (\tau) \tag{5.22}$$

  iteratively find branched solutions of Eqs. (2.10) and (2.11)[14]
end

---

[14]There are four non-zero solutions to Eqs. (5.11) and (5.9) for $T < T_b$. The two that satisfy $\lim_{w \to 0} |x^o{}_b| = a$; $\lim_{w \to 0} \Lambda_b = 0$ are centered over the peaks in $p_T(R)$ and are the ones of interest. The other two (which satisfy $\lim_{w \to 0} |x^o{}_b| = 0.80\, a$; $\lim_{w \to 0} \Lambda_b = 1.21 a$) are uninteresting and are stable only because they continue to overlap both peaks in $p_T(R)$ as $w \to 0$. In practice, the values of χ and γ used in Eqs. (5.21) and (5.22) will be offset to ensure that the appropriate solutions are found.

Although the real $p_T(R)$ will not have the simple form of Eq. (5.2), its packet branching will be governed by similar self-consistency considerations. In the multidimensional case, the positions of the branched solutions will be offset from $R^o{}_\alpha(\tau)$ by a vector of magnitude $\sim \chi \, \Lambda_\alpha(\tau_b)$ in the direction of the eigenvector of the Hessian of $\tilde{H}_{\gamma \Lambda_\alpha(\tau_b), T(\tau_b)}(R)$ which satisfies (5.20).[15] Approximately the same values of χ, γ and ρ given by Eqs. (5.17)-(5.19) can be used.[14] High accuracy is not required since χ and γ only provide initial values for iterative solution. Similarly, the precise value of ρ is not critical: if it is too large, branching will not be detected at the earliest possible point; if it is too small, some unnecessary attempts will be made to find iterative solutions that do not exist.

Other branch-testing algorithms can also be devised. For example, instead of examining $\partial^2 \tilde{H}_{\Lambda,T} / \partial R^2$ at $\Lambda = \gamma \Lambda_0$ by Eq. (5.20), we can examine $\partial^4 \tilde{H}_{\Lambda,T} / \partial R^4$ at $\Lambda = \Lambda_0$. The computational efficiency of the alternative methods must be compared.

The packet occupation probability for each branched solution, p_b, is smaller than the probability for the merged solution, p_o, since the total probability is split between the two branched solutions. Using Eqs. (2.10), (5.3), and (5.14)-(5.16) we get

$$p_o = 1.08 \tag{5.23}$$

$$p_b = 0.56 \tag{5.24}$$

However, $2 p_b > p_o$, reflecting the fact that the branched solutions provide a better fit to $p_T(x)$ than the merged solution. (Formally, this is the reason that the merged solution is replaced.) These discontinuities in the p_α's at branch points are schematically represented in Fig. 2.

---

[15]More than one eigenvector of the Hessian will satisfy (5.19) if a packet branches into more than two subpackets. The bifurcation procedure can be extended to accommodate this case.

6 The Molecular Conformation Problem

The packet annealing method depends on the feasibility of evaluating the effective potential $\tilde{H}_{\Lambda,T}(R)$ from Eq. (2.4). For some objective functions, the integrations will require as much computation as a local grid-search and will be impractical. However, for some H(R) of simple structure, (2.4) can be analytically approximated. Prediction of molecular conformation involves this type of H(R).

The probability distribution for a system of n atoms in thermal equilibrium with a heat bath at temperature T is proportional to the Gibbs distribution, Eq. (1.1), where R is the set of coordinates needed to specify the atom positions [$R=\{\bar{r}_i \in R^3; i=1, n\}$] and H is the energy function (Hamiltonian). For many problems, H(R) has the form

$$H(R) = \sum_{i=1, j>i}^{n} h_{ij}(|\bar{r}_i - \bar{r}_j|) + \sum_{i=1, j>i, k>j}^{n} h_{ijk}(|\bar{r}_i - \bar{r}_j|, |\bar{r}_j - \bar{r}_k|, |\bar{r}_i - \bar{r}_k|) + \ldots$$

(6.1)

That is, H is the sum of interatomic 2-body, 3-body,... potentials. For many important problems, n~ $O(10^2 - 10^4)$. The h_{ij}, h_{ijk},... are relatively simple functions that represent the interactions between different types of atoms and have lower, but not upper bounds (e.g., $-h_{ij}^{max} < h_{ij} < \infty$); that is, there are limits on attractive but not on repulsive potentials. Since only a small number (<10) of different atom types are usually involved, there are only a relatively small number of independent h_{ij}, h_{ijk} ... The complexity of the minimization problem arises from the summation and the fact that the $n(n-1)/2$ $|\bar{r}_i - \bar{r}_j|$ terms are not completely independent.[16]

---

[16] H(R) is invariant under rigid-body translations and rotations so it only depends on $3n(n-1)/2 - 6$ independent coordinates, and the global minimum is a 6-dimensional hypersurface. For pedagogical simplicity, we ignore this complication for this discussion.

For simplicity, we consider the problem of finding the lowest energy state of a "microcluster" of identical atoms that interact by a two-body potential:

$$H(R) = \sum_{i=1, j>i}^{n} h(|\bar{r}_i - \bar{r}_j|) \tag{6.2}$$

Some examples of low energy states are shown in Fig. 5. This problem, itself important in chemistry (Phillips, 1986; Duncan and Rouvray, 1989; Pool, 1990), is a model for the related but more complex problem of predicting the stable conformations of biological macromolecules such as proteins and nucleic acids (e.g., Brooks et al., 1988, for review). For inert atoms, h represents the VanderWaal's interaction and is often taken as a Lennard-Jones 6-12 potential:[17]

$$h(r) = E_o \left[\left(\frac{\sigma}{r} \right)^{12} - 2 \left(\frac{\sigma}{r} \right)^{6} \right] \tag{6.3}$$

E_0 and σ set the objective (energy) and spatial scales, respectively. This potential is positive and unbounded as $r \geq 0$, corresponding to the "hard core" repulsive interaction between adjacent atoms, and has an attractive well of depth E_0 with minimum at $r = \sigma$, the atom pair equilibrium distance (see Fig. 6; $\lambda=0$, T=1).

Most if not all local minima of Eqs. (6.2) and (6.3) for small microclusters ($n \leq 13$) have been identified by exhaustive search of the potential energy surface (Hoare and McInnes, 1976, 1983). It has been suggested that the number of local minima grows roughly like $\exp(-2.52 + 0.36n + 0.029n^2)$ (Hoare, 1979), indicating that identification of the global minimum for even moderately large n is a problem of significant

---

[17] Most of the discussion is valid for general h(r). The Lennard-Jones form is only used for numerical calculations.

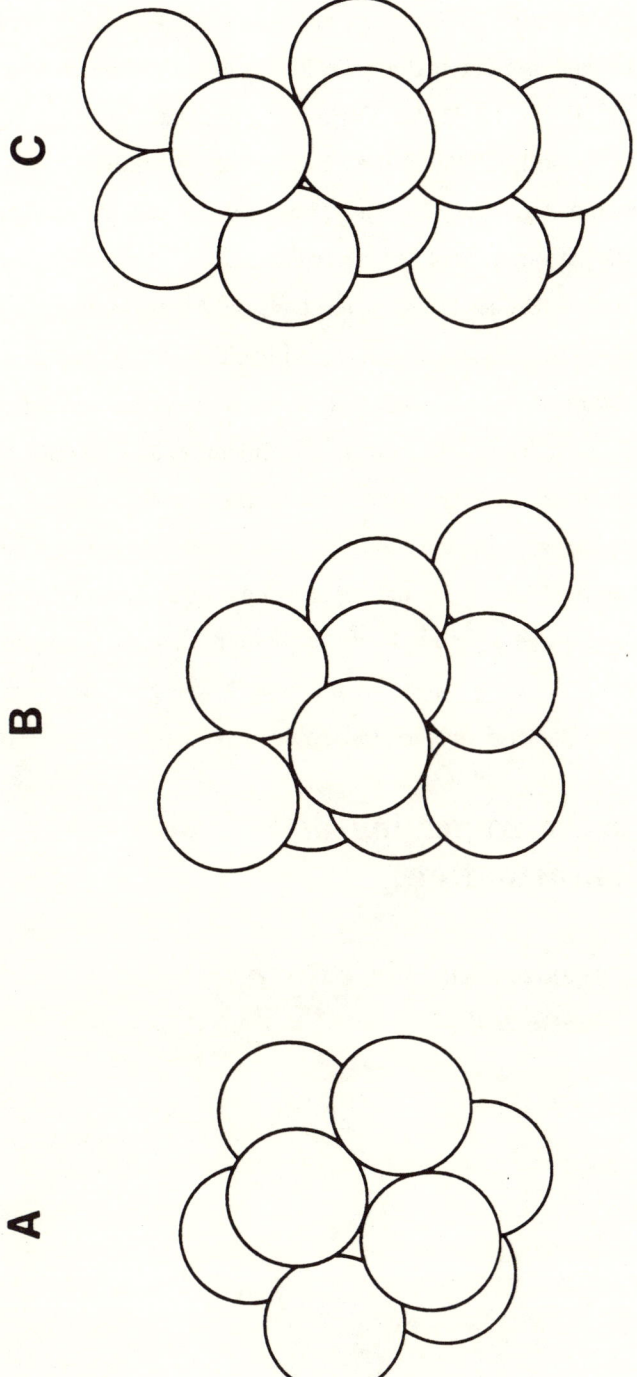

Fig. 5. Global and Local Microcluster Energy Minima. Three conformations of 13 Lennard-Jones atoms corresponding to minima of Eqs. (6.2) and (6.3) are shown. (A) Global minimum with icosahedral symmetry; (B) and (C) higher energy metastable local minima.

complexity. Putative global minima in the range n ≤ 46 have been identified by studying polyhedral growth sequences (Hoare and Pal, 1971; Farges et al., 1985), by Monte Carlo and simulated annealing methods (Freeman and Doll, 1986; Wille, 1987) and by molecular dynamics (Honeycutt and Andersen, 1987). These results provide standards for evaluating the performance of new algorithms.

Hoare and McGinnes (1976, 1983) have compared the number of local minima for microclusters interacting with either a Lennard-Jones 6-12 potential or with a slightly smoother $\alpha = 3$ Morse potential [$h(r) = \{1 - \exp[-\alpha(1-r)]\}^2 - 1$]. They find that even the minor relative smoothing of the Morse potential causes a large reduction in the numbers of local minima. For example, for n = 13, the Lennard-Jones potential supports 988 local minima while the Morse potential supports only 36 minima. This supports the intuitive idea that the complexity of the minimization problem will be reduced during the early stages of the annealing algorithm when highly averaged effective energy functions are being used.

7 Packet Annealing of Microcluster Conformation

7.1 Approximating the Effective Potential

Eqs. (2.4) and (6.2) yield

$$\tilde{Z}_{\Lambda,T}(R) \equiv e^{-\tilde{H}_{\Lambda,T}(R)} =$$

$$C(\Lambda) \int \exp\left\{-\sum_{i=1,j>i}^{n} h(|\bar{r}_i' - \bar{r}_j'|)/k_B T - \frac{1}{\Lambda^2}\sum_{i=1}^{n} |\bar{r}_i' - \bar{r}_i|^2\right\} \prod_{i=1}^{n} d\bar{r}_i'$$

(7.1)

In general, this complex integral can not be exactly evaluated, but it can be evaluated in two special cases:

If h is quadratic,

$$h(r) = \frac{g}{2} r^2 \quad \text{[quadratic case]} \tag{7.2}$$

then (7.1) can be analytically evaluated (see Appendix A) to give

$$e^{-\tilde{H}_{\Lambda,T}(R)} \propto \prod_{i=1,j>i}^{n} e^{-h_{\sqrt{n}\Lambda,T}(|\bar{r}_i - \bar{r}_j|)/k_B T} \quad \text{[quadratic case]} \tag{7.3}$$

where

$$\tilde{z}_{\lambda,T}(|\bar{r}|) \equiv e^{-\tilde{h}_{\lambda T}(|\bar{r}|)} \equiv (\sqrt{\pi}\,\lambda)^{-3} \int e^{-h(|\bar{r}'|)/k_B T} e^{-|\bar{r}' - \bar{r}|^2/\lambda^2} d\bar{r}' \tag{7.4}$$

That is

$$\tilde{H}_{\Lambda,T}(R) = \sum_{i=1,j>i}^{n} \tilde{h}_{\sqrt{n}\Lambda,T}(|\bar{r}_i - \bar{r}_j|) + c_q(n,\Lambda,T) \quad \text{[quadratic h]}$$

(7.5)

Alternatively, if $h(r)$ is a bounded function, then at high temperature T, where $|h(r)|/k_B T \ll 1$, we can approximate $\exp[-h(|\bar{r}_i' - \bar{r}_j'|)/k_B T] \approx 1 - h(|\bar{r}_i' - \bar{r}_j'|)/k_B T$ and evaluate (7.1) to get

$$\tilde{H}_{\Lambda,T}(R) \approx \sum_{i=1, j>i}^{n} \tilde{h}_{\sqrt{2}\Lambda,T}(|\bar{r}_i - \bar{r}_j|) + c_{ht}(n) \qquad [h(r)/k_B T \ll 1]$$

(7.6)

where we have again used definition (7.4).

Comparing (7.5) and (7.6) we see that in both cases

$$\tilde{H}_{\Lambda,T}(R) \approx \sum_{i=1, j>i}^{n} \tilde{h}_{\lambda,T}(|\bar{r}_i - \bar{r}_j|) + \text{constant} \qquad (7.7)$$

where $\lambda = f(n, \Lambda, T)$; $\lim_{\Lambda \to 0} f = 0$

Thus, it seems reasonable to use an approximation of this sort in the general case. The fact that $f(n,\Lambda,T)$ is unknown is unimportant since Λ will be determined self-consistently by homologs of Eqs. (2.11) and (2.12). This "factorization approximation" corresponds to averaging over atom positions in pairs and assuming that the effects of cross-correlations between pairs can be absorbed into Λ without changing the functional form of the effective interaction.[18] Even in cases where Eq. (7.5) is a poor approximation, it satisfies Eq. (2.6), the boundary condition as $\Lambda \to 0$, up to an unimportant constant and provides a natural method for spatial averaging of H(R).

The factorization approximation to $\tilde{H}_{\Lambda,T}(R)$ is actually a special case of the more general "effective harmonic approximation" which is applicable when anisotropic packets are employed (i.e., when Λ is a symmetic matrix rather than a single number). In this approximation, Eq.

---

[18] The factorization approximation is analogous to the two-body term in the Mayer cluster expansion of the partition function in statistical mechanics (Feynman, 1972, for pedagogical review), and it may be possible to represent its corrections in terms of 3-body, 4-body... integrals. However, it is unlikely that it would be computationally feasible to include more than a 3-body correction term.

(7.7) is replaced by a generalized form with distinct λ_{ij}. The central point is that, even with anisotropic packets, the effective potential $\tilde{H}$ can be approximated in terms of the two-body effective potentials $\tilde{h}$. A full exposition of the effective harmonic approximation will be deferred.

7.2 Evaluation of $\tilde{h}_{\lambda,T}(r)$

The three-dimensional integral in Eq. (7.4) can be partially evaluated in spherical coordinates and reduced to[19]

$$\tilde{z}_{\lambda,T}(r) = \frac{2 e^{-r^2/\lambda^2}}{r \lambda \sqrt{\pi}} \int_0^\infty e^{-[h(r')/k_B T + r'^2/\lambda^2]} \sinh(2 r r'/\lambda^2) \, r' \, dr'$$

$$= \frac{\lambda e^{-r^2/\lambda^2}}{r \sqrt{\pi}} \int_0^1 e^{-h[\lambda \sqrt{\log(1/v)}]/k_B T} \sinh\left[\frac{2 r \sqrt{\log(1/v)}}{\lambda}\right] dv \qquad (7.8)$$

The behavior of $\tilde{h}_{\lambda,T}(r)$ for the Lennard-Jones potential [Eq. (6.3)], calulated by numerical integration of (7.8) for varying λ (left panels) and varying T (right panels) is shown in Fig. 6. As expected, $\tilde{h}_{\lambda,T}(r)$ becomes progressively smoother as λ increases. The suppression of the repulsive singularity at $r = 0$ and the convexification of $\tilde{h}_{\lambda,T}(r)$ for large λ (e.g., see $\lambda = 1.5$, T=1 panel) deserves special note. This phenomenon occurs because the spatial average in Eq. (7.4) is over the probability density, not over the singular objective function itself. The convexification at large λ reflects the fact that, to low spatial resolution, the most probable conformation of two

---

[19] In spherical coordinates Eq. (7.4) becomes

$$\tilde{z}_{\lambda,T}(|r|) = \frac{e^{-|r|^2/\lambda^2}}{(\sqrt{\pi} \lambda)^3} \int_0^\infty e^{-h(r')/k_B T - r'^2/\lambda^2} r'^2 \int_{-1}^1 e^{2r'|r|\cos\theta/\lambda^2} \int_0^{2\pi} d\phi \, d\cos\theta \, d r'$$

The ϕ and $\cos\theta$ integrals are evaluated to yield the first equality in (7.8). The second form, which is more convenient for numerical computation, is obtained by the substitution $r'^2 = -\lambda^2 \log v$.

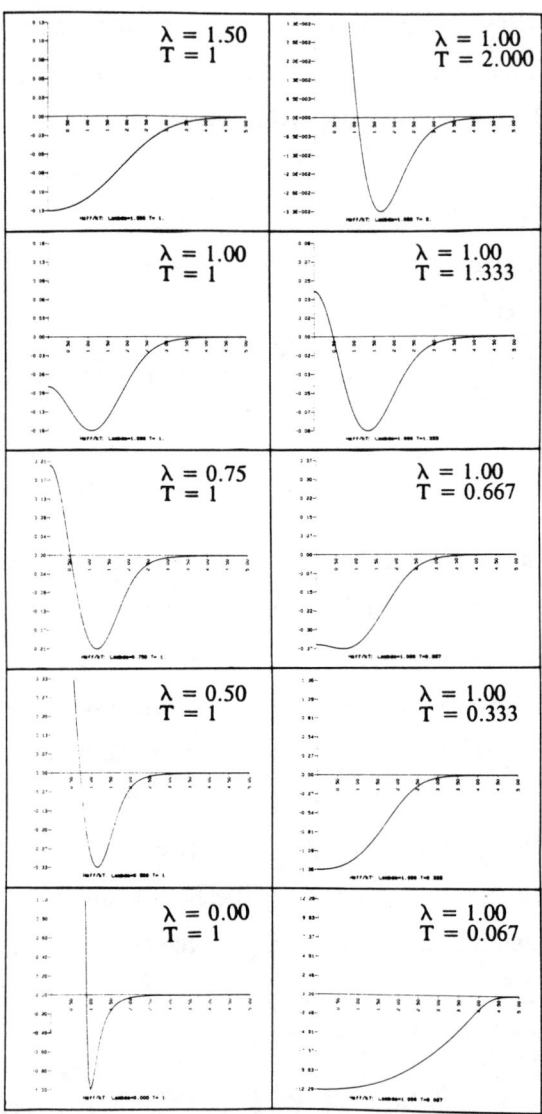

Fig. 6. $\tilde{h}_{\lambda,T}$. This is calculated from the 6-12 Lennard Jones potential with $E_0=1$, $\sigma=1$ [Eq. (6.3)] using Eq. (7.8). Effective potentials for varying λ, constant T (left panels) and varying T, constant λ (right panels are shown (ordinate = $\tilde{h}$, abscissa = r). The lower left panel ($\lambda = 0$, $\tau = 1$) is identical to h(r).

atoms occurs when they are together $\tilde{h}_{\lambda,T}(r)$ also broadens as T is lowered while keeping λ fixed. This is because lower energy atoms (with lower thermal velocities) do not penetrate and sample much of the repulsive core. Thus, their averaged interaction tends to be more attractive. Conversely, at higher temperatures, the bounded attractive part of the potential becomes less significant than the unbounded repulsive part so the minimum of $\tilde{h}_{\lambda,T}(r)$ is found at increasingly large distances. The panels in Fig. 6 do <u>not</u> represent a typical packet annealing progression of $\tilde{h}_{\lambda(\tau),T(\tau)}(r)$ since both λ and T will change simultaneously during the process.

7.3 Critical λ and T for Convexification

Convexification of $\tilde{h}_{\lambda,T}(r)$ implies convexification of $\tilde{H}_{\Lambda,T}(R)$ and provides a natural condition for choosing λ_{hi} for the initialization step (see Sec. 4). We select $\lambda_{hi} > \lambda_c$ (T$_{hi}$), where λ_c (T$_{hi}$) is the value at which convexification first occurs. λ_c is determined implicitly by (see Appendix B)

$$\int_0^\infty [e^{-h(u\lambda_c)/kT} - 1]\, \frac{d}{du}\, [u^3\, e^{-u^2}]\, du = 0 \qquad (7.9)$$

λ_c decreases with decreasing T (see Fig. 6). Applying Laplace's approximation to (7.9) as T $\to$ 0, we find that

$$\lim_{T \to 0}\, \lambda_c/\sigma = \sqrt{\frac{2}{3}} \qquad (7.10)$$

where σ is defined in Eq. (6.3). Convexification does not occur for T > T$_c$. T$_c$ is determined (see Appendix B) by

$$\int_0^\infty [e^{-h(u)/k T_c} - 1]\, u^2\, du = 0 \qquad (7.11)$$

Numerically solving Eq. (7.11) for the Lennard-Jones 6-12 potential, we get[20]

$$T_c = \frac{3.27 \, E_o}{k_B} \qquad (7.12)$$

8 Numerical Testing

We have not yet tested an implementation of the entire packet annealing algorithm. However, preliminary tests to demonstrate the feasibility of computing and minimizing $\tilde{H}_{\Lambda,T}(R)$ have been completed. Tables of $\tilde{h}_{\lambda,T}(r)$ for discrete values of λ, T and r were pre-calculated and values for arbitrary r were determined by cubic-spline interpolation. This allowed $\tilde{h}_{\lambda,T}(R)$ and its first and second derivatives to be rapidly evaluated. $\tilde{H}_{\Lambda,T}(R)$ was evaluated using the factorization approximation and local minimizations were performed using the conjugate gradient method.

A partial implementation was tested with $5 \le n \le 24$ by tracing a single trajectory as λ was decreased in small steps from $\lambda > \lambda_c(T)$ to $\lambda = 0$ at fixed $T < T_c$. Execution required relatively little computation and was carried out on a microcomputer. The overhead for computing the $\tilde{h}_{\lambda,T}$ tables was minimal; most time was used by the conjugate gradient minimizer.[21]

---

[20]The physical interpretation of T_c is not understood. It may be related to the physical critical point temperature, but the analysis contains no homolog to the physical critical point pressure. The physical critical temperature for argon is $151^{\circ}K$ (Kauzmann, 1966). Using the argon value $E_o \sim 120^{\circ}K \, k_B$ (Allen and Tildesley, 1987), $T_c \sim 390^{\circ}K$.

[21]Computations took from a few minutes (N=5) to about 20 hrs (N=24) on an IBM PS/2 Model 80. This probably represents an overestimate of computational effort since λ was decremented in conservatively small steps (50 steps/run) and no effort was made to optimize the conjugate gradient minimizer.

The minima identified by this procedure are compared with the best minima found by other means in Table 1. The fact that even this crude method finds many of the putative global minima correctly is encouraging. One of the computed minima (n = 23) had lower energy than the lowest minimum identified by studying polyhedral growth sequences (Hoare and Pal, 1971), although a lower value has recently been identified by both geometric (Farges et al., 1985) and simulated annealing (Wille, 1987) methods.

The 3-dimensional conformations were graphically examined as λ was reduced. As expected, for $\lambda > \lambda_c$, all atoms were condensed and located at the same point. As λ was decreased, the overlap between atoms decreased as they were "pushed" apart by the increasingly hard core of the potential (re Fig. 6, left panels). Spatial symmetry breaking (conformations going from higher to lower symmetry) was observed at some values of λ. In the complete algorithm, these events would be associated with trajectory branching. However, only a single, randomly selected, broken-symmetry trajectory was followed in this partial implementation. This may be the reason that the global minimum was not identified in some cases.

9 Discussion

The packet annealing method is a synthesis of both temperature-annealing and spatial-averaging methods which uses variable coarse-graining of both spatial and objective function values. It employs effective objective functions that are related to the effective energy functions in renormalization group methods (Wilson, 1975). Simulated annealing (Kirkpatrick, 1983) effectively provides coarse-graining in objective-function but not in spatial values. Conversely, methods that provide spatial coarse-graining alone have also been proposed (e.g., Zakharov, 1970). Levitt (1983) has shown that

Table 1. Lowest Minima Found by Partial Implementation

Number of Atoms	Previous Minimum Energy	Partial Implementation Energy
5	9.104	=
6	12.303	=
7	16.505	=
8	19.822	19.766
9	24.113	=
10	28.420	=
11	32.765	=
12	37.967	=
13	44.327	=
14	47.845	=
15	52.322	=
16	56.815	55.345
17	61.318	61.095
18	66.531	66.285
19	72.659	=
20	77.177	=
21	81.685	=
22	86.148	=
23	92.844	91.348
24	97.349	93.654

The magnitudes of the energies for the lowest energy states identified for microclusters of 5 to 24 atoms interacting with the Lennard-Jones potential [Eq. (6.3) with $E_0 = 1$] are listed. Minimum energies were obtained from studies of polyhedral growth sequences (Hoare and Pal, 1971) except for N = 17 (Freeman and Doll, 1985), N = 23 (Farges et al, 1985) and N = 24 (Wille, 1987). The partial implementation values were obtained by tracking a single packet as Λ was reduced [ignoring constraint (2.11)] at fixed $k_BT = 0.033\ E_0$. Runs were initiated with $\lambda(0) > \lambda_c$ [re Eq. (7.9)] so that $\tilde{h}_{\lambda(0),T}$ was convex. The energies obtained as lim $\lambda(\tau) \to 0$ are listed.

smoothing the repulsive hard-core part of the Lennard-Jones potential accelerates the rate at which the global minimum can be found in molecular dynamics simulations of protein condensation. This type of modification, which smooths the energy surface by allowing the atoms to pass-through each other, emerges naturally in the packet annealing method (e.g., see Fig. 6). Piela et al (1989) have recently proposed a differential method based on the diffusion equation for linear smoothing of the objective function. Their method, if recast as an integral method, corresponds to averaging the objective function itself with a Gaussian test-function and is similar to the high-temperature approximation Eq. (7.6). Like the partial implementation described in Sec. 8, this method has had partial success when used for global minimization of microcluster conformations (Kostrowicki et al, submitted).

These approaches can be related to penalty function methods (Gill et al., 1981) for constrained optimization if we view the hard-core repulsive part of the potential as a constraint which keeps atoms from overlapping. Increasing the coefficient of the penalty function is analogous to decreasing λ in the packet annealing method. Sha (Sha and Blank, 1987; Sha, 1989) has shown that penalty function methods can give good results in global optimization of 2-dimensional component layouts. Components have large overlaps at the start of the procedure and gradually separate as the penalty function coefficient increases. This is analogous to the progressive separation of atoms in the partial implementation (see Sec. 8).

Novel features of the packet annealing method are that spatial averages are performed over the Gibbs density rather than over the objective function and that T and Λ are coordinately varied. The essential point is that the method identifies and focuses computational effort on the spatial scales $\Lambda_\alpha(T)$ that dominate behavior at each temperature T. The use of the Gibbs density is heuristically motivated and provides a natural regularization of positive singularities that appear in many physical-model objective (energy) functions. It has the attractive feature that spatial averages are not unduly affected by large positive excursions of the objective function. As in statistical mechanics, there is little discrimination between regions where $H(R) \gg k_B T$; thus, computational effort is focused on discriminations in the important low energy regions.

The primary obstacle to implementation is the evaluation of the effective energy function (Eq. 2.4). In one sense, (2.4) is simply a method for recasting the global minimization problem: if it could be evaluated exactly, the objective function could be convexified and global minimization would be trivial.[22] The significance of Eq. (2.4) is that it (and the associated algorithms) provides a structure for novel approximation methods such as the effective harmonic expansion. Only the framework of the method has been developed and much remains to be done. In particular, the convergence properties and practical computational complexity of the method must be evaluated.

As we have pointed out, the packet annealing method will not converge to the global minimum in all cases. In particular, global minima which occur in regions where the Gibbs distribution is, on average, small will not be covered by packets and will be overlooked (e.g., see Fig. 1F). However, it seems likely that the global minima for many physically-motivated problems will be located in regions where the Gibbs distributions are, on average, large. This is particularly plausible for objective functions that represent extensive physical properties (e.g., energy) that are sums of large

---

[22] This can be recognized by applying Laplace's approximation to Eq. (2.4) at very low temperature. In this limit, $\tilde{Z}_{\Lambda,T}(R)$ is approximately Gaussian and $\tilde{H}_{\Lambda,T}(R)$ is convex.

numbers of small, partially-independent terms. In the spirit of the Central Limit Theorem, we expect that isolated anomalies like that shown in Fig. 1F will become increasingly rare as the number of terms increases. It will be interesting to determine if there are any classes of problems for which the packet annealing algorithm can be guaranteed to find the global minimum.

The solutions obtained by packet annealing may still be of physical interest even when they are not global minima. Like simulated annealing, packet annealing models the behavior of a cooled physical system as it "seeks" low-energy states. Isolated anomalous global minima which are likely to be missed by both forms of computational annealing, are also likely to be kinetically inaccessible to a physically cooled system if the system condenses in a time period that is short compared to the period required for stochastic sampling of the entire energy surface. It has been suggested that this may be the case for some proteins and macromolecules that condense down kinetically preferred pathways to metastable states (local minima) (Levinthal, 1968). Since it is the physically-selected minima (whether global or not) that are generally of primary interest, the potential neglect of anomalous global minima may not be a significant disadvantage.

Acknowledgments

This study was initiated during a Sabbatical with the group of Dr. P. Kollman (UCSF) during 1988-1989. I thank Dr. Kollman, his research group, Dr. I. Kuntz (UCSF), Dr. W. Murray (Stanford) and Dr. P. Pardalos (Penn State) for stimulating discussions and Drs. H. Scheraga and J. Kostrowicki (Cornell) for bringing their results to my attention and for sharing their manuscript prior to publication. In particular, I thank Lisa Miller for her excellent preparation of the manuscript.

Appendix A: Effective Potential for Quadratic Energy Function

Using Eq. (7.2) and

$$\sum_{i=1, j>i}^{n} |\bar{r}_i' - \bar{r}_j'|^2 = n \sum_{i=1}^{n} |\bar{r}_i' - \bar{r}_o'|^2 \qquad (A.1)$$

where

$$\bar{r}_o' \equiv \frac{1}{n} \sum_{i=1}^{n} \bar{r}_i' \qquad (A.2)$$

Eq. (7.3) can be rewritten as

$$\tilde{Z}_{\Lambda,T}(R) = C(\Lambda) \int_{-\infty}^{\infty} e^{-\phi} \prod_{i=1}^{n} d\bar{r}_i' \qquad (A.3)$$

where

$$\phi \equiv \frac{gn}{2k_BT} \sum_{i=1}^{n} |\bar{r}_i' - \bar{r}_o'|^2 + \sum_{i=1}^{n} \frac{|\bar{r}_i' - \bar{r}_i|^2}{\Lambda^2}$$

$$C(\Lambda) \equiv (\sqrt{\pi}\,\Lambda)^{-3n}$$

We release constraint (A.2) and treat $\bar{r}_o'$ as an independent variable by introducing a Dirac δ function and integrating over $\bar{r}_o'$:

$$\tilde{Z}_{\Lambda,T}(R) = C(\Lambda) \int_{-\infty}^{\infty} e^{-\phi} \, \delta\left(\bar{r}_o' - \frac{1}{n}\sum_{i=1}^{n} \bar{r}_i'\right) d\bar{r}_o' \prod_{i=1}^{n} d\bar{r}_i'$$

(A.4)

Using

$$\delta(\bar{x}) = \frac{1}{(2\pi)^3} \int_{-\infty}^{\infty} e^{i\bar{\alpha}\cdot\bar{x}} \, d\bar{\alpha}$$

we have

$$\tilde{Z}_{\Lambda,T}(R) =$$

$$\frac{C(\Lambda)}{(2\pi)^3} \int_{-\infty}^{\infty} \exp\left\{-\phi + i\bar{\alpha} \cdot \left(\bar{r}_0' - \frac{1}{n}\sum_{i=1}^{n} \bar{r}_i'\right)\right\} d\bar{\alpha}\, d\bar{r}_0' \prod_{i=1}^{n} d\bar{r}_i' \tag{A.5}$$

The integrand in (A.5) can now be factored into a product of terms of the form $\exp[-a\,(\bar{r}_i' - b)^2]$ and the $\bar{r}_i'$ integrals can be analytically evaluated. The α and $\bar{r}_0'$ integrals then take a similar form and can be evaluated to yield

$$\tilde{Z}_{\Lambda,T}(R) = \left(\frac{g\,n\,\Lambda^2}{2\,k_B T} + 1\right)^{-\frac{3(n-1)}{2}}$$

$$\exp\left\{-\left[\frac{g}{(g\,n\,\Lambda^2 + 2\,k_B T)} \sum_{i=1, j>i}^{n} |\bar{r}_i - \bar{r}_j|^2\right]\right\} \tag{A.6}$$

By similar means, we calculate $\tilde{z}_{\lambda,T}(r)$ from Eqs. (7.2) and (7.4):

$$\tilde{z}_{\lambda,T}(|\bar{r}|) = \left(\frac{g\,\lambda^2}{2\,k_B T} + 1\right)^{-3/2} \exp\left\{-\left[\frac{g\,|\bar{r}|^2}{(g\,\lambda^2 + 2\,k_B T)}\right]\right\} \tag{A.7}$$

Comparing (A.6) and (A.7) gives

$$\tilde{Z}_{\Lambda,T}(R) = \left(\frac{g\,n\,\Lambda^2}{2\,k_B T} + 1\right)^{3(n-1)(n-2)/4} \prod_{i=1, j>i}^{n} \tilde{z}_{\sqrt{n}\,\Lambda,T}(|\bar{r}_i - \bar{r}_j|) \tag{A.8}$$

Taking the logarithm of both sides yields Eq. (7.5).

Appendix B: Convexification of $\tilde{h}_{\lambda,T}(r)$

Examination of Fig. 6 shows that $\tilde{h}_{\lambda,T}(r)$ will be convex when

$$\frac{\partial^2 \tilde{h}_{\lambda,T}(r)}{\partial r^2}\bigg|_{r=0} \geq 0 \quad \text{[convexity condition]} \tag{B.1}$$

Since

$$\frac{\partial^2 \tilde{h}_{\lambda,T}(r)}{\partial r^2} = -k_B T \left\{ \frac{\partial^2 \tilde{z}_{\lambda,T}(r)}{\partial r^2}/\tilde{z}_{\lambda,T}(r) - \left[\frac{\partial \tilde{z}_{\lambda,T}(r)}{\partial r}/\tilde{z}_{\lambda,T}(r)\right]^2 \right\} \tag{B.2}$$

and [from Eq. (7.4)]

$$\frac{\partial \tilde{z}_{\lambda,T}(r)}{\partial r}\bigg|_{r=0} = 0 \tag{B.3}$$

(B.1) is equivalent to

$$\frac{\partial^2 \tilde{z}_{\lambda,T}(r)}{\partial r^2}\bigg|_{r=0} \leq 0 \quad \text{[convexity condition]} \tag{B.4}$$

Expanding Eq. (7.8) about r=0 yields

$$\tilde{z}_{\lambda,T}(r) = \frac{4}{\sqrt{\pi}} \int_0^\infty e^{-h(u\lambda)/k_B T} e^{-u^2} \left[1 + \frac{r^2}{\lambda^2}\left(\frac{2}{3}u^2 - 1\right) + O(r^4)\right] u^2\, du \tag{B.5}$$

Substitution of (B.5) into (B.4) yields

$$\int_0^\infty e^{-h(u\lambda)/k_B T} e^{-u^2} u^2 \left(\frac{2}{3} u^2 - 1\right) du \leq 0 \quad \text{[convexity condition]}$$

(B.6)

The u terms in the integrand of (B.6) can be rewritten as

$$e^{-u^2} u^2 \left(\frac{2}{3} u^2 - 1\right) \propto - \frac{d}{du} (u^3 e^{-u^2})$$

(B.7)

and a constant can be subtracted from the exp[-h] term [so that it approaches 0 as $u\lambda/\sigma \to \infty$; σ defined in Eq. (6.3)] without affecting the value of the integral. This yields Eq. (7.9).

From Fig. 6 we see that $\tilde{h}_{\lambda,T}(r)$ tends towards convexity as λ increases. The highest temperature, T_C, for convexification is determined by evaluating Eq. (7.9) for $\lambda/\sigma \to \infty$. In this limit, the only contribution to the integral occurs for $u \ll 1$ and the exp[-u^2] term can be ignored. After rescaling $u\lambda \to u$, Eq. (7.9) reduces to Eq. (7.10).

References

Allen, M.P. and D.J. Tildesley (1987). Computer Simulation of Liquids. (Clarendon Press: Oxford), p. 9.

Brooks, C.L.,III, M. Karplus, and B. M. Pettitt. (1988). Proteins: A Theoretical Perspective of Dynamics, Structure, and Thermodynamics. (J. Wiley and Sons: New York).

Duncan, M.A. and D. H. Rovray (1989). Microclusters, Sci. Am. 261:110-115.

Farges, J., M.F. DeFeraudy, B. Raoult, and G. Torchet. (1985). Cluster models made of double icosahedron units. Surface Sci. 156:370-378.

Feynman, R.P. (1972). Statistical Mechanics. (W.A. Benjamin: Reading, Mass.) pp. 105-110.

Freeman, D.L. and J.D. Doll. (1985). Quantum Monte Carlo study of the thermodynamic properties of argon clusters: the homogeneous nucleation of argon in argon vapor and "magic number" distributions in argon vapor. J. Chem. Phys. 82:462-471.

Gill, P.E., W. Murray, and M.H. Wright. (1981) Practical Optimization. (Academic Press: London), pp. 207-219.

Hoare, M.R. (1979). Structure and dynamics of simple microclusters. Adv. Chem. Phys. 40:49-135.

Hoare, M.R. and J. McInnes (1976). Statistical mechanics and morphology of very small atomic clusters. Faraday Discussions Chem. Soc. 61:12-24.

Hoare, M.R. and J. McInnes (1983). Morphology and statistical statics of simple microclusters. Advan. Phys. 32:791-821.

Hoare, M.R. and P. Pal. (1971). Physical cluster mechanics: statics and energy surfaces for monatomic systems. Adv. Phys. 20:161-196.

Honeycutt, J.D. and H. C. Andersen. (1987) Molecular dynamics study of melting and freezing of small Lennard-Jones clusters. J. Phys. Chem. 91:4950-4963.

Kauzmann, W. (1966). Kinetic Theory of Gases. (W.A. Benjamin: New York). p. 31.

Kawai, H., T. Kikuchi, and Y. Okamoto (1989). A prediction of tertiary structures of peptide by the Monte Carlo simulated annealing method. Protein Eng. 3:85-94.

Kirkpatrick, S., C.D. Gelatt, Jr., and M.P. Vecchi. (1983). Optimization by simulated annealing. Science 220:671-680.

Kostrowicki, J., L. Piela, B.J. Cherayil and H.A. Scheraga. Performance of the diffusion equation method in searches for optimum structures of clusters of Lennard-Jones atoms. (submitted)

Levitt, M. (1983). Protein folding by restrained energy minimization and molecular dynamics. J. Mol. Biol. 170:723-764.

Phillips, J.C. (1986). Chemical bonding, kinetics, and the approach to equilibrium structures of simple metallic, molecular, and network microclusters. Chem. Rev. 86:619-634.

Piela, L., J. Kostrowicki, and H.A. Scheraga. (1989). The multiple-minima problem in the conformational analysis of molecules. Deformation of the potential energy hypersurface by the diffusion equation method. J. Phys. Chem. 93:3339-3346.

Pool, R. (1990). Clusters: strange morsels of matter. Science 248:1186-1188.

Sha, L. (1989). A macrocell placement algorithm using mathematical programming techniques. Ph.D. Thesis, Department of Electrical Engineering, Stanford University.

Sha, L. and T. Blank. (1987). ATLAS-a technique for layout using analytic shapes. Digest of Technical Papers. IEEE Conf. on Computer-Aided Design, pp. 84-86.

Wille, L.T. (1986). Searching potential energy surfaces by simulated annealing. Nature 324:46-48.

Wille, L.T. (1987). Minimum-energy configurations of atomic clusters: new results obtained by simulated annealing. Chem. Phys. Lett. 133:405-410.

Wilson, K.G. (1975). The renormalization group: critical phenomena and the Kondo problem. Rev. Mod. Phys. 47:773-840.

Zakharov, V.V. (1970). The method of integral smoothing in many extremal and stochastic problems. Eng. Cybernetics 4:637-642.

Mixed-Integer Linear Programming Reformulations for Some Nonlinear Discrete Design Optimization Problems

I.E. Grossmann[1], V.T. Voudouris[1] and O. Ghattas[1,2]

Abstract

This paper deals with a special class of nonlinear discrete design optimization problems which involve nonlinear separable objective functions and bilinear constraints. These constraints involve products of design and state variables in which the former are restricted to take discrete values Two special cases are identified for which advantage can be taken of the discrete nature of the design variables to reformulate these problems as MILP models which can be solved to global optimality. The computational expense can be reduced with SOS1 sets and a simple solution strategy that is proposed. The application of the MILP reformulations is applied to multiproduct batch plant problems in chemical engineering and

---

[1] Engineering Design Research Center, Carnegie Mellon University, Pittsburgh, PA 15213.

[2] The authors gratefully acknowledge financial support from the Engineering Design Research Center.

to structural design problems in civil engineering. Numerical results and comparisons with other methods are also presented.

1. INTRODUCTION

Many problems in engineering design give rise to nonconvex nonlinear programming (NLP) problems (e.g. see Floudas and Pardalos, 1990). Furthermore, quite often due to manufacturing constraints, design variables are restricted to take discrete values for selecting standard sizes which gives rise to mixed-integer nonlinear programs (MINLP) (e.g. see Papalambros and Wilde, 1988; Grossmann, 1990). These problems in many cases have a continuous relaxation that corresponds to a nonconvex NLP. Due to the difficulty in solving these problems, many design models reported in the literature have assumed continuous sizes, and used ad-hoc rounding procedures. It is the purpose of this paper, to show that important classes of discrete design optimization problems that involve separable objective functions and bilinearities in the constraints, can in fact be reformulated as mixed-integer linear programs (MILP), and therefore solved rigorously to global optimality.

This paper will be organized as follows. In Section 2 we will present basic NLP and MINLP formulations that arise in discrete design optimization problems. In Section 3 we will consider two special cases of bilinear constraints that arise in many design applications. We will show that advantage can be taken of the discrete nature of the design variables in order to reformulate these problems as MILP models. Section 4 will compare the proposed formulations with other linearization schemes and briefly discuss computational aspects. Sections 5, 6 and 7 will present the application of the reformulations to multiproduct batch plant problems in chemical engineering and to structural design problems in civil engineering. These problems have traditionally been formulated as continuous optimization problems, and thereby neglected the fact that in most practical applications only standard sizes are available. Finally, Section 8 will present some numerical results.

2. BASIC FORMULATIONS

Consider the following MINLP problem with separable objective function and with linear and nonlinear constraints:

$$\min C = \sum_{i=1}^{n} f_i(x_i)$$
$$\text{s.t.} \quad Az \leq b$$
$$g(x,z) \leq 0 \quad \quad (P1)$$
$$x \in X_D, \ z \in Z$$

where $f_i: \mathbf{R}^1 \to \mathbf{R}^1$ and $g: \mathbf{R}^{n+m} \to \mathbf{R}^q$ are continuous functions, $X_D \subset \mathbf{R}^n_+$ is a set of discrete values, $Z \subset \mathbf{R}^m$ is a set of continuous values, and $A \in \mathbf{R}^{r \times m}$, $b \in \mathbf{R}^r$. In the context of a design optimization problem C is a cost function, x is the vector of discrete design variables which are restricted to choices of standard sizes in the set X_D, and z is the vector of continuous state variables.

Since the MINLP problem (P1) is in general difficult to solve, it is common to consider the continuous relaxation of problem (P1) which leads to the NLP problem:

$$\min C = \sum_{i=1}^{n} f_i(x_i)$$
$$\text{s.t.} \quad Az \le b$$
$$g(x,z) \le 0 \quad \text{(P2)}$$
$$x \in X, \; z \in Z$$

where X is the convex hull of X_D.

The common approach is then to find an optimal solution to problem (P2) with a standard NLP solver (e.g. MINOS or SQP algorithm) and round up the variables x_i, i=1...n, to the next highest discrete value in X_D. Clearly the difficulty is that this might lead to a suboptimal solution, or to a solution that is infeasible. Furthermore, to complicate matters, the functions $f_i(x)$, i=1..n, and $g(x,z)$ are often nonconvex which can give rise to several local optima in problem (P2).

Another way to circumvent the problem of getting non-discrete design variables is to reformulate problem (P2) as an MINLP problem with 0-1 variables. That is, let $DS(i)=\{d_{i1}, d_{i2}, ... d_{iN(i)}\}$ be the set of discrete values for each design variable x_i. Furthermore, let y_{is}, s=1,...N(i), be 0-1 variables defined as follows:

$$y_{is} = \begin{cases} 1 & \text{if } x_i = d_{is} \\ 0 & \text{otherwise} \end{cases} \quad (1)$$

Then, since each variable x_i can be expressed as:

$$x_i = \sum_{s=1}^{N(i)} d_{is} y_{is} \qquad (2)$$

$$\sum_{s=1}^{N(i)} y_{is} = 1 \qquad (3)$$

problem (P2) can be reformulated as the MINLP model:

$$\min C = \sum_{i=1}^{n} f_i(x_i)$$

$$\text{s.t.} \quad Az \leq b \qquad \text{(P3)}$$

$$g(x,z) \leq 0$$

$$x_i = \sum_{s=1}^{N(i)} d_{is} y_{is} \qquad i=1,\ldots n$$

$$\sum_{s=1}^{N(i)} y_{is} = 1 \qquad i=1,\ldots n$$

$$x \in X, \ z \in Z, \ y_{is} = \{0,1\}$$

Problems (P1) and (P3) can be solved in principle with a branch and bound method (Gupta, 1980), Generalized Benders Decomposition (Geoffrion, 1972) or with the Outer-Approximation method (Duran and Grossmann, 1986). However, the inherent difficulty is that due to possible nonconvexities in the nonlinear functions, these algorithms may not converge to the global optimum. The next section will show, however, that for special cases of the nonlinear constraints $g(x,z)$ that involve bilinearities, problem (P3) can be reformulated as an MILP problem and solved to global optimality.

3. MILP REFORMULATIONS FOR SPECIAL CASES

Consider the two following particular cases for the nonlinear constraints $g(x,z)$:

a) Case 1: $\quad g_{ij} = \alpha_{ij} x_i v_j - \beta_{ij} \leq 0$, $j \in J(i)$, $i=1..n$ \hfill (4)

where v is a subvector of $z^T = [u,v]^T$ and $\alpha_{ij} \neq 0$, $\beta_{ij} \neq 0$.

b) Case 2: $\quad g_{ij} = \alpha_{ij} x_i v_j - \beta_{ij} w_j \leq 0$, $j \in J(i)$, $i=1..n$ \hfill (5)

where v and w are subvectors of $z^T = [u,v,w]^T$ and $\alpha_{ij} \neq 0$, $\beta_{ij} \neq 0$.

For simplicity we consider here the case of inequalities, although (4) and (5) could also involve equality constraints. Case 1 is clearly a particular case of Case 2, but as will be shown below it leads to a simpler reformulation which is worth considering. Also, as will be shown later in the paper, Cases 1 and 2 arise in multiproduct batch design problems, while Case 2 arises in structural design problems.

For the MILP reformulation consider first the objective function C in (P3). By introducing the binary variables y_{is} as in (1) subject to the constraints in (3), then by defining

$$c_{is} = f_i(d_{is}) \tag{6}$$

it is clear that C can be expressed by the linear combination

$$C = \sum_{i=1}^{n} \sum_{s=1}^{N(i)} c_{is} \, y_{is} \tag{7}$$

Consider now Case 1. From (4) it follows that for $x_i > 0$,[3]

$$\alpha_{ij} \, v_j \leq \frac{\beta_{ij}}{x_i} \quad j \in J(i) \, , \, i=1..n \tag{8}$$

In order to remove the nonlinearity in the right-hand side of (8), the inverse of the design variable x_i will be represented by a linear combination of inverse values of the discrete sizes; that is,

$$\frac{1}{x_i} = \sum_{s=1}^{N(i)} \frac{y_{is}}{d_{is}} \quad i=1,..n \tag{9}$$

Then by substituting (9) into (8) yields the linear inequalities,

$$\alpha_{ij} \, v_j \leq \sum_{s=1}^{N(i)} \frac{\beta_{ij}}{d_{is}} y_{is} \quad j \in J(i) \quad i=1,..n \tag{10}$$

Hence, from (7), (10), (3) and by expressing the linear constraints $Az \leq b$ as $[A_1, A_2] \begin{bmatrix} u \\ v \end{bmatrix} \leq \begin{bmatrix} b_1 \\ b_2 \end{bmatrix}$, problem (P3) can be reformulated for Case 1 as the binary MILP problem:

---

[3] If $x_i > 0$ does not hold, Case 2 applies.

$$\min C = \sum_{i=1}^{n} \sum_{s=1}^{N(i)} c_{is} \, y_{is}$$

$$\text{s.t.} \quad [A_1 \, A_2] \begin{bmatrix} u \\ v \end{bmatrix} \leq \begin{bmatrix} b_1 \\ b_2 \end{bmatrix} \quad \quad (R1)$$

$$\alpha_{ij} v_j - \sum_{s=1}^{N(i)} \frac{\beta_{ij}}{d_{is}} y_{is} \leq 0 \quad j \in J(i) \,, \, i=1..n$$

$$\sum_{s=1}^{N(i)} y_{is} = 1 \quad i=1,..n$$

$$u \in U \,, \, v \in V \quad \quad y_{is} = \{0,1\}$$

where $U \times V = Z$ and c_{is} is given by (6). Note that the interesting feature in this formulation is that it has fewer variables and fewer constraints than the MINLP model (P3).

Consider now Case 2. Substituting (2) into (5) leads to the bilinear constraints

$$\alpha_{ij} \sum_{s=1}^{N(i)} d_{is} y_{is} v_j - \beta_{ij} w_j \leq 0 \quad j \in J(i) \,, \, i=1,..n \quad (11)$$

In order to remove the bilinear terms $y_{is} v_j$, define the continuous variables v_{ijs} such that

$$v_j = \sum_{s=1}^{N(i)} v_{ijs} \quad j \in J(i) \,, \, i=1,..n \quad (12)$$

$$v_j^L y_{is} \leq v_{ijs} \leq v_j^U y_{is} \quad j \in J(i), \, s=1, N(i), \, i=1..n \quad (13)$$

where v_j^L, v_j^U are valid lower and upper bounds. Then, the constraints in (11) can be replaced by the linear inequalities

$$\alpha_{ij} \sum_{s=1}^{N(i)} d_{is} v_{ijs} - \beta_{ij} w_j \leq 0 \qquad j \in J(i), \; i=1,..n \qquad (14)$$

A proof for the equivalence of the constraints in (14), (12) and (13) with the inequalities in (5) for discrete values in x_i is given in the Appendix. Hence, from (7), (12)-(14), (3) and by expressing the linear inequalities $Az \leq b$ as

$$[A_1 \; A_2 \; A_3] \begin{bmatrix} u \\ v \\ w \end{bmatrix} \leq \begin{bmatrix} b_1 \\ b_2 \\ b_3 \end{bmatrix} \qquad (15)$$

problem (P3) can be reformulated for Case 2 as the binary MILP problem:

$$\min \; C = \sum_{i=1}^{n} \sum_{s=1}^{N(i)} c_{is} y_{is}$$

$$\text{s.t.} \; [A_1 \; A_2 \; A_3] \begin{bmatrix} u \\ v \\ w \end{bmatrix} \leq \begin{bmatrix} b_1 \\ b_2 \\ b_3 \end{bmatrix} \qquad (R2)$$

$$\alpha_{ij} \sum_{s=1}^{N(i)} d_{is} v_{ijs} - \beta_{ij} w_j \leq 0 \qquad j \in J(i), \; i=1,..n$$

$$v_j^L y_{is} \leq v_{ijs} \leq v_j^U y_{is} \qquad j \in J(i), \; s=1..N(i), \; i=1..n$$

$$v_j = \sum_{s=1}^{N(i)} v_{ijs} \qquad j \in J(i), \ i=1..n$$

$$\sum_{s=1}^{N(i)} y_{is} = 1 \qquad i=1,..n$$

$$u \in U, \ v \in V, \ w \in W \quad y_{is} = \{0,1\}$$

where $U \times V \times W = Z$ and c_{is} is given by (6). Note that in this case the number of variables and constraints is larger than in the MINLP model (P3).

4. REMARKS

The proposed linearization of the bilinear constraints in (11) with (12)-(14) can also be applied to the case when (11) is given by equality constraints. This follows trivially from the proof in the Appendix. Also, given the reformulation of the objective function in (6) and (7), no assumption is required on the form of the cost function $f_i(x_i)$. In fact this function can be discontinuous, which is not an uncommon occurrence in practice. The linearizations in (12)-(14), however, are not unique. Other alternatives include the linearizations by Glover (1975) and by Torres (1991). As will be shown below the former requires a larger number of constraints and may yield a weaker LP relaxation. The latter, which is only applicable to inequalities, requires fewer constraints, but may also yield a weaker LP relaxation.

The bilinear constraints in (11) can be linearized with the following formulation proposed by Glover (1975):

$$v_j^L y_{is} \leq v_{ijs} \leq v_j^U y_{is}$$
$$v_{ijs} \geq v_j - v_j^U(1 - y_{is}) \quad j \in J(i), \; s=1, N(i), \; i=1..n$$
$$v_{ijs} \leq v_j - v_j^L(1 - y_{is}) \tag{16}$$

which requires $4 \sum_{i=1}^{n} |J(i)||N(i)|$ inequalities. In contrast the proposed linearization scheme in (12) and (13) only requires

$2 \sum_{i=1}^{n} |J(i)||N(i)| + \sum_{i=1}^{n} |J(i)|$ constraints. Furthermore, while a point (v_{ijs}, v_j, y_{is}) satisfying (12) and (13) satisfies the inequalities in (16), the converse may not be true. For instance, assume a non-integer point y_{is} such that $v_{ijs} = v_j^U y_{is}$. Using (3) it follows from (16) that

$$v_j^L + (v_j^U - v_j^L) y_{is} \leq v_j \leq v_j^U \tag{17}$$

while (12) yields $v_j = v_j^U$. Thus, the inequalities in (16) may produce a weaker LP relaxation.

For the case when the bilinear constraints in (11) are only inequalities, Torres (1991) has shown that it is sufficient to consider the following constraints from (16):

$$v_j^L y_{is} \leq v_{ijs}$$
$$v_{ijs} \geq v_j - v_j^U(1 - y_{is}) \quad j \in J(i), \; s=1, N(i), \; i=1..n \tag{18}$$

which requires $\sum_{i=1}^{n} |J(i)|$ fewer constraints than the proposed linearization in (12) and (13). However, the above inequalities also can produce a weaker LP relaxation. For instance, setting $v_{ijs} = v_j^L y_{is}$ for a non-integer point y_{is} yields,

$$v_j \leq v_j^U - (v_j^U - v_j^L) y_{is} \qquad (19)$$

while (12) yields $v_j = v_j^L$. In summary, the proposed linearizations in (12) and (13) are tighter since they exploit the convexity condition in (3), while the ones by Glover (1975) and Torres (1991) do not.

As for the computational requirements, the reformulations (R1) and (R2) correspond to MILP problems that can be solved to global optimality with branch and bound methods such as the ones implemented in SCICONIC, MPSX, ZOOM and LINDO. Furthermore, the constraints in (3) correspond to special ordered sets of type 1 (SOS1; e.g. see SCICONIC, 1990) whose structure can be exploited to reduce the number of nodes that must be examined in the branch and bound enumeration. While problem (R1) is somewhat smaller in size than the nonlinear model in (P3), problem (R2) is potentially much larger and has the additional difficulty that lower and upper bounds v_j^L, v_j^U, must be supplied which can have a great impact in the integrality gap of the LP relaxation.

The issue of size in problem (R2) can be addressed in several ways. One is to generate cutting planes in the MILP in order to strengthen the LP relaxation (e.g. see Van Roy and Wolsey, 1987). The other one is to apply Benders decomposition so as to greatly reduce the size of the LP subproblems (see Sahinidis, 1990). In our experience, however, we have found that the greatest source of computational difficulty in the proposed MILP models (particularly in (R2)) lies in their tendency to predict small sizes in the relaxed LP creating many infeasible nodes in the branch and bound tree. To circumvent this problem we have devised a simple but rigorous solution strategy for fixing subsets of 0-1 variables in (R1) and (R2) that consists of the following steps:

<u>Step 1</u>. (Optional). Obtain an upper bound. The relaxed LP is solved to compute sizes x_i^R from (9) or (2). The MILP model is solved with binary variables y_{is} fixed to zero for which $d_{is} < x_i^R$.

<u>Step 2</u>. Predict valid lower bounds x_i^L for each size x_i. In simple models these can be obtained analytically. In more complex models these can be obtained by maximizing (9) or minimizing (2)) with the relaxed LP model. Note that if step 1 is used, those x_i^R with zero value yield valid lower bounds.

<u>Step 3</u>. Obtain the global optimum by solving the MILP model with binary variables y_{is} fixed to zero for which $d_{is} < x_i^L$, and with the upper bound obtained in step 1.

This procedure can obviously be made more effective if the MILP problems in steps 1 and 3 are solved with SOS1 sets.

5. APPLICATION TO SIMPLE BATCH PROCESS DESIGN

In order to illustrate the application of the reformulation (R1), consider the problem of sizing multiproduct batch plants with one unit per stage (see Fig. 1) and operating with single product campaigns (see Grossmann and Sargent, 1978). In these plants it is assumed that each product requires all the processing stages in the same sequence.

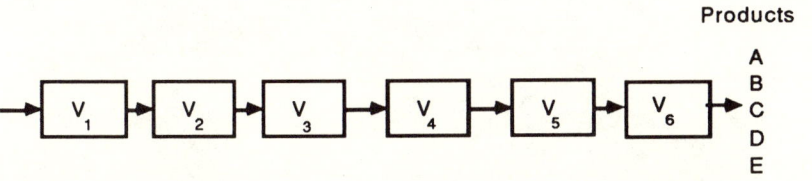

Fig. 1. Multiproduct batch plant with one unit per stage

The following parameters and variables are required to formulate the design optimization problem:

Parameters:
- N   the number of products j=1..N
- M   the number of stages i=1..M
- T_{Lj} the cycle time of product j
- Q_j demand of product j
- S_{ij} size factor for product j in stage i
- H   horizon time
- γ_i, δ_i cost coefficients for unit in stage i with $0 < \gamma_i < 1$

Variables:
- V_i size of unit in stage i
- B_j batch size of product j

If the sizes V_i are assumed to be continuous, the problem of finding an optimal design as given by the NLP:

$$\min \quad C = \sum_{i=1}^{M} \delta_i V_i^{\gamma_i}$$

(B1)

$$\text{s.t.} \quad V_i \geq S_{ij} B_j \quad j=1,..N, \quad i=1,..M$$

$$\sum_{j=1}^{N} \frac{Q_j}{B_j} T_{Lj} \leq H$$

$$B_j, V_i > 0$$

In problem (B1) the objective function represents the investment cost that is to be minimized. The first set of constraints simply states that at each stage i, the size V_i must be sufficiently large for all products j. Finally, the second constraint states that the total time for production, as given by the number of batches (Q_j/B_j) by the corresponding cycle time (T_{Lj}), should not exceed the allotted time H.

Problem (B1) corresponds to an NLP that involves nonconvexities in the objective function and in the second constraint. However, as shown by Grossmann and Sargent (1978) this problem can be transformed into a geometric programming problem with posynomial terms.

Assume, however, that the sizes V_i are only available in discrete values $DV(i) = \{V_{i1}, V_{i2}...V_{iN(i)}\}$. Rather than formulating (B1) as an MINLP according to problem (P3), we will show that it can be reformulated as an MILP. First, note that the first set of constraints in (B1), which contains the design variable V_i, is linear while the second constraint is nonlinear. In order to remove the nonlinearity in the latter, let

$$B_j = \frac{1}{b_j} \quad j=1..N \tag{20}$$

with which problem (B1) reduces to

$$\min \ C = \sum_{i=1}^{M} \delta_i V_i^{\gamma_i}$$

$$\text{s.t.} \ -V_i b_j + S_{ij} \le 0 \quad j=1..N, \ i=1..M \tag{B2}$$

$$\sum_{j=1}^{N} Q_j T_{Lj} b_j \le H$$

$$V_i > 0, \ b_j > 0$$

Note that the first set of constraints are now of the same form as equation (4) for Case 1. If we introduce the 0-1 variables y_{is} as in (3) to select the discrete size V_{is}, then by letting

$$\frac{1}{V_i} = \sum_{s=1}^{N(i)} \frac{y_{is}}{V_{is}} \tag{21}$$

as in (9), following a similar treatment for the derivation of problem (R1), the reformulated MILP corresponds to:

$$\min \ C = \sum_{i=1}^{M} \sum_{s=1}^{N(i)} c_{is} \, y_{is}$$

(RB)

$$\text{s.t.} \quad b_j \geq \sum_{s=1}^{N(i)} S_{ij} \frac{y_{is}}{V_{is}} \qquad j=1..N \quad i=1..M$$

$$\sum_{s=1}^{N(i)} y_{is} = 1 \qquad i=1..M$$

$$\sum_{j=1}^{N} Q_j T_{Lj} b_j \leq H$$

$$b_j \geq 0, \qquad y_{is} = \{0,1\}$$

where $c_{is} = \delta_i(V_{is})^{\gamma_i}$. The important feature of this model is that it readily allows the treatment of discrete sizes which have been commonly treated as continuous variables in previous work.

6. APPLICATION TO STRUCTURAL DESIGN

In order to illustrate the reformulation (R2), consider the least weight design of a truss consisting of a number of specified bars with fixed nodal locations (see Fig. 2) that is subject to a number of different loading conditions, and for which constraints on stresses, nodal displacements and bar elongations are specified (see Bremicker et al., 1990; Haftka et al., 1990; Ghattas and Grossmann, 1991).

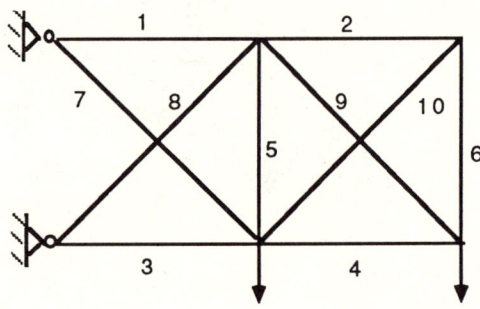

Fig. 2. Truss with 10 bars and 2 loads.

In order to find a design with minimum weight the following parameters and variables must be considered:

Parameters:

M	number of bars i=1,M
N	number of degrees of freedom[4] j =1,N
L	number of loading conditions ℓ=1,L
ℓ_i	length of bar i
E_i	modulus of elasticity of bar i
ρ_i	density of bar i
$P_{\ell j}$	jth component of load at condition ℓ
b_{ji}	direction cosine relating the force in bar i with the degree of freedom j
σ_i^L, σ_i^U	maximum stresses in compression and tension in bar i
v_i^L, v_i^U	limits on elongations in bar i
d_j^L, d_j^U	limits of displacements for degree of freedom j

Variables:

A_i	cross sectional area of bar i
$s_{i\ell}$	force in bar i for condition ℓ
$\sigma_{i\ell}$	stress in bar i for condition ℓ
$v_{i\ell}$	elongation of bar i for condition ℓ
$d_{j\ell}$	displacement at degree of freedom j for condition ℓ

---

[4]Number of degrees of freedom is given by $N = N_N e - b$, where N_N is the number of nodes, e the degrees of freedom at each node, and b is the number of support fixities.

If the cross sectional areas are assumed to be continuous, the optimal design for a truss with minimum weight is given by the NLP:

$$\min \quad C = \sum_{i=1}^{M} \rho_i \ell_i A_i \qquad \text{(ST)}$$

s.t.
(a) Equilibrium equations
$$\sum_{i=1}^{M} b_{ji} s_{i\ell} = p_{\ell j} \qquad j=1,N \quad \ell=1,L$$
(b) Compatibility equations
$$\sum_{j=1}^{N} b_{ji} d_{j\ell} = v_{i\ell} \qquad i=1,M \quad \ell=1,L$$
(c) Hooke's law
$$\frac{E_i}{\ell_i} A_i v_{i\ell} = s_{i\ell} \qquad i=1,M \quad \ell=1,L$$
(d) Stress equations
$$\sigma_{i\ell} = \frac{E_i}{\ell_i} v_{i\ell} \qquad i=1,M \quad \ell=1,L$$
(e) Bounds
$$\sigma_i^L \leq \sigma_{i\ell} \leq \sigma_i^U \qquad i=1,M \quad \ell=1,L$$
$$v_i^L \leq v_{i\ell} \leq v_i^U \qquad i=1,M \quad \ell=1,L$$
$$d_j^L \leq d_{j\ell} \leq d_j^U \qquad j=1,N \quad \ell=1,L$$
$$A_i \geq 0 \qquad i=1,M$$
$$s_{i\ell} \in \mathbf{R}^1 \qquad i=1,M, \quad \ell=1,L$$

Note that except for Hooke's law equations in (c), which involve bilinear terms of the form in equation (5), the above model is linear in the objective function and in the remaining constraints.

It should be noted that the above model is commonly formulated by eliminating the variables $v_{i\ell}$ and $d_{i\ell}$ by substituting the compatibility equations into Hooke's law equations, and the result in the equilibrium equations. Although this reduces the number of variables and constraints, it actually increases the number of bilinear terms.

Assume now that the cross section areas A_i are specified with discrete values $DA(i)=\{A_{i1}, A_{is2},...A_{iN(i)}\}$. To reformulate problem (ST) as an MILP, we introduce the 0-1 variables y_{is} as in (3) to select the discrete sizes A_{is}. By then letting

$$A_i = \sum_{s=1}^{N(i)} A_{is} \, y_{is} \qquad (22)$$

as in (2), substituting into Hooke's law equations yields

$$\frac{E_i}{\ell_i} \sum_{s=1}^{N(i)} A_{is} \, v_{i\ell} \, y_{is} = s_{i\ell} \qquad i=1,M \quad l=1,L$$

By defining the variables $v_{i\ell s}$ as in (12) to (14), and by setting

$$c_{is} = \rho_i \, \ell_i \, A_{is} \qquad (23)$$

then by analogy to problem (R2) the MILP reformulation of (ST) yields:

$$\min \quad C = \sum_{i=1}^{M} \sum_{s=1}^{N(i)} c_{is} \, y_{is}$$

(RST)

$$\text{s.t.} \quad \sum_{i=1}^{M} b_{ji} \, s_{i\ell} = p_{\ell j} \qquad j=1..N, \ \ell=1,L$$

$$\sum_{j=1}^{N} b_{ji} d_{j\ell} = \sum_{s=1}^{N(i)} v_{i\ell s} \qquad i=1..M, \; \ell=1,L$$

$$\frac{E_i}{\ell_i} \sum_{s=1}^{N(i)} A_{is} v_{i\ell s} = s_{i\ell} \qquad i=1..M, \; \ell=1,L$$

$$v_{i\ell}{}^L y_{is} \leq v_{i\ell s} \leq v_{i\ell}{}^U y_{is} \qquad s=1,N(i), \; \ell=1,L \\ i=1,M$$

$$\sum_{i=1}^{N(i)} y_{is} = 1 \qquad i=1,M$$

$$\sigma_{i\ell} = \frac{E_i}{\ell_i} \sum_{s=1}^{N(i)} v_{i\ell s} \qquad i=1..M, \; \ell=1,L$$

$$\sigma_i{}^L \leq \sigma_{i\ell} \leq \sigma_i{}^U \qquad i=1..M, \; \ell=1,L$$

$$d_j{}^L \leq d_{j\ell} \leq d_j{}^U \qquad j=1..N, \; \ell=1,L$$

$$s_{i\ell} \in \mathbf{R}^1 \quad i=1,M, \; \ell=1,L$$
$$v_{i\ell s} \in \mathbf{R}^1 \quad s=1,N(i), \; \ell=1,L \; i=1,M$$
$$y_{is} = \{0,1\} \quad s=1,N(i), \; i=1,M$$

where the variable $v_{i\ell}$ has been substituted from the equation

$$v_{i\ell} = \sum_{s=1}^{N(i)} v_{i\ell s} \qquad (24)$$

and c_{is} is a parameter given by (23). The importance of model (RST) is that it allows the rigorous removal of bars with which one can optimize the topology, as well as the sizes of the bars. Further discussion on this model can be found in Ghattas and Grossmann (1991).

7. APPLICATION TO COMPLEX BATCH PROCESS DESIGN

As a third application we will consider the optimal design of multipurpose batch plants with multiple production routes, a problem that has been considered recently by Faqir and Karimi (1990).

As opposed to the problem considered in Section 5, in this problem we are given a number of stages with a number of potential units of identical type. Also, not all products require all production stages and therefore potential production routes are specified for each product as shown in Fig. 3. Faqir and Karimi (1989) formulated this design problem as an MINLP problem which involves bilinearities in the constraints which cannot be transformed into a geometric programming problem.

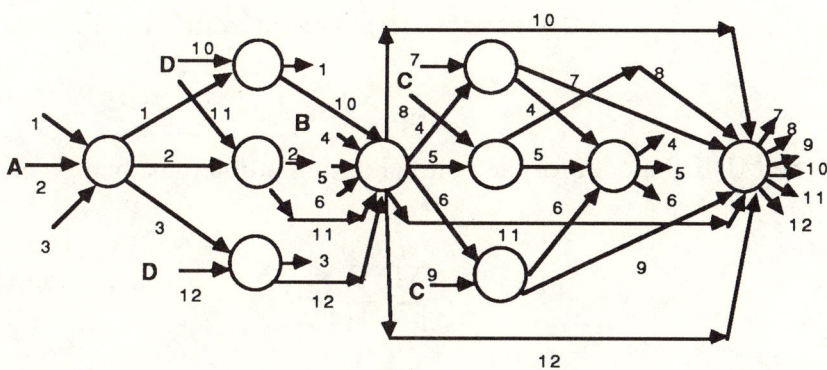

Fig.3. Complex multiproduct batch plant

The following parameters and variables are involved in this formulation.

Parameters

    N   the number of products j=1,..N

M the total number of potential units
NR the number of production routes r=1,..NR
E_r index set for units i involved in route r
R_j index set for routes r that produce product j
T_{Lr} cycle time of route r
Q_j demand of product j
S_{ri} size factor for route r in unit i
V_{is} discrete size s for unit i, i=1, N(i) ($V_{i1} = 0$)
H horizon time
γ_i, δ_i cost coefficients for unit i with $0 < \gamma_i < 1$

Variables

V_i size of unit i
B_r batch size in route r
q_r amount produced in route r
θ_r production time spent with route r
y_{is} 0-1 variable to denote selection of size s in unit i

The MINLP model for the optimal design problem is then given by

$$\min \quad C = \sum_{i=1}^{M} \sum_{s=1}^{N(i)} \delta_i V_i^{\gamma_i} \quad \text{(MB)}$$

s.t.

(a) Volume requirements for units

$$V_i \geq S_{ri} B_r \quad i \in E_r, \quad r=1,NR$$

(b) Production in each route

$$q_r = \frac{B_r \theta_r}{T_{Lr}} \quad r=1,NR$$

(c) Demand constraint
$$\sum_{r \in R_j} q_r = Q_j \qquad j=1,N$$

(d) Definition of sizes
$$V_i = \sum_{s=1}^{N(i)} V_{is}\, y_{is}\,; \qquad \sum_{s=1}^{N(i)} y_{is} = 1, \ i=1,M$$

(e) Horizon constraints
$$h(\theta_1, \theta_2,\ldots\theta_r,\ldots) \leq H$$
$$V_i \geq 0 \quad i=1,M \quad y_{is} = \{0,1\} \quad s=1,N(i) \quad i=1,N$$
$$B_r,\, q_r,\, \theta_r \geq 0 \qquad r=1,NR$$

where $h(\theta_1, \theta_2\ldots\theta_{NR})$ are linear functions that define the allocation of times for production (see Faqir and Karimi, 1990).

As can be seen, problem (MB) is an MINLP that involves nonconvexities in the objective function and in the constraints (b). In order to reformulate (MB) as an MILP, substitute B_r from constraints (b) into the constraints in (a) which then leads to

$$- V_i \theta_r + S_{ri} T_{Lr} q_r \leq 0 \quad i \in E_r,\ r=1,NR \qquad (25)$$

which has exactly the same form as equation (5) for Case 2.

By setting the cost coefficients $c_{is} = \delta_i V_i^{\gamma_i}$, and by applying equations (12) and (13) as in model (R2) for the variables θ_{ris}, the resulting MILP model is

$$\min\ C = \sum_{i=1}^{M} \sum_{s=1}^{N(i)} c_{is}\, y_{is}$$

(RMB)

s.t. $\sum_{s=1}^{N(i)} V_{is} \theta_{ris} \geq S_{ri} T_{Lr} q_r \quad i \in E_r, \ r=1, NR$

$\theta_r = \sum_{s=1}^{N(i)} \theta_{ris} \quad i \in E_r, \ r=1, NR$

$\theta_{ris} - \theta_r^U y_{is} \leq 0 \quad s=1, N(i), \ i \in E_r \quad r=1, NR$

$\sum_{s=1}^{N(i)} y_{is} = 1 \quad i=1, M$

$\sum_{r \in R_j} q_r = Q_j \quad j=1, N$

$h(\theta_1, \theta_2, ..\theta_r) \leq H$

$\theta_r, q_r \geq 0 \quad r=1, NR$

$\theta_{ris} \geq 0 \quad s=1, N(i), \ i \in E_r, \ r=1, NR$

$y_{is} = \{0,1\} \quad s=1, N(i), \ i=1, M$

where a simple choice of the upper bound θ_r^U is H, the total horizon time.

8. NUMERICAL RESULTS

Batch process design

Consider the problem by Voudouris and Grossmann (1991) of a multiproduct batch plant which produces five different products A,B,C,D and E. The plant consists of six stages involving one piece of equipment (see Fig. 1). The demands for the five products are: 250,000 tons per year for A, 150,000 tons per year for B, 180,000 tons per year for C, 160,000 tons per year for D and 120,000 tons per year for E. Size factors and cycle times are given for each product and the specified time horizon is 6000 hours. The equipment at any stage are available in 5 discrete sizes; namely 3000,3750,4500,5860 and 7325 liters. In this case the equipment cannot be removed from the process train and so the value of 0 liters is not allowed.

The problem can be formulated as the NLP (B1) where the equipment volumes are relaxed to be continuous variables. The optimal design found by this formulation has an investment cost of $231,489.6, and the corresponding optimal volumes are [6017.6, 3483.6, 3960.9, 4823.4, 4646.5, 3885.5] liters. Rounding up to the next available size gives [7325, 3750, 4500, 5860, 5860, 4500] liters and a capital investment of $255,886.2. In this case there is no need to check for the feasibility of the proposed solution since rounding up gives always a feasible design. The CPU time required with GAMS/MINOS 5.2 (Brooke et al, 1988; Murtagh and Saunders, 1985) was 2.3 seconds in a VAX-6420. The NLP involved 11 continuous variables and 31 constraints.

The MINLP formulation of the problem was convexified with exponential transformations and solved using the Outer Approximation algorithm implemented in DICOPT (Kocis and Grossmann, 1989). It involved 30 0-1 variables, 11 continuous variables and 43 constraints. For the NLP's the solver used was MINOS 5.2 whereas for the MILP's SCICONIC (1990) was used. The problem converged in 19.9 seconds in a VAX-6420: 2.3 seconds or 12% of the total time were required for the NLP subproblems and 17.6 seconds or 88% of the total time were required for the MILP master problems. The optimal solution obtained was [5860, 3750, 3750, 5860, 4500, 4500] liters for the equipment in each stage and the capital investment required was $238,650.24.

By formulating the problem as the MILP in problem (RB), 30 0-1 variables, 5 continuous variables and 37 constraints were involved. Using SCICONIC on a VAX-6420, 3.1 seconds were required to solve the MILP problem to optimality; using ZOOM (Marsten, 1986) in the same computer required 23.2 seconds. The globally optimal solution found using the MILP formulation had a capital investment of $238,650.24 with equipment sizes [5860, 3750, 3750, 5860, 4500, 4500] liters. This solution is the same as the one obtained by the MINLP formulation. Note that the solution found by the NLP formulation followed by the rounding of the equipment sizes does not yield an optimal design as its capital investment is 7.2% higher than the global optimum solution. It is also worth noting that the CPU time required by the MILP formulation when solved with SCICONIC is only slighly higher than the CPU time required by the continuous NLP model when solved with MINOS.

Structural Design

Consider the 5 bar fan stress shown in Fig. 4 which is subject to one load of 100,000 lbs (Ghattas and Grossmann, 1991). The modulus of elasticity is 1×10^7 psi, the density is 0.1 lb/in^3 for each bar and the maximum stress is 20,000psi in compression or tension. Also, for each bar 6 discrete sizes are assumed, [0,2,4,6,8,10] in^2, where the zero value allows for the possible removal of the bars. Also constraints were specified on maximum displacement and elongation; these were not binding at the solution. The problem was first formulated as the NLP in (ST) which led to a minimum weight of 146.25 lb with cross-section areas of A=[6.988,0,0,0,6.525] in^2. By rounding up the sizes to [8,0,0,0,8] and resolving the NLP a feasible solution was obtained with a weight of 172.97 lb. The total CPU-time required with GAMS/MINOS was 1.94 sec on VAX-6420 (1.43 sec first NLP, 0.51 sec second NLP). The NLP involved 23 continuous variables and 18 constraints.

The problem was also formulated as an MINLP which required 30 0-1 variables, 23 continuous variables and 28 constraints. With the outer-approximation method implemented in DICOPT (Kocis and Grossmann, 1989) the problem converged in 6.4 sec using MINOS for the NLP subproblems and ZOOM for the MILP master problems. Generalized Benders decomposition was also applied but it failed to converge to the optimum. Finally, a branch and bound method

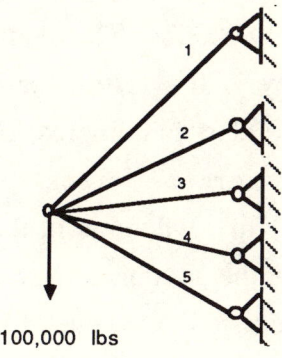

Fig. 4. Five-bar fan truss

was also used for the MINLP which was formulated as in (P1) requiring 5 discrete variables, 23 continuous variables and 18 constraints. The method required 11.5 sec enumerating 10 nodes which were solved with MINOS. By formulating the problem as an MILP in problem (RST) 30 0-1 variables, 43 continuous variables and 83 constraints were involved. The computer code SCICONIC solved the MILP problem to optimality in 3.5 sec while ZOOM required 9.3 sec. The optimal solution of the MILP led to a minimum weight of 172.18 lb with areas A = [8,0,0,2,6] in^2 which was the same solution that was found with DICOPT and with the branch and bound method. Note that in this case the rounded solution is [8,0,0,0,8] with weight 172.97 lb. This solution does not correspond to the optimum design although the difference in the weight is only 0.5% higher. It is also worth noting that the rounded solution yields a different topology than the global optimum.

As an additional example, consider the truss shown in Fig. 2. which consists of 10 bars and is subjected to two loading conditions of 100,000 lbs. The modulus of elasticity is 1×10^7 psi, the density is 0.1 lb/in^3 for each bar and the maximum stress is 25,000psi in compression or tension. For each bar 11 discrete sizes are assumed, [0,1,2,3,4,5,6,7,8,9,10] in^2, where as in the example above, the zero value allows for the possible removal of the bars. Also constraints were specified for the displacements and elongations. By formulating the problem as the MILP (RST), 110 0-1 variables, 148 continuous variables and 268 constraints

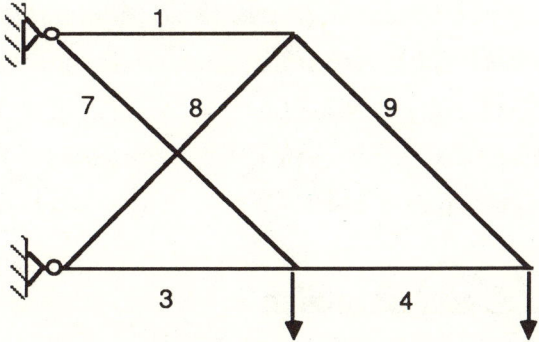

Fig. 5. Optimal topology for truss in Fig. 2.

were involved. The computer code SCICONIC solved the MILP problem to optimality in 602 sec on an HP-9000/720 (without SOS1 sets). On a VAX-6420 the problem could not be solved to optimality in 900 sec without the SOS1 sets; when they were included, the problem was solved to global optimality in 135 sec. As for the solution strategy described in the remarks section, the problem was solved also with SCICONIC on the VAX-6420 in 75.2 sec (step1: 6 sec LP, 7.5 sec MILP; step 2: 30.2 sec LP's; step3: 31.5 sec MILP with SOS1). The optimal solution of the MILP led to a minimum weight of 1636.4. lb with areas A = [8,0,8,4,0,0,6,6,6,0] in^2 which yields the configuration shown in Fig. 5. When this problem was solved as an MINLP in (P3), DICOPT (Kocis and Grossmann, 1989) failed to converge to the global optimum from several starting points due to the nonconvexities; Generalized Benders decomposition did not converge after 900 sec due to the large number of infeasible NLP subproblems. It should be noted, that when the problem was solved as a continuous NLP, the model required 48 continuous variables and 38 constraints. With zero initial guesses for all variables MINOS failed to converge; with nonzero guesses (all areas with size 5), the rounded solution was the same as the one of the MILP requiring a total of 6.1 sec.

Complex batch process design

Consider the design of a multipurpose batch plant with multiple production routes that is shown in Fig.3 (Faqir and Karimi, 1990). Assume that this plant must produce four products A, B, C and D satisfying the demands of 300000, 250000, 180000 and 200000 tons per year, respectively. The time horizon in which these demands have to be satisfied is 6000 hours. As seen in Fig. 3, there are a total of 10 potential equipment units available which are placed in 6 different processing stages. Each product must be manufactured with different processing stages which then gives rise to a total of 12 potential production routes. The equipment are available in 6 discrete sizes; namely 0, 500, 1000, 2000, 2500 and 3000 liters. Note that the nonexistence of an equipment is represented by a volume of 0 liters. There is a constant processing time characterizing every task of a product in a specific equipment group.

Faqir and Karimi (1990) have developed a special purpose strategy for solving the associated MINLP problem which requires a significant amount of preprocessing and user interaction. Formulating the problem as the MILP in (RMB), 50 0-1 variables, 225 continuous variables and 234 constraints are involved. The computer code SCICONIC was used in order to solve the problem through GAMS. The optimal solution obtained requires a capital investment of $124,500 which is the same solution as the one reported by Faqir and Karimi (1990). The CPU requirements on a VAX-6420 were 305.2 seconds. It has to be mentioned here that exploiting the fact that the logical constraints in (RMB) can be treated as special ordered sets of type 1 (SOS1), the CPU time was further reduced to 148.4 seconds.

9. CONCLUSIONS

This paper has considered a special class of nonlinear discrete design optimization problems which involve nonlinear separable objective functions in the design variables and bilinear constraints that are given by products of design and state variables, where the former are restricted to take discrete values. Two special cases for the constraints were identified for which it was shown that the discrete nature of the design variables can be exploited to reformulate these problems as MILP models. The solution of these models can be expedited through the use of SOS1 sets, and with a simple solution strategy that relies on deriving valid lower bounds on the sizes. The application of the MILP reformulations was applied to multiproduct batch plant problems in chemical engineering and to structural design problems in civil engineering. These represent novel design optimization models that can explicitly handle discrete sizes, and therefore avoid the common heuristic rounding procedures for discrete nonlinear programming models. Numerical results have been presented which show that the proposed models not only produce global optimum solutions, but are computationally competitive when compared to nonlinear formulations with continuous sizes.

REFERENCES

Bremicker, M., P.Y. Papalambros and H.T. Loh, "Solution of Mixed-Discrete Structural Optimization Problems with a New Sequential Algorithm", *Computers and Structures*, **37**, 451-461 (1990).

Brooke, A., D. Kendrick and A. Meeraus, "GAMS A User's Guide", Scientific Press, Palo Alto (1988).

Duran, M.A. and I.E. Grossmann, "An Outer-Approximation Algorithm for a Class of Mixed-Integer Nonlinear Programs", *Mathematical Programming*, **36**, 307-339 (1986).

Faqir, N.M. and I.A. Karimi, "Design of Multipurpose Batch Plants with Multiple Production Routes", Proceedings of Foundations of Computer Aided Process Design (Eds. Siirola et al.), Elsevier, Amsterdam, 451-468 (1990).

Floudas, C.A. and P.M. Pardalos, "A Collection of Test Problems for Constrained Global Optimization Problems", Lecture Notes in Computer Science (Eds. G.Goos and J.Hartmanis), Springer-Verlag, Berlin (1990).

Geoffrion, A.M., "Generalized Benders Decomposition", *Journal of Optimization Theory and Applications*, **10**, 237-260 (1972).

Ghattas, O. and I.E. Grossmann, "MINLP and MILP Strategies for Discrete Sizing Structural Optimization Problems", to appear in Proceedings of ASCE 10th Conference on Electronic Computation, Indianapolis (1991).

Glover, F., "Improved Linear Integer Programming Formulations of Nonlinear Integer Problems", *Management Science*, **22**, 455-460 (1975).

Grossmann, I.E., "Mixed-Integer Nonlinear Programming Techniques for the Synthesis of Engineering Systems", *Research in Engineering Design*, **1**, 205-228 (1990).

Grossmann, I.E. and R.W.H. Sargent, "Optimum Design of Multipurpose Chemical Plants", *Ind.Eng.Chem. Process Design and Dev.*, **18**, 343 (1978).

Gupta, O.K., "Branch and Bound Experiments in Nonlinear Integer Programming", Ph.D. Thesis, Purdue University (1980).

Haftka, R.T., Z. Gurdal and M.P. Kamat, "Elements of Structural Optimization", Kluwer Academic Publishers, Dordecht (1990).

Kocis, G.R. and I.E. Grossmann, "Computational Experience with DICOPT Solving MINLP Problems in Process Systems Engineering", *Computers and Chemical Engineering*, **13**, 307-315 (1989).

Marsten, R., "User's Manual for ZOOM/XMP", University of Arizona (1986).

Murtagh, B.A. and M.A. Saunders, "MINOS User's Guide", Systems Optimization Laboratory, Department of Operations Research, Stanford University (1985).

Papalambros, P.Y. and D.J. Wilde, "Principles of Optimal Design", Cambridge University Press, Cambridge (1988).

Sahinidis, N.V., "Mixed-Integer Optimization Techniques for Planning and Scheduling of Chemical Processes", Ph.D. thesis, Carnegie Mellon University, Pittsburgh (1990).

SCICONIC/VM User Guide (Version 2.11), Scicon Ltd., Milton Keynes (1990).

Torres, F.E., "Linearization of Mixed-Integer Products", *Mathematical Programming*, **49**, 427-428 (1991).
Van Roy, T.J. and L.A. Wolsey, "Solving Mixed Integer Programs by Automatic Reformulation", *Operations Research*, **35**, 45-57 (1987).

Voudouris, V.T. and I.E. Grossmann, "MILP Formulations for the Optimal Design of Batch Processes", submitted for publication (1991).

APPENDIX. On the equivalence for the linearization of case 2.

Proposition. The constraints,

$$\alpha_{ij} \sum_{s=1}^{N(i)} d_{is} v_{ijs} - \beta_{ij} w_j \leq 0 \qquad j \in J(i), \ i=1,..n \qquad (A1)$$

$$v_j = \sum_{s=1}^{N(i)} v_{ijs} \qquad j \in J(i), \ i=1,..n \qquad (A2)$$

$$\sum_{s=1}^{N(i)} y_{is} = 1 \qquad (A3)$$

$$v_j^L y_{is} \leq v_{ijs} \leq v_j^U y_{is} \qquad j \in J(i), \ s=1, \ N(i), \ i=1..n \qquad (A4)$$

are equivalent to the inequalities,

$$\alpha_{ij} x_i v_j - \beta_{ij} w_j \leq 0, \ j \in J(i), \ i=1..n \qquad (A5)$$

for

$$x_i = \sum_{s=1}^{N(i)} d_{is} y_{is}, \ \sum_{s=1}^{N(i)} y_{is}=1, \ i=1,..n; \qquad y_{is}=0,1 \qquad (A6)$$

<u>Proof.</u> Let $y_{it(i)} = 1$ and $y_{is}=0$ for $s \neq t(i)$, $s=1,..N(i)$, $i=1,...n$. From (A6) it follows that $x_i = d_{it(i)}$, with which (A5) becomes,

$$\alpha_{ij} d_{it(i)} v_j - \beta_{ij} w_j \leq 0, \ j \in J(i), \qquad i=1..n \qquad (A7)$$

Also, from (A4) it then follows that $v_{ijs} = 0$, $j \in J(i)$, $s = 1,...N(i)$, $i = 1,...n$, $s \neq t(i)$. Hence, from (A2),

$$v_j = v_{ijt(i)} \qquad j \in J(i), \qquad i=1..n \qquad (A8)$$

and $\alpha_{ij} d_{it(i)} v_{ijt(i)} - \beta_{ij} w_j \leq 0, \ j \in J(i), \qquad i=1..n \qquad (A9)$

Substituting (A8) into A(9) leads to

$$\alpha_{ij} d_{it(i)} v_j - \beta_{ij} w_j \leq 0, \ j \in J(i), \qquad i=1..n \qquad (A10)$$

which is identical to (A7).

Mixed-Integer Nonlinear Programming on Generalized Networks [*]

Soren S. Nielsen[†]   Stavros A. Zenios[†]

Abstract

An algorithm for solving mixed-integer network problems is developed and implemented. The algorithm is based on Generalized Benders Decomposition, retaining a network structure in the subproblems. The master problems are solved inexactly as binary linear feasibility problems using Branch-and-Bound and Lagrangean relaxation. Vector and parallel computing on a shared memory vector multiprocessor reduces significantly the solution times for the master problem.

Computational results are reported for mixed-integer linear and non-linear network problems obtained from applications in portfolio management (Markowitz models) and the placement of electrical circuits on printed boards (quadratic assignment problems). The experiences on an Alliant FX/4 vector multiprocessor are also discussed.

Key Words: Generalized Benders decomposition, Branch-and-Bound, Mixed-integer network problems, Lagrangean relaxation.

---

[*]Research partially supported by NSF grants SES-91-00216 and CCR-8811135, AFOSR grant 91-0168 and AT&T contract CH292700MK.
[†]Decision Sciences Department, The Wharton School, University of Pennsylvania, Philadelphia, PA 19104.

1 Introduction

Network problems constitute an important and widely applicable class of mathematical programming problems. Numerous applications and very efficient solution procedures have long been known for linear, pure or generalized, networks (Kennington and Helgason [1980]). More recently, the development of efficient solution algorithms for non-linear networks have further broadened the scope and applicability of network formulations. The success of network algorithms can primarily be attributed to the special basis structure and sparsity that is characteristic of these problems. This, in turn, facilitates the solution of realistic problems, which are often very large, within a reasonable time frame. Applications today include transportation and distribution planning, financial cash-stream management, air traffic control, hydroelectric power scheduling, and the estimation of social accounting matrices. See Dembo, Mulvey and Zenios [1989] for a survey of nonlinear network methods and related models.

In this paper we discuss the solution of mixed-integer nonlinear programs with generalized network constraints, [MINLGN]. Such problems arise in situations where there is a fixed charge associated with the presence of non-zero flow on a network arc (for example due to transaction costs in financial applications, or infrastructure construction costs in transportation problems). The problem is defined as follows:

[MINLGN]
Minimize
$$f(x) + dy \qquad (1)$$
$x \in R^n, y \in \{0,1\}^{n_c}$

Subject to
$$\sum_{j \in \delta_i^+} x_{ij} - \sum_{k \in \delta_i^-} m_{ki} x_{ki} = b_i \qquad i \in \mathcal{N} \qquad (2)$$

$$l_{ij} \leq x_{ij} \leq u_{ij} \qquad (i,j) \in \mathcal{E} \setminus \mathcal{C} \qquad (3)$$

$$l_{ij} y_{ij} \leq x_{ij} \leq u_{ij} y_{ij} \qquad (i,j) \in \mathcal{C} \qquad (4)$$

where

$\mathcal{N} = \{1, 2, 3, ..., m\}$ is the set of nodes,

$\mathcal{E} = \{(i,j) \mid i, j \in \mathcal{N}\}$ is the set of arcs of cardinality n,

$\mathcal{C} \subseteq \mathcal{E}$ is the set of arcs whose flows are of the "nothing-or-range" type, of cardinality n_c,

$f : R^n \to R$ is a twice continuously differentiable convex function,

$d = (d_{ij}), (i,j) \in \mathcal{C}$, is a vector of fixed costs for the presence of non-zero flow on (i,j),

$x = (x_{ij}), (i,j) \in \mathcal{E}$, is the vector of flow over the arcs of the network,

$y = (y_{ij}), (i,j) \in \mathcal{C}$, is a vector of binary variables associated with the arcs in $\mathcal{C}$,

m_{ij} is the multiplier on arc (i,j), $\forall (i,j) \in \mathcal{E}$,

$b = (b_i), i \in \mathcal{N}$, is the vector of supply/demand at node i,

$\delta_i^+ = \{j \mid (i,j) \in E\}$, is the set of outgoing arcs from node i, $\forall i \in \mathcal{N}$,

$\delta_i^- = \{j \mid (j,i) \in E\}$, is the set of incoming arcs into node i, $\forall i \in \mathcal{N}$,

$l = (l_{ij}), (i,j) \in \mathcal{E}$, is the vector of lower bounds,

$u = (u_{ij}), (i,j) \in \mathcal{E}$, is the vector of upper bounds.

If all the multipliers $\{m_{ij}\}$ are equal to 1 the network is pure, otherwise it is a generalized network. Use of the term linear or nonlinear depends on whether the cost function $f(x)$ is linear or nonlinear. When the set $\mathcal{C}$ of arcs controlled by binary variables is empty the [MINLGN] model simplifies to the non-linear generalized network structure of Ahlfeld et al. [1987].

We will solve this optimization problem using Generalized Benders Decomposition (GBD), decomposing the problem constraints into the two sets (2)–(3) and (4). This results in an algorithm that solves a series of *subproblems*, which are nonlinear, generalized network problems, and *master problems*, which are mixed-integer linear programming problems. The subproblems are solved using the nonlinear network optimizer GENOS of Mulvey and Zenios, [1987]. The master problems are solved inexactly as pure binary feasibility problems using branch-and-bound and Lagrangean relaxation.

Generalized Benders decomposition provides a suitable framework for the solution of [MINLGN] since the network structure of the optimization problem is preserved in the subproblems. The master problems are binary linear problems containing as many binary variables as the original [MINLGN] problem and one continuous variable. The sequence of master problems are solved as pure binary feasibility problems, which allows restarting the branch-and-bound phase from the previous branch-and-bound tree.

Benders Decomposition (Benders [1962]) is one of the earliest decomposition methods for solving linear problems with a small number of complicating variables. It was generalized by Geoffrion [1972] to non-linear problems. Benders Decomposition was used with great success in the design of a production and distribution system by Geoffrion and Graves [1974], who also used the idea of solving the mixed-integer master problems inexactly. Jörnstein [1980] solves a distribution and assignment model, and Bienstock and Shapiro [1988] solve a stochastic programming problem using Benders decomposition coupled with preprocessing techniques to improve convergence. Magnanti and Wong [1981] give theoretical results on chosing proper Benders cuts when multiple alternatives exist, and comment on model formulation. Paules and Floudas [1989] effectively use Generalized Benders Decomposition for chemical process synthesis problems.

Lagrangean relaxation — which we adopt here in solving the master problems — has been successfully applied in a number of problem areas. Geoffrion [1974] gives a systematic development of the concept of dualizing constraints and develops the general Lagrangean relaxation theory. Shapiro [1979] is a survey of Lagrangean techniques for discrete optimization. Fisher [1981, 1985] gives overviews of the application of Lagrangean relaxation covering basic theory and its use in a branch-and-bound framework. Subgradient optimization for solving the Lagrangean dual is explained. These references provide the background on which we build and implement our solution procedure for [MINLGN].

The rest of the paper is organized as follows. Section 2 is an overview of generalized Benders decomposition as specialized for [MINLGN]. In Section 3, we discuss details of system integration and implementation. Section 4 reports numerical results, and Section 5 concludes the report.

2 Specialization of Generalized Benders Decomposition

This section presents the components of the generalized Benders decomposition (GBD) framework as specialized for solving [MINLGN].

2.1 Generalized Benders Decomposition

The algorithm is stated as follows:

Generalized Benders Decomposition

GBD-0 **Initialization**: Set $k = 1$. Let y^k be an initial integer *proposal*. Let $\mathcal{C}_\mathcal{F} = \emptyset$ and $\mathcal{C}_\mathcal{I} = \emptyset$ be the index sets of cuts generated from feasible and infeasible subproblems, respectively. Let $z_{lo} = -\infty$ and $z_{up} = \infty$ be lower and upper bounds on the objective function, respectively.

GBD-1 **Subproblem**: Solve the non-linear generalized network

$$\underset{x \in R^n}{\text{Minimize}} \quad f(x) + dy^k \tag{5}$$

$$\text{Subject to} \sum_{j \in \delta_i^+} x_{ij} - \sum_{k \in \delta_i^-} m_{ki} x_{ki} = b_i \quad i \in \mathcal{N} \tag{6}$$

$$l_{ij} \le x_{ij} \le u_{ij} \quad (i,j) \in \mathcal{E} \setminus \mathcal{C} \tag{7}$$

$$l_{ij} y_{ij}^k \le x_{ij} \le u_{ij} y_{ij}^k \quad (i,j) \in \mathcal{C} \tag{8}$$

If the problem has no feasible solution, proceed to step GBD-2. Otherwise, let $\mathcal{C}_\mathcal{F} = \mathcal{C}_\mathcal{F} \cup \{k\}$ and let $\underline{\pi}^k \in R_+^{n_c}$ and $\overline{\pi}^k \in R_+^{n_c}$ be the shadow prices of the left and right inequalities, respectively, of constraints (8). Let x^k be the optimal solution and let $z_{up} = \min\{z_{up}, f(x^k) + dy^k\}$. Proceed to step GBD-3.

GBD-2 **Infeasibility minimizing subproblem**: Solve the linear generalized network problem

$$\underset{x \in R^n, s^p, s^n \in R_+^{n_c}}{\text{Minimize}} \quad es^p + es^n \tag{9}$$

$$\text{Subject to} \sum_{j \in \delta_i^+} x_{ij} - \sum_{k \in \delta_i^-} m_{ki} x_{ki} = b_i \quad i \in \mathcal{N} \tag{10}$$

$$l_{ij} \le x_{ij} \le u_{ij} \quad (i,j) \in \mathcal{E} \setminus \mathcal{C} \tag{11}$$

$$l_{ij} y_{ij}^k \le x_{ij} + s_{ij}^p - s_{ij}^n \le u_{ij} y_{ij}^k \quad (i,j) \in \mathcal{C} \tag{12}$$

where e is a vector of all 1, $s^p = (s_{ij}^p)$, $s^n = (s_{ij}^n)$, $\forall (i,j) \in \mathcal{C}$.

Let $\mathcal{C}_\mathcal{I} = \mathcal{C}_\mathcal{I} \cup \{k\}$. Let $\underline{\pi}^k \in R_+^{n_c}$ and $\overline{\pi}^k \in R_+^{n_c}$ be the shadow prices of the left and right inequalities, respectively, of constraints (12). Let (x^k, s^{pk}, s^{nk}) be the optimal solution.

GBD-3 **Master problem**: Solve the mixed-integer program

$$\underset{\mu \in R, y \in \{0,1\}^{n_c}}{\text{Minimize}} \quad \mu \tag{13}$$

Subject to

$$\mu \ge f(x^\ell) + dy - \underline{\pi}^\ell (ly - x^\ell) - \overline{\pi}^\ell (x^\ell - uy) \quad \ell \in \mathcal{C}_\mathcal{F} \tag{14}$$

$$0 \ge es^{p\ell} + es^{n\ell} - \underline{\pi}^\ell (ly - x^\ell) - \overline{\pi}^\ell (x^\ell - uy) \quad \ell \in \mathcal{C}_\mathcal{I} \tag{15}$$

Let (μ^k, y^k) be the optimal solution; set $\mu^k = -\infty$ if $\mathcal{C}_\mathcal{F} = \emptyset$. Let $z_{lo} = \max\{z_{lo}, \mu^k\}$.

GBD-4 **Termination**: If $z_{lo} = z_{up}$ then (y^k, x^k) is the optimal solution. Otherwise let $k = k + 1$ and proceed to step GBD-1.

The network structure of the problem [MINLGN] is directly present in the subproblems of step GBD-1. In step GBD-2 we choose to minimize the sum of infeasibilities because this preserves the network structure of the infeasibility minimizing subproblem: The slack variables s_{ij}^p and s_{ij}^n correspond to network arcs between nodes i and j. The master problem stated above is a mixed-integer optimization problem. Instead, we will solve a pure binary feasibility problem as explained in the sequel.

2.2 Pure Binary Feasibility Master Problems

The bottleneck of GBD is the solution of the master problems. This is especially true when the subproblems have a special structure which can be exploited for efficient solution, such as network structure. For this reason it is desirable to avoid solving the master problems to optimality, but rather settle with an approximate solution. Doing this raises two problems:

1. The decomposition algorithm may lose finite convergence unless care is taken to avoid the situation where the master problems return the same integer proposal when that proposal is not optimal.

2. The termination criterion of step GBD-4 is not applicable since the objective function value of the master problem is not a lower bound on the optimal objective function value.

We propose a remedy to both problems which recovers the finiteness of the algorithm and yields a valid termination criterion. The master problem of step GBD-3 is replaced by the pure binary *Revised Master Problem*:

[RMP] Solve with respect to $y \in \{0,1\}^{n_c}$
$$f(x^\ell) + dy - \underline{\pi}^\ell(ly - x^\ell) - \overline{\pi}^\ell(x^\ell - uy) \leq z_{up} \quad \ell \in \mathcal{C}_\mathcal{F} \quad (16)$$
$$es^{p\ell} + es^{n\ell} - \underline{\pi}^\ell(ly - x^\ell) - \overline{\pi}^\ell(x^\ell - uy) \leq 0 \quad \ell \in \mathcal{C}_\mathcal{I} \quad (17)$$

Solving this system with respect to y corresponds to finding a solution to the master problem with objective value $\mu \leq z_{up}$.

The revised master problem (RMP) is a pure binary feasibility problem. In Section 2.3 we discuss the procedures used for solving RMP. For simplicity the problem is restated as follows:

[RMP] Solve $Ay \leq r$ for $y \in \{0,1\}^{n_c}$,

where A is the $(|\mathcal{C}_\mathcal{I}| + |\mathcal{C}_\mathcal{F}|) \times n_c$ constraint coefficient matrix and r is the right hand side of the revised master problem.

Let RMP^k denote the revised master problem at iteration k, and let $\mathcal{F}(\text{RMP}^k)$ denote the set of solutions to this problem. Since RMP^k is obtained from RMP^{k-1} by adding a constraint and (possibly) decreasing z_{up}, we have

$$\mathcal{F}(\text{RMP}^k) \subseteq \mathcal{F}(\text{RMP}^{k-1}) \text{ for } k > 1.$$

This implies that the branch-and-bound tree upon termination of RMP^{k-1} can be used as the initial tree for finding a solution to RMP^k. This continued use of the same branch-and-bound tree decreases computing time, but also eliminates the problems stated above:

1. *Finite convergence*: This version of GBD is finite since the initial branch-and-bound tree used to solve RMP^k does not include any nodes not present in (or generated from) the initial tree of RMP^{k-1}. Thus, integer solutions previously returned cannot be generated again, and the tree will be exhausted after a finite number of GBD iterations.

2. *Termination*: We may terminate the algorithm when the branch-and-bound tree is exhausted, i.e., when no feasible solution is found to the revised master problem. The solution which resulted in the lowest subproblem objective value is then optimal.

The idea of solving the master problem inexactly is not new. Geoffrion and Graves [1974] also solved the master problems inexactly and had very good results. They do not use a partially enumerated tree as a starting point for subsequent solves.

2.2.1 Improving the Integer Solutions

The integer solutions resulting from solving the revised master problems may result in poor Benders cuts since, presumably, optimal integer solutions to the master problem will generate better cuts. The algorithm allows for continuing the search for a better solution after a feasible solution to the revised master problem has been found. After a feasible solution has been found the final tree is saved for later use. Using a lower bound on the optimal solution value (lower bounds are discussed in Section 2.2.2) and an *aggression factor* $\alpha \in (0, 1)$, a target $\hat{\mu}$ is established:

$$\mu = \alpha z_{lo} + (1 - \alpha) z_{up}.$$

A solution to the master problem with objective function $z \leq \hat{\mu}$ is then sought. The right hand side of RMP is adjusted corresponding to this new upper bound on the objective value (using $\hat{\mu}$ in place of z_{up}), and the branch-and-bound algorithm is resumed. If a solution to this feasibility problem is found, it is used as an integer proposal for the next subproblem, otherwise μ is a new lower bound on the optimal solution value of [MINLGN], and z_{lo} is updated. In this case it is necessary to re-solve the problem using an aggression factor closer to 1, in order to guarantee finding an optimal solution.

The aggression factor α may be either constant or dynamically adjusted under algorithm control. Section 4.4.2 contains numerical results of the effect of varying the aggression factor on solution times and the number of Benders iterations required.

2.2.2 Lower Bounds on the Optimal Solution Value

It is desirable to have a lower bound on the optimal solution value of [MINLGN]. When the master problems are not solved to optimality, the objective values may not be lower bounds. We therefore employ the following scheme: Any time a cut of the form

$$\sum_{j=1}^{n_c} a_{ij} y_j \leq b_i + \mu$$

is added to the master problem (i.e., whenever a feasible subproblem is solved), we update the lower bound as

$$z_{lo} = \max\{z_{lo}, \sum_j a_{kj}^- - b_k\},$$

where $a_{ij}^- = \min\{0, a_{ij}\}$. This lower bound is clearly much weaker than the one which could be obtained from solving the master problem to optimality.

2.3 Solving the Master Problems

The basic approach to solving RMP is branch-and-bound, Garfinkel and Nemhauser [1972], using Lagrangean relaxation and subgradient optimization to obtain bounds for pending nodes. A *merit function* based on the bounds is employed to estimate the relative attractiveness of nodes for algorithm consideration. Lagrangean relaxation and subgradient optimization are discussed in Sections 2.4 and 2.5. Here we discuss the branch-and-bound algorithm. First we need some definitions.

Following Crowder, Johnson and Padberg [1983], we call constraint i *blatantly infeasible* if

$$\sum_j a_{ij}^- > r_i,$$

GENERALIZED NETWORKS

where $a_{ij}^- = \min\{0, a_{ij}\}$. A set of constraints is called blatantly infeasible if any of the constraints are blatantly infeasible. Constraint i of the problem is called *inactive* if

$$\sum_j a_{ij}^+ \leq r_i,$$

where $a_{ij}^+ = \max\{0, a_{ij}\}$.

The initial branch-and-bound tree consists of the node with all variables free, or a previously saved tree. The Branch-and-Bound procedure is as follows:

Branch-and-Bound Algorithm

BB-1 If the tree is empty, terminate: RMP is infeasible.

BB-2 Remove the node with the largest merit function value from the tree, making it the *current node*, CN.

BB-3 If CN is blatantly infeasible, return to step BB-1.

BB-4 If any coefficient a_{ij} satisfies

$$a_{ij} > b_i - \sum_k a_{ik}^-$$

fix y_j at the value 0. If any coefficient a_{ij} satisfies

$$-a_{ij} > b_i - \sum_k a_{ik}^-$$

fix y_j at the value 1. Whenever a variable is fixed, remove it from the problem by substitution of its value. Repeat steps BB-3 and BB-4 until no more variables can be fixed this way.

BB-5 If no free variables remain in CN, terminate: The current node is feasible.

BB-6 Select a free variable y_j as branching variable. In CN, fix y_j at $v \in \{0, 1\}$. Create a node identical to CN except $y_j = 1 - v$. If the new node is not blatantly infeasible, establish a lower bound on the minimal infeasibility of it (see point 4 below). If this bound is positive, the node is infeasible and is discarded. Otherwise it is inserted into the branch-and bound tree. Proceed to step BB-4.

Some explanations are in order about this procedure:

1. In step BB-6, a branching variable is chosen such that the infeasiblity of CN assuming all other free variables set to the value 0 is minimized. See, e.g., Garfinkel and Nemhauser [1974].

2. In step BB-3 and whenever a variable is fixed in step BB-4, inactive constraints could be detected and removed from the problem.

3. If at any point, a variable y_j has only positive coefficients (disregarding inactive constraints), that variable may be set to 1, and similarly, if the variable has only negative coefficients, it may be set to 0. Note, however, that this is only valid for the duration of the current branch-and-bound algorithm: Since the final tree of this branch-and-bound session is to be used as a starting tree for the next RMP, the variables thus fixed must be released from their fixed values if the conditions mentioned are not satisfied by the new cut.

4. Whenever a new node is created in step BB-6, Lagrangean relaxation is used to establish a lower bound on the mimimal infeasibility obtainable from the node (See Section 2.4). If this value is strictly positive, the node is infeasible and is discarded. Otherwise, the value is used to form the merit of the node, and the node is inserted into the list of pending nodes.

2.4 Lagrangean Relaxation

When a new node is generated during the branch-and-bound algorithm, a bound on the infeasibility of solutions obtainable from the node is established using subgradient optimization on a Lagrangean relaxation (LR) of the node problem. The purpose of LR is twofold:

1. If the node is infeasible, LR may prove it and the node can then be discarded.

2. If LR fails to prove the node infeasible, it may still yield a measure of the potential for the node to produce a feasible solution, thus guiding the Branch-and-Bound search towards a feasible solution.

When a node is created for insertion in the tree, a subset of the variables in the node are fixed at either 0 or 1. The remaining variables are free. We will assume that fixed variables have had their values substituted. Denoting the remaining p free variables by $y \in \{0,1\}^p$, we may write the problem represented by the node as the feasibility problem:

[FP] Solve $Dy \leq q$ for $y \in \{0,1\}^p$.

D is the $(m \times p)$ submatrix of the constraint matrix A obtained by removing columns corresponding to fixed variables. q is the right hand side r adjusted by subtracting from it the columns of A corresponding to variables fixed at 1.

To apply Lagrangean relaxation we need an optimization problem. We therefore reformulate the problem as the following infeasibility minimization problem:

$$[\text{IM}] \quad \underset{y \in \{0,1\}^p, s \in R^m_+}{\text{Minimize}} \quad es$$

$$\text{Subject to} \quad Dy \leq q + s$$

The feasibility problem has a solution if and only if $v(\text{IM}) = 0$, where $v(\text{IM})$ denotes the optimal objective value of problem IM. The goal of using Lagrangean relaxation is to establish a lower bound on $v(\text{IM})$. If a *strictly positive* lower bound on $v(\text{IM})$ can be found, then the feasibility problem has no solution.

Lagrangean relaxation has its primary advantage if the constraint set can be partitioned into two constraint sets, one of which is relatively easy to solve. In our application, the matrix (D) arises from the application of Generalized Benders Decomposition and cannot be expected to possess any special structure. We therefore dualize the entire constraint set to obtain the *Lagrangean*

$$L(u) = \underset{y \in \{0,1\}^p, s \in R^m_+}{\text{Minimize}} \quad es + u(Dy - (q+s))$$

$L(u)$ is defined for $u \geq 0$. Since, if any $u_i > 1$, $L(u) = -\infty$, we may restrict the domain of L to $\{u \in R^m : 0 \leq u \leq 1\}$. This in turn implies that any optimal solution to the minimization problem embedded in $L(u)$ has $s = 0$. Substituting this into the expression above, we obtain the following equivalent definition for L (for $0 \leq u \leq 1$):

$$L(u) = \underset{y \in \{0,1\}^p}{\text{Minimize}} \quad u(Dy - q)$$

The optimal solution of the minimization problem is obtained trivially for a given value of u. Thus, a closed form expression for L is obtained:

$$L(u) = \sum_{i=1}^{m} v_i^- - uq$$

where $v = uD$, and $v_i^- = \min\{0, v_i\}$.

We need some standard results from duality theory (see e.g. Geoffrion [1974], Fisher [1985], or Nemhauser and Wolsey [1988]).

1. For any $u \geq 0$, $L(u) \leq v(\overline{\text{IM}}) \leq v(\text{IM})$, where $\overline{\text{IM}}$ is the LP-relaxation of IM.

2. $v(\text{LD}) = v(\overline{\text{IM}})$, where LD is the *Lagrangean dual* defined as

$$[\text{LD}] \quad \underset{0 \leq u \leq 1}{\text{Maximize}} \quad L(u)$$

The implication is that by solving the Lagrangean Dual to optimality, we obtain the same bound on the problem as if we had solved the LP-relaxation of the problem. If, while solving LD, we find a vector u such that $L(u) > 0$, we can conclude that the node is infeasible and terminate the optimization.

One apparent weakness of the Lagrangean approach for this application should be pointed out. Since the objective function of $\overline{\text{IM}}$ is always non-negative, 0 is a lower bound on $v(\overline{\text{IM}})$, and, indeed, $L(0) = 0$. In other words, for a node with feasible LP-relaxation, $v(\text{LD}) = 0$. This means that we cannot directly use $v(\text{LD})$ as a feasibility measure for nodes not proven infeasible. To remedy this situation, we further restrict the domain of L to

$$\{u \in R^m \mid 0 \leq u \leq 1, \max_{i=1}^{m}\{u_i\} = 1\}.$$

This will have no effect on proving nodes infeasible, but will provide us with a meaningful (though heuristic) measure of the chances that a node is feasible, which is desirable for guiding the branch-and-bound search.

2.5 Subgradient Optimization

Subgradient optimization is used to solve the Lagrangean Dual at each step of the branch-and-bound algorithm. Assuming an initial *stepsize* t_0 and initial *multipliers* $u \in R_+^m$, the procedure is:

SO-0 **Initialization**: Set $k \leftarrow 1$, $u^k = u_0$ and $t^k = t_0$.

SO-1 **Calculate Lagrangean**: Calculate $L(u^k)$ and let y^k be the corresponding integer values.

SO-2 **Update Multipliers**: Set $u^{k+1} = \max\{u^k + t^k(Dy^k - q), 0\}$ where $t^k = \alpha \frac{-L(u^k)}{|Dy-q|^2}$.

SO-3 **Termination test**: If $L(u^k) > 0$, terminate: The Lagrangean Dual has a positive optimum. If $k \geq k_{max}$, terminate on iteration limit, otherwise let $k \leftarrow k + 1$ and proceed from step SO-1.

The updating formula for t^k used in step SO-2 is given in Fisher [1985]. The max operation is applied element-wise. The parameter a employed in step SO-2 should be between 0 and 2. It is here initialized to 2 and halved if no improvement of $L(u)$ has occurred for a number of iterations. The termination criterion of step SO-3 could be more advanced than just an iteration limit, e.g. terminate if no improvement of $L(u)$ has occurred during a number of iterations. The multipliers u are initialized to the final value of u from the previous application of subgradient optimization, or set to 1 if no such value exists (which happens whenever a new Benders cut is added to the master problem). The "feasibility estimate" $\max_{k=1}^{m}\{\frac{L(u^k)}{\max_i\{u_i^k\}}\}$ is returned to the branch-and-bound routine.

2.6 Problem Preprocessing

In several network models, strong constraints on the binary variables may be derived directly from the problem as presented by the user, and we automatically generate a class of such constraints for inclusion in the master problem constraint sets. Consider, for example, a network node with only 3 incident arcs:

1. Arc 1 is incoming with a flow of either zero or in the range (10,100),
2. Arc 2 is incoming with a flow of either zero or in the range (40,60),
3. Arc 3 is outgoing with a flow in the range (5,45).

This problem may be represented using a binary variable for each of the incoming arcs, y_1 and y_2. It is easily seen that $y_1 + y_2 = 1$ in any feasible solution. This constraint may therefore be included in the master problem constraint set.

Constraints thus implied by the flow conservation constraints of network problems are discovered automatically. For each node j define the set $\mathcal{E}_j \subseteq \mathcal{E}$ as the set of outgoing arcs *not* controlled by a binary variable, and $\mathcal{C}_j \subseteq \mathcal{C}$ as the set of incoming arcs controlled by a binary variable. Without loss of generality we assume that no controlled outgoing arcs or un-controlled incoming arcs exist, and that all multipliers $m_{ij} = 1$. Now define
$$L_j = \sum_{(j,k) \in \mathcal{E}_j} l_{jk}$$
and
$$U_j = \sum_{(j,k) \in \mathcal{E}_j} u_{jk}.$$

These are the minimum and maximum total flows allowable on controlled arcs adjacent to node j. Then for each $j \in \mathcal{N}$, the constraints

$$\sum_{(i,j) \in C_j} u_{ij} y_{ij} \geq L_j$$

and

$$\sum_{(j,k) \in C_j} l_{jk} y_{jk} \leq U_j$$

must be satisfied in any feasible solution to the problem. The first constraint is trivially satisfied unless

$$\sum_{(i,j) \in C_j} u_{ij}^- < L_j$$

and the second unless

$$\sum_{(i,j) \in C_j} l_{ij}^+ > U_j,$$

where $u_{ij}^- = \min\{0, u_{ij}\}$ and $l_{ij}^+ = \max\{0, l_{ij}\}$.

For the example given above, the two constraints generated are: $100y_1 + 60y_2 \geq 5$ and $10y_1 + 40y_2 \leq 45$. By using coefficient reduction (Crowder, Johnson and Padberg [1983]), the former constraint is sharpened to $5y_1 + 5y_2 \geq 5$, a change which improves the bounds obtained from Lagrangean relaxation.

The algorithm automatically generates all such constraints which are not trivially satisfied and includes them in the master problem constraint set. Experience with one set of problems (backboard wiring problems) indicates that this preprocessing can make a substantial difference in solution time (Section 4.4.3). For related preprocessing techniques, see van Roy and Wolsey [1987].

3 Implementation

The various algorithmic approaches discussed in Section 2 were integrated in the software system GENOS/MIP. The main issues of the implementation are explained in this section.

3.1 Systems Integration

The GENOS/MIP system is written in Fortran 77. It uses a branch-and-bound algorithm, and the generalized network optimizer GENOS of

Mulvey and Zenios [1987], as subparts for solving the master and subproblems of Benders Decomposition, respectively. A driver alternates control between GENOS and the MIP part of the algorithm.

Solving a subproblem is a two-stage process. First, GENOS is called to find a feasible solution to the problem with controlled arcs corresponding to binary variables equal to 0 removed. This network is augmented with explicit slack arcs so that GENOS will find a feasible solution even if the current integer proposal renders the problem infeasible. The objective is to minimize flow on the slack arcs. If there is flow on the slack arcs, the current configuration is infeasible, and a Benders "infeasibility cut" is generated. Otherwise stage two is initiated; solving the network with the current configuration, and with the original objective function. From the result a Benders "feasibility cut" is generated.

Control then returns to the driver, which initiates the next master problem. Based on the previous branch-and-bound tree, a solution satisfying the current collection of Benders cuts and corresponding to a sufficiently good objective function value is sought as explained in Section 2.3.

3.2 Branch-and-bound Data Structures

The branch-and-bound algorithm relies upon two central data structures: the constraint matrix and the branch-and-bound tree. The constraint matrix is represented in the usual column-wise packed format. Coefficients from new Benders cuts are merged *in place* with existing coefficients.

The branch-and-bound tree is implemented as a priority queue of pending nodes, i.e. a linked list of pending nodes sorted in order of merit. The node at the head of the list is thus the next to be considered. The information stored for each node is the status of each binary variable: Whether it is fixed or free and, if fixed, at what value. No explicit tree-structure is thus present.

3.3 Parallel MIP Solution

A version of GENOS/MIP was written to allow parallel computing in the MIP phase on an Alliant FX/4 computer with 4 processors. The Alliant Fortran 77 compiler will (guided by the compiler directives CNCALL or NODEPCHK, and compiler options like -DAS) schedules DO-loops for execution across all 4 processors. This feature is used to start 4 independent processes, each executing on its own processor.

The steps BB-1 through BB-6 of the Branch-and-Bound algorithm (Section 2.3) are executed asynchronously in parallel on the 4 proces-

Problem	No. of obj. function coefficients	No. of binary vars	No. of nodes	No. of arcs	Obj. value
Mark100	4712	100	2	100	-0.001329
Back8	1008	64	18	80	429
Back10	2610	100	22	120	1069
Back20	37050	400	42	440	3894
Back34	216720	1224	72	1294	7926

Table 1: Test Problem Characteristics. All test problems have a quadradic objective function.

sors. A table of pending branch-and-bound nodes is shared among all processes. Synchronization is only necessary while accessing the node table, i.e., during steps BB-1 and BB-2, where processes remove a node from the node table, and step BB-6, where a new node is inserted into the table. In particular, compute intensive parts of the algorithm, such as the subgradient optimization, are performed in parallel on different nodes. The only potential bottleneck is insertion of nodes into the table, since new nodes are inserted such that the node table is kept sorted on node merits. Program profiling revealed (Section 4.5) that node insertion is no serious bottleneck. Access to the node table is synchronized using a semaphore and the associated lock/unlock primitives.

4 Numerical Experiments

A number of numerical experiments were performed to evaluate the algorithm, its robustness and performance. In this section, we present the test problems, and give results from their solution. A number of internal parameters and strategies are varied and the results presented, and results from the implementation on a parallel computer are given. The goal is to give a summary of the experiments and to highlight the features of the implementation.

4.1 Test Problems

Table 1 summarizes the characteristics of the test problems.

1. **Mark100**. This is a Markowitz model that creates a portfolio minimizing a weighted combination of risk (as measured by the covariance of the portfolio) and expected return. Minimum trading

requirements of 5% introduce integrality constraints. The network formulation is described in Dembo et al. [1989].

2. **Back***n*. These problems are Backboard Wiring Problems, Steinberg [1961]. n electrical components must be placed on a backboard such as to minimize the total length of electrical wires connecting the components. The problems are thus quadratic assignment problems. Back8, Back10 and Back20 are smaller versions of the full problem, Back34. Back34 was solved by Mawengkang and Murtagh [1986]. The objective function is minimizing total (Euclidean) wire length. It is non-convex, and numerous local minima exist. The algebraic statement of Back34 in the GAMS modelling language (Brooke, Kendrick and Meeraus [1988]) is found in Appendix B.

The best objective values known, either from the literature or from our own results, are stated in Table 1. The objective function values stated in Table 2 are the actual solutions obtained for the runs given. These values may differ for one of two reasons: Because multiple locally optimal solutions exist (which is the case for the backboard wiring problems), or because our runs were terminated due to time limits or exhaustion of the node table.

4.2 Solution of Test Problems

The test problems were solved on an Alliant FX/4. Results given in Table 2 include objective function value, total time (excluding input/output) and the time spent in the branch-and-bound algorithm.

1. **Mark100**. The Markowitz model was solved without much difficulty. The algorithm is apparently well suited for this important problem structure. It would be of interest to examine whether this still holds true as the number of binary variables increases.

2. **Back8 and Back10**. These problems are relatively easy and were solved repeatedly in testing system parameters (Section 4.4). Results given here are for a typical execution.

3. **Back20 and Back34**. These problems proved difficult and were not solved to optimality. The objective function values and times reported in Table 2 are for the best solutions obtained. For both problems, quite good solutions were found quickly, and finding improved solutions took increasingly long time. Figure 1 shows the best objective function value as a function of CPU time (Alliant

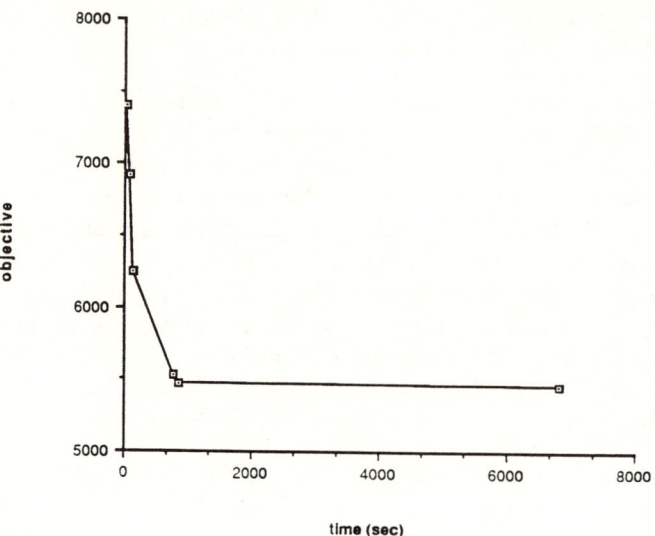

Figure 1: Back34: Objective function value vs. CPU time (Alliant FX/4).

FX/8, 4 processors in parallel) for the Back34 problem. The non-determinacy introduced by using parallel computing caused the algorithm to return different solutions in successive runs. The solutions found in most cases had objective values which were very close, although the actual solutions could differ substantially. This also indicates that starting the algorithm from different starting points probably would not have a big impact on the value of the solutions returned.

4.3 Comment on Problem Suitability

It was attempted to solve a portfolio dedication model using GENOS/MIP, but the Benders Decomposition scheme failed to converge after 200 iterations. Almost all of the subproblems were infeasible, and it was evident from inspection of the successive integer proposals generated that very little progress was being made. It is instructive to investigate why this happened.

The network model underlying the portfolio dedication model is shown in Appendix A. The purchase of a financial instruments, such as a bond, is modelled as flow on an arc leading from the supply node

Problem	Objective Value	Total Time (sec.)	MIP Time (sec).	No. of Benders Cuts
Back8	467	22	18	13
Back10	1076	44	40	6
Back20	3894	3357	3305	15
Back34	5452	32978	32696	21
Mark100	-0.001329	74	67	16

Table 2: Solution of Test Problems on Alliant FX/4.

to a bond node. Only unit flow (or no flow) is allowed on such arcs. Payments from a purchased bond are modelled using arcs leading from the bond node to a number of nodes corresponding to payment dates for a stream of liabilities. Arcs between payment date nodes represent short-term reinvestment of surplus cash at a conservative interest.

The multipliers on the bond arcs and the upper bounds on the payment arcs are such that when a bond is purchased, payments are given exactly (i.e., the multiplyer of an bond arc equals the sum of the payments resulting from holding the bond). The objective of the model is to satisfy liabilities (demands) at a number of the payment dates at a minimum portfolio cost.

We have no formal explanation why the decomposition fails to converge, but we provide here some observations. Under an infeasible integer proposal, if a liability at a payment date cannot be satisfied, the resulting Benders cut reflects that flows on controlled arcs having cash-flows due on that date must be increased, but the cut fails to reflect that the majority of the resulting increased bond cash-flow will be diverted to other payment dates. As a result, even for subsequent integer proposals which satisfy the new cut, the same liability is likely to remain unfunded. The problem would be eliminated if the underlying network model were able to represent proportionality of cash-flows. This, however, is a *blending problem*, which cannot be fit into our framework. Although the dedication model with proportionality constraints can be solved by Benders Decomposition, it does not seem effective to formulate the model as a mixed-integer network.

4.4 Testing Internal Implementation Strategies

A number of system parameters can be fine-tuned for better performance on specific large-scale problems. In this section, we investigate the relationship between parameter settings and algorithm performance,

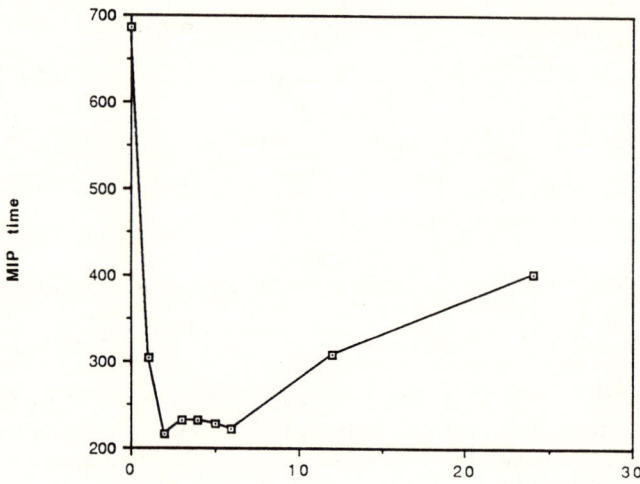

Figure 2: Varying subgradient Iteration Limit: MIP time and Benders Cuts.

exemplified by the Back8 and Back10 problems. The parameters are: Iteration limit on the subgradient optimization, the aggression factor of Section 2.2.1, using or not using automatic preprocessing (Section 2.6), and, finally, which node merit function to use.

4.4.1 Subgradient Iteration Limit

Whenever the Lagrangean relaxation routine is executed, it is terminated when either a positive value of the Lagrangean is found, or the number of subgradient iterations reaches a limit. Figure 2 shows the effect of varying the iteration limit on the Back10 problem on solution time and the number of Benders cuts generated for Back10.

It is quite remarkable that the fastest execution results from a very small limit, about 2. The added precision of the bounds provided by Lagrangean relaxation when using a higher limit does not offset the added computational expense.

4.4.2 The Aggression Factor

Once a solution to the binary feasibility problem has been found, the algorithm tries to find a better one, i.e., one with a better objective

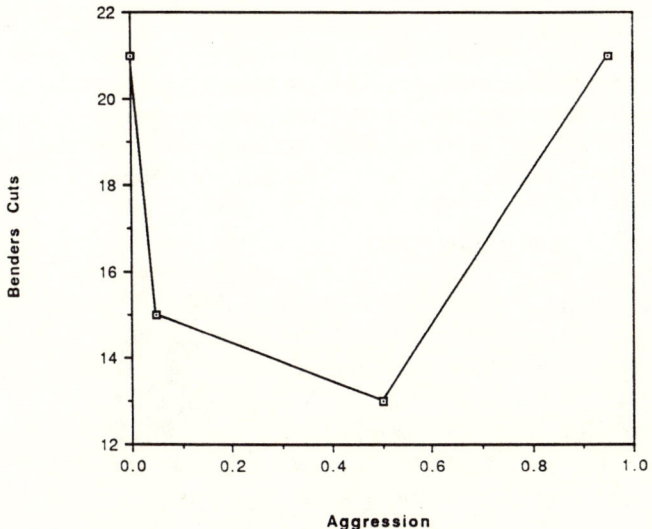

Figure 3: Varying the Aggression Factor: MIP time and Benders Cuts.

function in terms of the original (minimization) master problems. This translates into finding a feasible solution to a more tightly constrained binary feasibility problem.

A more aggressive solution strategy, i.e., a higher value of the "aggression factor" $\alpha \in (0,1)$, results in a more tightly constrained feasibility problem. It is not clear whether such a problem should be easier or more difficult. Also, being more aggressive results, presumably, in better Benders cuts. A series of experiments were conducted to clarify the effect of varying the aggression factor. Runs were performed on the Back10 problem on an Alliant FX/4 (with no concurrent branch-and-bound) and results are depicted in Figure 3.

The best times are for either very low or for very high aggression factors. For very low aggression factors, it is relatively easy to find an improved feasible solution (for the case of the aggression factor being 0, no improved solution is sought after the initial feasible one). For very high aggression factors, the problem is very tightly constrained. It is then likely that no feasible solution exists, and that this can be easily proved by the algorithm.

4.4.3 Preprocessing

The Back8 and Back10 problems were originally solved without using the preprocessing step of the algorithm. Solution time for the Back8 problem was 99 seconds, whereas the Back10 solve had to be terminated after 10,000 seconds! Compare these to the solution times of 22 and 44 seconds respectively from Table 2, where preprocessing was employed.

The number of infeasible subproblems that had to be solved was in both cases about the same as the number of cuts the preprocessing would have generated, but the cuts were very dense (in contrast to the preprocessing cuts). The time it took the algorithm to generate these cuts was not big, but it appears that the structure of the cuts made the subsequent branch-and-bound much more difficult (at least in the Back10 case) than the preprocessing cuts.

When the other test problems were solved without preprocessing the difference in solution time was not significant. In no case was a problem solved faster without preprocessing.

4.4.4 Merit Function

Nodes generated during the branch-and-bound phases are inserted into the list of pending nodes (and subsequently expanded) in order of their "merit". Two different merit functions were tried, viz.

$$m_1 = -LD$$

and

$$m_2 = -L,$$

where $L \leq 0$ is the value returned by the Lagrangean relaxation application, and D is the depth of the node in the tree (i.e. the number of variables fixed). These were both felt to be relevant for measuring the attractiveness of a node: L is a (heuristic) measure of the potential feasibility of the node, and a higher value of D means that the node is closer to having all variables fixed, i.e. being feasible.

For the problems tested, Back8 and Back10, m_2 resulted in solution times 20% - 25% lower than m_1. The number of pending nodes in the branch-and-bound tree was somewhat higher (10% - 20%) for m_2 than for m_1, which means that the gain in speed is at the expense of having to store a larger node table.

4.5 Vector and Parallel Computations

The Branch-and-Bound part of GENOS/MIP was parallellized on an Alliant FX/4 with 4 processors. The processors operate indepently of

Problem	No conc.	1 proc.	2 proc.	4 proc.	Speed-up
Back8	98	90	62	49	2.0
Back10	569	509	192	192	3.0

Table 3: Parallellization on the Alliant FX/4, MIP times (in seconds). Shown is the speed-up factor from using no concurrency or vectorization to using 4 processors with vectorization.

each other and communicate through shared memory. Each processor has vector capabilities.

A series of runs were conducted on problems Back8 and Back10. First, the program was run with no vectorization or concurrency, using the Alliant compiler option for global scalar optimization. Then the program was compiled allowing for vectorized and concurrent execution of DO loops. The number of processors working concurrently in the parallel branch-and-bound algorithm was 1, 2 and 4. Time spent in the branch-and-bound phase are given in Table 3. Total times are not reported since they are consistently only a few percents higher. The speed-up factor is the ratio of the execution time with no concurrency with that of the time on 4 processors. Due to the non-determinism of parallel computation, all times are averages of several (4-8) runs.

The least speed-up for Back8 was 1.0, the largest 2.5. The reason for this considerable variation is that small differences in the synchronization may result in different integer solutions being found, leading to completely different Benders cuts and subsequent master problems to solve.

The added overhead of synchronizing access to the node table (see Section 3.3) was measured to be 5% in one case with 4 processors. While this is not prohibitive, a different way of sharing the node table would be needed if the number of processors were much higher.

5 Conclusions

In this paper, we have reported on the development and implementation of an algorithm for solving mixed-integer nonlinear network problems. Solution of a number of test problems derived from real applications shows that Generalized Benders Decomposition is a viable algorithm for solving mixed-integer non-linear (or linear) programming problems. Care must be taken, though, to exploit any problem specific knowledge (in this case network conservation constraints) to include suitable preprocessing constraints. Lagrangean relaxation proved a very robust tool for generating branch-and-bound bounds.

The bottleneck of the algorithm remains the master problems. Much time seems to be saved by not solving these to optimality, but the increased number of Benders cuts needed, and hence the increased number of subproblems as well as master problems that must be solved, must be weighted carefully against the savings of solving the master problems inexactly. More research needs to be put into this area, specifically whether this trade-off can be performed under algorithm control. The idea of solving the master problems as infeasibility problems seems viable. Additional work is needed in order to determine whether this approach is superior to solving them as optimization problems.

Results from using multiple processors on the Alliant FX/4 indicate that a concurrent evaluation of the branch-and-bound tree can save significant amounts of computation time.

Acknowledgements: The authors would like to acknowledge the comments of M. Guignard-Spielberg and A. H. G. Rinnooy Kan on an earlier draft of this paper. The research partially supported by NSF grants SES-91-00216 and CCR-8811135, AFOSR grant 91-0168 and AT&T contract CH292700MK.

References

D.P. Ahlfeld, R.S. Dembo, J.M. Mulvey and S.A. Zenios. *Nonlinear Programming on Generalized Networks*, ACM Transactions on Mathematical Software, Vol. 13, pp. 350-367, 1987.

Daniel Bienstock and J.F. Shapiro. *Optimizing Resource Acquisition Decisions by Stochastic Programming*, Management Science, Vol. 34(2), pp. 215-229, Feb. 1988.

J.F. Benders. *Partitioning Procedures for solving Mixed Variable Problems*, Num. Math. 4, pp. 238-252, 1962

A. Brooke, D. Kendrick and A. Meeraus *GAMS: A Users' Guide*, The Scientific Press, Redwood City, CA. 1988.

Harlan Crowder, Ellis L. Johnson and Manfred Padberg. *Solving Large-*

Scale Zero-One Linear Programming Problems, Operations Research, Vol. 31(5), pp. 803-834, September-October 1983.

Ron S. Dembo, John M. Mulvey and Stavros A. Zenios. *Large Scale Nonlinear Network Models and their Application*, Operations Research, Vol. 37(3), pp. 353-372, 1989.

Marshall L. Fisher. *The Lagrangian Relaxation Method for Solving Integer Programming Problems.* Management Science, Vol. 27(1), pp. 1-16, January 1981.

Marshall L. Fisher. *An Applications Oriented Guide to Lagrangian Relaxation.* Interfaces, Vol. 15(2), pp. 10-21, 1985.

R.S. Garfinkel and G.L. Nemhauser. *Integer Programming*, John Wiley and Sons, New York, 1972.

A.M. Geoffrion. *Generalized Benders Decomposition.* J. Optim. Theory Appl. 10, pp. 237-260, 1972.

A.M. Geoffrion. *Lagrangean Relaxation for Integer Programming*, Mathematical Programming Study 2, pp. 82-114, 1974.

A.M. Geoffrion and G.W. Graves. *Multicommodity Distribution System Design by Benders Decomposition.* Management Science, Vol. 20(5), pp. 822-844, 1974.

J.L. Kennington and R.V. Helgason. *Algorithms for Network Programming*, John Wiley and Sons, New York, 1980.

Kurt O. Jörnstein. *A Maximum Entropy Combined Distribution and Assignment Model Solved by Benders Decomposition.* Transportation Science, Vol 14(3), pp. 262-276, 1980.

Gary R. Kocis and Ignacio E. Grossmann. *Relaxation Strategy for the Structural Optimization of Process Flowsheets.* Department of Chemical Engineering, Carnegie-Mellon University, Pittsburgh, PA 15213, 1986.

T.L. Magnanti and R.T. Wong. *Accelerating Benders Decomposition: Algorithmic Enhancement and Model Selection Criteria.* Operations Research, Vol. 29, pp. 464-484, 1981.

H. Mawengkang and B.A. Murtagh. *Solving Nonlinear Integer Programs with Large-Scale Optimization Software*, Annals of Operations Research 5, pp. 425-437, 1986.

J.M. Mulvey and S.A. Zenios. *GENOS 1.0: A Generalized Network Optimization System. User's Guide.* Report 87-12-03, Decision Sciences Department, The Wharton School, University of Pennsylvania, Philadelphia, PA 19104, 1987.

George L. Nemhauser and Laurence A. Wolsey. *Integer and Combinatorial Optimization.* John Wiley and Sons, Inc., 1988.

G.E. Paules and C.A. Floudas. *APROS: A Discrete-Continuous Optimizer for the Automatic Solution of Mixed-Integer Nonlinear/Linear Programming Problems.* Operations Research, Vol 37(6), pp. 902-915, 1989

Tony J. van Roy and Laurence A. Wolsey. *Solving Mixed Integer programming Problems using Automatic Reformulation.* Operations Research, Vol 35(1), pp. 45-57, January-February 1987.

Jeremy F. Shapiro. *A Survey of Lagrangean Techniques for Discrete Optimization.* Annals of Discrete Mathematics 5, 113-138.

Leon Steinberg. *The Backboard Wiring Problem: A Placement Algorithm.* SIAM Review, Vol 3(1), January 1961.

Appendix A
Portfolio Dedication Model

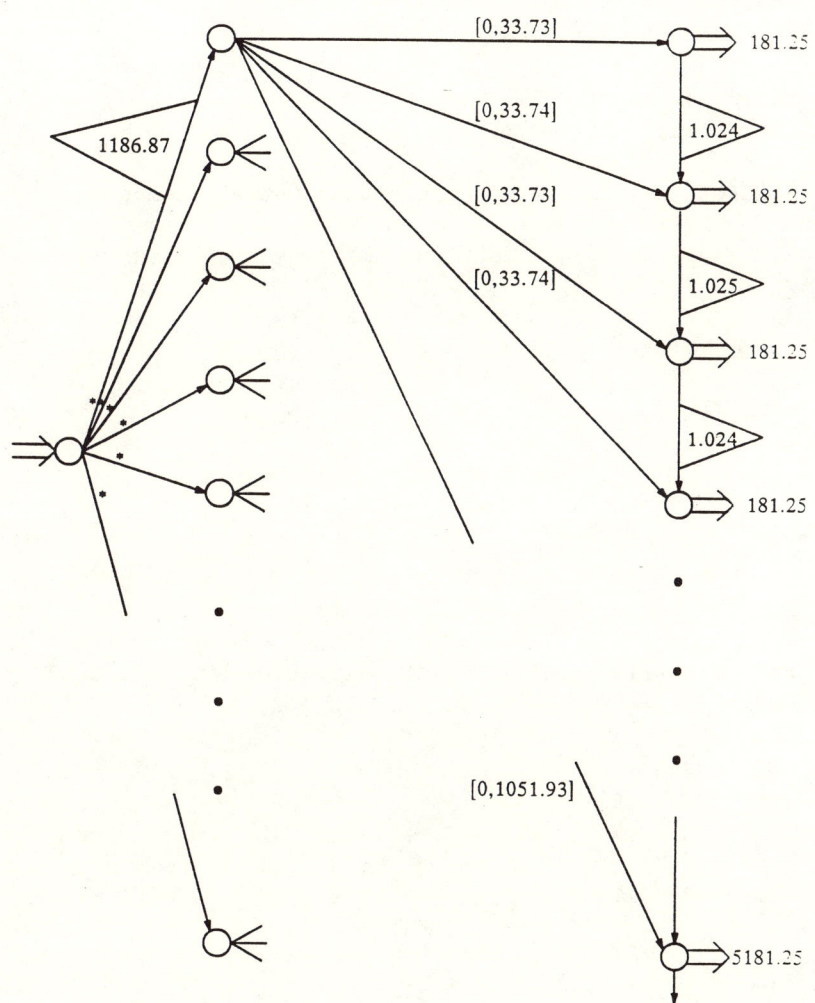

* indicates binary variable

Appendix B
Backboard Wiring Problem in GAMS

```
set components /e1 * e34/;
set locations  /loc1 * loc36/;

alias (components, c1, c2);
alias (locations,  l1, l2);

table conn(c1, c2) Number of connections between components
```

	e1	e2	e3	e4	e5	e6	e7	e8	e9	e10	e11	e12
e1				2	1	7	9		4	75	7	12
e2							4	16		8		
e3						4	16	20				
e4					29	5	18	47	23	2	4	
e5						18	12	25			4	
e6							4	2		1	23	2
e7									14	72	7	8
e8									10	71	2	
e9										14		
e10											11	1
e11												316

+	e13	e14	e15	e16	e17	e18	e19	e20	e21	e22	e23	e24
e1	22	7	1					23				
e2	16					6		4				
e3	20							4				
e4	48		4					25				
e5	25		3					18		3		
e6	19						2	19				
e7	39	8	40	8		8	4	7				
e8						41						
e9	18											
e10	17		1			17		15				
e11	33	8	2				8	34			6	
e12	157	25	4			1						
e13		11	6			6		5	8	3	10	
e14				3		1	1	21		1		2
e15				19		2	2	12				
e16						6		1				
e17						40						

	e25	e26	e27	e28	e29	e30	e31	e32	e33	e34
e18							26			
e19							13	9		7
e20								11	4	36
e21									36	6

+	e25	e26	e27	e28	e29	e30	e31	e32	e33	e34
e7				28	8					
e8			7	8						
e10										
e11			10			6				
e12	22		1							
e13			9	11	2			1		
e14			5			3	2	5	5	4
e15				7	3					
e19				27	16	3		20		4
e20			16	18	9	10	1	28	6	2
e21	8		2							
e22			4							
e23		12	9							
e24	26		5							
e25		35	2							
e26			4							
e27										
e28					10	22	4	6	4	12
e29						19	12			
e30							19	4	5	8
e31									3	13
e32									18	24
e33										20
e34										

```
parameter rowtotal(components);
rowtotal(c1) = sum(c2, conn(c1,c2));
display rowtotal;

parameter dist(l1, l2)   Euclidean distances between locations

* Calculate distances:
parameter xpos(locations), ypos(locations);
ypos(locations) = 1 +
                ( 1 $ (ord(locations) gt 9)) +
                ( 1 $ (ord(locations) gt 18)) +
                ( 1 $ (ord(locations) gt 27)) ;
xpos(locations) = ord(locations) - ypos(locations) * 9 + 9;
dist(l1,l2)     =
            sqrt(sqr(xpos(l1)-xpos(l2))+sqr(ypos(l1)-ypos(l2)));

display xpos, ypos, dist;

binary variables x(components, locations);
variable wirelength;

equations
    sum1    Each component goes to one location,
    sum2    Each location  receives at most one component
    obj     Objective function definition ;

sum1(locations) ..sum(components,x(components,locations)) =l= 1;
sum2(components)..sum(locations, x(components,locations)) =e= 1;

obj    ..  wirelength =e=
            sum((c1,c2,l1,l2),
                conn(c1,c2)*dist(l1,l2)* x(c1,l1) * x(c2,l2));

model backboard /all/;
solve backboard minimizing wirelength using midnlp;
```

Global Minima in Root Finding

Angelo Lucia*   Jinxian Xu*

Abstract

Two chemical process examples are used to illustrate that conventional trust region methods for finding roots on the real line can terminate at a singular point, corresponding to a nonzero-valued local minimum of the least squares objective function. Furthermore, it is shown that this behavior is the result of an exchange of a pair of zero-valued global minima for a local minimum that accompanies a bifurcation of a pair of real roots into the complex plane.

It is observed that these nonzero-valued local minima on the real line frequently correspond to saddlepoints of the complex absolute value function. Thus, when viewed in the complex plane, these stationary points possess directions of negative curvature which can be used to construct a computational path to a global minimum (or root). Many geometric illustrations are used to present the fundamental observations on which an extended trust region method is based.

1. Introduction

Trust region (or dogleg) methods for solving nonlinear algebraic equations were introduced by M.J.D. Powell in 1970 and have been very popular in a variety of disciplines over the last twenty years due to their global convergence properties. However, what is not made clear, either in the original paper by Powell or in any of the many modifications that have appeared since, is that the dogleg strategy can converge to a nonzero-valued local minimum of the least squares objective function, which does not correspond to a root, but rather a singular point, of the function under consideration. This is true even for single variable problems.

---

*Department of Chemical Engineering, Clarkson University, Potsdam, NY 13699-5705.

In a recent manuscript largely concerned with the periodic and chaotic behavior of Newton's method in the complex plane, Lucia and Guo (1991) observe that nonzero-valued local minima on the real line can correspond to saddlepoints of the complex absolute value function. Consequently, when viewed in the complex plane, there are directions of negative curvature that can be used to construct a computational path from a strict local minimum (a singular point) to a zero-valued global minimum (a root).

The objective of this paper is to provide further numerical and geometric support for the extension of the dogleg strategy to the complex plane given in Lucia and Guo (1991). In particular, common chemical process applications are used to illustrate the exchange of a pair of global minima for a local minimum on the real line that accompanies a bifurcation into the complex plane and that these real local minima often correspond to saddlepoints in the complex absolute value function. Finally, termination of the complex dogleg strategy at a global minimum is illustrated by extensive numerical experimentation; however no proof of global convergence is given.

2. Background and Motivation

The problem in which we are interested is that of finding a root to

$$F(X, p) = 0 \qquad (1)$$

where F is some nonlinear algebraic equation of a single unknown variable X and p is a scalar parameter. For the moment, assume that F, X and p are real.

There are many ways to solve Eq. 1 for a root, including Newton's method, dogleg or trust region methods, and continuation method, and each has its own set of advantages and disadvantages. Newton's method can behave periodically or chaotically, trust region methods can terminate at stationary points of the objective function $\phi = |F(X, p)|$ that are not roots, and continuation methods can be computationally expensive. This manuscript is concerned exclusively with the numerical performance of trust region methods, and some familiarity with the basic concepts of trust region methods is assumed.

Figures 1 and 2 are plots of the function $\phi = |F(V, T)|$, where $F(V, T)$ is the Soave-Redlich-Kwong (SRK) equation of state, where V represents a volume root of the SRK equation and where T, the temperature, is the

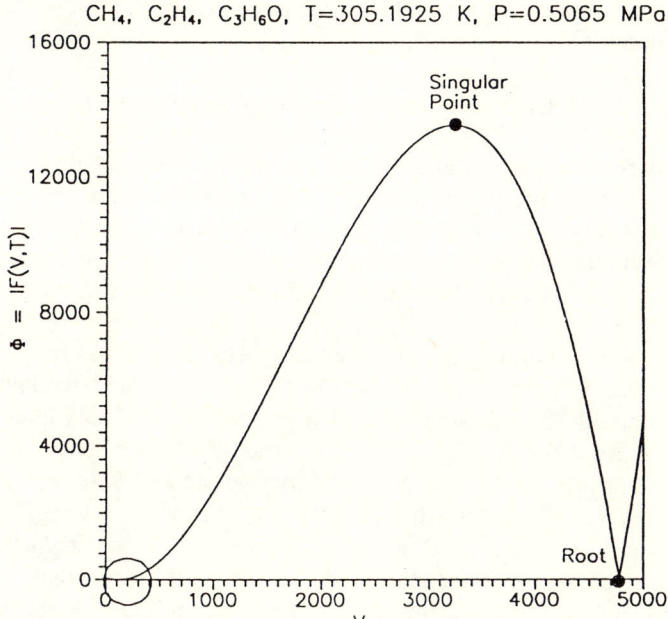

Figure 1: Real Stationary Points of $|F(V,T)|$

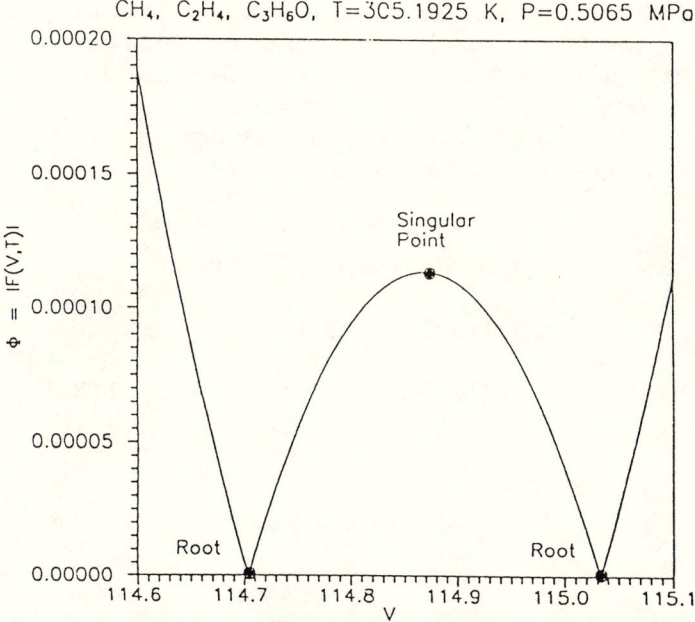

Figure 2: Real Stationary Points of $|F(V,T)|$

parameter. The particular form of the equation of state that we have used for this illustration is

$$F(V, T) = V^3 - [RT/p]V^2 + [a\alpha - bRT - pb^2]V - (a\alpha b/p) = 0, \qquad (2)$$

where p is the pressure, R is the universal gas constant and where a, b, and α depend on the critical constant associated with the species in the mixture and the mixing rules. We refer the reader to Walas (p. 53, 1985), noting that for the example that follows, the Soave parameters, k_{ij}, in the mixing rules were assumed to be zero. Note also that Eq. 2 is a cubic equation.

The particular mixture used in generating Figs. 1 and 2, as well as any subsequently related figures, was a ternary mixture of 30 mol % methane, 40 mol % ethylene and 30 mol % acetone at a pressure of 0.5065 MPa. At a temperature of 305.1925 K, Eq. 2 has three real roots (see Table 1), and Figs. 1 and 2 clearly show that the three roots correspond to zero-valued global minima of $\phi = |F(V, T)|$, while the two singular points where $F'(V, T) = 0$ are maxima. Note that two of the roots are quite close in numerical value. When all of the roots are real, we have observed that dogleg strategies have no difficulty finding a global minimum since these are the only minima that are present.

On the other hand, Table 1 shows that at a slightly higher temperature of 305.1928 K, two of the roots have bifurcated into the complex plane. Moreover, Figs. 3 and 4, and in particular Fig. 4, show that the objective that function $\phi = |F(V, T)|$ now has only two real minima, one local and one global, that correspond to a singular point and the vapor root of F(V, T) respectively. An exchange of a local minimum for two global minima has

Table 1
Roots and Singular Points of the SRK Equation[*]

Temperature (K)	Volume Root (cc)	Singular Point (cc)
305.1925	114.704 + 0i	114.868
	115.033 + 0i	3224.190
	4778.851 + 0i	
305.1928	114.868 + 0.0704i	114.868
	114.868 − 0.0704i	3224.193
	4778.856 + 0i	

[*] 0.3 CH_4, 0.4 C_2H_4, 0.3 C_3H_6O at 0.5065 MPa.

GLOBAL MINIMA IN ROOT FINDING

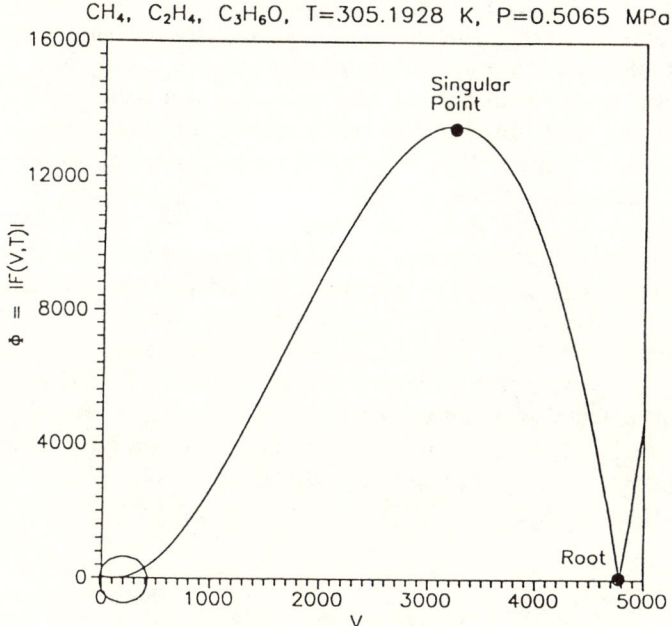

Figure 3: Real Stationary Points of |F(V,T)|

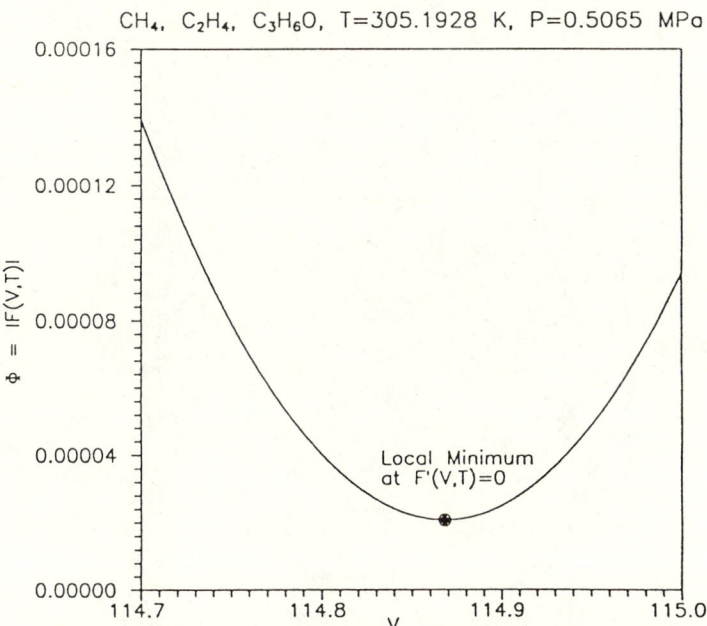

Figure 4: Real Stationary Points of |F(V,T)|

accompanied the bifurcation of the liquid root, and the value of the objective function at the singular point is nonzero for any temperature above the bifurcation temperature. In addition, Figs. 3 and 4 show that any local nonlinear programming algorithm, such as a trust region method, will terminate at this local minimum if the starting point is between zero and the other singular point at V = 3224.193 cc, and is doomed to fail if it is a root (or global minimum) that is sought.

In a recent paper, Lucia and Guo (1991) make the key observation that nonzero-valued local minima of $\phi = |F(V, T)|$ on the real line frequently correspond to saddlepoints of the complex absolute value function. This is clearly illustrated in Figs. 5 and 6 for the real local minimum shown in Fig. 4. Thus, perturbation from this saddlepoint along the imaginary axis represents a direction of negative curvature and movement toward a global minimum (see Fig. 5). This is the primary feature that we have exploited in developing an extension of the dogleg strategy to the complex plane.

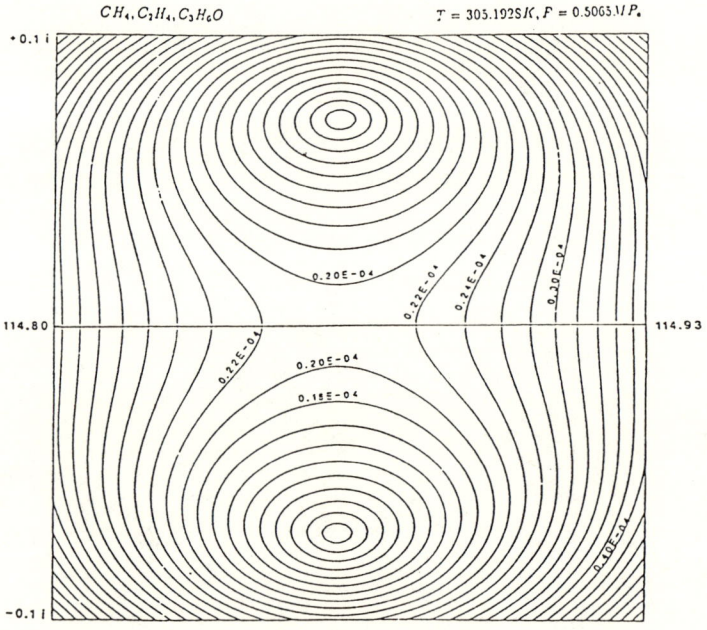

Figure 5: Level Sets for $\phi = |F(V,T)|$ where F is the SRK Equation

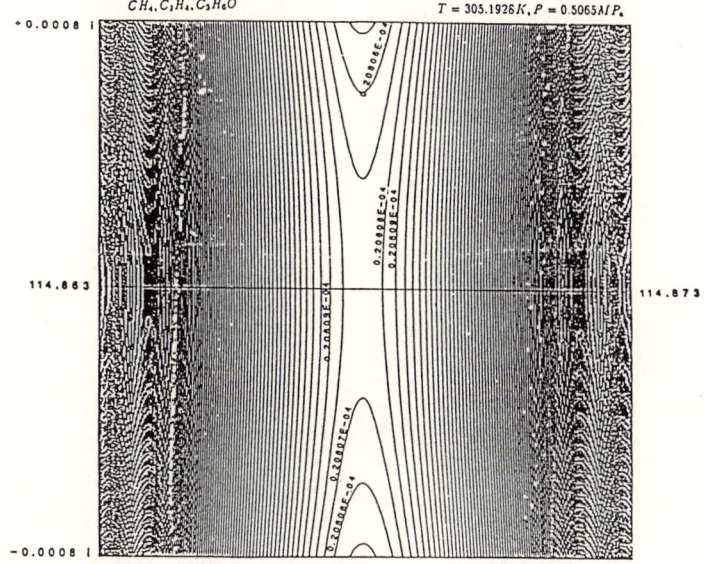

Figure 6: Level Sets for $\phi = |F(V,T)|$ where F is the SRK Equation

3. A Trust Region Algorithm

The steps of our trust region algorithm for functions defined on the complex plane are as follows. Some familiarity with traditional trust region methods is assumed.

1. Choose p and X°, initial and maximum trust region radii Δ° and Δ_{max} respectively and a convergence tolerance $\epsilon > 0$. Set the iteration counter $k = 0$.

2. Calculate $F(X^k, p)$, $F'(X^k, p)$ and the objective function value $\phi^k = |F(X^k, p)|$, where $|\cdot|$ denotes the complex absolute value function.

3. If $\phi^k \leq \epsilon$, stop. Otherwise, go to step 4.

4. Calculate the approximate Cauchy (or gradient) direction

$$\phi' = \text{Re}[F(X^k, p)]F'(X^k, p) \qquad (3)$$

where

$$F'(X^k, p) = \begin{cases} \text{Re}[F'(X^k, p)] - \text{Im}[F'(X^k, p)], & \text{if } 0.1 \, \text{Re}[F'] < \text{Im}[F'] \\ F'(X^k, p), & \text{otherwise} \end{cases}$$

where $\text{Re}[\,\cdot\,]$ and $\text{Im}[\,\cdot\,]$ denote the real and imaginary part of the argument in brackets. If $|\phi'| < \epsilon$ and $\phi^k > \epsilon$, go to step 15.

5. If $|\phi'| \geq \Delta^k$, calculate $\alpha = \Delta^k/|\phi'|$, set $\phi' = \alpha\phi'$, set $\beta^k = 0$ and go to step 9.

6. Calculate the Newton step, s_N, by the rule

$$s_N = F(X^k, p)/F'(X^k, p) \tag{4}$$

7. If $|s_N| \leq \Delta^k$, set $\beta^k = 1$ and go to step 9.

8. Calculate $\beta^k = [-b + \sqrt{(b^2 - 4ac)}]/2a$, where

$$a = |\phi'|^2 + |s_N|^2 - 2\{\text{Re}[\phi']\text{Re}[s_N] + \text{Im}[\phi']\text{Im}[s_N]\}$$

$$b = -2|\phi'|^2 + 2\{\text{Re}[\phi']\text{Re}[s_N] + \text{Im}[\phi']\text{Im}[s_N]\}$$

and

$$c = |\phi'|^2 - (\Delta^k)^2.$$

9. Calculate the direction d^k by the rule

$$d^k = (1 - \beta^k)(-\phi') - \beta^k s_N \tag{5}$$

10. Set $X^{k+1} = X^k + d^k$ and evaluate $F(X^{k+1}, p)$, $F'(X^{k+1}, p)$ and $\phi^{k+1} = |F(X^{k+1}, p)|$.

11. If $\phi^{k+1} \geq \phi^k$, set $X^{k+1} = X^k$, $F(X^{k+1}, p)$, and $F'(X^{k+1}, p) = F'(X^k, p)$.

12. Compute the error function

$$e^k = \phi^{k+1} - 0.1\, \phi^k \tag{6}$$

13. Adjust the trust region radius using the following rules

(i) If $\phi^{k+1} < \phi^k$ and $|s_N| < \Delta_{max}$, $\Delta^{k+1} = \text{argmin}(10\Delta^k, \Delta_{max})$

(ii) If $\phi^{k+1} < \phi^k$ and $e^k > 0$, $\Delta^{k+1} = \Delta^k$

(iii) If $\phi^{k+1} \geq \phi^k$, $\Delta^{k+1} = 1/2\, \Delta^k$

14. Set $k = k + 1$ and go to step 3.

15. Set $X^{k+1} = X^k + [0 + (\Delta_{max})i]$ and compute $F(X^{k+1}, p)$, $F'(X^k, p)$ and $\phi^{k+1} = |F(X^{k+1}, p)|$. Set $\Delta^k = \Delta_{max}$ on only the first of successive passes and go to step 11.

For the most part, the steps of the algorithm are self explanatory, with the exception of step 4. Because the complex absolute value function does not satisfy Cauchy-Riemann conditions (see p. 239 in Greenberg, 1978), it is nondifferentiable except at $F(X, p) = 0 \pm 0i$. As a consequence, ϕ' does not necessarily represent a descent direction of $\phi = |F(X, p)|$ due to rotations in the complex plane, and descent is needed for convergence to a global minimum of $F(X, p)$. To circumvent these difficulties, we use the approximate gradient (or Cauchy) direction defined by Eq. 3, which contains an embedded safeguard against rotations in the complex plane due to a dominant imaginary part and the fact that $i^2 = -1$. Step 4 also contains a saddlepoint test. Note that rotations are avoided by the rules for calculating $F'(X^k, p)$, while the simultaneous test on $|\phi'|$ and ϕ^k clearly identifies a nonzero-valued stationary point. Step 8 provides a <u>real</u> value of β^k for computing the dogleg direction. Step 15 makes use of the observation that any real nonzero-valued local minimum frequently corresponds to a saddlepoint of the complex absolute value function, move is likely to the unknown variable in the direction of negative curvature (i.e., in the imaginary direction) and continues the steps of the algorithm at the appropriate place.

4. Numerical Support

In this section, we present two common chemical process examples and several geometric illustrations that show that the dogleg algorithm given in section 3 terminates at a global minimum. These examples include the (polynomial) Soave-Redlich-Kwong (SRK) equation of state used in the motivation section, and a continuous stirred tank reactor (CSTR) which involves a transcendental function and a large number of complex roots.

4.1. Volume Roots of the SRK Equation

The SRK equation is a cubic equation of state and thus an example of a chemical process application in which the roots of a polynomial function are required. Furthermore, Figs. 3, 4, 5 and 6 clearly illustrate the main point that we have tried to stress. That is, while nonzero-valued minima of the (real) absolute value function can occur at singular points, perturbation into the complex plane frequently provides a path to a global minimum in the complex absolute value function.

Figures 7 and 8 show the basins of attraction for the dogleg strategy presented in section 3 for the specific problems defined in Table 1. For each case, 250,000 initial values on an equally spaced grid of 5000 x 5000 cc in volume (in increments of 10 cc) were used. Note that all initial values terminated at a global minimum, despite the fractal nature of the basin boundaries for the case in which two of the roots are complex (see Fig. 8). The convergence tolerance, initial and maximum trust region radii in these numerical experiments were 10^{-8}, 100 cc, and $|V^k|$ respectively.

To illustrate the importance of both our observation that local minima on the real line correspond to saddlepoints of the complex absolute value function and the utility of steps 4 and 15 in the algorithm, consider the case where the temperature is 305.1928 K and the initial value is specified as $V° = 1669.53$ cc. For this particular initial value, as well as others in the fractal boundary, the saddlepoint in $\phi = |F(V, T)|$ at $V = 114.868$ cc is usually encountered reasonably quickly. In this case, it took 44 iterations and the same number of function and gradient evaluations. Conventional trust region algorithms terminate here. However, step 4 in our complex dogleg strategy correctly identifies the singular point as a nonzero-valued stationary point of ϕ, step 15 initiates the subsequent perturbation of that stationary point into the complex plane, and the algorithm finds the root (global minimum) at $V^* = 114.868 + 0.0704i$ in 12 additional iterations and 24 additional function and gradient evaluations. Similar behavior has been observed for other starting points, particularly those in the fractal basin boundaries, for a wide variety of temperature parameter values, even for those in which the imaginary part of the root is quite large in magnitude.

4.2. Temperature Roots of a CSTR Mass/Energy Balance Equation

Our second example is a problem concerned with finding the temperature roots of the combined mass and energy balance equations for a CSTR, and was taken directly from Smith (p. 234, 1970). The nonlinear function whose temperature roots we seek is given by

$$F(T, \Delta H) = [(\rho C_p/C_o \Delta H)](T-T_o)$$

$$+ \theta A \exp(-E/RT)/[1 + \theta A \exp(-E/RT)] = 0, \qquad (7)$$

GLOBAL MINIMA IN ROOT FINDING

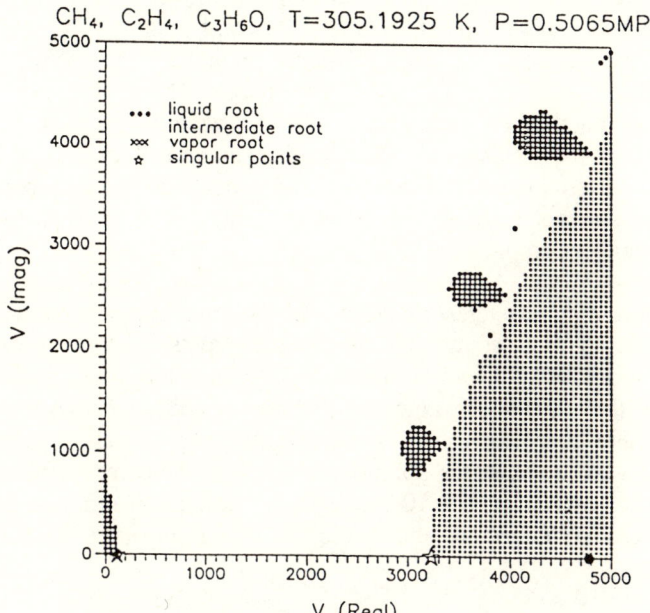

Figure 7: Basins of Attraction for Dogleg Method for SRK Equation with Three Real Roots

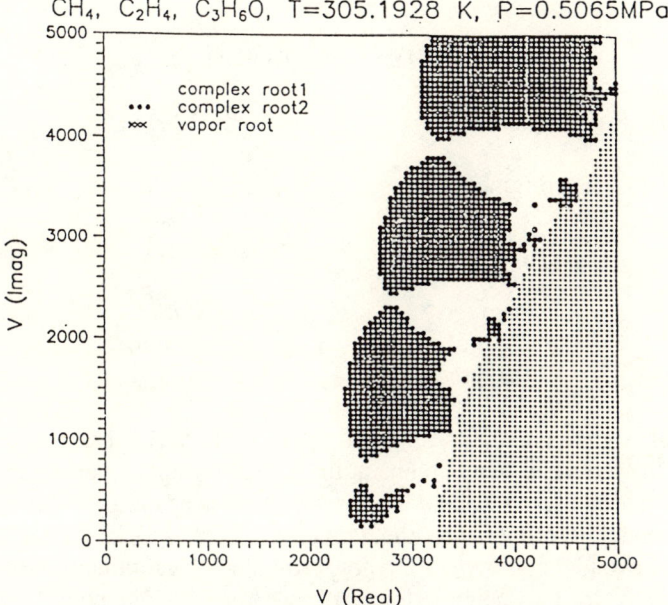

Figure 8: Basins of Attraction for Dogleg Method for SRK Equation with One Real Roots

where ρ is the fluid density, C_p is the specific heat, C_o and T_o are the inlet concentration and temperature of the reactant respectively, ΔH is the heat of reaction, θ is the average residence time for the reactor, A is a frequency factor, and E is the activation energy. Note that Eq. 7 is a transcendental function of temperature and the heat of reaction, ΔH, has been selected as the parameter. The specific numerical data used in this illustration is given in Table 2 and sets of roots and singular points for two particular parameter values are given in Table 3. Note again the roots can be either real or complex.

One of the more interesting aspects of the transcendental nature of Eq. 7 is that there is a large number of complex roots. More specifically, Lucia and Guo (1991) found a total of 159 complex roots. However, regardless of the parameter value, only three of the roots have physical significance, and there are only two (real) singular points by continuity arguments. Figure 9 shows the solution structure in the complex plane for a parameter value of $\Delta H = -35510.3$ cal/gmol. The three roots of physical significance are contained in the circle on the right hand side of that figure.

Figures 10, 11 and 12 show the stationary points of the absolute value function for the solutions given in Table 3. Note that when the three roots

Table 2

Numerical Data for CSTR Example

Quantity	Value
ρ (gmol/cc)	1.0
C_p (cal/gmol)	1.0
C_o (gmol/l)	3.0
T_o (K)	298.
θ (sec)	3000.
A (l/gmol sec)	4.48×10^6
E (cal/gmol)	$-15000.$

of physical significance are all real, $\phi = |F(T, \Delta H)|$ has three global minima at the roots and two maxima at the singular points given by $F'(T, \Delta H) = 0$, as shown in Figs. 10 and 11. On the other hand, bifurcation of a pair of real temperature roots into the complex plane is accompanied by an exchange of two global minima for a nonzero-valued local minimum at the real singular point in the neighborhood of the bifurcation point. This is clearly illustrated in Fig. 12. Note that despite the transcendental

Table 3

Roots and Singular Points for CSTR Example

$-\Delta H$ (cal/gmol)	Temp. Root (K)	Singular Point (K)
35,958	299.633 + 0i	339.730
	380.540 + 0i	380.859
	381.117 + 0i	
35,510.3	299.609 + 0i	340.450
	380.390 + 7.854i	379.215
	380.390 − 7.854i	

nature of Eq. 7, the behavior of the absolute value function $\phi = |F(T, \Delta H)|$ is qualitatively similar to the behavior of $\phi = |F(V, T)|$ for the polynomial case of the SRK equation.

Figure 13 again illustrates that nonzero-valued local minima on the real line often correspond to saddlepoints of the complex absolute value function. In particular, while the singular point at 379.215 K corresponds to a real local minimum of $\phi = |F(T, \Delta H)|$ (see, Fig. 12), it clearly possesses negative curvature in the direction of the imaginary axis.

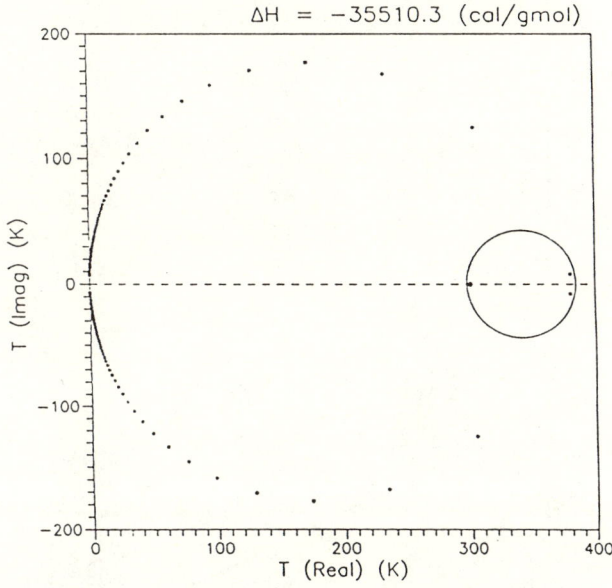

Figure 9: Solution Structure for CSTR with First-Order Kinetics

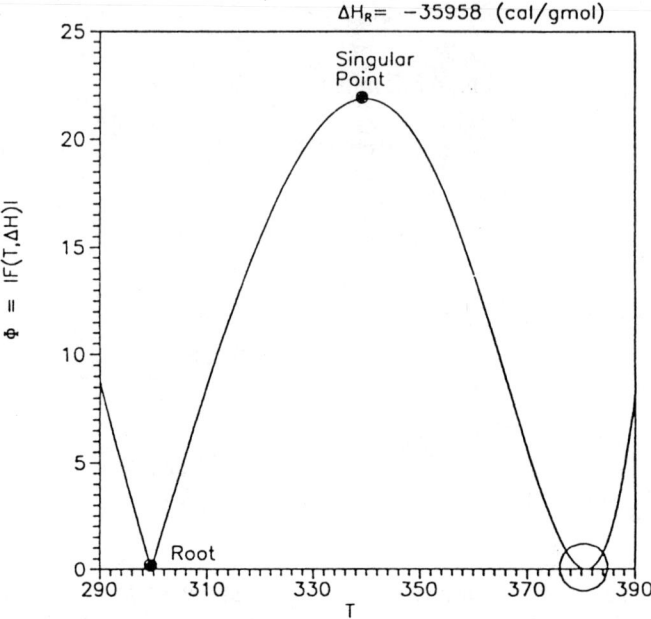

Figure 10: Real Stationary Points of $|F(T,\Delta H)|$

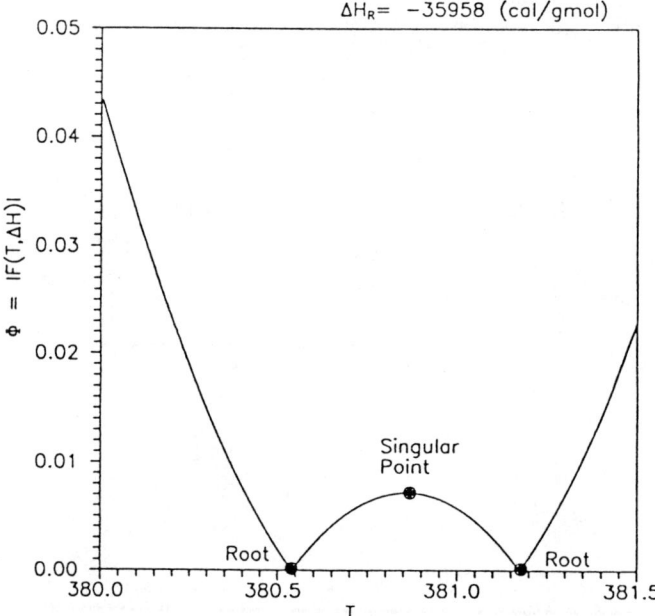

Figure 11: Real Stationary Points of $|F(T,\Delta H)|$

GLOBAL MINIMA IN ROOT FINDING

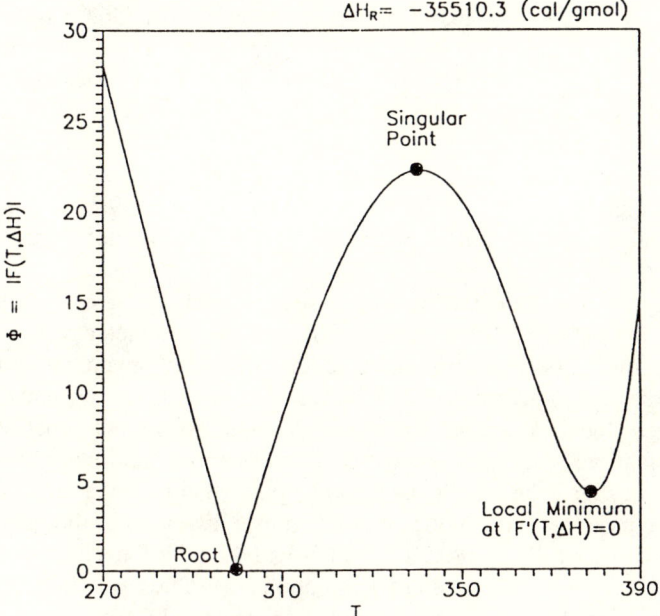

Figure 12: Real Stationary Points of |F(T,ΔH)|

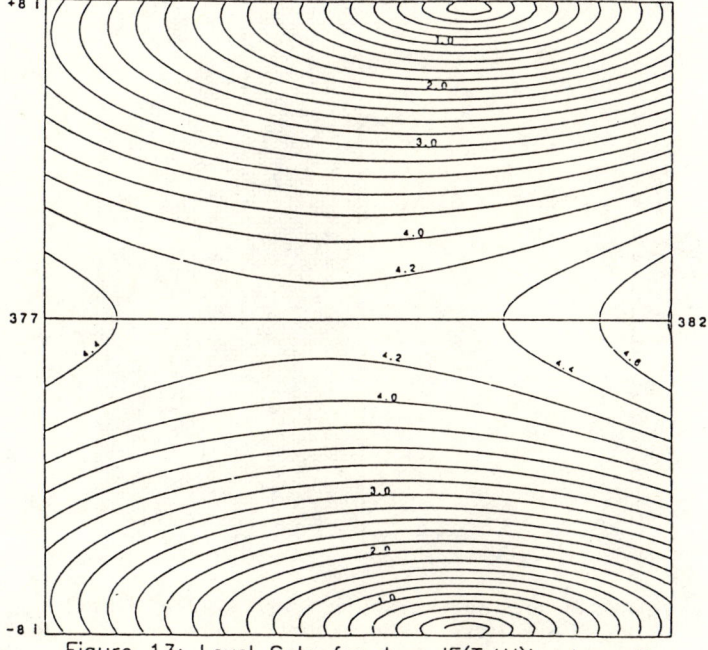

Figure 13: Level Sets for $\Phi = |F(T,\Delta H)|$, where F is the Mass/Energy Equation for CSTR

Figure 14 is a plot of the basins of attraction for the dogleg algorithm given in section 3, and was generated by constructing 5 by 5 K (in increments of 0.05 K) grid of initial temperature values in the neighborhood of the singular point at 340.450 K. For this large set of starting points, Fig. 14 clearly shows that the dogleg strategy in the complex plane always terminates at a global minimum of ϕ or a root of F(T, ΔH). Note, however, that some of these initial values converge to physically meaningless complex roots. For example, for an initial value of T° = 339.4 + 4.5i, the algorithm terminates at the root T* = 304.097 − 124.338i in 4 iterations.

Finally, to illustrate the utility of steps 4 and 15 in our algorithm, consider the initial value T° = 370 + 0.1i for the parameter value ΔH = 35,510.3 cal/gmol. For this initial value the algorithm given in section 3 encounters the singular point at 379.215 K, which corresponds to a nonzero-valued local minimum of ϕ, in 53 iterations and function and gradient evaluations. Step 4 correctly identifies the stationary point as nonzero-valued, step 15 makes the appropriate perturbation in the direction of negative curvature (i.e., along the imaginary axis), and the algorithm terminates at the root T* = 380.390 + 7.854i in 12 additional iterations and 19 additional function and gradient evaluations. Values of 1000 K, $|T^k|$, and 10^{-5} were used for Δ, Δ_{max} and ϵ respectively for all numerical experiments in this subsection.

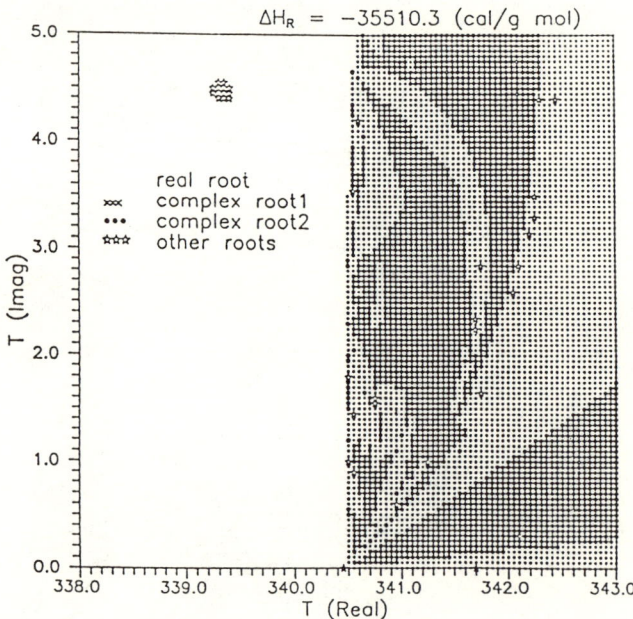

Figure 14: Basins of Attraction for Dogleg Method on CSTR Mass/Energy Equation with One Real Root

5. Conclusions

Two common chemical engineering applications were used to provide numerical and geometric support for an extension of the dogleg strategy to the complex plane recently proposed by Lucia and Guo (1991). This extension was based on the key observation that local minima in the absolute value function on the real line frequently correspond to saddlepoints of the complex absolute value function. Extensive numerical experiments indicate that our dogleg algorithm always terminates at a global minimum or root. However, no proof of global convergence was presented. We are currently attempting to prove global convergence results for the algorithm using an isomorphism from C to R^2 to get around the difficulty associated with the nondifferentiability of the complex absolute value function in the context of Cauchy-Riemann conditions. Extensions to multivariable problem in C^n are also being pursued.

Acknowledgements. This work was supported by the National Science Foundation under Grant No. CTS-8922316.

Notation

A	= frequency factor
C_p	= specific heat
C_o	= inlet concentration
E	= activation energy
e	= error function
F(X, p)	= complex function
ΔH	= heat of reaction
p	= pressure, parameter value
T, T_o	= temperature, inlet temperature
V	= volume
X	= unknown complex variable
x	= concentration

Greek Symbols

β	= dogleg parameter
Δ	= trust region radius
θ	= average residence time
ρ	= density
φ	= absolute value function

Subscripts

max = maximum
N = Newton's method

Superscripts

k = iteration counter
* = global minimum

Literature Cited

Greenberg, M.D., <u>Foundations of Applied Mathematics</u>, Prentice Hall, Engelwood Cliffs, NJ (1978).

Lucia, A. and X. Guo, "Newton's Method and Process Models in the Complex Domain," submitted to the AIChE J. (1991).

Powell, M.J.D., "A Hybrid Method for Nonlinear Equations," in <u>Numerical Methods for Nonlinear Algebraic Equations</u>, P. Rabinowitz, ed., Gordon and Breach, London (1970).

Smith, J.M., <u>Chemical Engineering Kinetics</u>, McGraw-Hill, New York (1970).

Walas, S.M., <u>Phase Equilibria in Chemical Engineering</u>, Butterworth, Stoneham, MA (1985).

HOMOTOPY-CONTINUATION ALGORITHM FOR GLOBAL OPTIMIZATION

Amy C. Sun[*]   Warren D. Seider[*]

May, 1991

ABSTRACT

A continuation algorithm is proposed to locate all of the stationary points that satisfy the inequality constraints on a single homotopy path. For these constraints, the Mangasarian equation replaces the complementary slackness equations. The resulting Kuhn-Tucker equations are solved using a differential-arclength continuation algorithm that implements Newton and affine homotopy formulations and a toroidal mapping function. Himmelblau's function is minimized, with and without inequality constraints. All of the stationary points are located, but some of the homotopy paths do not return to their starting points. In one case, the path terminates at a real bifurcation point. The proposed method, using a Newton homotopy formulation, is shown to have promise for increasing the reliability in locating the global minimum of the Gibbs free energy when solving multiphase equilibrium problems. Results are presented for the calculation of liquid-liquid equilibrium.

INTRODUCTION

In chemical processes, nonlinear programs (NLPs) arise in the models for many components of the process flowsheet. It is usually important, although difficult, to locate all of their stationary points, and especially the global minimum. Often the NLP is for the minimization of the Gibbs free energy to calculate the composition of a process stream in phase and chemical equilibrium, an especially difficult problem when the phase distribution at equilibrium is uncertain. Other NLPs involve the minimization of the

---

[*]Chemical Engineering Department, University of Pennsylvania, Philadelphia, PA 19104-6393

annualized cost, often for chemical reactors that exhibit multiple steady states. The sources of the nonlinearities and solution methods for these and many other NLPs are reviewed by Seider et al. (1991).

Efficient and reliable algorithms are needed to locate the global optimum, and in some cases all of the stationary points. During the past decade, first- and second-order methods have been used successfully to locate one or more stationary points. Two of the most popular are the reduced-gradient and successive quadratic programming methods. However, their solutions are not guaranteed to be globally optimal, and probably of greater consequence, they are not designed to seek the global optimum.

In one approach to locate the global optimum, Floudas et al. (1989) use Benders decomposition. These authors introduce transformations that render both the primal and reduced master subproblems convex. They apply this approach for the calculation of phase and chemical equilibrium (Paules and Floudas, 1989; see the section on Phase Equilibrium). In a related approach, Kocis and Grossmann (1987, 1988) introduce an outer approximation/equality relaxation algorithm to solve problems in process synthesis.

Another potential approach involves the use of continuation methods to find the roots of nonlinear equations. For polynomial functions, where the number of roots are known, Garcia and Zangwill (1981) prove that all of the solutions can be found using homotopy-continuation methods, and an algorithm is devised to solve polynomial functions by Watson et al. (1987). For nonpolynomial functions, where the number of roots are unknown, Seader et al. (1990) apply mapped-continuation methods to trace all of the solutions, but not specifically for the stationarity conditions associated with NLPs that have inequality constraints. While Lundberg et al. (1990) use continuation to minimize objective functions with continuous penalty functions, to our knowledge, no path-following algorithms have been developed to locate all of the stationary points for NLPs with inequality constraints.

In this paper, a new continuation algorithm is introduced, and results are presented for the optimization of Himmelblau's function (Reklaitis et al., 1983), with and without inequality constraints, and for the calculation of liquid-liquid equilibrium. First, the phase equilibrium problem and the mapped-continuation methods are reviewed briefly.

PHASE EQUILIBRIUM

The phase equilibrium problem is shown schematically in Figure 1. A feed stream, containing C chemical species with flow rates, n_{jf}, in kmol/s of species j, enters a vessel. The vessel is sufficiently large to enable an equilibration of P phases at the temperature and pressure in the vessel. Each of the phases is isolated and exits with a flow rate n^ℓ, where $n^\ell = \sum_{j=1}^{C} n_{j\ell}$, and $n_{j\ell}$ is in kmol/s of species j in phase ℓ.

At equilibrium, the Gibbs free energy is minimized subject to mass balance constraints. This is stated in NLP1 as follows:

$$\min_{\underline{n}} G = \sum_{\ell=1}^{P} \sum_{j=1}^{C} n_{j\ell} \bar{G}_{j\ell} = \sum_{\ell=1}^{P} \sum_{j=1}^{C} n_{j\ell} (G_{j\ell}^\circ + RT \ln \{\gamma_{j\ell} \frac{n_{j\ell}}{n^\ell}\})$$

ST: $n_{jf} = \sum_{\ell=1}^{P} n_{j\ell} \qquad \forall j$

$n_{j\ell} \geq 0 \qquad \forall j, \ell$

NLP1

Here, G is the Gibbs free energy, $\bar{G}_{j\ell}$ is the partial molal Gibbs free energy ($\partial G_{j\ell}/\partial n_{j\ell}$), $G_{j\ell}^\circ$ is the Gibbs free energy of formation (from its elements) of species j in phase ℓ in the standard state, and $\gamma_{j\ell} = \gamma_{j\ell}\{\underline{n}_\ell, T\}$ is the activity coefficient, a nonlinear function of $\underline{n}_\ell$ and T. Note that when the nonnegativity constraints become active ($n_{j\ell} = 0$), $\ln n_{j\ell}$ approaches $-\infty$, yielding an unbounded G.

In solving NLP1, it is often difficult to determine P at which G is a global minimum. Consider a mixture that has three phases at equilibrium. If P is incorrectly set at 2, when NLP1 is solved, a constrained minimum in G is obtained; constrained because potential three-phase solutions are excluded from the search. Even when P is set at 3, often $n_{j3} \to 0$, $n^3 \to 0$, and a local minimum is located with $P = 2$. When a mixture contains two phases at equilibrium, two additional problems arise. With P set at 3, when three phases exist at $G_{\min}$, a local minimum is obtained. Finally, when P is set at 2, the first-order stationarity conditions often have multiple solutions and local minima can be computed. This is illustrated for a system in liquid-liquid equilibrium (LLE) in the Results section.

Of the many methods proposed to solve NLP1, a Newton-like method, known as the Rand method, has been widely used. Gautam and Seider (1979a) review the evolution of the Rand method and evaluate it for several chemical and phase equilibrium problems. In one system, containing 40 mole % ethylene glycol, 30% lauryl alcohol, and 30% nitromethane, at 295.2K and 1.013 bar, three phases exist at equilibrium, as illustrated in Figure 2. Note the level contours of ΔG^m, the Gibbs free energy of mixing ($\Delta G^m = G - \sum_{j=1}^{C} n_{jf} G_j^\circ$), and A, B, and C, the phases at equilibrium, which are the points of tangency for the plane ABC. With 50 random guesses for the concentrations of the three phases, during 39 minimizations, $n_{j3} \to 0$, $j = 1,2,3$, and $n^3 \to 0$. In the Rand method, these variables must be eliminated to avoid a singular Hessian matrix. Hence, Gautam and Seider (1979b) introduced a method for reinitiating a phase. Subsequently, Michelsen (1982) applied the Gibbs tangent plane analysis to check the stability of the phases at G_{min} and, when unstable, computed the composition of trial phases closer to the global G_{min}. The Michelsen algorithm is particularly effective for locating small phases near phase boundaries and in the close proximity of critical points. However, it is not guaranteed to identify phase instability, and consequently, may not proceed to locate the global G_{min}.

In a recent development, Paules and Floudas (1989) have applied the algorithm of Floudas et al (1989) to seek the global minimum Gibbs free energy. NLP1 is augmented with binary variables to represent the existence or nonexistence of the phases, resulting in the following mixed-integer nonlinear program:

$$\min_{\underline{n}} G = \sum_{\ell=1}^{P} \sum_{j=1}^{C} n_{j\ell} \, [G_{j\ell}^\circ + RT \ln \{\gamma_{j\ell} \frac{n_{j\ell}}{n_\ell}\}]$$

$$\text{ST:} \quad n_{jf} = \sum_{\ell=1}^{P} n_{j\ell} \qquad \forall j \qquad \text{MINLP}$$

$$\left. \begin{array}{l} n_{j\ell} \leq n_{jf} y_{j\ell} \\ y_{j\ell} \leq y_\ell \\ n_{j\ell} \geq 0 \\ y_{j\ell}, y_\ell \in \{0,1\} \end{array} \right\} \quad \forall j, \ell$$

where $y_{j\ell}$ is unity when species j exists in phase ℓ and y_ℓ is unity when phase ℓ exists.

In Benders decomposition, the binary variables are the complicating variables. These are fixed (corresponding to a specific phase distribution) and the resulting NLP (primal problem) is solved for an upper bound on the minimum of G. At this solution, the constraints of the reduced master problem (RMP) are augmented. Then, the RMP is solved to locate a new lower bound for G and its associated binary variables. The nonconvexities in the objective function, however, can lead to a local G_{min}. To improve the reliability in locating the global G_{min}, Paules and Floudas introduce variable transformations in the nonconvex expressions for the liquid phase activity coefficients. These are included in the set of complicating variables and render both the primal and reduced master problems convex. The resulting subproblems are solved for their global solutions, but the global optimum of the MINLP cannot be guaranteed due to "the potential transfer of nonconvexities through the Kuhn-Tucker multipliers." The reliability and efficiency of this potentially attractive approach for global optimization needs to be compared with the methods involving stability analysis and the homotopy-continuation method proposed herein.

NONLINEAR PROGRAMMING

In the next two sections, strategies are developed to more reliably locate the global optimum of the nonlinear program:

$$\text{Optimize}_{\underline{x}} \quad f\{\underline{x}\} \quad \text{NLP}$$

$$\text{ST:} \quad \underline{h}\{\underline{x}\} = 0$$
$$\underline{g}\{\underline{x}\} \geq 0$$

This constrained optimization problem can be transformed into an unconstrained problem by forming the Lagrangian, L, and applying the stationarity conditions, $\underline{\nabla}L = 0$. The result is a system of NLEs known as the Kuhn-Tucker (KT) necessary conditions:

$$\nabla_{\underline{x}} f\{\underline{x}\} + \underline{\pi}^T \nabla_{\underline{x}} \underline{h}\{\underline{x}\} + \underline{\lambda}^T \nabla_{\underline{x}} \underline{g}\{\underline{x}\} = 0 \quad \text{KT-1}$$
$$\underline{h}\{\underline{x}\} = 0 \quad \text{KT-2}$$
$$g_i \lambda_i = 0 \quad \forall i \quad \text{KT-3}$$

which can be summarized as:

$$\underline{F}\{\underline{z}\} = 0 \quad \text{KT-S}$$

where $\underline{z} = [\underline{x}, \underline{\pi}, \underline{\lambda}]^T$ and $\underline{\pi}$ and $\underline{\lambda}$ are vectors of Lagrange and Kuhn-Tucker multipliers.

CONTINUATION METHODS

When solving $\underline{F}\{\underline{z}\} = 0$, the homotopy-continuation methods provide a smooth transition between an approximation to the solution, $\underline{G}\{\underline{z}\} = 0$, and the desired solution. In many formulations, a convex, linear homotopy function is utilized:

$$\underline{H}\{\underline{z}, t\} = t\underline{F}\{\underline{z}\} + (1-t)\underline{G}\{\underline{z}\} \quad (1)$$

where t is a real, scalar parameter and $\underline{G}\{\underline{z}\}$ is a vector of residuals that can easily be solved for $\underline{z}$. Note that t can be a scaled parameter in the equation set. The continuation algorithms solve Eqn. (1) beginning with t = 0 and track the homotopy path as t is increased to unity. Three choices for $\underline{G}\{\underline{z}\}$ are widely used:

$$\underline{G}\{\underline{z}\} = \underline{F}\{\underline{z}\} - \underline{F}\{\underline{z}^\circ\} \qquad \text{Newton homotopy} \qquad (2a)$$

$$\underline{G}\{\underline{z}\} = \underline{z} - \underline{z}^\circ \qquad \text{Fixed-point homotopy} \qquad (2b)$$

$$\underline{G}\{\underline{z}\} = \text{diag}\{\nabla_{\underline{z}}\underline{F}\{\underline{z}^\circ\}\}(\underline{z} - \underline{z}^\circ) \qquad \text{Affine homotopy} \qquad (2c)$$

Here, $\underline{z}^\circ$ is a vector of guessed values. While the Newton homotopy is scale-invariant, several $\underline{z}$ may satisfy $\underline{G}\{\underline{z}\} = 0$. This can present difficulties in locating all of the solutions of $\underline{F}\{\underline{z}\} = 0$. With the fixed-point homotopy, only one solution of $\underline{G}\{\underline{z}\} = 0$ exists. Garcia and Gould (1980) and Wayburn and Seader (1987) recommend the affine homotopy because it is scale-invariant, under linear transformations of $\underline{z}$, and has a unique solution of $\underline{G}\{\underline{z}\} = 0$. For the existence and uniqueness of a homotopy path, regular conditions are required; that is, the Jacobian of $\underline{H}\{\underline{z},t\}$, for given t, must be of full rank for $\forall \underline{z}$ and zero must be a regular value of $\underline{H}\{\underline{z},t\}$. Furthermore, since the homotopy-continuation paths often include singular points, such as limit points, it is important to trace the paths with sufficient accuracy, while bypassing the singular points. One effective algorithm, introduced by Klopfenstein (1961) utilizes a differentiated form of Eqn. (1) suggested by Davidenko (1953):

$$\begin{bmatrix} \dfrac{\partial \underline{H}}{\partial \underline{z}} & \dfrac{\partial \underline{H}}{\partial t} \end{bmatrix} \begin{bmatrix} \dfrac{d\underline{z}}{dp} \\ \dfrac{dt}{dp} \end{bmatrix} = 0 \qquad (3a)$$

where p is the arclength parameter defined by:

$$\left(\dfrac{d\underline{z}}{dp}\right)^T \left(\dfrac{d\underline{z}}{dp}\right) + \left(\dfrac{dt}{dp}\right)^2 = 1 \qquad (3b)$$

Allgower and Georg (1980) present a differential-arclength homotopy-continuation algorithm that integrates the ODEs along the arclength. A predictor step is calculated in the direction of the tangent vector along its arclength. Then, $\underline{z}$ and t are returned to the homotopy path with Newton correction steps taken orthogonal to the tangent vector. This strategy is implemented by Wayburn and Seader (1984) with special attention to the methods for step-size adjustment and parameter-variable exchange to avoid singularities at the limit points. The turning point algorithm is further extended by Kovach and Seider (1987) to bypass the limit points when multiple solutions are encountered as a second liquid phase is introduced on the trays of a heterogeneous azeotropic distillation tower.

Many homotopy paths exhibit turning points and pass t = 1 more than once. Using the Newton homotopy, Wayburn and Seader (1987) examine the relationship between the starting guesses and the path trajectories. When a unique starting point is used, multiple solutions are found by tracing the path as t varies over the entire real domain. To locate all of the solutions, Kuno and Seader (1988) formalize a starting-point criterion, using the fixed-point homotopy, which must be satisfied for the homotopy path to pass through all of the solutions. According to their criterion, $\underline{z}_c^o$ is chosen such that:

$$\underline{z}_c^o \in \{ \underline{z}^o : \underline{z}^o \in \mathbf{R}^n, \min_{\underline{z}^o} N\{\underline{z}^o\} \} \tag{4}$$

where $N\{\underline{z}^o\}$ is the number of real roots of $\underline{F}\{\underline{z}\} - \underline{z} + \underline{z}^o = 0$. This criterion, which seeks to reduce the number of solutions of $\underline{H}\{\underline{z},t\} = 0$ as t becomes large, applies for several engineering problems, but is not sufficient to guarantee that all of the solutions can be found. Note that when $\underline{z}^o$ is selected such that $N\{\underline{z}^o\} = 0$, the homotopy path has finite bounds on t and the likelihood of finding all of the solutions on a single path is increased. However, for complex equations, it is important to recognize that Eqn. (4) can be difficult to solve for $N\{\underline{z}^o\}$. In a more rigorous approach to obtaining a single connected path, Diener (1987) introduces a theorem concerning the connectivity of the solutions along the homotopy path:

Let $\underline{H}\{\underline{z},t\}: \mathbf{R}^{n+1} \rightarrow \mathbf{R}^n$ be twice differentiable, and its Jacobian be denoted by $\underline{J}\{\underline{z},t\}$. If

$$\sup\{\|(\underline{J}\{\underline{z},t\}\underline{J}\{\underline{z},t\}^T)^{-1}\|\} \leq K < \infty \tag{5}$$

then the inverse mapping $\underline{H}^{-1}\{\underline{0}\}$ is a connected one-dimensional submanifold in $\mathbf{R}^{n+1}$

By this theorem, when $(\underline{J}\,\underline{J}^T)^{-1}$ has a finite norm, all of the solutions to $\underline{H}\{\underline{z},t\} = 0$ are connected. That is, in addition to the regularity condition, the connectiveness of the path requires a bounded trajectory emanating from $\underline{z}^o$. A simpler criterion is presented by Lin (1988), who notes that when condition (5) is satisfied:

$$\det\{\underline{J}\{\underline{z},t\}\underline{J}\{\underline{z},t\}^T\} \neq 0 \tag{6}$$

That is, to have a finite inverse norm, $\underline{J}\,\underline{J}^T$ must be nonsingular. For a one-dimensional system, using fixed-point homotopy, Eqn. (6) becomes:

$$\det \{\underline{J}\,\underline{J}^T\} = \left(\frac{\partial H}{\partial \underline{z}}\right)^2 + \left(\frac{\partial H}{\partial t}\right)^2 = [t\frac{dF}{dz} + (1-t)]^2 + [F\{z\} - (z - z_c^o)]^2 \quad (7)$$

Clearly, when $N\{\underline{z}^o\} = 0$, the second term remains finite and a connected path is assured. For two-dimensional problems, when $\partial H_1/\partial z_1 = \partial H_1/\partial z_2 = \partial H_2/\partial z_1 = \partial H_2/\partial z_2 = 0$, $\det \{\underline{J}\,\underline{J}^T\} = 0$, regardless of $\partial H_1/\partial t$ and $\partial H_2/\partial t$ and $\underline{z}_c^o$. Hence for n = 2, and in general for n ≥ 2, the starting point criterion of Kuno and Seader is less applicable. For these problems, it is important to address the existence of a connected path through all of the solutions from any starting point. Lin (1988) shows that the paths are connected when:

(1) t approaches zero as z_i grows unbounded and z_i/F_i approaches zero. Then, the homotopy path is connected at $-z_i$ and $t = 0$.

(2) $\underline{z}$ approaches $\underline{z}^k$ as t approaches infinity. Then, the homotopy path is connected at $t = -\infty$ and $\underline{z} = \underline{z}^k$.

This is illustrated schematically in Figure 3 for $F\{x\} = x^2 - 3x + 2$. Accordingly, Seader et al. (1990) connect the paths in a mapped-continuation algorithm. To avoid very large or very small iteration variables, two mapping functions are included that scale all of the variables between -1 and 1:

Boomerang: $$y_i' = \frac{2y_i}{1+y_i^2} \quad (8a)$$

Toroidal: $$y_i' = \frac{y_i}{\sqrt{(1+y_i^2)}} \quad (8b)$$

Note that $-1 \le y_i' \le 1$ when $-\infty \le y_i \le \infty$, where $\underline{y} = [\underline{z}\ t]^T$. They show that these mapped-continuation techniques find all five of the steady-state solutions for an adiabatic, continuous, stirred-tank reactor (CSTR).

In this work, using the fixed-point homotopy and the toroidal mapping function, all nine of the stationary points of the Himmelblau function:

$$f\{\underline{x}\} = (x_1^2 + x_2 - 11)^2 + (x_1 + x_2^2 - 7)^2 \quad (9)$$

were located, beginning at $\underline{x}^o = \underline{z}^o = [-6,6]^T$. Figure 4 shows a contour map and tabulates the stationary points. Note the existence of four global minima at stationary points 1, 3, 7, 9, with $f\{\underline{x}\} = 0$. Three saddle points are at 2, 6, and 8, and the global maximum is at point 5. Figures 5a and 5b are the solution diagrams for x_1 and x_2, respectively. Considerable computational work is required to trace the homotopy path, with its many turning points. Yet, all of the solutions, including the global optima, are located reliably.

CONSTRAINED OPTIMIZATION

As illustrated above, the general nonlinear program, NLP, can be expressed using the Kuhn-Tucker necessary conditions, KT. This problem is commonly solved using reduced gradient and successive quadratic programming methods, which attempt to maintain feasibility, with respect to the inequality constraints, $g\{\underline{x}\} \geq 0$, by either assuming an active constraint set or penalizing the constraint violations. The KT equations are often difficult to solve because KT-3, the equations of complementary slackness:

$$g_i \lambda_i = 0 \qquad \forall i \qquad (10)$$

are not continuously differentiable at the origin, where $g_i = \lambda_i = 0$. This problem can be overcome by applying the Mangasarian Theorem (Mangasarian, 1976), which is stated:

> For N inequality constraints, $g\{\underline{x}\} \geq 0$, if a monotonically increasing function θ exists such that $\theta\{0\} = \theta'\{0\} = 0$, then
>
> $$M_i = \theta\{|g_i\{\underline{x}\} - \lambda_i|\} - \theta\{g_i\{\underline{x}\}\} - \theta\{\lambda_i\} = 0 \qquad \forall i \qquad (11)$$
>
> is satisfied if and only if the solution: (1) satisfies the equations of complementary slackness, $g_i \lambda_i = 0$, (2) is feasible, $g_i \geq 0$, and (3) has the proper sign of the Kuhn-Tucker multipliers, $\lambda_i \geq 0$.

One such function is $\theta\{\chi_i\} = \chi_i|\chi_i|$. Using this formulation, Seider et al. (1991) have found the radius of convergence of the homotopy methods to be extended greatly, with solutions that are guaranteed to be feasible and to have the proper sign for the KT multipliers. Vasudevan et al. (1989) apply the theorem with $\theta\{\chi_i\} = \chi_i^3$ to solve an optimal-fuel orbital problem.

With the equations of complementary slackness (10) replaced by continuously differentiable functions (11), all points satisfying the first-order necessary conditions can be found using the continuation algorithm, and subsequently, the global optimum can be determined. The algorithm implemented in this work is outlined below:

(1) Continuous and differentiable homotopy functions are formulated by writing the first-order necessary conditions and incorporating them

into the affine homotopy formulation. That is, for the NLP, the KT equations become:

$$t \nabla_x L\{\underline{x},\underline{\lambda},\underline{\pi}\} + (1 - t)\text{diag}\{\underline{\nabla^2 L}\{\underline{x}^o,\underline{\lambda}^o,\underline{\pi}^o\}\}(\underline{x} - \underline{x}^o) = 0 \quad (12a)$$

$$t M_i\{\underline{x},\lambda_i\} + (1 - t)\nabla_{\lambda_i} M_i\{\underline{x}^o,\lambda_i^o\}(\lambda_i - \lambda_i^o) = 0 \quad \forall i \quad (12b)$$

$$t \underline{h}\{\underline{x}\} + (1 - t)\text{diag}\{\nabla_{\underline{\pi}}\underline{h}\{\underline{x}^o\}\}(\underline{\pi} - \underline{\pi}^o) = 0 \quad (12c)$$

where

$$\nabla_x L\{\underline{x},\underline{\lambda},\underline{\pi}\} = \nabla_x f\{\underline{x}\} + \underline{\lambda}^T \nabla_x \underline{g}\{\underline{x}\} + \underline{\pi}^T \nabla_x \underline{h}\{\underline{x}\} \quad (12d)$$

$$M_i\{\underline{x},\lambda_i\} = (g_i\{\underline{x}\} - \lambda_i)^2 - |g_i\{\underline{x}\}|g_i\{\underline{x}\} - |\lambda_i|\lambda_i \quad \forall i \quad (12e)$$

Here, $L\{\underline{x},\underline{\lambda},\underline{\pi}\}$ is the Lagrangian, $\underline{h}\{\underline{x}\}$ is a vector of equality constraints, $\underline{\pi}$ is the corresponding vector of Lagrange multipliers, $\underline{g}\{\underline{x}\}$ is a vector of inequality constraints, $\underline{\lambda}$ is the corresponding vector of Kuhn-Tucker multipliers, and $M_i\{\underline{x},\lambda_i\}$ is the Mangasarian equation (11), expressed with $\theta\{\chi_i\} = \chi_i|\chi_i|$, $\forall i$.

(2) The homotopy path is traced for Eqn. (12) from an arbitrary starting point using the predictor-corrector algorithm, with up to 10 Newton correction steps, as suggested by Allgower and Georg.

(3) The variables are mapped using the toroidal function. Note that the algorithm by Seader et al. calculates the Euler step in the mapped space. The algorithm herein maps the variables after a complete step has been taken. When $|t'|$ or $|x'_i|$ approach unity (within 0.05%), the sign is changed, and path tracing is continued. The multipliers are not monitored because λ_i/M_i does not approach infinity as λ_i becomes unbounded.

(4) When the solution is bracketed at $t = 1.0$ (i.e., $t^k > 1$ and $t^{k-1} < 1$ and vice versa, where k is the step counter), Newton-Raphson iterations locate the exact solution before another step is taken.

(5) When the path returns to the starting point, the algorithm stops; otherwise, the algorithm returns to Step 2.

RESULTS

Himmelblau Function

For the minimization of the Himmelblau function, three cases are examined. All involve an inequality constraint, one linear and two quadratic, as shown in Figure 6. All of the stationary points are located in each case.

Case a. For this case, $g\{\underline{x}\} = x_1 - x_2 \geq 0$. Plots 7a and 7b show the solution paths for x_1 and x_2. All five solutions lie on one homotopy path, starting from $\underline{x}^\circ = [6,5]^T$, $\lambda^\circ = 0$. Note that the Kuhn-Tucker multiplier, λ, remains zero since all of the stationary points lie within the feasible region. After the first minimum is found, the continuation variable remains within $t = [0.95, 1.05]$, while x_1 and x_2 move to the remaining four solutions. Similar narrow solution paths are seen in the next two cases. This behavior is discussed below.

Case b. For this case, $g\{\underline{x}\} = -(x_1 - 3)^2 - (x_2 - 2)^2 \geq 0$. This constraint crosses the minimum at $\underline{x}^* = [3,2]^T$, as shown in Figure 6b. Clearly, both the constraint and the objective function have zero gradients at $\underline{x}^*$ and the path terminates at this real bifurcation point, where the Jacobian of $\underline{H}$ contains a zero row and is singular. However, all three feasible solutions are found, beginning at $\underline{x}^\circ = [6,-2]^T$, $\lambda^\circ = 0$, as shown in Figures 8a and 8b. The solution for x_1 appears to be discontinuous in dx_1/dt at $t = 1.0523$ and $x_1 = 2.2877$. In an expanded view, however, the first derivatives can be seen to remain continuous.

Case c. For this case, $g\{\underline{x}\} = x_2 - x_1^2 \geq 0$. Beginning at $\underline{x}^\circ = [2,5]^T$, $\lambda^\circ = 0$, a saddle point, $\underline{x}^{1*} = [0.086678, 2.8843]^T$, $\lambda^{1*} = 0$, and two minima, $\underline{x}^{2*} = [1.6445, 2.7043]^T$, $\lambda^{2*} = 9.9925$, and $\underline{x}^{3*} = [-1.99085, 3.20714]^T$, $\lambda^{3*} = 10.0062$ are located. The global minimum for this constrained objective function is at $\underline{x}^{3*}$. Figures 9a, 9b, and 9c show the homotopy path for x_1, x_2, and λ, respectively. Note that $\underline{x}^{2*}$ is identified first, then $\underline{x}^{1*}$ and $\underline{x}^{3*}$, after which the homotopy path cycles through these last two solutions. The Kuhn-Tucker multiplier increases from zero as $\underline{x}^{2*}$ is approached and becomes very large before returning to zero. At $\lambda = 0$, t increases slightly, then decreases to 0.6548 before increasing and becoming nonzero at $t = 0.7334$. Then, λ increases past $t = 1$ ($\underline{x}^{3*}$), after which it returns to zero in a closed cycle. When $\lambda^\circ > 0$, the algorithm terminates prematurely.

In the three constrained cases, no asymptotic behavior is observed and all three paths do not return to $\underline{x}^\circ$. In addition, as noted in the last paragraph, t varies slightly as x_1 and x_2 trace the solution path, with the Jacobian becoming ill-conditioned near the sharp turning points. This is because the Mangasarian formulation affects the curvature of the homotopy paths. For this reason, the algorithm may have difficulties handling highly nonlinear objective functions and constraints. Furthermore, $M_i\{\underline{x},\lambda_i\}$ is not a C^2 function for $\theta\{\chi_i\} = \chi_i|\chi_i|$ since:

$$\frac{\partial^2 \theta_i}{\partial \chi_i^2} = 2\chi_i|\chi_i|^{-1} \tag{13}$$

For this reason, Diener's Theorem does not apply.

<u>Liquid-liquid Equilibrium</u>

For a binary mixture in liquid-liquid equilibrium (LLE), NLP1 becomes:

$$\min_{\underline{n}} G = \sum_{\ell=1}^{2}\sum_{j=1}^{2} n_{j\ell}\left(G_{j\ell}^{\circ} + RT \ln \left\{\gamma_{j\ell}\frac{n_{j\ell}}{n^{\ell}}\right\}\right)$$

$$\text{ST:} \quad n_{jf} = \sum_{\ell=1}^{2} n_{j\ell} \qquad \forall j \qquad \text{NLP2}$$

$$n_{j\ell} \geq 0 \qquad \forall j,\ell$$

An equivalent formulation, presented by Lin (1988), and modified slightly here, is:

$$\min_{x_{11},x_{12}} \quad \frac{\Delta G^m}{n^f RT} = \beta\left[x_{11}\ln x_{11}\gamma_{11} + x_{21}\ln x_{21}\gamma_{21}\right]$$

$$+ (1-\beta)\left[x_{12}\ln x_{12}\gamma_{12} + x_{22}\ln x_{22}\gamma_{22}\right]$$

NLP3

ST:
$$z_1 - \beta x_{11} + (1-\beta)x_{12} = 0$$
$$x_{11} + x_{21} - 1 = 0$$
$$x_{12} + x_{22} - 1 = 0$$

$$x_{j\ell} \geq 0 \qquad \forall j,\ell$$
$$1 - x_{j\ell} \geq 0 \qquad j=1, \ell=1,2$$
$$\beta \geq 0$$
$$1 - \beta \geq 0$$

where $\Delta G^m \,(= G - \sum_{j=1}^{2} n_{jf} G_{j\ell}^{\circ})$ is the Gibbs free energy of mixing, $x_{j\ell}$ is the mole fraction of species j in phase ℓ, $\beta\,(=n^1/n^f)$ is feed fraction in the first liquid phase, and z_j is the mole fraction of species j in the feed. Lin (1988) solved a set of NLEs, comparable to the KT equations, and found three solutions for an equimolar feed of n-butylacetate and water, using the NRTL equation for the activity coefficients, with $\tau_{12} = 3.00498$, $\tau_{21} = 4.69071$, and $\alpha_{12} = \alpha_{21} = 0.391965$.

Unfortunately, the affine homotopy does not apply due to the equality constraints, $\underline{h}\{\underline{x}\} = 0$. For these constraints, the homotopy function becomes:

$$H_i\{\underline{x},\underline{\pi},\underline{\lambda},t\} = t\,h_i\{\underline{x}\} + (1-t)\left(\frac{\partial h_i}{\partial \pi_i}\right)_{\underline{x}^\circ,\underline{\pi}^\circ,\underline{\lambda}^\circ}(\pi_i - \pi_i^\circ) \qquad (14)$$

The partial derivative is zero, and hence, the homotopy function and its first derivatives vanish at $t = 0$. That is, at $t = 0$, the Jacobian of $\underline{\underline{H}}$ is singular and a vector tangent to the homotopy path cannot be calculated.

Hence, the algorithm above was applied using the Newton homotopy, rather than the affine homotopy. Beginning with $x_{21}^\circ = 0.05$ ($x_{21}^\circ = 0.95$), $x_{12}^\circ = 0.95$ ($x_{22}^\circ = 0.05$), $\beta^\circ = 0.4$, $\underline{\pi}^\circ = [0,-5,-1]^T$, and $\lambda_i = 0.01$ ($\forall i$), the homotopy path for x_{22} is shown in Figure 10. The left-most branch is traced first, with the trivial solution at $t = 1$ ($x_{21} = x_{22} = 0.5$, $\beta = 0$). The branch is retraced in the opposite direction, past $t = 0$. It traverses a limit point and traces a new branch with three solutions. The global minimum is at the third solution ($x_{11} = 0.004557$, $x_{21} = 0.995443$, $x_{12} = 0.59199$, $x_{22} = 0.40801$). The other solutions have no physical significance at equilibrium. This branch terminates at $t = 1.1431$, close to the reversal point ($t = 1.0598$) of the first branch. At termination, the Newton correction step encounters a singular Jacobian. Figure 11 illustrates the homotopy path beginning with a different vector of Lagrange multipliers, $\underline{\pi}^\circ = [1,5,1]^T$. The three two-phase solutions are traversed, as well as the trivial solution.

Discussion. Multiphase equilibrium problems are difficult to solve reliably. It is essential to locate the global G_{min} to have the correct solution. Yet, local minima in G are important to know, as small changes in temperature, pressure, and feed composition can cause these solutions to become global minima. The efficiency of the solution method is also important, as equilibrium problems are solved repeatedly in the simulation of process flowsheets.

Based upon the limited results above, these methods are potentially very reliable, but may be inefficient in comparison with other methods. Far more testing is needed, especially near phase boundaries where small amounts of incipient phases are difficult to detect. The reliability and efficiency of the algorithm should be compared with the Michelsen algorithm, which checks for phase stability using the tangent-plane distance function, and the algorithm of Paules and Floudas.

CONCLUSIONS

It is concluded that:

(1) For the Himmelblau function, all feasible optima are found on a single homotopy path. However, some of the paths trace the solutions with small changes in the homotopy parameter, t, and for the constrained cases, the paths do not return to their starting points.

(2) For the Himmelblau function, Case b, the proposed algorithm approaches the stationary point at $\underline{x}^* = [3,2]^T$, which is a real bifurcation point, and terminates.

(3) Further work is needed to obtain closed homotopy paths (i.e., paths that return to their starting points). The Mangasarian equations prevent asymptotic behavior along the homotopy paths in all of the examples presented. Hence, the connectivity conditions of Lin (1988) do not apply. Furthermore, Diener's Theorem does not apply, and hence, a single, connected path containing all of the stationary points is not guaranteed.

(4) The proposed algorithm, using the Newton homotopy (rather than the affine homotopy), successfully locates the global and local G_{min} when solving an LLE problem. The algorithm has great potential for improving the reliability of equilibrium calculations; extensive testing is needed.

(5) In one LLE case (Figure 10), termination occurs at a singular point. With modifications to the algorithm, this point should be by-passed.

ACKNOWLEDGMENT

Partial funding was provided by the Design Theory and Methodology Program of the NSF under Grant No. DMC-8613484 and is gratefully acknowledged.

NOMENCLATURE

C   number of chemical species
f   objective function
$\underline{F}$ vector of residuals of the Kuhn-Tucker equations
$\underline{g}$ vector of residuals of the inequality constraints
G   Gibbs free energy

$\underline{G}$	vector of residuals of the Kuhn-Tucker equations whose solution is known or easily found
$\bar{G}_{j\ell}$	partial molal Gibbs free energy of species j in phase ℓ
$G^\circ_{j\ell}$	Gibbs free energy of formation (from the elements) of species j in phase ℓ in the standard state
ΔG^m	Gibbs free energy of mixing
$\underline{h}$	vector of residuals of the equality constraints
$\underline{H}$	vector of Homotopy functions
$\underline{\underline{J}}$	Jacobian matrix of $\underline{H}$
L	Lagrangian
$\underline{M}$	vector of Mangasarian functions, Eqn. (11 and 12e)
n^f	kmol/s of the feed
$n_{j\ell}$	kmol/s of species j in phase ℓ
n^ℓ	kmol/s of phase ℓ
N	number of real roots of H as $t \to \infty$
p	arclength parameter
P	number of phases
P	pressure
R	universal gas constant
t	homotopy parameter
T	temperature
$\underline{x}$	vector of problem variables
$x_{j\ell}$	mole fraction of species j in phase ℓ
$\underline{x}^\circ$	value of $\underline{x}$ at starting point of the homotopy path
y	binary variables
$\underline{y}$	vector of $\underline{z}$ and t; i.e., $\underline{y} = [\underline{z}\ t]^T$
$\underline{y}'$	value of $\underline{y}$ in the mapped space
$\underline{z}$	$[\underline{x},\underline{\pi},\underline{\lambda}]^T$
z_j	mole fraction of species j in the feed
$\underline{z}^\circ$	value of $\underline{z}$ at starting point of homotopy path
$\underline{z}^\circ_c$	value of $\underline{z}$ at starting point given by Eqn. (4)

Greek Symbols

α interaction coefficients in NRTL equation

β fraction of feed in the first liquid

$\gamma_{j\ell}$ activity coefficient of species j in phase ℓ

θ an arbitrary function that is monotonically increasing and has zero value and zero derivative at the origin

$\boldsymbol{\lambda}$ vector of Kuhn-Tucker multipliers

$\boldsymbol{\pi}$ vector of Lagrange multipliers

τ interaction coefficients in NRTL equation

χ independent variable of the θ function

Superscript

o   starting point

Subscript

'   mapped space

LITERATURE CITED

Allgower, E., and K. Georg, "Simplicial and Continuation Methods for Approximating Fixed Points and Solutions to Systems of Equations," *SIAM Rev.*, 22, 28 (1980).

Davidenko, D., "On a New Method of Numerically Integrating a System of Nonlinear Equations," *Dokl. Akad. Nauk. USSR*, 88, 601 (1953).

Diener, I.,"On the Global Convergence of Path-following Methods to Determine All Solutions to a System of Nonlinear Equations," *Mathematical Programming*, 39, 181-188 (1987).

Floudas, C.A., A. Aggrawal, and A. Ciric, "Global Optimum Search for Nonconvex NLP and MINLP Problems," *Comput. Chem. Eng.*, 13, 10, 1117 (1989).

Floudas, C.A., and V. Visweswaran, "A Global Optimization Algorithm (GOP) for Certain Classes of Nonconvex NLPs - Theory," *Comput. Chem. Eng.*, 14, 12, 1397-1417 (1990).

Garcia, C.B., and F.J. Gould, "Relations between Several Path Following Algorithms and Local and Global Newton Methods," *SIAM Review*, 22, 3, 263-274 (1980).

Garcia, C.B., and W.I. Zangwill, *Pathways to Solutions, Fixed Points and Equilibria*, Prentice-Hall, Ch.18 (1981).

Gautam, R., and W.D. Seider, "Calculation of Phase and Chemical Equilibria, Part I: Local and Constrained Minima in Gibbs Free Energy," *AIChE J.*, 25, 6 (1979a).

Gautam, R., and W.D. Seider, "Calculation of Phase and Chemical Equilibria, Part II: Phase Splitting," *AIChE J.*, 25, 6 (1979b).

Klopfenstein, R.W., "Zeros of Nonlinear Functions," *J. Assoc. Comput. Mach.*, 8, 366 (1961).

Kocis, G.R., and I.E. Grossmann, "Relaxation Strategy for the Structural Optimization of Process Flowsheets," *Ind. Eng. Chem. Res.*, 26, 1869 (1987).

Kocis, G.R., and I.E. Grossman, "Global Optimization of Non-convex Mixed-integer Nonlinear Programming (MINLP) Problems in Process Synthesis," *Ind. Eng. Chem. Res.*, 27, 1407 (1988).

Kovach, III, J.W., and W.D. Seider, "Heterogeneous Azeotropic Distillation: Homotopy-continuation Methods," *Comput. Chem. Eng.*, 11, 6, 593-603 (1987).

Kuno, M., and J.D. Seader, "Computing All Real Solutions to Systems of Nonlinear Equations with Global Fixed-point Homotopy," *Ind. Eng. Chem. Res.*, 27, 7, 1320-1329 (1988).

Lin, W.-J., "Application of Continuation and Modeling Methods to Phase Equilibrium, Steady-state, and Dynamic Process Calculations," PhD Thesis, University of Utah, 1988.

Lundberg, B.N., A.B. Poore, and B. Yang, "Smooth Penalty Functions and Continuation Methods for Constrained Optimization," *Lectures in Applied Mathematics*, 26, 389-412 (1990).

Mangasarian, O.I.,"Equivalence of the Complementarity Problem to a System of Nonlinear Equations," *SIAM J. Appl. Math.*, 31, 1, 89-91 (1976).

Michelsen, M.L., "The Isothermal Flash Problem, I: Stability," *Fluid Phase Equil.*, 4, 1 (1982).

Paules IV, G.E., and C.A. Floudas, "A New Optimization Approach for Phase and Chemical Equilibrium Problems," AIChE Annual Mtg. (1989).

Reklaitis, G.V., A. Ravindran, and K.M. Ragsdell, *Engineering Optimization: Methods and Application*, Wiley (1983).

Seader, J.D., M. Kuno, W.-J. Lin, S. A. Johnson, K. Unsworth, and J.W. Wiskin, "Mapped Continuation Methods for Computing All Solutions to General Systems of Nonlinear Equations," *Comput. Chem. Eng.*, 14, 1, 71-85 (1990).

Seider, W.D., D.D. Brengel, and S. Widagdo, "Nonlinear Analysis in Process Design - Journal Review," *AIChE J.*, 37, 1-38 (1991).

Vasudevan, G., L.T. Watson, and F. H. Lutze, "A Homotopy Approach for Solving Constrained Optimization Problems," *Proc. Amer. Cont. Conf.*, Pittsburgh, 780-785 (1989).

Watson, L.T., S.C. Billups, and A.P. Morgan, "Algorithm 652, HOMPACK: A Suite of Codes for Globally Convergent Homotopy Algorithms," *ACM Trans. Math. Software*, 13, 281-310 (1987).

Wayburn, T.L., and J.D. Seader, "Solutions of Systems of Interlinked Distillation Towers by Differential Homotopy-Continuation Methods," *Proc. Conf. on Found. Computer-aided Process Design*, ed., A.W. Westerberg and H.H. Chien, CACHE, 765 (1984).

Wayburn, T.L., and J.D. Seader, "Homotopy Continuation Methods for Computer-Aided Process Design," *Comput. Chem. Eng.*, 11, 1, 7-25 (1987).

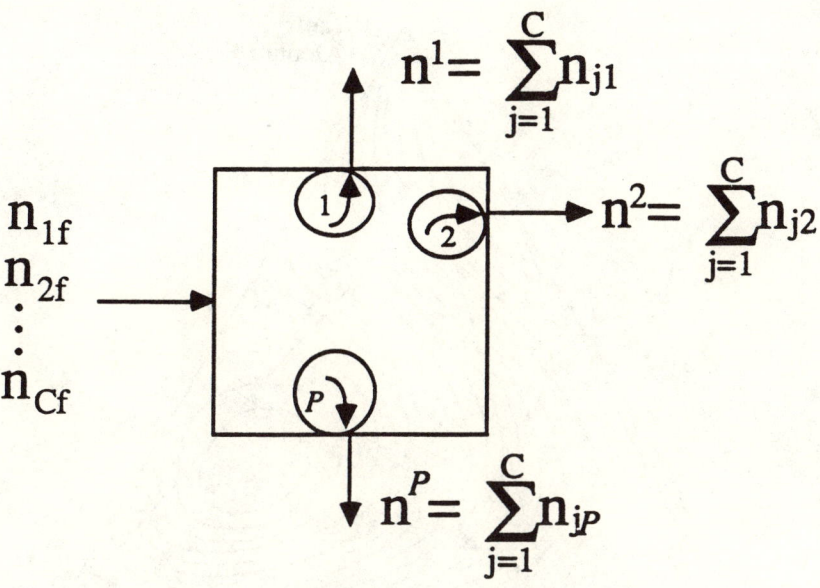

Figure 1. Vessel in phase equilibrium.

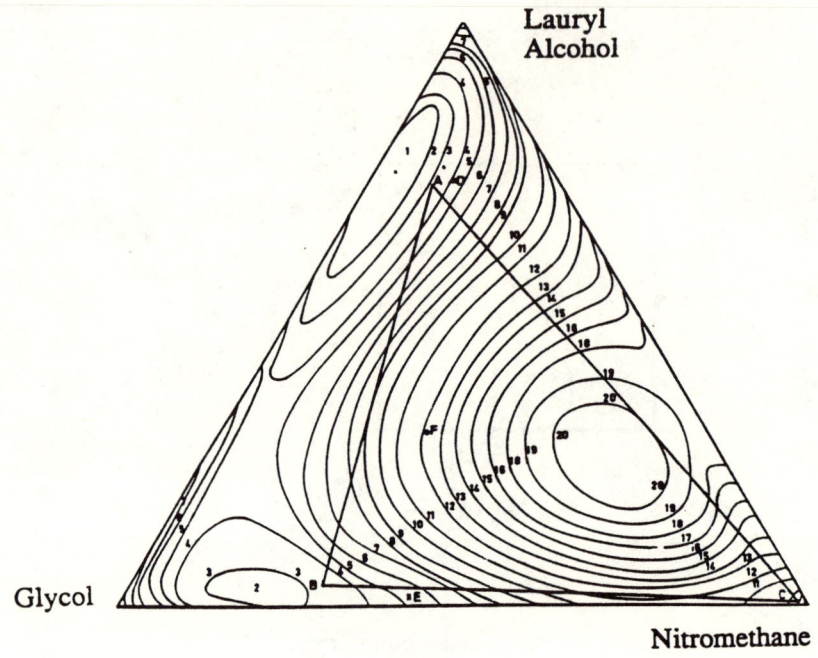

Figure 2. Contours of constant Gibbs free energy of mixing for a mixture of ethylene glycol, lauryl alcohol, and nitromethane at 295.2K and 1.013 bar. Activity coefficients computed using the van Laar equation (Gautam and Seider, 1979a).

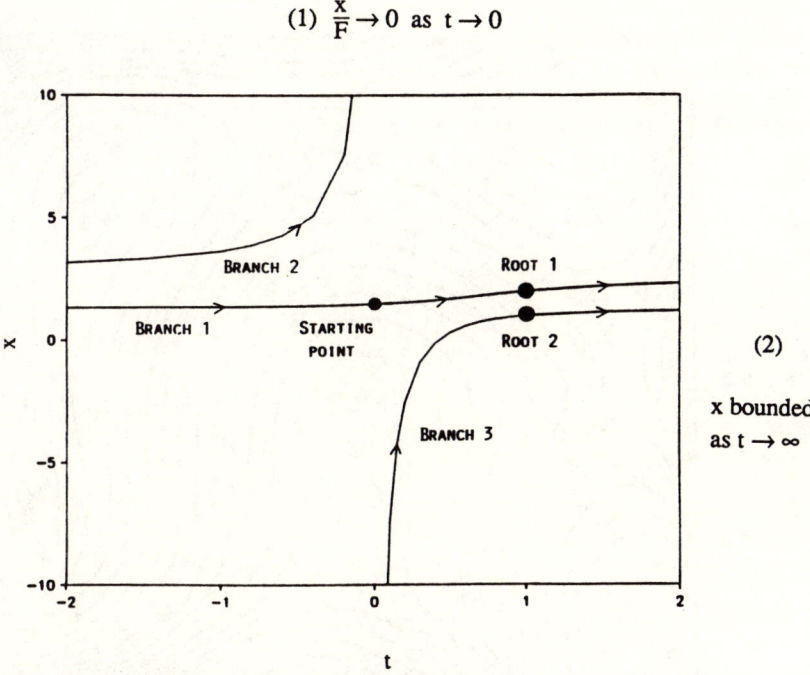

Figure 3. Fixed-point homotopy path for $F\{x\} = x^2 - 3x + 2$ with $x^o = 1.5$. Reprinted with permission from Lin (1988).

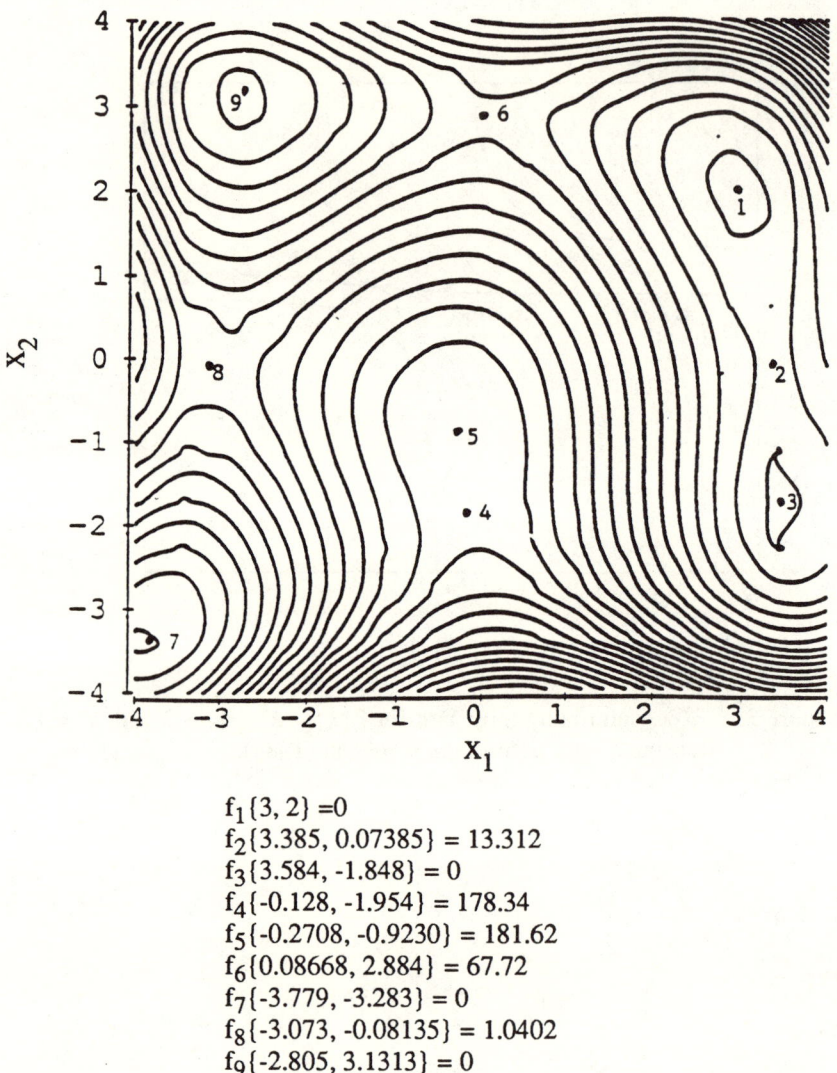

$f_1\{3, 2\} = 0$
$f_2\{3.385, 0.07385\} = 13.312$
$f_3\{3.584, -1.848\} = 0$
$f_4\{-0.128, -1.954\} = 178.34$
$f_5\{-0.2708, -0.9230\} = 181.62$
$f_6\{0.08668, 2.884\} = 67.72$
$f_7\{-3.779, -3.283\} = 0$
$f_8\{-3.073, -0.08135\} = 1.0402$
$f_9\{-2.805, 3.1313\} = 0$

Figure 4. Level contours and stationary points of Himmelblau's function.

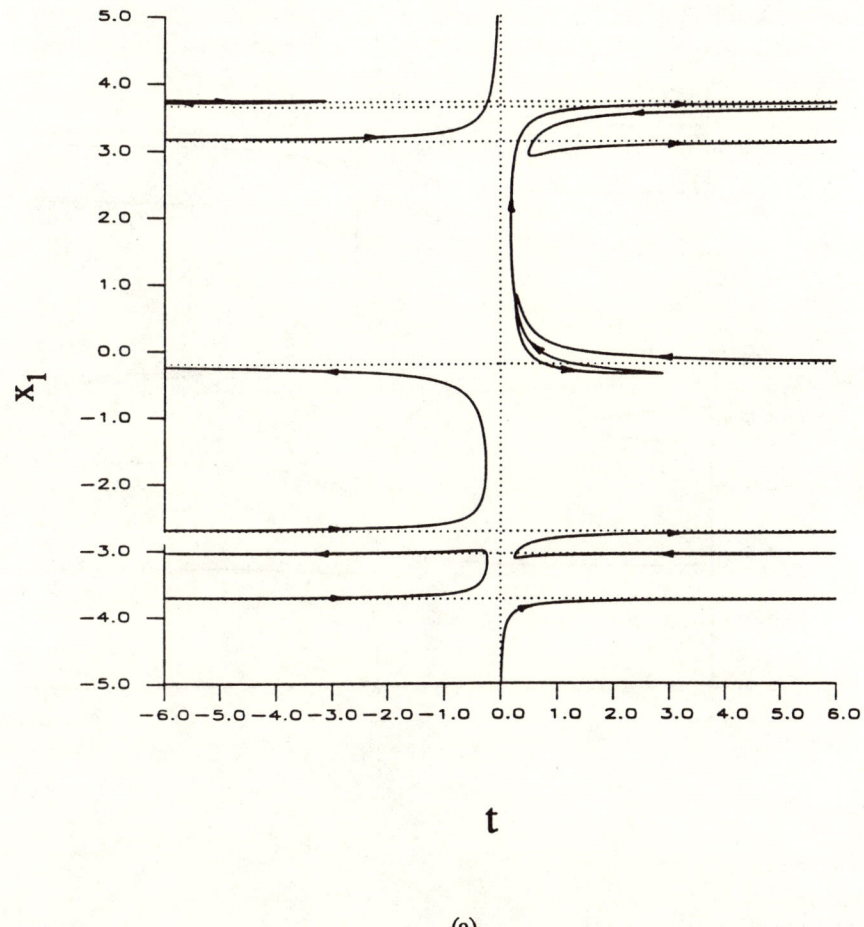

(a)

Figure 5. Solution diagrams for the unconstrained Himmelblau function.

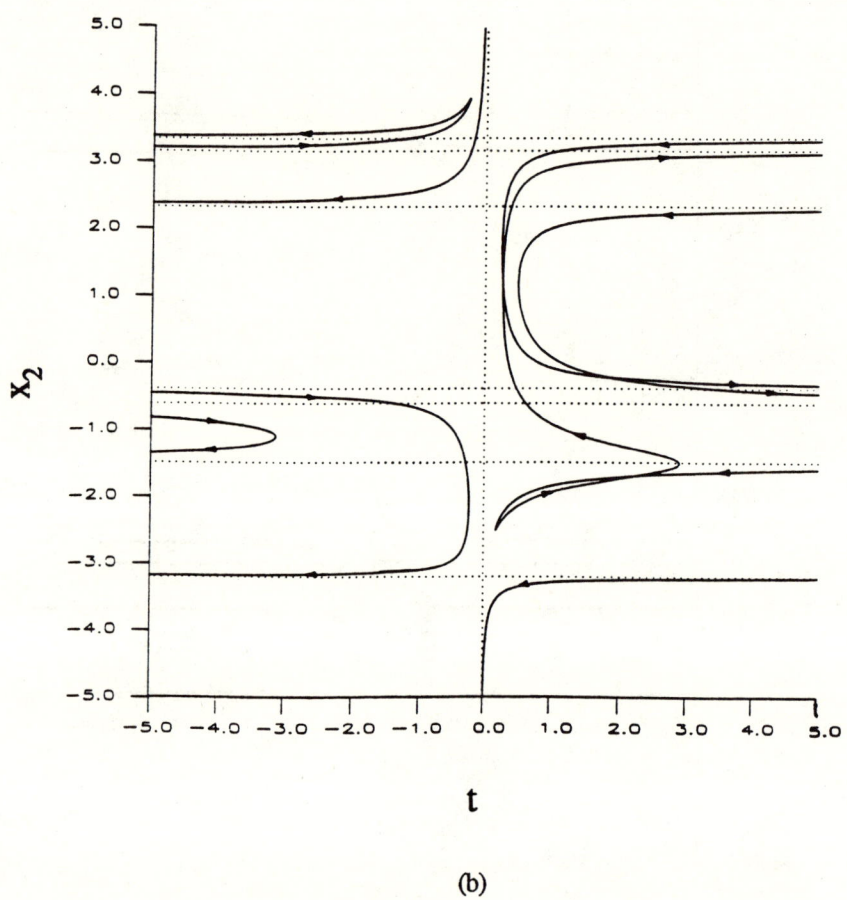

(b)

Figure 5. (cont) Solution diagrams for the unconstrained Himmelblau function.

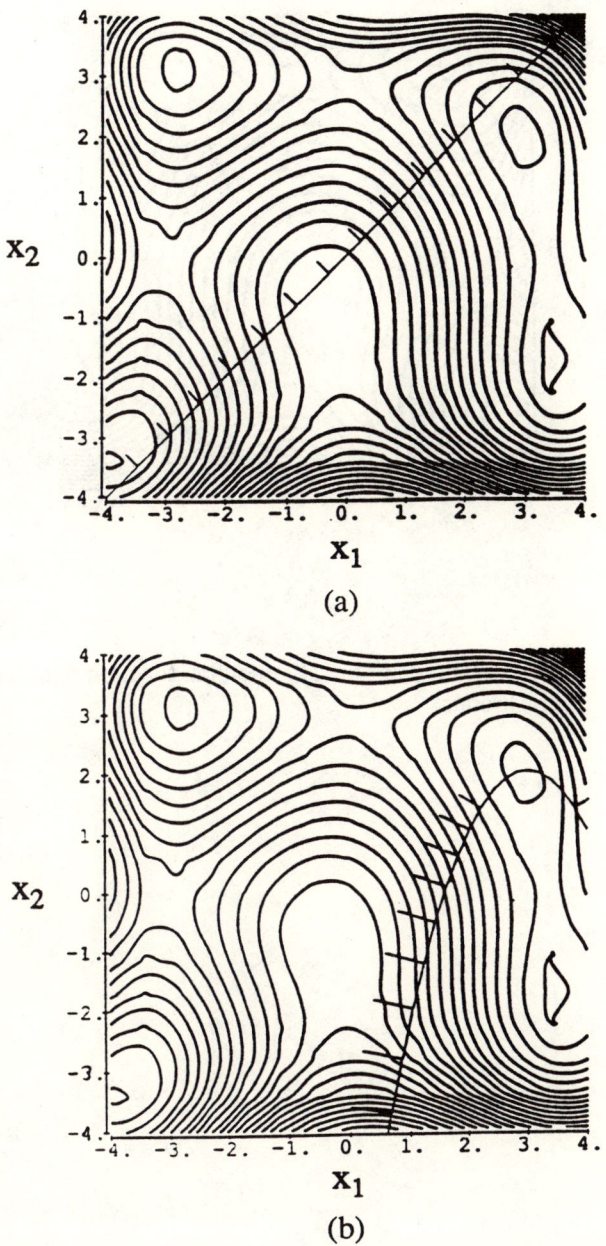

Figure 6. Level contours of the Himmelblau function with inequality constraints.

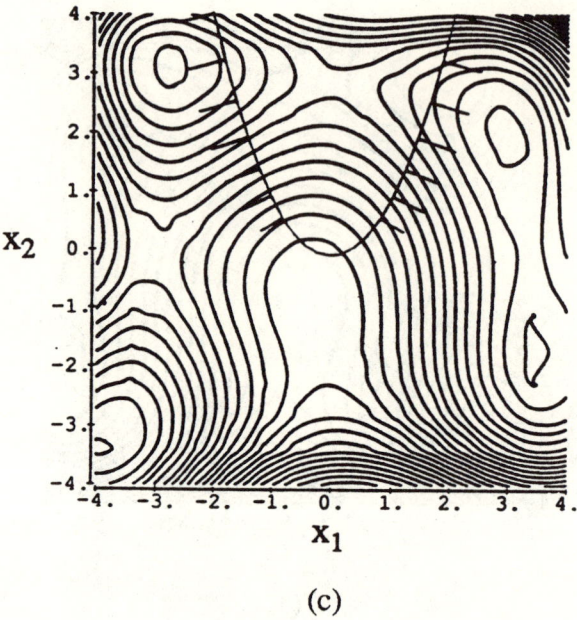

(c)

Figure 6. (cont) Level contours of the Himmelblau function with inequality constraints.

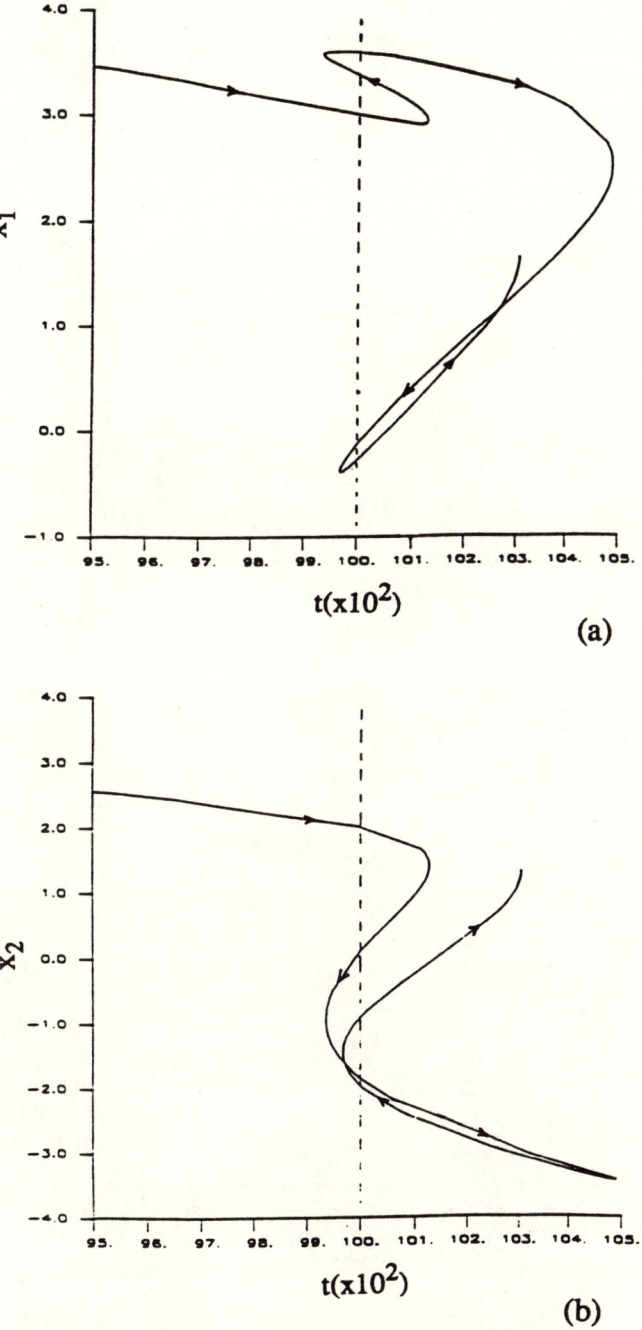

Figure 7. Solution path for Case a.

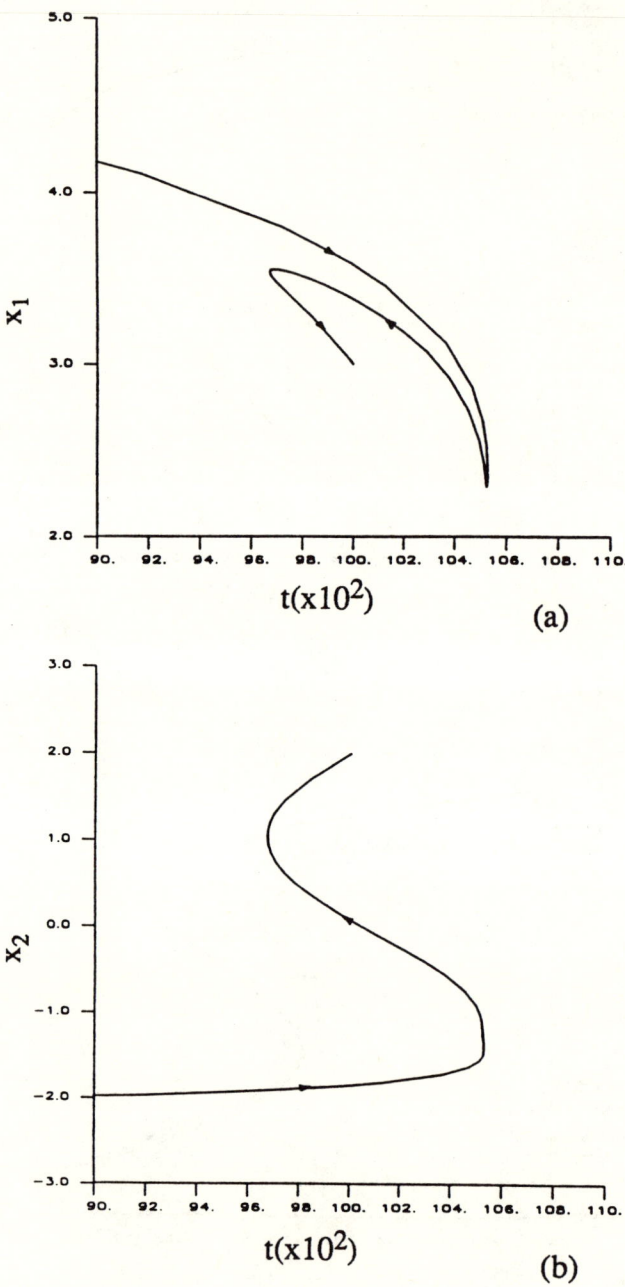

Figure 8. Solution path for Case b.

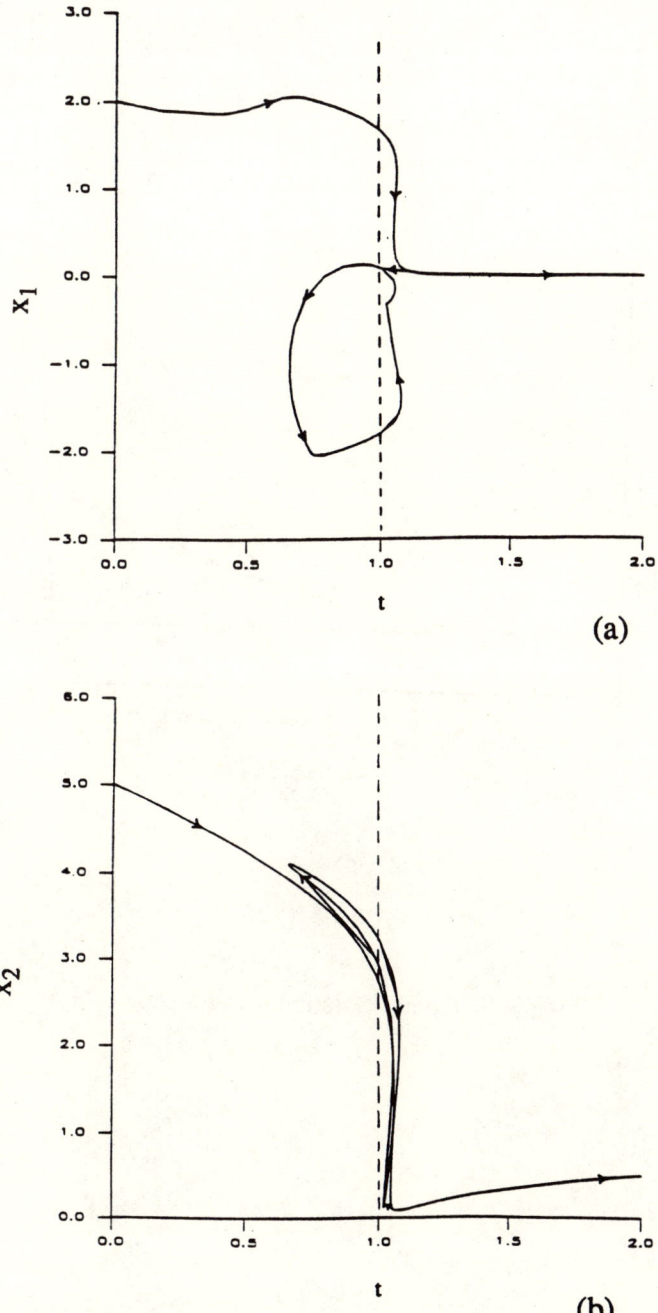

Figure 9. Solution path for Case c.

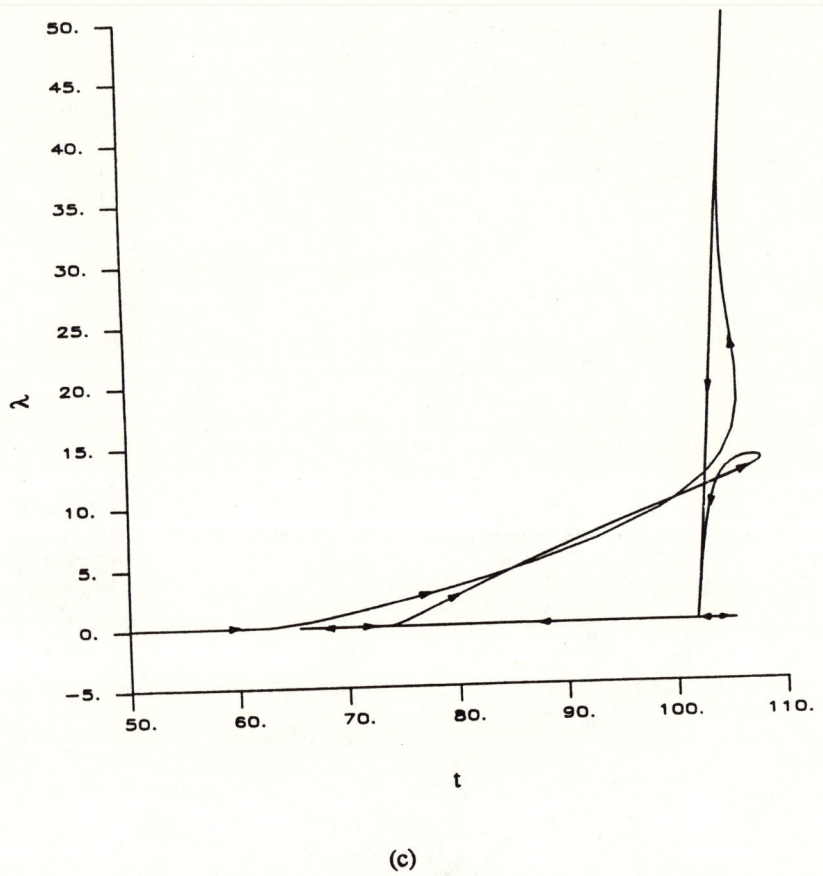

(c)

Figure 9. (cont) Solution path for Case c.

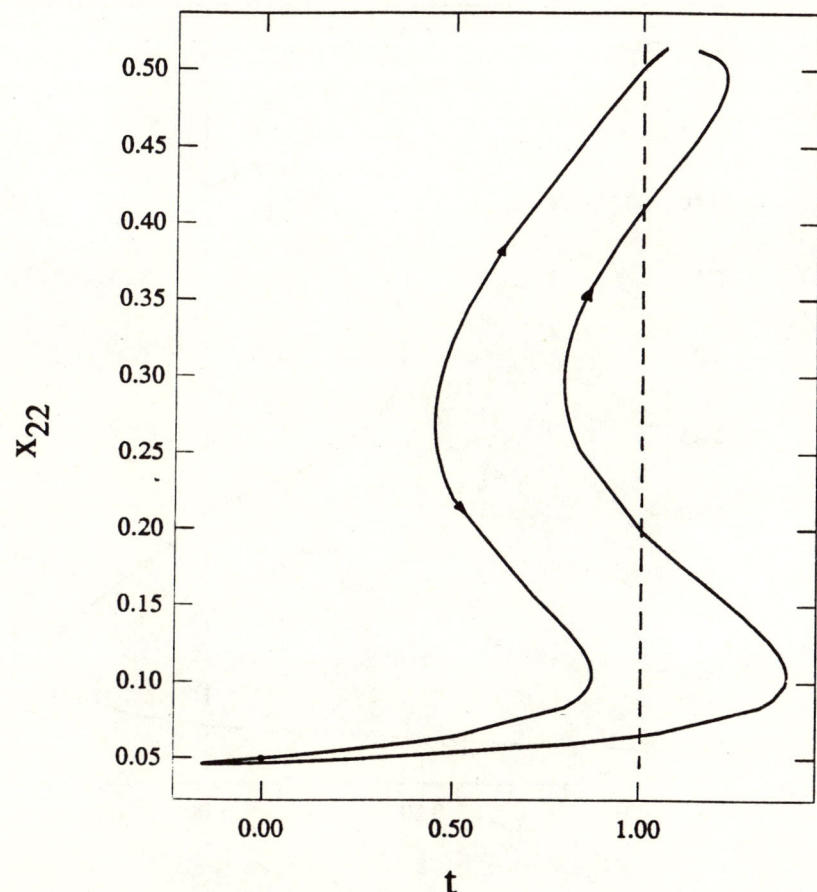

Figure 10. Homotopy path for LLE problem beginning with $\underline{x}^\circ = [0,5,-1]^T$.

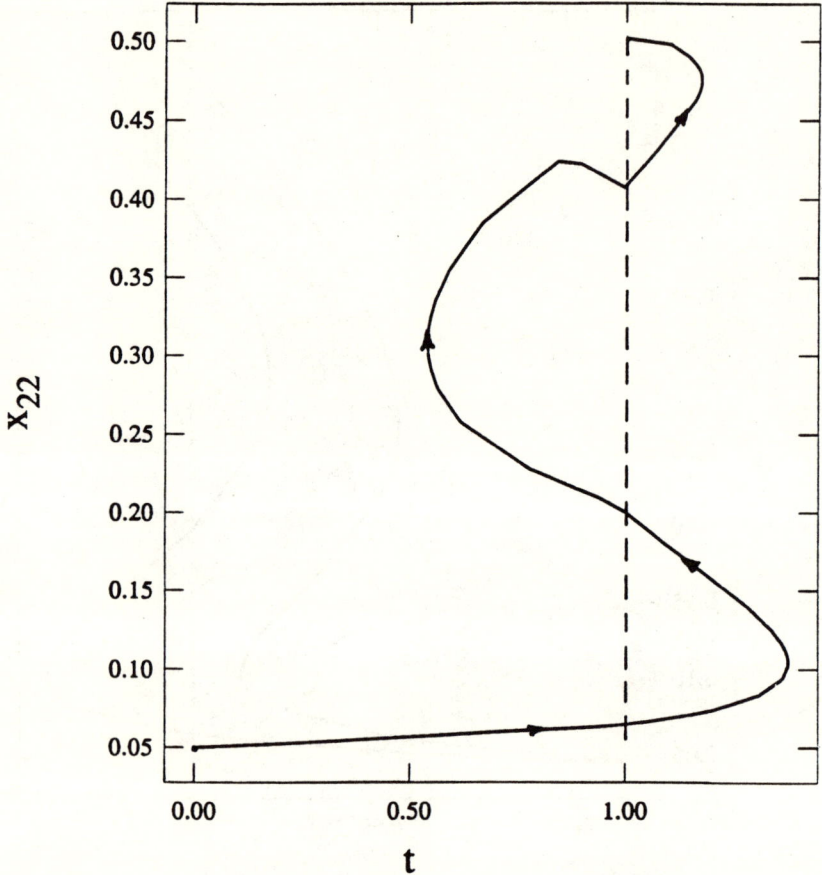

Figure 11. Homotopy path for LLE problem beginning with $\underline{x}^\circ = [1,5,1]^T$.

SPACE-COVERING APPROACH AND MODIFIED FRANK-WOLFE ALGORITHM FOR OPTIMAL NUCLEAR REACTOR RELOAD DESIGN

Zhian Li[*]   P. M. Pardalos[+]   S. H. Levine[*]

January 1991

Abstract

A global optimization method has been developed for a constrained black-box type optimization problem involved in the optimal nuclear reactor reload design. In this method, the problem domain was partitioned into sub-domains and the Modified Frank-Wolfe Algorithm was employed to find a local maximum for each of the sub-domains if it is feasible. This method has been verified in the first step of an optimization study of the Beaver Valley nuclear reactor Unit-2, cycle-2 (Pennsylvania, USA) reload design and an optimal solution has been obtained with reasonable computer time (about 9 hours on an IBM 3090 computer). The results of this study indicate that the proposed method can be used in constrained black-box type global optimization problems such as optimal nuclear reactor reload design. In such a case, the analytic expressions of the objective function and the constraint functions can not be obtained or are too complex, and the feasible domain may not be connected as well. The great advantages of the proposed method are: (1) it does not require the objective function, as well as the constraint functions, to be differentiable over the entire domain, (2) it does not require the feasible region to be connected either, and (3) the method is accurate and efficient because the Linear Programming it uses is fast and the sub-domains can be made sufficiently small. It should be pointed out that the computation time was mainly used in the evaluations of the objective function and the constraint functions. For the test case, the computation time needed can be reduced to about 3 hours on the same computer by modifying the evaluation model of the objective function and the constraint functions as well as their gradients without significant loss of accuracy.

---

[*]Nuclear Engineering Department, The Pennsylvania State University, University Park, PA16802.
[+]Computer Science Department, The Pennsylvania State University, University Park, PA16802.

Introduction

The safe and economic operation of a nuclear power reactor is directly related to the nuclear reactor reload design. Therefore, the goal of nuclear reactor fuel management has been to make nuclear reactor fuel reload design optimal and automatic. Attending this goal involves a few sequential optimal calculations. Among them, the first step is to obtain the optimal End-Of-Cycle(EOC) state k_∞ distribution.

The primary objective of an optimal nuclear reactor reload design is to minimize the nuclear reactor fuel inventory and/or the cost, while satisfying the energy production and the safety requirement imposed on the reactor operation. Many studies have been made and various approaches and optimization techniques have been used in these studies. *Melice* [1] and *Huang* [2] used a Lagrange multiplier method in their studies and a Dynamic Programming method was used by *Wall et al.* [3]. Recently, *Suh and Levine* [4] developed an optimized automatic reload program for pressurized water reactor using linear programming by a direct search manner. *J. A. Stillman et. al* [5] also used linear programming algorithm in searching optimal End-Of-Cycle k_∞, but a backward depletion scheme was used to link the EOC state to the Beginning-Of-Cycle state. In an earlier study by *G. H. Hobson et. al* [6], a Monte Carlo integer programming scheme was used in attempting to find global optimum of the EOC state k_∞ distribution. However, because the Monte Carlo method is a random sampling approach, it is extremely inefficient and the quality of the solution is uncertain. A recent attempt to find the global optimal solution was made by *Morita et al.* [7] . In their study, a large number of loading patterns have been analyzed and the best candidate is chosen based on the results of their analysis. However, because of the complexity of the problem, a global optimal solution can not be guaranteed and greatly depends on the initial guess which is totally depended on the experience of the reload designer.

Most recently, *Levine et al.* [8] has summarized the whole strategy of the nuclear reactor fuel reload design in a paper. In this paper, they pointed out that: "The optimum core depends on the end-of-cycle (EOC) state. This final state should have the maximum allowed k_{eff} for the fissionable material in the core at the EOC state. Any movement of the fuel in the EOC core will either reduce the k_{eff} or violate the constraints." Here k_{eff} is the effective neutron population multiplication factor. If it goes below 1.0, the reactor will shut down automatically. If it is greater than 1.0, the reactor power will go up. For a steady state reactor, it is maintained at 1.0 by adjusting the soluble boron concentration in the reactor or some other control mechanism (see, J. R. Lamarsh [14]).

OPTIMAL NUCLEAR REACTOR RELOAD DESIGN 595

This statement given by *Levine* has clearly defined the objective of the nuclear reactor fuel reload optimization problem, that is, the objective of the nuclear reactor fuel reload optimization is to find an optimal reactor core End-Of-Cycle state. The goal of this study is, therefore, to develop a method which can be used to find this optimal EOC state accurately and efficiently.

In the first step of the optimization problem of optimal nuclear reactor reload design, the goal is to determine the physical characters of this optimal EOC state in this analysis. Because neither the objective function nor the the constraint functions can be expressed analytically, only the values of the objective function and the constraint functions can be evaluated by a sophisticated reactor physics model, this optimization problem is a black-box type optimization problem. In general, it is very difficult to solve this type of optimization problem because very little is known for both, the objective function and the constraint functions. In addition, the feasible domain may not be connected because the constraint functions may break the feasible domain into pieces. This added difficulty makes the global optimization problem even more complex. A new approach for the black-box type global optimization problem has been developed in our study on the optimal nuclear reactor reload design. In this approach, the problem domain is partitioned into sub-domains and a moving window moves from one sub-domain to another to cover the entire domain. Then the *Modified Frank-Wolfe Method* [9] is used to perform the search for a local optimum within the moving window.

In this analysis, a great effort has been made to reduce the number of sub-domains to be searched based on the analysis on the constraint functions. If the problem has n variables and we divide each coordinate into m intervals, the total number of sub-domains is m^n. For our problem, n = 26, and we chose m = 5. The number of sub-domains to be searched is about 1.5×10^{18}. The analysis shows that the conservation constraints on the variables make some of the sub-domains identical with each other. Therefore, the actual number of sub-domains to be searched becomes 1178. This significant reduction of the number of sub-domains has made the approach practical and very efficient.

In order to test the proposed approach, the EOC state of the Beaver Valley nuclear reactor Unit-2, Cycle-2 (Pennsylvania, USA) reload design has been analyzed. The results show that the value of the objective function SB (SB denotes the Soluble Boron concentration in the reactor which indicates the reactor core reactivity; if this value becomes 0.0, the reactor will stop) may reach 240 ppm (Part Per Million boron, a measurement of soluble boron concentration) by the proposed method, compared to 4 ppm as the original design data and 84 ppm as the result by a direct search method used by Suh (see Suh, 1990). This could give about an extra 70 Effective Full Power Days for the reactor to operate. It should be mentioned that in this approach only the EOC state is optimized, whereas in the other two cases the core is depleted from the Beginning Of Cycle(BOC) to EOC. The computer time in solving the problem is about 9 hours of cpu time on an IBM 3090 600S. The results also show that the computer time was mainly used in the evaluations of the objective and the constraint functions. It has been proved that the computer time needed can be reduced to about 3 hours of cpu time on the same machine by modifying the evaluation model used for these evaluations without significant loss of accuracy.

It should be pointed out that to attend an optimal nuclear reactor reload design involves several stages and the practical availability of the fuel assemblies required to make the load optimal. This study is only to find the optimal EOC state. To attend the actual optimal reload design obtained, we need to get a priority scheme of the k_∞ distribution from the result of the optimization, load the actual fuel assemblies according to the priority scheme, find an optimal Haling power distribution, and load burnable poison into the core to match the optimal Haling distribution. However, those details are beyond the scope of this paper.

Problem Statement

In a nuclear reactor, the nuclear fuel is made in a form of fuel assembly. The reactor is loaded by various types of fuel assemblies. The number of fuel assemblies in a nuclear reactor may vary from reactor to reactor. The Beaver Valley nuclear Unit-2 reactor has 157 fuel assemblies. To make the nuclear reactor reload design simpler, the common method is to design a 1/8 symmetric core. Figure 1 has shown the 1/4 of the Beaver Valley reactor core with 1/8 symmetry. In this figure, the assembly identification number 28 and 27 represent the pure water assemblies and the baffle assemblies. Those assemblies are not nuclear fuel assemblies and their positions are fixed. The dimension n, in our problem, is therefore reduced to 26 (as shown in Figure 1). The problem becomes that how to arrange the 26 fuel assemblies in such a way that any movement of the assembly positions will either reduce the reactor life-time (the energy production) or violate the constraints, so that the design will be optimal (The definition of optimal reload design given by Levine. see Levine, 1991).

To achieve the goal design, there are two major stages. The first is to find the optimal EOC state. And the second is to rearrange these fuel assemblies so as to achieve the optimal EOC state. The studies made here provide the first stage of the optimal nuclear reactor fuel reload design. That is to find an optimal EOC state under the constraints of given nuclear fuel inventory, energy production, and the upper and lower limits of the normalized power.

Let SB denote the soluble boron concentration at the End-Of-Cycle (EOC) state in the reactor and KINF denote the k_∞ distribution vector of the fuel assemblies which uniquely governs the End-Of-Cycle state of a nuclear reactor. The general form of the optimization problem can be written as:

$$Global \quad \underset{KINF \in S^n}{Max} \quad SB(KINF) \quad (1)$$

subject to

$$NP(KINF) \leq C_1^n$$

$$NP(KINF) \geq C_2^n$$

$$W \bullet KINF = TK$$

$$KINF \in S^n$$

Where SB(KINF) is the objective function, NP is the assembly Normalized Power distribution vector and $NP(KINF) \leq C_1^n$ and $NP(KINF) \geq C_2^n$ are the upper and lower limits on the power assembly normalized powers. $W \bullet KINF = TK$ is the total nuclear fuel inventory constraint with W as the fuel assembly volume weighting vector, and S^n is the problem domain.

—— REACTOR MATERIAL PICTURE ——

I J-1	2	3	4	5	6	7	8	9	10	11	12	13	14	15	16	17	18
1	1	2	2	3	3	4	4	5	5	6	6	7	7	8	8	27	28
2	2	9	9	10	10	11	11	12	12	13	13	14	14	15	15	27	28
3	2	9	9	10	10	11	11	12	12	13	13	14	14	15	15	27	28
4	3	10	10	16	16	17	17	18	18	19	19	20	20	27	27	27	28
5	3	10	10	16	16	17	17	18	18	19	19	20	20	27	28	28	28
6	4	11	11	17	17	21	21	22	22	23	23	24	24	27	28	28	28
7	4	11	11	17	17	21	21	22	22	23	23	24	24	27	28	28	28
8	5	12	12	18	18	22	22	25	25	26	26	27	27	27	28	28	28
9	5	12	12	18	18	22	22	25	25	26	26	27	28	28	28	28	28
10	6	13	13	19	19	23	23	26	26	27	27	27	28	28	28	28	28
11	6	13	13	19	19	23	23	26	26	27	28	28	28	28	28	28	28
12	7	14	14	20	20	24	24	27	27	27	28	28	28	28	28	28	28
13	7	14	14	20	20	24	24	27	28	28	28	28	28	28	28	28	28
14	8	15	15	27	27	27	27	27	28	28	28	28	28	28	28	28	28
15	8	15	15	27	28	28	28	28	28	28	28	28	28	28	28	28	28
16	27	27	27	27	28	28	28	28	28	28	28	28	28	28	28	28	28
17	28	28	28	28	28	28	28	28	28	28	28	28	28	28	28	28	28
18																		

FIGURE 1. 1/4 NUCLEAR REACTOR LOAD MAP
BEAVER VALLEY REACTOR, 1/8 SYMMETRY

For a given fuel assembly, the k_∞ of the fuel assembly is directly related to the assembly BurnUp (BU denotes the integrated assembly energy production). For convenience, here we use assembly burnup as the control variable instead of KINF. Thus, the problem can be written as:

$$\text{Global} \quad \underset{BU \in S^n}{Max} \quad SB(BU) \qquad (2)$$

subject to

$$NP(BU) \leq C_1^n$$

$$NP(BU) \geq C_2^n$$

$$W \bullet BU = TBU$$

$$BU \in S^n$$

Where SB(BU) is the objective function as a function of burnup, BU, $NP(BU) \leq C_1^n$ and $NP(BU) \geq C_2^n$ are again the upper and lower limits of the normalized power as a function of assembly burnup BU and $W \bullet BU = TBU$ is the total reactor burnup constraints for given fuel assemblies and W is the fuel assembly volume weighting vector, and S^n is the problem domain.

In general, to solve this problem, we need to know the analytical expressions of the objective function and the constraint equations. Unfortunately, neither the objective function nor the constraint equations can be written analytically. Only the values of the objective function and the constraint functions can be obtained by sophisticated reactor physics models [4]. We conclude that the optimization problem of optimal reactor reload design, therefore, is a complicated black-box type optimization problem. In general, it is difficult to solve the black-box type optimization problem because very little is known about the objective function as well as the constraint functions.

Modified Frank-Wolfe Algorithm

A Modified Frank-Wolfe Algorithm was proposed by *Pardalos* [9]. The algorithm is described as follows.

Consider the problem

$$global \; \underset{x \in \Omega}{Max} \; f(x) \qquad (3)$$

where the objective function is continuously differentiable, and the feasible domain Ω is a nonempty bounded polyhedron in R^n. Suppose we have a number of "starting points" $\alpha_1, ..., \alpha_m \in \Omega$ and let $g = \nabla f(x)$ be the gradient vector. Then for each $x = \alpha_i$, $i = 1, ..., m$, we have the following algorithm:

Algorithm:

1. Initial point $x_0 = x_s \in \Omega$,

2. Given x_k, compute the gradient $g_k = \nabla f(x_k)$,

3. Solve the linear programming problem

$$\underset{x \in \Omega}{Max} \; g_k^T x \qquad (4)$$

to obtain a solution s_{k+1}. Let x_{k+1} be the max point of f(x) on the line $[x_k, s_{k+1}]$.

4. If $f(x_{k+1}) = f(x_k) \pm \varepsilon$, stop (ε is a given tolerance); x_{k+1} is a local maximum. If not, $x_k \leftarrow x_{k+1}$, and go to step 2.

It has been proved that this method will guarantee the convergence to a local maximum with the assumption that $f(x)$ is continuously differentiable. The two major advantages of the this algorithm are:

1. It does not require the analytic expression of the objective function nor the constraints. Only the gradients $\nabla f(x)$ are needed;

2. Linear programming is used in solving the optimization problem for each "starting point". This will allow a large number of search points to be evaluated within limited computer time.

These major advantages made the algorithm particularly suitable for the black-box type optimization problem as involved in the optimal nuclear reactor reload design in which the objective function and/or the constraints can not be expressed analytically or are too complicated. However applying the Modified Frank-Wolfe Algorithm from only one initial point can only guarantee the convergence to a local maximum. In addition, the feasible domain may not be connected because the constraint functions may break the feasible domain into pieces. To solve these problems, a new approach was proposed. In this approach, we first partition the feasible domain into sub-domains. Then, a moving window is used from one sub-domain to another to cover the entire feasible domain. Within the moving window, the Modified Frank-Wolfe Algorithm is applied to find the local maximum for each sub-domain. Next, we will describe and discuss the approach in details.

Space Covering Approach

Consider the following global optimization problem.

$$global \; \underset{x \in \Omega}{Max} \; f(x) \qquad (5)$$

subject to

$$L_1(x) \leq C_1^n$$

$$L_2(x) \geq C_2^n$$

$$x \in \Omega$$

We partition the feasible domain Ω into sub-domains Ω_1, Ω_2,..., Ω_m. Assume $f(x)$, $L_1(x)$, $L_2(x)$ are continuously differentiable within each feasible sub-domain Ω_i. Applying the Modified Frank-Wolfe Algorithm to each sub-domain Ω_i, we can generate N local maxima s_1, s_2,..., s_N. Because s_i may not be distinct or may not exist under the constraints $L_1(x) \leq C_1^n$ and $L_2(x) \leq C_2^n$, we have $N \leq m$. Thus, if each Ω_i is sufficiently small (depends on the accuracy requirement of the problem), and let $f(s) = \max_{1 \leq i \leq N} f(s_i)$ denote the maximum of the solutions obtained, we conclude that $f(s)$ is an ε-approximation of the global maximum of $f(x)$, $x \in \Omega$, subject to $L_1(x) \leq C_1^n$ and $L_2(x) \geq C_2^n$. Regarding the quality of f(s), we have the following.

Assume that $s_1, \ldots, s_N$ are local maxima that have been found and let $f(s) = \max_{1 \leq i \leq N} f(s_i)$. If L is the Lipschitz constant of $f(x)$ and $V_i(r_i)$ denotes the sphere with center s_i and radius $r_i = \dfrac{f(s) - f(s_i) + \varepsilon}{L}$ ($\varepsilon > 0$, given tolerance) and if $\bigcup_{i=1}^{N} V_i \supseteq \Omega$, then $0 \leq f(s^*) - f(s) \leq \varepsilon$, where $f(s^*)$ is the global maximum.

It can be shown (see Vavasis, 1990) that, in the worst case, no algorithm can find in finite number of steps the global maximum in a black-box type optimization model.

Figure 2 has illustrated the schematic feasible domain of the optimization problem and the Space-Covering Approach for fictitious 2 dimensional case.

OPTIMAL NUCLEAR REACTOR RELOAD DESIGN

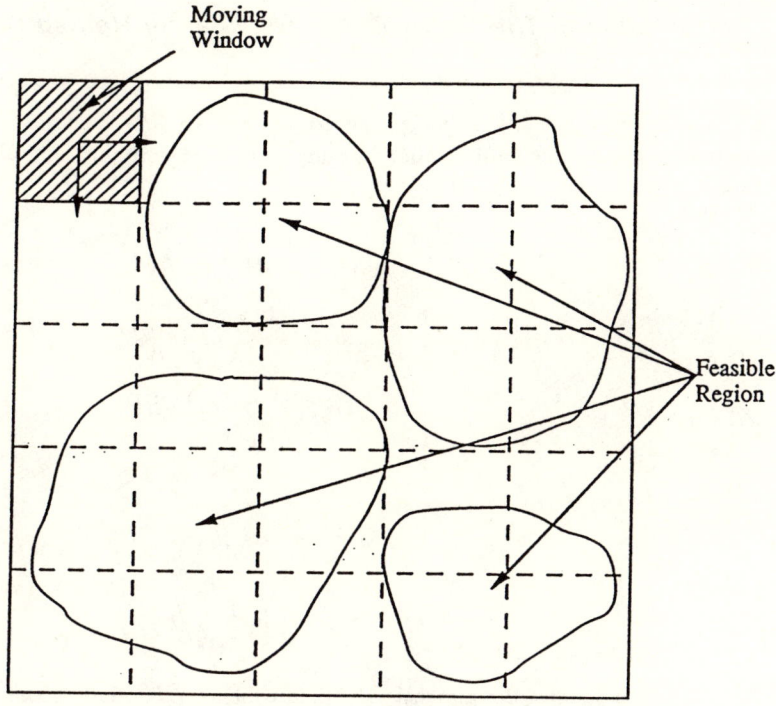

Figure 2. Problem Domain, Feasible Regions and the Space-Covering Approach (Moving Window)

Formulation of The Optimal Nuclear Reactor Reload Design Problem

To apply the algorithm proposed above, we first partition the feasible domain of the given optimization problem into sub-domains call $D_1, ..., D_m$. Each $D_l \in [D_1,...,D_m]$ is defined by:

$$D_l \begin{bmatrix} BU_i^j \leq BU_i \leq BU_i^{j+1} \\ \\ BU_i^{max} \leq MAXBU \\ \\ BU_i^{min} \geq MINBU \\ \\ BU_i^{j+1} - BU_i^j = DBU \\ \\ BU_i = BU_0, BU_1,...,BU_n \\ \\ i=1,...,n, \\ j=1,...,MNI \end{bmatrix} \quad (6)$$

Here n denotes the dimension of the problem, BU_i denotes the i th component of the control variable, MAXBU and MINBU are the upper and lower bounds of the control variables, DBU is the size of the moving window in each dimension to cover the D_l.

In order to apply the Modified Frank-Wolfe Algorithm to each sub-domain D_l, we linearize the objective function and the constraint functions as follows:

$$SB = SB^0 + \frac{\partial SB}{\partial BU}(BU - BU^0)^T$$
$$NP = NP^0 + \frac{\partial NP}{\partial BU}(BU - BU^0)^T \quad (7)$$

with

$$\frac{\partial SB}{\partial BU} = \left[\frac{\partial SB}{\partial BU_1},...,\frac{\partial SB}{\partial BU_n}\right] \quad (8)$$

$$\frac{\partial NP}{\partial BU} = \begin{bmatrix} \dfrac{\partial NP_1}{\partial BU_1} & \cdots & \dfrac{\partial NP_1}{\partial BU_n} \\ \vdots & \cdots & \vdots \\ \dfrac{\partial NP_n}{\partial BU_1} & \cdots & \dfrac{\partial NP_n}{\partial BU_n} \end{bmatrix} \quad \bullet \quad (9)$$

Then, the optimization problem becomes the following:

$$\underset{BU \in v_i}{Max} \; SB(BU) = \frac{\partial SB}{\partial BU} BU^T + SB^0 - \frac{\partial SB}{\partial BU}(BU^0)^T \quad (10)$$

subject to

$$\frac{\partial NP}{\partial BU}(BU)^T \leq NP_{max} - NP^0 + \frac{\partial NP}{\partial BU}(BU^0)^T$$

$$\frac{\partial NP}{\partial BU}(BU)^T \geq NP_{min} - NP^0 + \frac{\partial NP}{\partial BU}(BU^0)^T$$

$$W \bullet BU^T = C \qquad \bullet \quad (11)$$

$$BU \leq BU_{i+1} \leq BU_{max}$$

$$BU \geq BU_i \geq BU_{min}$$

$$BU \geq 0$$

where BU^0, is the value of the control variables at the "starting point" of each sub-polyhedron. SB^0 and NP^0 are the associated values of the objective function and constraint functions. All other symbols have the same meaning as defined in equation (2).

Because BU^0, SB^0 and NP^0 are known constants, we rearrange equation (10), (11) and get the following Modified Frank-Wolfe Algorithm forms.

$$\underset{BU \in D_i}{Max} \; SB(BU) = \frac{\partial SB}{\partial BU} BU^T + CONSTANT \quad (12)$$

subject to

$$\frac{\partial NP}{\partial BU}(BU)^T \leq CONSTANT$$

$$\frac{\partial NP}{\partial BU}(BU)^T \geq CONSTANT$$

$$W \bullet BU^T = C \qquad \bullet \qquad (13)$$

$$BU \leq BU_{i+1} \leq BU_{max}$$

$$BU \geq BU_i \geq BU_{min}$$

$$BU \geq 0$$

A FORTRAN program has been written for the optimal nuclear reactor reload design problem based on the proposed algorithm. The program will (1) move a window to a start point and generate a sub-polyhedron with its associated boundaries. (2) calculate the gradients for both the objective function and the constraints by calling a two-dimensional reactor physics model ADMARC which will calculate the values of the objective function as well as the constraints. (3) Having obtained the gradients and the values of the objective function and the constraint functions, the Modified Frank-Wolfe Algorithm is used to search for a local maximum if there exists one. A linear programming subroutine ZX3LP from IMSL package is called to solve the linear programming problem formulated by the Modified Frank-Wolfe Algorithm. If the current sub-polyhedron is not feasible, the program will skip the local maximum search and go to step (4). If the current sub-polyhedron is feasible, the program will perform the local maximum search by iteration. Having found the local maximum for the current sub-polyhedron, the program will then (4) move the window to its next adjacent sub-polyhedron and test whether the new sub-polyhedron is identical with any of the previous sub-polyhedrons. If yes, go to step (4). If not, the program will go to step (2) and repeat the procedure of a local maximum search. After covering all the feasible area defined by the problem by moving the window, the program will sort the results of all feasible solutions and print out 5 best solutions for alternative selection. Figure 3 shows the flowing diagram of the FORTRAN program.

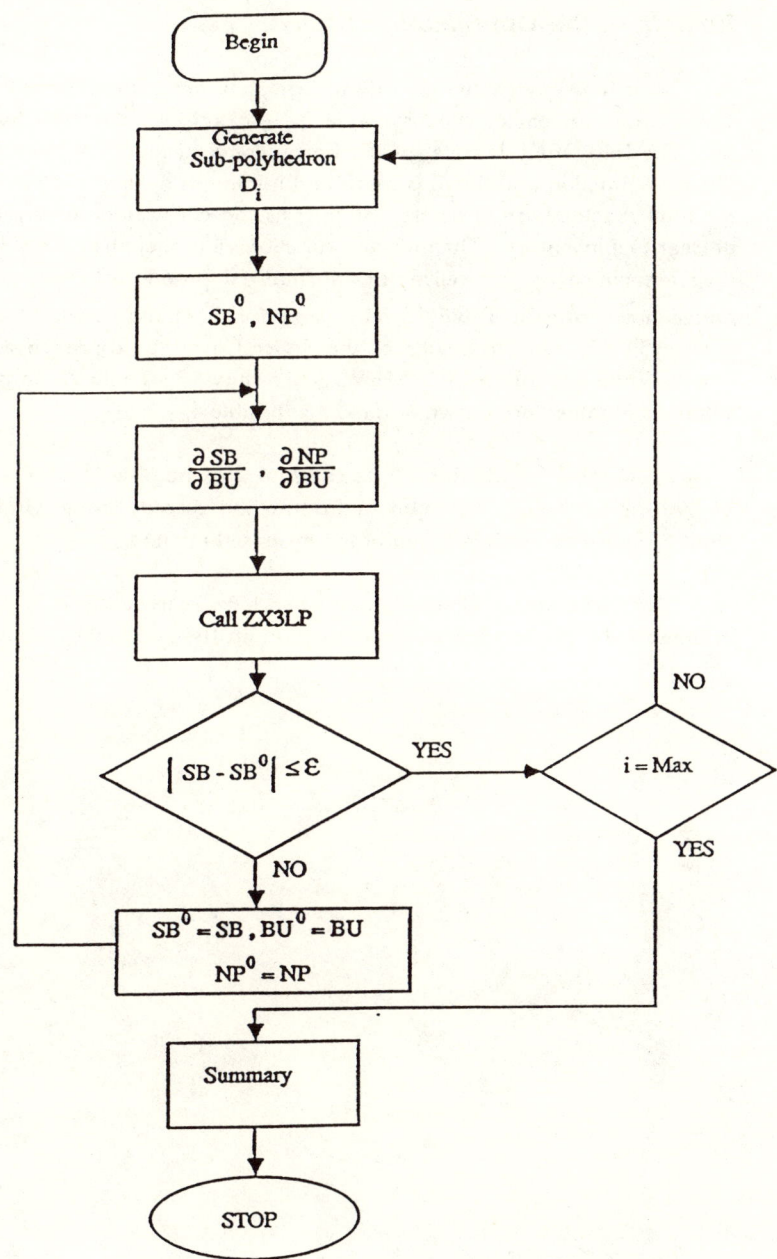

Figure 3. Flowing Chart of the SCMFWA method

Results of the Optimization Study

In order to verify the algorithm developed here, the Beaver Valley Unit-2, Cycle-2 nuclear reactor (Pennsylvania, USA) End-Of-Cycle state has been studied. The End-Of-Cycle state soluble boron concentration has been chosen as the objective function and the maximum and minimum normalized power as well as the total reactor burnup (energy output) as the constraints for given amount of nuclear fuel inventory. The objective function values, control variable values and their associated k_∞ as well as the normalized power distributions for original design are shown in Table 1. The comparison of the results of direct search method(DSM) and the results of the Space-Covering Approach and Modified Frank-Wolfe Algorithm (SCAMFWA) are shown in Table 2. Some alternative solutions obtained are shown in table 3 and table 4.

From table 1 and table 2, one can find that the objective function is 4 ppm of the original design, 84 ppm of the method directly using Modified Frank-Wolfe Algorithm, and 240 ppm of the proposed method.

The total number of sub-domains had been searched is 1178. The computer time used is about 9 hours of cpu time on an IBM 3090 600S computer.

Table 1: Original Design Parameters

Original Design

SB   4 ppm

BU

33328	16720	26874	33392	17309	25926	15667	25151
	31374	29043	17135	31339	26336	14748	24668
		16982	30823	30957	16552	19560	
			31055	16796	14564	25688	
				21866	25979		

k_∞

0.9270	1.0918	0.9997	0.9267	1.0865	1.0062	1.1390	0.9766
	0.9377	0.9850	1.0885	0.9763	1.0034	1.1358	0.9813
		1.0902	0.9729	0.9342	1.1161	1.0541	
			0.9397	1.1157	1.1395	0.9743	
				1.0315	0.9724		

Normalized Powers

0.996	1.249	1.093	1.009	1.314	1.192	1.237	0.488
	0.988	1.091	1.279	1.104	1.150	1.150	0.408
		1.280	1.052	1.004	1.284	0.787	
			0.979	1.282	1.112	0.438	
				0.918	0.486		

Table 2: Comparison of The Results From DSM and SCAMFWA

```
DSM  Method      SB = 84 ppm
   33328   16720    24874   33392   14874   25926   15667   25151
           33374    29043   17135   31339   26336   13123   26668
                    14982   32823   32798   14552   19560
                            31055   14796   15264   27688
   BU                               19866   28979
   0.9270 1.0918 1.0135 0.9267 1.1062 1.0062 1.1241 0.9766
          0.9257 0.9850 1.0885 0.9763 1.0034 1.1497 0.9677
                 1.1065 0.9622 0.9291 1.1314 1.0541
                        0.9397 1.1315 1.1570 0.9610
   $k_\infty$                   1.0505 0.9527
   0.918   1.162   1.064   0.984   1.346   1.198   1.246   0.485
           0.905   1.041   1.237   1.093   1.161   1.183   0.399
                   1.259   1.008   0.985   1.343   0.805
                           0.978   1.347   1.179   0.440
   Normalized Powers               0.975   0.487

SCAMFWA Method   SB = 240 ppm
   36953   11144   19534   40267   16176   20843   12607   32026
           38144   22615   10874   29221   19711    9653   31543
                   10414   37692   36521   10567   26435
                           37930    9921    9168   34563
   BU                              28005   32854
   0.9062 1.1329 1.0542 0.8884 1.0957 1.0428 1.1466 0.9331
          0.8981 1.0296 1.1346 0.9915 1.0517 1.1792 0.9367
                 1.1412 0.9318 0.9072 1.1590 1.0027
                        0.8993 1.1685 1.1902 0.9187
   $k_\infty$                   0.9920 0.9288
   0.979   1.359   1.246   0.957   1.349   1.345   1.345   0.437
           0.958   1.213   1.354   1.129   1.295   1.255   0.363
                   1.355   0.905   0.888   1.347   0.700
                           0.796   1.245   1.119   0.357
   Normalized Powers               0.749   0.387
```

Table 3:   4 Alternative Optimal Solutions

```
    1.     SB = 232 ppm
  36953  11519  19553  40267  16357  21420  14150  32026
         37928  22543  11250  29487  19711   7873  31543
                10821  37198  35876  10301  26435
                       37430   9921  11286  32563
  BU                          28560  33854
```

0.9062 1.1306 1.0541 0.8884 1.0942 1.0384 1.1355 0.9331
       0.8990 1.0301 1.1323 0.9899 1.0517 1.1935 0.9367
              1.1382 0.9347 0.9109 1.1606 1.0027
                     0.9021 1.1685 1.1690 0.9305
k_∞ 0.9883 0.9231

```
  0.982  1.360  1.252  0.960  1.349  1.345  1.345  0.445
         0.964  1.219  1.354  1.130  1.308  1.303  0.373
                1.356  0.914  0.893  1.347  0.709
                       0.797  1.222  1.059  0.358
  Normalized Powers            0.723  0.367
```

```
    2.     SB = 231 ppm
  35203  11935  17999  40267  16116  20368  11031  32026
         38249  23088  11025  29599  19711  11883  31543
                10574  37539  37832  10628  26435
                       37681   9921   8314  32563
  BU                          28091  32854
```

0.9162 1.1278 1.0615 0.8884 1.0962 1.0465 1.1572 0.9331
       0.8976 1.0262 1.1337 0.9893 1.0517 1.1602 0.9367
              1.1400 0.9327 0.8998 1.1586 1.0027
                     0.9007 1.1685 1.1986 0.9305
k_∞ 0.9915 0.9288

```
  1.000  1.355  1.264  0.960  1.349  1.346  1.346  0.430
         0.958  1.210  1.353  1.122  1.284  1.201  0.353
                1.354  0.909  0.878  1.348  0.695
                       0.805  1.261  1.152  0.372
  Normalized Powers            0.763  0.397
```

Table 4: 4 Alternative Optimal Solutions (continued)

```
3.    SB  = 229 ppm
     36703    13503    18047    40267    16278    20917    12678    32026
              34270    22735    11384    29417    19711     9873    31543
                       11094    37621    38255     9961    26435
                                37930     9921    10140    32563
     BU                                  27911    32854
     0.9077  1.1167  1.0611  0.8884  1.0949  1.0422  1.1461  0.9331
             0.9205  1.0287  1.1314  0.9903  1.0517  1.1774  0.9367
                     1.1362  0.9322  0.8975  1.1624  1.0027
                             0.8993  1.1685  1.1812  0.9305
```
k_∞
```
                                     0.9921  0.9288
     0.991   1.355   1.280   0.965   1.349   1.346   1.346   0.438
             1.016   1.232   1.353   1.125   1.295   1.254   0.364
                     1.354   0.903   0.866   1.348   0.702
                             0.790   1.229   1.096   0.363
     Normalized Powers               0.741   0.382

4.    SB  = 228 ppm
     37453     9859    24690    40267    14850    21294    14272    32026
              27233    26092    10298    30103    19711     9873    31543
                       10107    33962    37822    12994    26435
                                37930    11420     8064    34563
     BU                                  13625    32854
     0.9077  1.1167  1.0611  0.8884  1.0949  1.0422  1.1461  0.9331
             0.9205  1.0287  1.1314  0.9903  1.0517  1.1774  0.9367
                     1.1362  0.9322  0.8975  1.1624  1.0027
                             0.8993  1.1685  1.1812  0.9305
```
k_∞
```
                                     0.9921  0.9288
     0.991   1.355   1.280   0.965   1.349   1.346   1.346   0.438
             1.016   1.232   1.353   1.125   1.295   1.254   0.364
                     1.354   0.903   0.866   1.348   0.702
                             0.790   1.229   1.096   0.363
     Normalized Powers               0.741   0.382
```

Conclusions

A new approach, using Space-Covering technique and the Modified Frank-Wolfe Algorithm, has been developed for the constrained black-box type global optimization problem involved in the optimal nuclear reactor reload design. In this type of problem the analytical expressions of the objective function as well as the constraint functions can not be obtained or are too complex and the feasible domain may not be connected.

The results obtained from the analysis on a real world problem, the Beaver Valley Unit-2, Cycle-2, show that the proposed algorithm is accurate and efficient. The main computation time used is the evaluations of the the objective and the constraints. The maximum objective function value obtained is 240 ppm compared to 84 ppm of the result from direct search method and 4 ppm for the original design. However, the latter two results involve depleting the core from BOC to EOC whereas in this method only the EOC state is optimized. The computation time used is about 9 hours of cpu time on an IBM 3090 600S. It has been proved by *Suh* [13] that the computation time for the evaluations of the objective function and the constraint functions can be reduced to 1/3 to 1/4 of the time used in this study if 1 mesh point is used for each fuel assembly in the reactor physics model instead of 4 mesh points as used in this study without significant loose of accuracy. The detailed discussion of the computation time reduction is beyond the scope of this paper.

The major advantages of the proposed method here are:

1. The Space-Covering technique and the Modified Frank-Wolfe Algorithm does not require the analytic expressions neither for the objective function nor for the constraint functions. Only the values of them at some points are needed. This is very important for the black-box type optimization problem in which the analytic expressions of the objective function and the constraint functions can not be obtained or are too complex.

2. The method developed does not require either the objective function or the constraint function to be globally differentiable. Local differentiability is sufficient for the algorithm to solve the problem. This is particularly helpful for the constrained optimization problem in which the feasible domain may not connected.

3. The accuracy of the solution depends on the size of the sub-polyhedrons generated as well as the quality of the gradients of the objective function and the constraint functions.

4. The optimal solution is not unique in the optimization problem in nuclear reactor reload design.

In conclusion, the proposed approach is very efficient for the constrained black-box type global optimization problem involved in the optimal nuclear reactor fuel reload design. The accuracy of the proposed algorithm depends on the size of the sub-domains. It can also be used in similar problems involved in other areas. Although the sub-domains are equally sized in this study, the algorithm does not require the sub-domain to be equally sized.

References

1. M. Melice, Pressurized Water Reactor Optimal Core Management and Reactivity Profiles, Nuclear Science Engineering, 37: p. 451-477 (1969).

2. H. Huang, Optimal Fuel Management by Maximizing the End-Of-Cycle Multiplication Factor, Ph. D Thesis, Penn State University (1981).

3. I. Wall et al., Optimization of Refueling Schedule for Light Water Reactors, Nuclear Science Engineering, 22: p. 285-297 (1965).

4. J. S. Suh, S. H. Levine, Optimized Automatic Reload Program for Pressurized Water Reactors Using Direct Search Optimization Techniques, Nuclear Science Engineering, 105: p. 371-382 (1990).

5. J. A. Stillman, Y. A. Chao, T. J. Downar, The Optimum Fuel and Power Distribution For A PWR Burnup Cycle. Proc. 1988 Intl. Reactor Physics Conference, ANS, Jackson Hole, IV-247 (1988).

6. J. H. Hobson, P. J. Turinsky, Automatic Optimization Of Core Loading Patterns To Maximize Cycle Energy Production Within Operation Constraints, Proc. Topic Meeting, Advances in Fuel Management, ANS, Pinehurst, 509 (1986).

7. T. Morita, Y. A. Chao, A. J. Federowicz, and P. J. Duffey, LPOP: Loading Pattern Optimization Program, Trans. Am Nucl. Soc., 52: p. 41-42 (1986).

8. S. H. Levine, et al. Using Optimization Techniques for Finding Potential Superior Reload Configurations for the TMI-1 cycle-9, To be published.

9. P. M. Pardalos, Parallel Search Algorithm in Global Optimization, Applied Mathematics and Computation, 29: p. 219-229 (1989).

10. S. A. Vavasis, Black-box Complexity of Local Minimization, TR 90-1132 (1990).

11. S. H. Levine, A Colloquium on Optimization Techniques for Applications to PP&L Programs (1988)

12. Yuriji. G. Evtushenko, Numerical Optimization Techniques, Optimization Softwear, Inc., Publications Division, New York, (1985).

13. J. S. Suh, Optimized Automatic Reload Programs For PWRs Using The Haling Distribution, Ph. D Thesis, Penn State University (1989).

14. J. R. Lamarsh, Introduction To Nuclear Reactor Theory, Mass.: Addison Wesley Pub. Co., (1966).

A GLOBAL OPTIMIZATION APPROACH TO SOFTWARE TESTING

Roberto Barbagallo[†], Maria Cristina Recchioni[†]
Istituto Nazionale di Alta Matematica "F.Severi"
Piazzale Aldo Moro 5, 00185 Roma (Italy)

Francesco Zirilli
Dipartimento di Matematica "G.Castelnuovo"
Universitá di Roma "La Sapienza", 00185 Roma (Italy)

1. Introduction

In today software industry the cost of the software production is highly dependent on software testing. That is the activity of testing that the software modules produced actually perform the functions that they are supposed to perform. Despite its great importance software testing is a mainly empirical activity [1]. In a more rational approach we can divide the software testing activity in two branches: the structural and the functional testing [2]. The structural testing is independent of the code and consists of checking the flow-chart of the software modules; the functional testing is dependent on the code and consists of checking that the desidered functions are actually performed. In this paper we restrict our attention to the problem of the structural testing of the codes represented by flow-charts that satisfy the rules of the Warnier's methodology [3],[4]. A Warnier's flow-chart is made of two basic structures: the "loop" and the "alternative" structures (see Fig.1.1,Fig.1.2). In section 4 ten Warnier's flow-charts used to test our work are shown. Given a Warnier's flow-chart there are many walks to go through it from the beginning to the end. Each one of these walks will be considered as a possible test case. Given $n \geq 2$ we want to answer the following question:
find the n best possible test cases.
In this paper first of all we give a precise meaning to the above statement, so that the question proposed is translated in a combinatorial

---

[†] The research of these authors has been made possible through the support and the sponsorship of the ITALSIEL S.P.A. (Italy) to the graduate fellowship program of the Istituto Nazionale di Alta Matematica "Francesco Severi"

optimization problem. Moreover we present an optimization algorithm of the same type as those in [5],[6],[7] to solve the combinatorial optimization problem considered.

In section 2 we give the mathematical formulation of the question posed above as a combinatorial optimization problem.

In section 3 we give the optimization algorithm. This algorithm is based on a Markov chain obtained from a discretized version of a Fokker Planck partial differential equation in a singular limit [8].

In section 4 we describe ten Warnier's flow-charts on which the algorithm has been tested and the numerical results obtained.

Fig.1.1 A loop structure

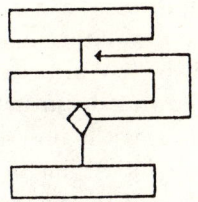

Fig.1.2 An alternative structure

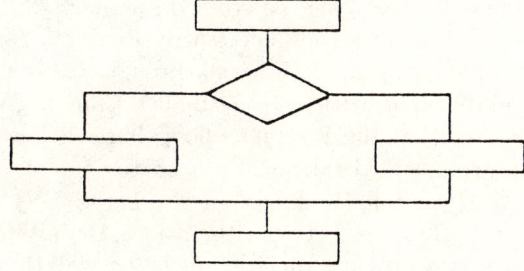

2. Software testing as a global optimization problem

The testing of complex computer codes starting from their code lines using deductive analysis is in practice impossible. In a similar way for these codes a complete input-output analysis of their behaviour is impossible. So that their testing often consists of a partial input-output analysis. The input cases, also called test cases, should be chosen to make as reliable as possible the testing procedure. However since the analysis of each input case has a cost (i.e. computer time, manpower,...) only a certain number of input cases is affordable to test a given code.

Here we give a mathematical formulation to the problem of maximizing the reliability of the testing procedure subject to the constraint that only a given number n of input cases can be used. That is the computer code will be represented by a walkthrough of the Warnier's flow-chart (see Definition 2.2). Let W be a Warnier's flow-chart (see for example Fig.2.1), we will use the following definitions:

Definition 2.1: A branch of W is the line that connects two blocks of a Warnier's flow-chart (see the dotted line of Fig.2.1).

Definition 2.2: A walkthrough of W is a connected set of branches that goes from the initial block to the final block following the flow described by W. For example only in the presence of a loop structure the same branch can appear more than once in a walkthrough. Two walkthroughs are considered different if they contain different branches or the same branches in a different order (see the crossed line in Fig.2.1).

The number of the branches (not necessarily distinct) contained in a walkthrough is called length of the walkthrough. If we restrict our attention to walkthroughs with length smaller than a given constant then it is easy to see that the Warnier's flow-chart W has only a finite number of walkthroughs. We define Ω to be the set of walkthroughs of W, and $\Omega_k \subseteq \Omega$ be a finite subset of Ω. Let $x_1, x_2 \in \Omega$ we will define a "distance" $d(x_1, x_2)$ between them (see Definition 2.4). This notion of distance is a measure of how much the walkthroughs x_1, x_2 are different one from the other. So that the mathematical formulation of the software testing problem posed above is the following one:

Problem 2.3: Given a Warnier's flow-chart W, $n \geq 2$ and a distance function d such that $d : \Omega \times \Omega \to \mathbf{R}^+$ find the n-tuple $(x_1, x_2, ..., x_n) \in \Omega^n$ that solves the following optimization problem:

$$\max_{(\xi_1,\xi_2,...,\xi_n)\in\Omega^n} \min_{\substack{i,j=1,...,n \\ i \neq j}} d(\xi_i, \xi_j) \qquad (2.1)$$

In our work the distance function d is defined only on $\Omega_k \times \Omega_k$ as follows. Given W and Ω_k for $x \in \Omega_k$ let $L(x)$ be the length of x, moreover for x_1 and $x_2 \in \Omega_k$ let B_{x_1} and B_{x_2} be the set of the branches contained in x_1

and x_2 respectively and $Q_{x_1x_2} = B_{x_1} \triangle B_{x_2}$ be the symmetric difference between B_{x_1} and B_{x_2}.

Definition 2.4: Let $x_1, x_2 \in \Omega_k$, $L(x_i)$ $i = 1, 2$, $Q_{x_1x_2}$ be defined as above. We define the distance $d(x_1, x_2)$ as follow:

$$d(x_1, x_2) = |L(x_1) - L(x_2)| + \#\{Q_{x_1x_2}\} \qquad (2.2)$$

For example in Fig.2.1 if we choose $x_1 = (1, 2, 4, 6), x_2 = (1, 2, 4, 5, 2, 4, 6)$ then $Q_{x_1x_2} = (1, 2, 4, 5, 6)$ so that $d(x_1, x_2) = |4 - 7| + 5 = 8$. Let $n_k = \#\{\Omega_k\}$ and $\binom{n_k}{n} = \#\{\Omega_k\}$ when $n \leq n_k$. So that the optimization Problem 2.3 can be reduced to the problem of maximizing a function defined on a finite subset of **Z**.

Fig2.1 A Warnier's flow-chart

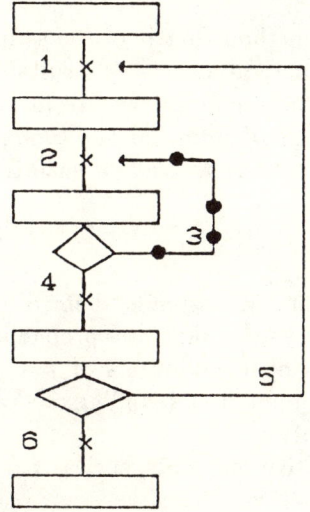

—●— branch
-×- walkthrough $X_1 = (1, 2, 4, 6)$

The distance (2.2) is a rough measure of how much two walkthroughs are different. The difference is measured considering the length and the branches of the walkthroughs. The validity of the formulation of the software testing problem as the combinatorial optimization problem must be tested exeperimentally. This testing activity will permit the refinement of the distance definition.

3. The Optimization Algorithm

Let $\mathbf{R}^+ = \{x \in \mathbf{R} \mid x \geq 0\}$, m be a natural number and $\mathbf{Z}_m$ be the set $\mathbf{Z}_m = \{-m, -m+1, ..., -1, 0, 1, ..., m\}$ and $f : \mathbf{Z}_m \to \mathbf{R}^+$ be a real function. In this section we consider the problem of finding a global minimizer of f, that is a point $x^* \in \mathbf{Z}_m$ such that

$$f(x^*) \leq f(x) \quad \forall\, x \in \mathbf{Z}_m. \tag{3.1}$$

Here we suggest a method that is of the same type of those suggested in [5],[6],[7]. This method associates a suitable Markov chain with the function whose global minimizer we are looking for. In particular the algorithm seeks a global minimizer of f along trajectories generated by a Markov chain. Let M be a constant such that:

$$\max_{i=-m,...,m-1} |f(i+1) - f(i)| < M \tag{3.2}$$

We construct a random variable defined on $\mathbf{Z}_m$, depending on a parameter $\epsilon > 0$, such that its probability distribution $u^\epsilon(x)$ is maximized at the global minimizers of f as ϵ goes to zero. A global minimizer of f is computed by sampling random variables corresponding to different values of ϵ.

Let v, r, h be a positive constants, such that:

$$v = 1 \tag{3.3}$$

$$0 < r < 0.5 \tag{3.4}$$

$$0 < h \leq \frac{1}{2\epsilon^2(1+r/2)} \tag{3.5}$$

and let $A^\epsilon \in \mathbf{R}^{(2m+1)\times(2m+1)}$ be the tridiagonal matrix defined by:

$$A_{ij}^\epsilon = \begin{cases} \frac{h\epsilon^2}{v^2} & i-j = 1 \\ 1 + \frac{h\epsilon^2}{v^2}(-1 + \frac{v}{\epsilon^2}(f(i+v) - f(i))) & i = j \\ \frac{h\epsilon^2}{2v^2}(-1 + \frac{v}{\epsilon^2}(f(i+v) - f(i))) & i-j = -1 \\ 0 & otherwise \end{cases} \tag{3.6}$$

for $i = -m+1, ..., m-1$, moreover:

$$A^\epsilon_{mj} = \begin{cases} 1 - \frac{h\epsilon^2}{v^2} & j = m \\ \frac{h\epsilon^2}{2v^2} & j = m-1 \\ 0 & \text{otherwise} \end{cases} \quad (3.7)$$

and

$$A^\epsilon_{-mj} = \begin{cases} 1 + \frac{h\epsilon^2}{v^2}(-1 + \frac{v}{\epsilon^2}(f(-m+v) - f(-m))) & j = -m \\ \frac{h\epsilon^2}{2v^2}(-1 + \frac{v}{\epsilon^2}(f(-m+v) - f(-m))) & j = -m+1 \\ 0 & \text{otherwise} \end{cases} \quad (3.8)$$

Let s be a constant such that:

$$\frac{2M}{s\epsilon^2} \leq r \quad (3.9)$$

and $f^* = f/s$. The matrix A^ϵ associated to f^* defined by (3.6), (3.7), (3.8) is a stochastic matrix, in fact:

$$\sum_{j=-m}^{m} A^\epsilon_{ij} = 1 \quad \forall i = -m, ..., m \quad (3.10)$$

and

$$0 \leq A^\epsilon_{ij} \leq 1 \quad \forall i, j = -m, ..., m \quad (3.11)$$

In the following to simplify the notation we will continue to use f instead of f^*.
Let $U^{0\epsilon} = U^0 \in \mathbf{R}^{2m+1}$ be:

$$U^0_i = \frac{1}{2m+1} \quad i = -m, -m+1, ..., m \quad (3.12)$$

then the stochastic matrix A^ϵ together with the initial distribution U^0 defines a Markov chain.
Let $U^{n,\epsilon} = (u^\epsilon_{n,-m}, ..., u^\epsilon_{n,m})^T \in \mathbf{R}^{2m+1}$ be the column vector defined by:

$$U^{n+1,\epsilon} = A^{\epsilon^T} U^{n,\epsilon} \quad n = 0, 1, 2, ... \quad (3.13)$$

where A^{ϵ^T} denotes the transpose matrix of A^ϵ. It is easy to see that $u^\epsilon_{n,i} > 0$ $i = -m, ..., m$ and $\sum_{i=-m}^{m} u^\epsilon_{n,i} = 1$.
Moreover $U^{n,\epsilon}$ is the probability distribution of the Markov chain (3.13) at the step n. We note that if $f(\xi)$ is the restriction to $\mathbf{Z}_m$ of a smooth function f defined on the interval $[-m, m] \subset \mathbf{R}$ the transition probability matrix A^ϵ of the Markov chain (3.13) can be interpreted as follow:

Lemma 3.1 The matrix A^ϵ defined in (3.6),(3.7),(3.8) is the discretization by finite differences of the Fokker Planck equation:

$$\frac{\partial u^\epsilon}{\partial t}(t,x) = \frac{\epsilon^2}{2}\frac{\partial^2 u^\epsilon}{\partial x^2}(t,x) + \frac{\partial}{\partial x}\left(\frac{df}{dx}u^\epsilon\right) \quad (t,x) \in \mathbf{R}^+ \times [-m,m] \quad (3.14)$$

with boundary conditions:

$$\left[\frac{\epsilon^2}{2}\frac{\partial u^\epsilon}{\partial x} + \frac{df}{dx}u^\epsilon\right]_{x=-m} = 0 \quad (3.15)$$

$$\left[\frac{\epsilon^2}{2}\frac{\partial u^\epsilon}{\partial x} + \frac{df}{dx}u^\epsilon\right]_{x=m} = 0 \quad (3.16)$$

Proof: Let $h > 0$ and $v > 0$ be the discretization step of the variable t and of the variable x respectively such that equations (3.3),(3.5) hold. Let $U^\epsilon(t,x)$ be a solution of (3.14),(3.15),(3.16) we will approximate u^ϵ_{nk}, $n = 0, 1, ..., k = -m, -m+1, ...m$, where u^ϵ_{nk} is the corresponding solution of the difference equation:

$$u^\epsilon_{n+1,k} = h\frac{\epsilon^2}{2}u^\epsilon_{n,k+1} + (1 + h\epsilon^2(-1 + \frac{1}{\epsilon^2}(f_{k+1} - f_k)))u^\epsilon_{nk} +$$

$$+h\frac{\epsilon^2}{2}(1 - \frac{2}{\epsilon^2}(f_k - f_{k-1}))u^\epsilon_{n,k-1} \quad n = 0,1,2,... \quad k = -m+1,...,m-1 \quad (3.17)$$

where $f_k = f(kv) \quad k = -m, -m+1, ..., m$.

The boundary conditions (3.15),(3.16) are discretized using forward-finite differences. In particular (3.15) becomes

$$\frac{\epsilon^2}{2}(u^\epsilon_{n,-m+1} - u^\epsilon_{n,-m}) + (f_{-m+1} - f_{-m})u^\epsilon_{n,-m} = 0 \quad (3.18)$$

which implies

$$u^\epsilon_{n+1,-m} = h\frac{\epsilon^2}{2}u^\epsilon_{n,-m+1} + \{1 + h\frac{\epsilon^2}{2}(-1 + \frac{2}{\epsilon^2}(f_{-m+1} - f_{-m}))\} \quad (3.19)$$

A similar formula can be obtained for (3.16).
Finally we identify $U^{n\epsilon} = (u^\epsilon_{n,-m}, ..., u^\epsilon_{n,m})^T$. This completes the proof.
We have:
Lemma 3.2 Let $U^\epsilon = (u^\epsilon_{-m}, ..., u^\epsilon_m)^T \in \mathbf{R}^{2m+1}$ be the vector:

$$u^\epsilon_i = \begin{cases} D(\epsilon)\exp(-(f_{-m}\frac{2}{\epsilon^2}))\prod_{j=-m}^{i-1} -\frac{2}{\epsilon^2}(f_{j+1} - f_j)) & i = -m+1, ..., m \\ D(\epsilon)\exp(-(f_{-m}\frac{2}{\epsilon^2})) & i = -m \end{cases}$$

$$(3.20)$$

where $D(\epsilon)$ is a normalization constant such that $\sum_{i=-m}^{m} u_i^\epsilon = 1$. Then the vector U^ϵ is the stationary distribution of the Markov chain (3.13), that is

$$\lim_{n \to \infty} U^{n,\epsilon} = U^\epsilon \tag{3.21}$$

Proof: It is easy to see that:

$$U^\epsilon = A^{\epsilon^T} U^\epsilon \tag{3.22}$$

Since A^ϵ is irreducible we have the thesis.
Without loss of generality we assume now that zero is the global minimum of f and that the point $-m$ is not a global minimizer. In the following Lemma 3.3 the function f is not rescaled by s.
Lemma 3.3 Let ϵ and s be defined as above, and $\{\xi_1, \xi_2, ..., \xi_p\} \subset \mathbf{Z}_m$ be the set of the global minimizers of f. We have

$$\lim_{s \to \infty} \lim_{\epsilon^2 \to 1/s^2} U^\epsilon = \delta \tag{3.23}$$

where $\delta \in \mathbf{R}^{2m+1}$ is the following probability distribution:

$$\delta_k = \begin{cases} 0 & k \in \mathbf{Z}_m \ \ k \notin \{\xi_1, \xi_2, ..., \xi_p\} \\ 1/p & k \in \{\xi_1, \xi_2, ..., \xi_p\} \end{cases} \tag{3.24}$$

Proof: For $i = -m, ..., m$ we have

$$u_i^\epsilon = D(\epsilon) \exp(-(f_m \frac{2}{\epsilon^2 s})) \prod_{j=-m}^{i-1} (1 - \frac{2}{\epsilon^2 s}(f_{j+1} - f_j)) =$$

$$= D(\epsilon) \exp(-(f_m \frac{2}{\epsilon^2 s})) \{ \prod_{j=-m}^{i-1} \exp(-\frac{2}{\epsilon^2 s}(f_{j+1} - f_j)) + O(\frac{1}{\epsilon^4 s^2}) \} \tag{3.25}$$

so that:

$$u_i^\epsilon = D(\epsilon) \exp(-(f_i \frac{2}{\epsilon^2 s})) + D(\epsilon) \exp(-(f_{-m} \frac{2}{\epsilon^2 s})) O(\frac{1}{\epsilon^4 s^2}) \tag{3.26}$$

We have:

$$\lim_{s \to \infty} \lim_{\epsilon^2 \to 1/s^2} \exp(-(f_i \frac{2}{\epsilon^2 s})) = \lim_{s \to \infty} \exp(-2s f_i) = \begin{cases} 1 & \text{if } f_i = 0 \\ 0 & \text{otherwise} \end{cases} \tag{3.27}$$

We observe that:

$$\lim_{s \to \infty} \lim_{\epsilon^2 \to 1/s^2} \sum_{i=-m}^{m} u_i^\epsilon = \lim_{s \to \infty} D(1/s) \sum_{i=-m}^{m} \exp(-2s f_i) \tag{3.28}$$

that is:

$$\lim_{s\to\infty}\lim_{\epsilon^2\to 1/s^2} D(\epsilon) = (\lim_{s\to\infty}\lim_{\epsilon^2\to 1/s^2} \sum_{i=-m}^{m} \exp(-(f_i\frac{2}{\epsilon^2 s})))^{-1} = 1/p \quad (3.29)$$

This concludes the proof.

The proposed optimization method consists of computing several trajectories of the Markov chain (3.13) for decreasing values of ϵ. To compute a trajectory we start from an initial value ϵ_0 of ϵ and from an arbitrary point $\xi_0 \in Z_m$, the algorithm for each trajectory is given by:

0. Set $k=0$, $\epsilon = \epsilon_0$, $\xi_k = \xi_0$.
1. Let ζ be a random number uniformly distributed in $[0,1]$.
2. If $0 \leq \zeta < A_{\xi_k,\xi_k-1}$ then $\xi_{k+1} = \xi_k - 1$ if $A_{\xi_k,\xi_k-1} \leq \zeta < A_{\xi_k,\xi_k-1} + A_{\xi_k,\xi_k}$ then $\xi_{k+1} = \xi_k$ otherwise $\xi_{k+1} = \xi_k + 1$.
3. If k is less than a certain fixed number then set $k = k+1$ and go to step 1, otherwise decrease ϵ of a constant factor.

The algorithm consists of following more trajectories simultaneously. At the end of some fixed periods established according to some heuristic rules, a comparison is made between the trajectories. Finally when the function values observed satisfy the stopping conditions the computation stop.

4. Test problems and numerical results

Let us give the description of the test problems on which the algorithm described in section 3 has been tested.

A test problem consists of a Warnier's flow-chart W, of the number of test cases required n and of $\Omega_k \subseteq \Omega$. In our exeperience Ω_k has been chosen as the set of walkthroughs such that each loop is repeated at most two times. The ten Warnier's flow-charts used to test our work are shown in Fig.4.1-4.10. Let us consider now the function $\tilde{f}$ to be optimized

$$\tilde{f}(x_1, x_2, ..., x_n) = \min_{\substack{i,j=1,...,n \\ i \neq j}} d(x_i, x_j) \quad \forall (x_1, x_2, ..., x_n) \in \Omega_k^n \quad (4.1)$$

in order to use the optimization algorithm of the section 3 we identify $\tilde{f}$ with a function f defined on Z_m for a suitable m. So that the computational cost of a function evaluation of f is made of two terms:
1. Given $z \in Z_m$ determine the corresponding n-tuple of Ω_k^n.
2. Given $(x_1, x_2, ..., x_n) \in \Omega_k^n$ compute the function $\tilde{f}$ defined in (4.1).

The computational cost of the first term depends on the choice of the correspondence between Ω_k^n and Z_m, that is in our numerical experience:

$$O(n^3/6) \quad (4.2)$$

elementary operations. Let m_L the maximum length of the walkthroughs of Ω_k, the computational cost of the second term is:

$$O(m_L^2 n^2) \tag{4.3}$$

elementary operations. So that the computational cost of a function evaluation is:

$$O(m_L^2 n^2) + O(n^3/6) \tag{4.4}$$

elementary operations. In table 4.1 we give some numerical results. The computational time needed to obtain the results shown in Table 4.1 consists, on the avarage, of a few seconds of C.P.U. time on a VAX 8530. Moreover the computational time required by the algorithm to find a global minimizer of f is smaller than the one needed to seek a global minimizer computing the function f in every point of Z_m. In Table 4.1 for each test problem we give: n_p the number of Warnier's flow-chart (i.e.Fig.4.1-4.10) corresponding to the test problem considered, n defined as above, n_k the number of walkthroughs in Ω_k and $l_k = \binom{n}{k}$ the number of the n-tuple in Ω_k, n_f the number of distinct function evaluations necessary to the algorithm of section 3 to find an optimizer, s_t the success indicator ($s_t = 1$ success, $s_t = 2$ failure, $s_t = 3$ failure due to machine overflow).

The algorithm has been coded in FORTRAN on a VAX/VMS VERSION V5.1 in double precision arithmetic.

We note that the algorithm goes through the same points many times before claiming success or failure so that n_f is only the number of the distinct function evaluations.

Table 4.1

n_p	n	n_k	l_k	n_f	s_t
1	2	18	153	18	1
1	3	18	816	173	1
1	4	18	3060	61	1
1	8	18	43758	386	1
2	2	84	3486	531	1
3	2	46	1035	143	1
3	3	46	15180	636	1
4	2	258	33153	859	1
5	3	10	120	61	1
5	4	10	210	71	1
6	4	32	35960	1021	1
6	5	32	201376	3056	1
6	6	32	906192	4067	1
7	3	39	9139	149	1
7	4	39	82251	9658	1
8	3	7	35	21	1
8	4	7	35	14	1
9	2	78	3003	431	1
9	3	78	76076	925	2
10	2	195	18915	603	1

Acknowledgements: We thank G.Militello, A.Raffone, M.Scarino of the ITALSIEL S.P.A. for many helpful discussions and for providing us with the test problems.

Fig.4.1

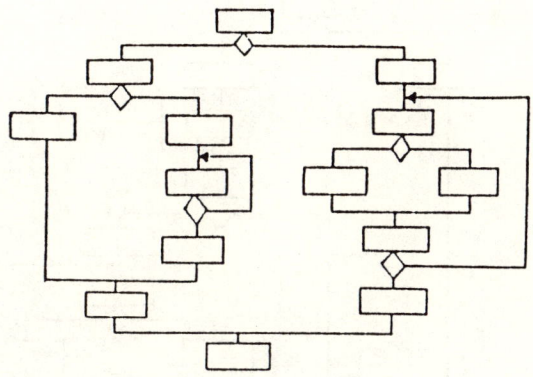

Fig.4.2

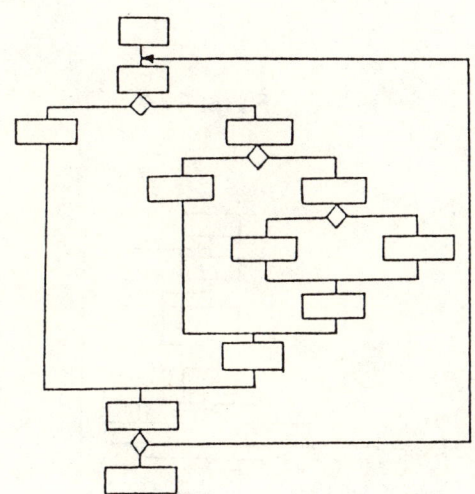

Fig. 4.3

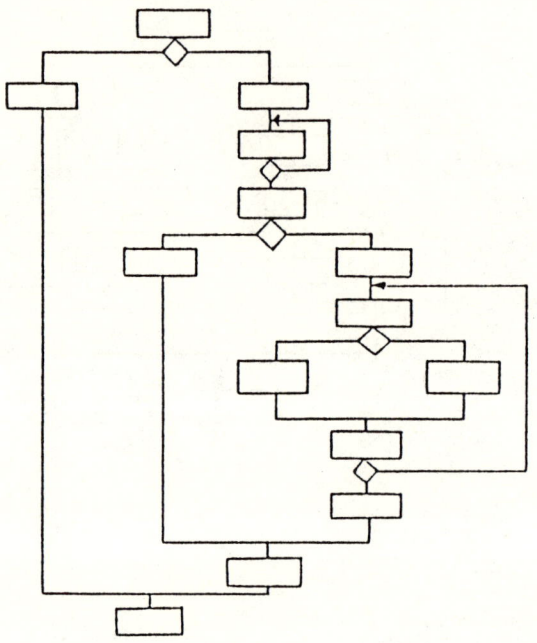

Fig. 4.4

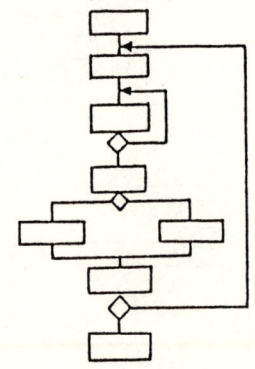

Fig.4.5

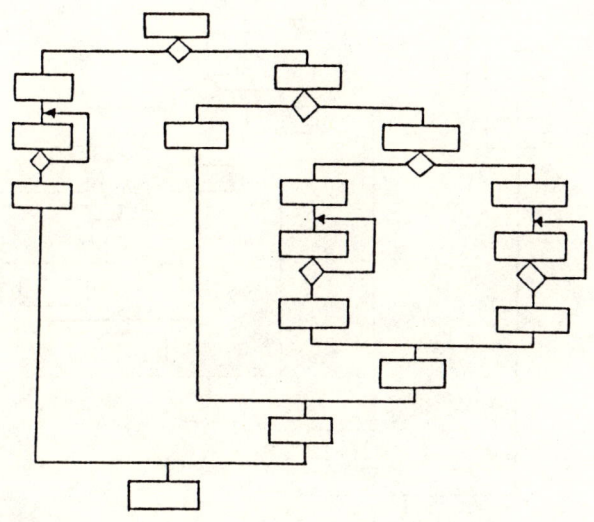

Fig. 4.6

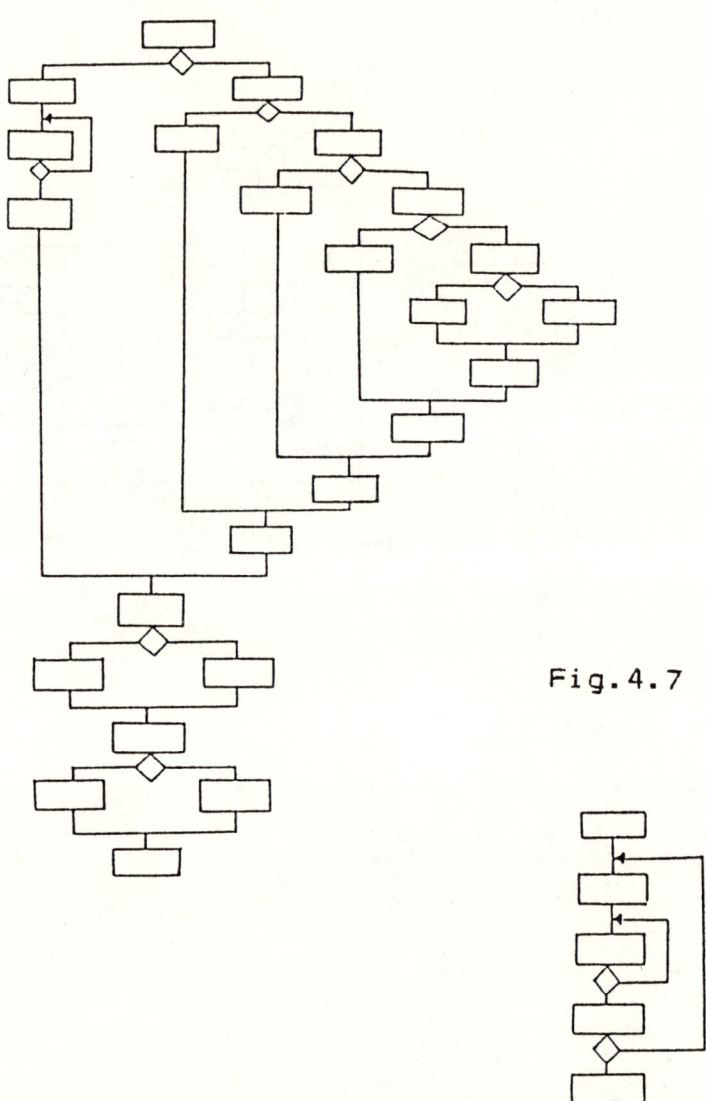

Fig. 4.7

Fig. 4.8

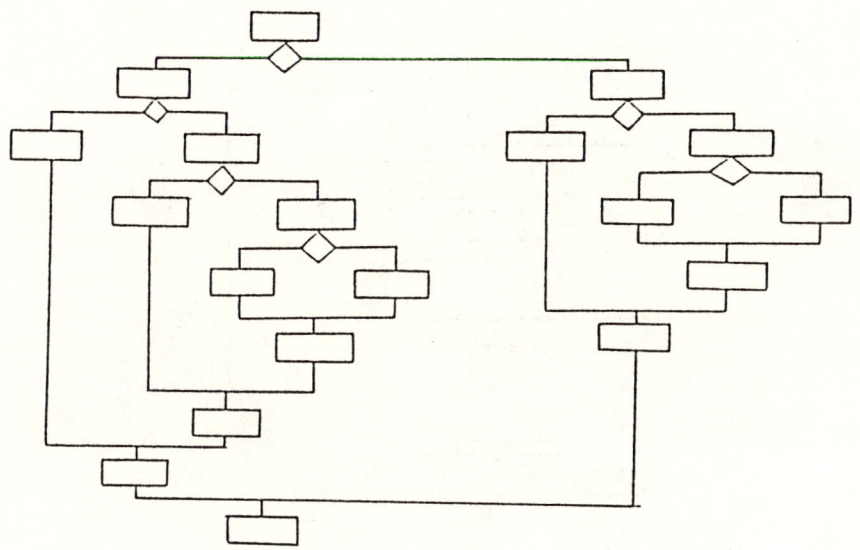

Fig. 4.9

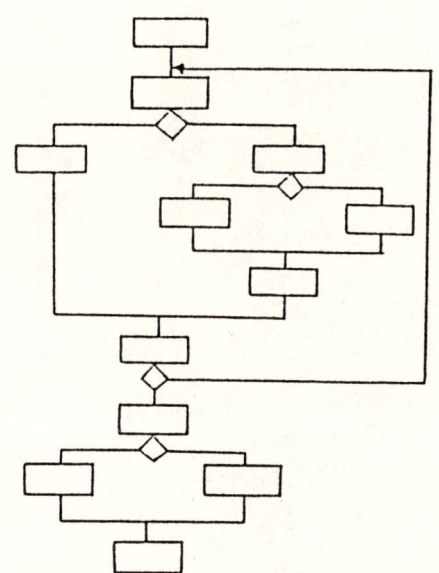

Fig. 4.10

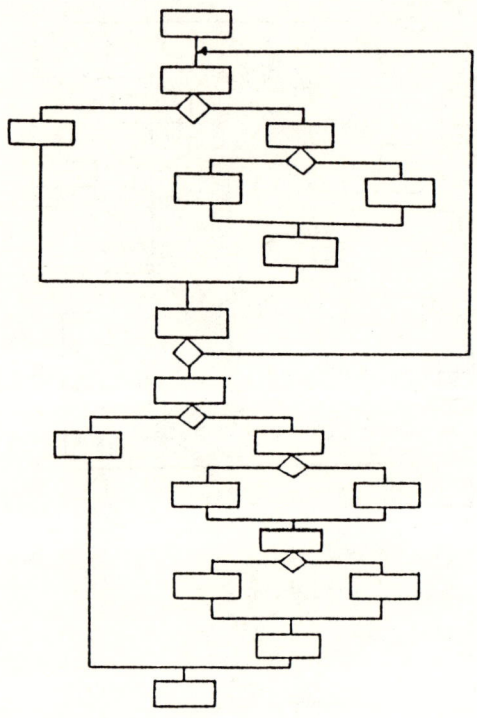

References

[1] W.Hetzel: *The complete guide to software testing*, Collins, (1985).
[2] G.J.Myer: *The art of software testing*, Wiley-Interscience Publication, New York, (1978).
[3] J.D.Warnier, B.M.Flanagan: *Entrainement à La Programmation*, Vol.I, *Construction des Programmes*, Les Editions d'organisation, Paris, (1972).
[4] J.D.Warnier:*Entrainement à La Programmation*, Vol.Ii, *Exploitation des donnèes*, Les Editions d'organisation, Paris, (1972).
[5] F.Aluffi-Pentini, V.Parisi, F.Zirilli: *Global Optimization and Stochastic Differential Equations*, J.Optimization Theory and Applications 47, (1986), 1-16.
[6] F.Aluffi-Pentini, V.Parisi, F.Zirilli: *A global optimization Algorithm using Stochastic Differential Equations*, Transactions on Mathematical Software of the A.C.M. 14, (1988), 345-365.
[7] F.Aluffi-Pentini, V.Parisi, F.Zirilli: *Algorithm 667 SIGMA - A Stochastic-Integration Global Minimization Algorithm*, Transaction on Mathematical Software 14, (1988), 366-380.
[8] A.Angeletti, C.Castagnari, F.Zirilli: *Asymptotic eingevalues degeneracy for a class of one dimensional Fokker-Planck operators*, J. Math. Phys. 26, (1985),678-691.